Plant Physiology

A TREATISE

EDITED BY
F. C. STEWARD

Professor Emeritus
Cornell University
Ithaca, New York

Volume VII: Energy and Carbon Metabolism

Coedited by

F. C. STEWARD   and   R. G. S. BIDWELL

Charlottesville, Virginia   *Department of Biology*
                                  Dalhousie University
                                  Halifax, Nova Scotia, Canada

1983

ACADEMIC PRESS

A Subsidiary of Harcourt Brace Jovanovich, Publishers

New York   London
Paris   San Diego   San Francisco   São Paulo   Sydney   Tokyo   Toronto

ACADEMIC PRESS, INC.
111 Fifth Avenue, New York, New York 10003

United Kingdom Edition published by
ACADEMIC PRESS, INC. (LONDON) LTD.
24/28 Oval Road, London NW1 7DX

Library of Congress Cataloging in Publication Data

Main entry under title:

Plant physiology.

   Vol 7-   edited by F.C. Steward & R.G.S. Bidwell.
   Includes bibliographies and indexes.
   Contents: v. 1. A. Cellular organization and
respiration. B. Photosynthesis and chemosynthesis.
2. v. -- v. 2. Plants in relation to water and
solutes -- [etc.] -- v. 7. Energy and carbon metabolism.
   Plant physiology. I. Steward, F. C. (Frederick)
II. Bidwell, R. G. S. (Roger Grafton Shelford)
QK711.S7   581.1   59-7689
ISBN 0-12-668607-6 (v. 7)

PRINTED IN THE UNITED STATES OF AMERICA

83 84 85 86   9 8 7 6 5 4 3 2 1

Contents

CHAPTER ONE

Cells and Their Organization: Current Concepts
by J. L. HALL 3

CHAPTER TWO

Energy Metabolism in Plants by D. T. DENNIS 163

Contributors to Volume VII

Numbers in parentheses indicate the pages on which the authors' contributions begin.

R. G. S. BIDWELL (287), *Department of Biology, Dalhousie University, Halifax, Nova Scotia B3H 4J1, Canada*

D. T. DENNIS (163), *Biology Department, Queen's University, Kingston, Ontario K7L 3N6, Canada*

DOROTHY F. FORWARD (459), *Department of Botany, University of Toronto, Toronto, Ontario M5S 1A1, Canada*

J. L. HALL (3), *Department of Biology, The University, Southampton S09 5NH, United Kingdom*

Foreword: The New Volumes

The first volume of "Plant Physiology: A Treatise" appeared in 1959, though planning for it commenced as early as 1957. The treatise was arranged in three sections: Cell Physiology, Nutrition and Metabolism, Growth and Development. As it developed, its planned six volumes extended to eleven tomes, the last of which appeared in 1972. Over this time span change was inevitable although, in a measure, the later volumes could update the first.

The decision to add additional volumes to the treatise (Volumes VII *et seq.*) was not made either hastily or lightly. The new volumes will summarize salient developments and thus epitomize the present status of the subject. In so doing, they will add to the historical record of Volumes I through VI in the belief that the continued use of the treatise has owed much to its historical perspective. This perspective, set initially in brief historical sketches, has permeated the work throughout. Not all topics dealt with in Volumes I–VI have been selected for updating. For our arbitrary selections or omissions the editors assume responsibility.

An early injunction to authors, variously fulfilled according to their individual styles, to write comprehensively but in a narrative text that, as necessary, rises above the mass of detail still applies. Thus the new volumes do not discard but will build on the existing ones, for the progress of science is always indebted to its past.

Therefore, as much as possible of the format of the treatise as published will be preserved. This is done even though the restless mood of the times clamors for the new look, and change often seems a virtue. But natural science, like nature, should surely be conservative, and any look that emerges will be seen in the content of the new chapters and the points of view summarized by the authors. Meanwhile editors and authors will strive to preserve that essential continuity with the past that is the basis of scholarship and is essential to an understanding of how and when progress was made.

Organization may survive, but, inevitably, authors change. Many who have served the treatise well no longer survive; others have moved into retirement, albeit with varying degrees of maintained professional activity. The present group of authors represents a blend of some who are new to the task and others who have previously served in this capacity. The plan is to add new volumes that cover selected topics in the general areas of cell biology, metabolism, and growth and development, invoking the aid of three coeditors. Each coeditor assumes, with the editor, primary responsibility for one area while maintaining a general vigilance for the integrity of the treatise as a whole. Moreover, coeditors bring not only their own expertise but experience in teaching of plant physiology. Indeed, to promote the continued usefulness of the treatise to teachers, research workers, and students is part of our collective aims.

But in the final analysis, the value of the new volumes will rest on the ability of the invited authors to add to the existing content of Volumes I–VI and to present an overview of the status of the subject as it now moves through the turbulent last quarter of the twentieth century and anticipates the problems of the twenty-first.

F. C. Steward

Plant Physiology: The Contemporary Scene

This is a sequel to a historical sketch, written prior to 1960 in Volume I of this treatise (pp. xv–xxvi), which examined the subject of plant physiology as it was then adjusting to the massive increase in the quantity and diversity of relevant knowledge. The second quarter of the twentieth century witnessed a very rapid rise, later accentuated by the postwar boom in science, in the ever-expanding flow of scientific publication. By the end of the third quarter of this century, scientific information was, and still is, being disseminated in vast quantity and in a variety of ways. In the years 1972 and 1976 the total number of published articles in biology was already of the order of 100,000 in each year and, of these, the number that related specifically to plant physiology was of the order of 7000. Similar estimates for the 1980s would be very much greater. Therefore, the impact of this ever-present and growing problem on those who attempt to present the subject as it now is, as in this treatise, needs to be understood.

The earlier brief historical sketch traced progress, qualitatively, by reference to events from Aristotelian dogma to mid-twentieth century experimental science. But, since then, some more quantitative analyses of the progress of science in general and of biology, even plant physiology, have been made. De Solla Price[1] showed that the exponential expansion of science, like that of the population from which the scientists came, has been in progress for a very long time, with manpower or publication increasing with a more or less stable doubling time of 10–15 years. But even a 15-year doubling time over three centuries (1660–1960) corresponds to an increase of 2^{20}, or an increase in the "size" of science "of the order of a millionfold." Even from 1945 to 1975 this general rate of increase would have produced a fourfold increase in science as a whole. But as Menard[2] showed, the journal *Plant Physiology*, unlike the older and more sedate *American Journal of Botany*, was dominated between 1930 and 1968 by such rapidly expanding research that 90% of the citations were less than 20 years old, and their doubling time

[1] Derek J. De Solla Price (1963). "Little Science, Big Science." Columbia University Press, New York and London.

[2] Henry W. Menard (1971). "Science: Growth and Change." Harvard University Press, Cambridge, Massachusetts.

was only 7 years. But plant physiology is, as Menard also showed (*loc. cit.,* Fig. 3.5), like molecular biology, nuclear astrophysics, marine biology, etc., a "transitional" or borderline field that remained quiescent until important problems suddenly became tractable because of discovery, or newly appreciated applications, in another related science. Such borderline fields have expanded in the last 30–40 years with a doubling time as short as 2.5 years. For plant physiology the stimulus of the "other science" clearly occurred about 1930 and came from biochemistry, enzymology, and later from the applications of radioactive procedures. But as also shown by Menard, "whereas older papers were still being cited as late as 1950, they were not so cited in 1968," and "this rejection of the mass of older literature has been observed in other fields." This trend to ignore all but contemporary works is obviously to be deplored. It reflects on modern workers in the field, the manner in which they were trained, and the organization of their scientific meetings and journals. Thus information *about* plants, as conveyed in the flood of currently published minutiae, only leads to understanding of plants inasmuch as it can be integrated and as the preoccupations of the present do not remain severed from the accumulated knowledge of the past.

Many groups of biologists now "hunt in packs" as they associate primarily with others of like interest, and the groups then proliferate seminars, symposia, or workshops, often with printed versions of their repetitive proceedings, that promote these often parochial interests. It is also a feature of the times that there are multiauthored, multivolume sets on an incredible array of topics, many of which would hitherto have seemed to be adequately covered by a single scholarly article, as in *Biological Reviews.* Any attempt to even list all of these as they may, in the last decade or so, affect plant physiology would be tedious and never up-to-date because the mail announces almost weekly some new and seemingly relevant publications. Even new journals devoted to special topics that are relevant to plant physiology now appear more frequently than single scholarly works did in the memory of some now living. In short, gaining access to, not necessarily digesting, all that is being or has been written, either on the whole subject of plant physiology or even its relatively small subdivisions, has become a superhuman task; a task which, incidentally, is not solved solely by computerized literature searches that, in the inadequacy of code words and recall mechanisms, are but poor substitutes for the now outmoded practice of browsing through the stacks of a well-stocked library.

But this being so, where are we? Obviously scholarship should not go the way of modern industry in which mass production has replaced quality and craftsmanship, and "built-in obsolescence" ensures that the wheels keep turning even though the world's resources are depleted and the wastes threaten to clog our civilization. Nor should we arrest the search for truth. Nevertheless, some balance needs to be struck between calling a halt to the proliferation of the trivial and the unnecessary duplication as publishing houses, like cereal firms, must seemingly have competing wares however brief their useful lives. Meanwhile, there is much to be said for the view that plant physiology today, as a branch of human knowledge, has unity and coherence only inasmuch as it can be comprehended by the individual scholar whose task or interest it is to know and communicate how plants are nourished, metabolize, grow, and reproduce. To this end, therefore, authors will show their wisdom and serve the users of this treatise inasmuch as they can rise above the temptation merely to cite what is miscellaneous information as they integrate it into broad concepts or interpretations. This task is not made easier by the conspicuous tendency in recent years to give overriding attention to certain topical special problems of plant physiology and virtually to ignore others equally as fundamental.

It is still the laudable objective of plant physiologists to explain how plants function in terms of known principles of chemistry and physics, but that does not mean that respect should be denied for ways that biological organization, through its evolutionary history, has erected environments and circumstances in which events still difficult to emulate *in vitro* proceed with baffling ease *in vivo*. And if in our search for causality one goes back to the beginning one may still ask: How did nature by locally reduced entropy first create complexity and, thereafter, systems that enabled "nature to feed on negative entropy"? Plant physiologists, however, are prone to interpret plants "as they are." But here lies a dilemma. How far can one understand the physics and chemistry of developed plants without also understanding how plants use inherited information and random molecules from their environments to create the complex organizations at all levels that become the milieux for the vital activities that occur? Biochemistry may describe in molecular terms the course of events as they finally occur provided that inheritance, morphology, and development have furnished the setting in which they may operate.

In the period covered by these supplementary volumes great problems, both terrestrial and cosmic, already face plant physiologists. World agriculture, with its finite resources, faces the feeding of an apparently unlimited population of humans. The technical approach has been to improve the genetics of food plants. The "Green Revolution" was a plant breeder's triumph, though it was more exuberantly heralded than such earlier triumphs as the breeding of wheats for the northern climates of

Canada. But, as yet, the "Green Revolution" has not achieved its full potential because genetics and plant breeding alone are not enough. To develop a "wonder rice," or any other "wonder" plant to answer all problems, it should be tested, during breeding, not only in one climatic environment but against all the combinations of environmental factors that it may encounter. Had the new "wonder plants" been so tested, as the breeding went along, for all their responses to the interactions of long and short days, of high and low day and night temperatures, as well as their respective demands on nutrition, this could have tempered expectations. Highly specialized and demanding single cultivars should not be expected to solve all the problems equally of the very diverse and often nutritionally deficient habitats in which a given crop may be expected to grow.

Although the volumes to come may deal with the physiological background of these problems, it is clear that enlightened plant physiology should be allowed to play its timely part in efforts to solve them. Obviously, a few supplementary volumes to "Plant Physiology: A Treatise," with their preselected topics, cannot deal with all that is touched on above. They may, however, as they analyze the present, set the stage for future developments and, as students are trained to be investigators, enable them to face realistically the problems of the century to come.

F. C. Steward

Preface to Volume VII

To supplement the original list of volumes in this treatise with a limited number of new ones raised the inevitable problem of where to begin. The general topics of metabolism and energy relations (biogenergetics) and the organization of the cells in which all occurs permeate the subjects to be updated. The decision to cover these fields first with the smallest number of authors and main topics (a device to ensure integrated accounts of their diverse aspects) led to some very long chapters. The field has been subdivided into two major categories: Energy and Carbon Metabolism (introduced by an updated account of cellular organization) and Nitrogen and Protein Metabolism. Although the subject matter is closely interrelated, the contributions are presented in two separate volumes (VII and VIII), each of which can stand alone and be of maximum use to research workers, teachers, and students.

As much as possible of the design and organization of the earlier volumes has been retained, including the Subject and Author Indexes and the Index to Plant Names. Continuity and consistency in the latter are happily ensured by the services of Dr. W. J. Dress, for whose help the editors are again grateful. In the necessarily speedy preparation of the Subject Index for Volume VII the aid of Mrs. Shirley Bidwell is acknowledged. It is a pleasure also to acknowledge the sympathetic understanding of the personnel of Academic Press in the more unusual problems the editors have had to face.

F. C. Steward

Note on the Use of Plant Names

The policy has been to identify by its scientific name, whenever possible, any plant mentioned by a vernacular name by the contributors to this work. In general, this has been done on the first occasion in each chapter when a vernacular name has been used. Particular care was taken to ensure the correct designation of plants mentioned in tables and figures which record actual observations. Sometimes, when reference has been made by an author to work done by others, it has not been possible to ascertain the exact identity of the plant material originally used because the original workers did not identify their material except by generic or common name.

It should be unnecessary to state that the precise identification of plant material used in experimental work is as important for the enduring value of the work as the precise definition of any other variables in the work. "Warm" or "cold" would not usually be considered an acceptable substitute for a precisely stated temperature, nor could a general designation of "sugar" take the place of the precise molecular configuration of the substance used; "sunflower" and "*Helianthus*" are no more acceptable as plant names, considering how many diverse species are covered by either designation. Plant physiologists are becoming increasingly aware that different species of one genus (even different varieties or cultivars of one species) may differ in their physiological responses as well as in their external morphology and that experimental plants should therefore be identified as precisely as possible if the observations made are to be verified by others.

On the assumption that such common names as lettuce and bean are well understood, it may appear pedantic to append the scientific names to them—but such an assumption cannot safely be made. Workers in the United States who use the unmodified word "bean" almost invariably are referring to some form of *Phaseolus vulgaris;* whereas in Britain *Vicia faba*, a plant of another genus entirely, might be implied. "Artichoke" is another such name that comes to mind, sometimes used for *Helianthus tuberosus* (properly, the Jerusalem artichoke), though the true artichoke is *Cynara scolymus*.

By the frequent interpolation of scientific names, consideration has also been given to the difficulties that any vernacular English name alone may present to a reader whose native tongue is not English. Even some American and most British botanists would be led into a misinterpretation of the identity of "yellow poplar," for instance, if this vernacular American name were not supplemented by its scientific equivalent *Lirodendron tulipifera,* for this is not a species of *Populus* as might be expected, but a member of the quite unrelated magnolia family.

When reference has been made to the work of another investigator who, in his published papers, has used a plant name not now accepted by the nomenclature authorities followed in the present work, that name ordinarily has been included in parentheses, as a synonym, immediately after the accepted name. In a few instances, when it seemed expedient to employ a plant name as it was used by an original author, even though that name is not now recognized as the valid one, the valid name, preceded by the sign =, has been supplied in parentheses: e.g., *Betula verrucosa* (= *B. pendula*). Synonyms have occasionally been added elsewhere also, as in the case of a plant known and frequently reported upon in the literature under more than one name: e.g., *Pseudotsuga menziesii* (*P. taxifolia*); species of *Elodea* (*Anacharis*).

Having adopted these conventions, their implementation rested first with each contributor to this work; but all outstanding problems of nomenclature have been referred to Dr. W. J. Dress of the L. H. Bailey Hortorium, Cornell University. The authority for the nomenclature now employed in this work has been the Bailey Hortorium's "Hortus Third." For bacteria Bergey's "Manual of Determinative Bacteriology" and for fungi Ainsworth and Bisbee's "Dictionary of the Fungi" have been used as reference sources; other names have been checked where necessary against Engler's "Syllabus der Pflanzenfamilien." Recent taxonomic monographs and floras have been consulted where necessary. Dr. Dress's work in ensuring consistency and accuracy in the use of plant names is deeply appreciated.

The Editors

PREAMBLE TO CHAPTER ONE

It is appropriate that Chapter 1 should deal with cells, for the need to understand their organization and functioning now permeates all aspects of plant physiology. Furthermore, the great advances made during the life of this treatise have permeated to the innermost structure of cells, from their inclusions, organelles, membranes, tubules, and particles virtually to the molecular level. It is in this period also that knowledge of the material universe has been so unbelievably advanced by modern physical science and interpretation. This knowledge and even the sight of bodies and conditions in outer space have indeed captured the popular imagination. Equally dramatic, however, has been the exploration of the inner world of cells and of protoplasm, where the crucial events of life occur; all this has been made possible by the development of tools and physical methods applicable to the infinitesimally small. But Chapter 1 does not merely document descriptively the submicroscopic morphology of cells. It also portrays the cell as the physical setting within which the dynamic molecular machinery of metabolism is made to work.

F.C.S.

1

CHAPTER ONE

Cells and Their Organization: Current Concepts[1]

J. L. HALL

[1] Abbreviations used in this chapter: CB, cytochalasin B; ER, endoplasmic reticulum; GERL, Golgi–endoplasmic reticulum–lysosome complex; MTOC, microtubule organizing center.

Plant Physiology
A Treatise
Vol. VII: Energy and Carbon Metabolism

I. Introduction: Plant Cells

All descriptions of plant cells, classical or modern, have recognized the great variation in cell size, shape, and function. Cells of higher plants may vary in size from 10 μm in length or diameter in the meristem to 300 μm or more in fibers. Some blue-green algal cells may be as small as 0.5 μm in diameter, whereas individual cells of certain coenocytes (e.g., *Nitella*) may be several centimeters in their longest dimension. There may be considerable variation in cell size within a particular organ. For example, during development of a root tip, very large changes in cell dimensions occur within a few millimeters of the meristem (Fig. 1). Considerable changes also occur in cell shape as differentiation proceeds because both size and shape are closely related to cellular function. Plant organs show division of labor because different cells and tissues are specialized to perform the various functions of protection, absorption, secretion, translocation, storage, support, and so on. Many of these adaptations are closely related to, and manifested in, modifications in the cell wall, which undergoes major alterations in size, form, and composition during differentiation. Although animal and bacterial cells may also exhibit somewhat similar extracellular structures, a more highly organized and relatively rigid cell wall is a major feature that characterizes plant tissues. However, this chapter assumes the wide variation in the general morphology of plant cells, which has been thoroughly described in the first volume of this treatise (29) and in a number of other studies (e.g., 36, 46), and it is concerned with concepts of the intracellular organization of plant cells that have resulted from the routine use of the electron microscope in biological research. Much of the economy of plant cells, however, is directed toward the synthesis and differentiation of the extracellular wall, and so the interaction of intracellular and extracellular processes is an important aspect of plant cell organization.

The examination of a wide range of plant and animal cells has revealed that, despite considerable diversity, a basic structural framework is present in all cells and is particularly marked in eukaryotes. We now know that this basic framework is needed for the essential biochemical

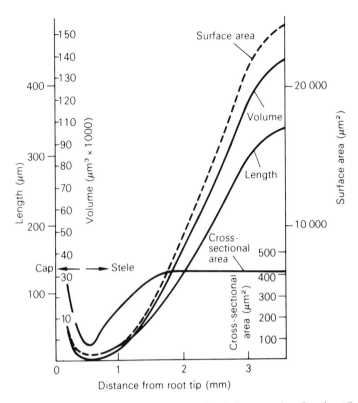

Fig. 1. The dimensions of cells along the axis of the root tip of maize (*Zea mays*). Redrawn from Clowes and Juniper (46).

mechanisms that all cells must possess to maintain life. The study of cell chemistry has likewise revealed similarities in many macromolecules (and in their subunits) that are present in all living cells. From these separate approaches, which have attained a high level of sophistication and produced a great wealth of information, it has been possible to construct a concept of unity at the cellular level (156): unity of plan, unity of function, and unity of composition. (Inasmuch as each living cell has similar basic structures as revealed by electron microscopy, it carries out similar metabolic activity and is composed of similar types of macromolecules built up from similar smaller units.) Nevertheless, in this search for "identities" among cells, there must be room for the difference in structure and organization that is consistent with obvious cellular diversity.

This development of our understanding of cellular structure and function has been very dependent on developments in physics and chemistry and is illustrated by a brief consideration of microscopy. Cell biology began with the first microscopes in the seventeenth century. Their increased use, improved design, and the application of staining techniques led to the discovery of a number of major cell components and the concept of the cell theory [see Brown (29)]. By about 1900, a fairly clear picture of plant cell structure had been established, and this picture remained more or less static until well into the twentieth century. Then a great boost to cytology was given by the introduction of electron microscopy, which in turn had a major impact on biochemistry and cellular physiology (206). The electron microscope, when it became applicable by the development of appropriate fixation and staining techniques, clearly revealed a high level of structural organization not apparent from light microscopy alone. Micrographs of a plant cell taken by light and electron microscopes are compared in Fig. 2. This advance was followed by the development of a range of preparative and staining techniques, such as freeze-etching, enzyme cytochemistry, and auto radiography and by the introduction of new instruments such as the scanning electron microscope and analytical attachments that contributed, and are still contributing, greatly to our knowledge of cellular organization. At the same time, though, the scope of light microscopy was increased by innovations such as interference and fluorescence microscopy and by new methods of specimen preparation often developed for electron microscopy (161). These methods added important additional tools to the cytologist's range of techniques for investigating cell structure and function. In conjunction with these momentous changes in microscopy, the refinement of cell fractionation procedures and associated techniques for the biochemical analysis of isolated cell fractions [see de Duve (55)] has revealed major relationships between structure and function in cells.

Since the discussion of plant cell structure in Volume I of this treatise in 1960, new organelles have been discovered, the existence of others clarified, and an inventory of the major subcellular components of cells has been established. A diagrammatic representation of plant cell structure as revealed by electron microscopy is shown in Fig. 3b, which may be compared with the model based on light microscopy (Fig. 3a). Extensive atlases of plant cell structure are now available (e.g., 99, 150, 200), and the concept of the unity of cellular structure is very apparent although numerous modifications of basic structural features exist. At the same time, the major specific functions of these cellular components have been unraveled, and the concept of cellular compartmentation in which the different organelles are specialized to perform specific meta-

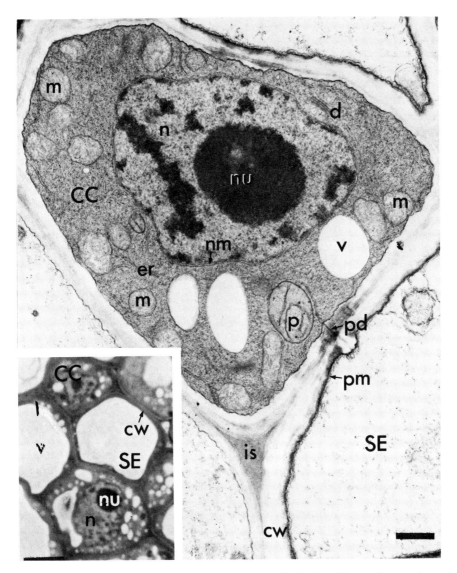

Fig. 2. Electron and light (inset) micrographs of cells in the phloem of pea (*Pisum sativum*) root. The inset is from a 1-μm thick resin-embedded section stained with toluidine blue. Scale bar = 5 μm. The electron micrograph is from an ultrathin section of a similar region stained with uranyl acetate and lead citrate, and it shows a sieve element (SE) and associated companion cell (CC). Scale bar =1 μm. For abbreviations, see Fig. 3.

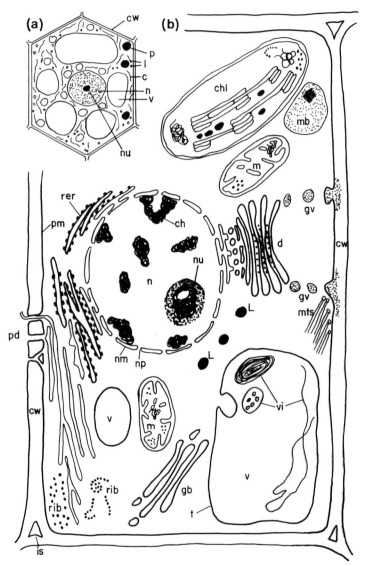

Fig. 3. Diagrammatic representation of a vacuolated plant cell based on light (inset) and electron microscopy. The inset is redrawn after Guilliermond (91) and illustrates the concept of cell structure before the introduction of electron microscopy. Abbreviations: c, chondriosome; ch, chromatin; chl, chloroplast; cw, cell wall; d, dictyosome or Golgi body; gv, Golgi vesicle; is, intercellular space; 1, lipid droplet; mb, microbody; m, mitochondrion; mt, microtubule; n, nucleus; nm, nuclear membrane; np, nuclear pore; nu, nucleolus; pd, plasmodesmata; pm, plasma membrane; p, plastid, rer, rough endoplasmic reticulum; rib, ribosomes; ser, smooth endoplasmic reticulum; t, tonoplast; v, vacuole; vi, vacuolar inclusions. Not to scale.

bolic tasks has become widely recognized. Membranes are vital components in this conceptual framework, not only acting as barriers that control permeation between the various compartments (or indeed eliminate permeation), but also providing a structural framework on which many multienzyme systems may exist and function. A cell is, therefore, to be envisaged as a complex community of separate organized systems existing "symbiotically" (19). Thus, before considering the separate organelles and other major components of plant cells, the structure and properties of biological membranes should be reviewed.

II. Cell Membranes

Living cells are bounded and separated from their environment by the cell membrane, commonly known as the plasma membrane or plasmalemma. Although the thickness of membranes is well below the resolution of the light microscope, the presence and properties of the plasma membrane and, indeed, of the vacuolar membrane or tonoplast were inferred by micromanipulative and permeability studies (29, 48) long before the introduction of the electron microscope, and much was established concerning both their chemical nature and structure. Electron microscopy showed that eukaryotic cells were highly structured systems further compartmentalized by membranes into various discrete organelles, each with characteristic structures and functions. The semipermeable or selectively permeable properties of membranes allow the maintenance of microenvironments within cells or organelles, with each environment presumably adapted to the special metabolic functions that it performs. In addition to these permeability properties, which allow for the selection of solutes and the regulation of their movement between a cell or organelle and its environment, membranes are intimately associated with a wide range of biochemical processes, as is evident in later chapters and volumes of this treatise. These events include the energy conversion processes in mitochondria and chloroplasts, protein synthesis, various secretory processes including those associated with cell wall formation, contractile processes, and reception of growth substances. In many of these situations membranes may be considered a cytoskeleton (221) that provides a framework for the ordered arrangement of multienzyme systems. A sequence of reactions can thus be more efficiently coordinated than would be possible if the enzymes were randomly distributed within the cell or its organelles.

What then is known about the composition and organization of these unique biological structures, which are so intimately associated with cellular function?

A. COMPOSITION

Cellular membranes consist largely of protein and lipid. Some animal membranes, such as those of red cells (erythrocytes), may also contain an appreciable proportion of carbohydrate (up to 10%) in the form of glycoproteins and glycolipids; evidence suggests that glycoproteins are absent or very scarce in plant membranes and that their role has been assumed by the glycolipids (16). Much of the information available on the composition of specific membranes has been derived from animal cells with the red cell membrane being particularly well understood. (That the red cell membrane may be atypical hardly needs to be stressed here.) The paucity of data concerning plant membranes is in part a reflection of the more limited research effort in this field, but it also reflects the greater difficulty in the fractionation of plant cells attributable to the tough cell wall and complicated by the presence of the vacuole. Chloroplast membranes, which may be isolated and purified more easily than most plant membranes, have been studied quite extensively, however. The range and structure of the lipids and proteins that have been characterized from cell membranes are described in a number of reviews (e.g., 16, 44a, 62, 90a, 104) and are not discussed in detail here. However, some of the more important features relating to membrane structure are described briefly.

The four major classes of lipids found in plant cell membranes are phospholipids, sulfolipids, glycolipids, and polyisoprenoids; cholesterol, a common constituent of animal membranes, does not seem to occur in plant membranes. Most of these lipids are amphipathic, that is, they contain both hydrophobic (nonpolar) and hydrophilic (polar) regions. In an aqueous (polar) environment these lipids tend to form aggregates in which their nonpolar regions are aligned to form a hydrophobic phase to minimize contact with the polar environment. An example that is particularly important in relation to membrane structure is the bimolecular layer (see Fig. 4), which has hydrophilic surfaces and a hydrophobic interior. The proportions of the different classes of lipids vary widely in different membrane types (see 105, 238), although the significance of these differences is uncertain. Certainly there is evidence that lipids have more than a purely structural role in forming a permeability barrier. They may play a major role in membrane selectivity, and certain enzyme activities may require specific membrane lipids (104, 105, 238).

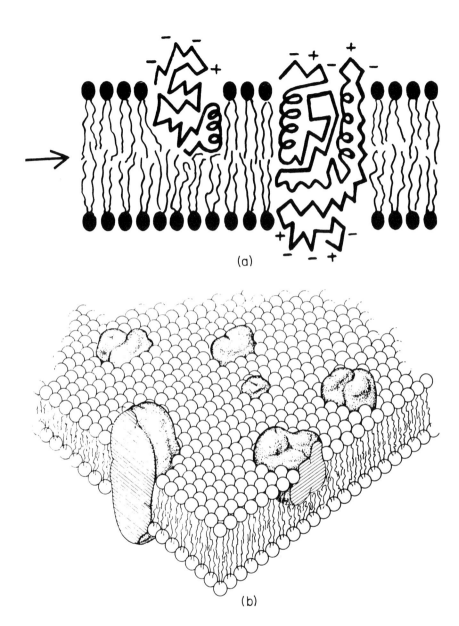

Fɪɢ. 4. Fluid mosaic model of membrane structure. (a) Schematic cross-sectional view. The phospholipids are arranged in a discontinuous bilayer with their ionic and polar heads in contact with water. The integral proteins are shown as globular molecules partially embedded in the membrane. The arrow marks the plane of cleavage to be expected in freeze-etching. (b) Schematic three-dimensional view. The solid bodies with stippled surfaces represent the globular integral proteins. Reproduced from Singer and Nicolson (271). Copyright 1972 by the American Association for the Advancement of Science.

TABLE I

CHEMICAL COMPOSITION OF VARIOUS CELL MEMBRANES

Membrane	Protein (%)	Lipid (%)	Carbohydrate (%)	Protein/Lipid ratio
Nerve myelin	18	79	3	0.23
Plasma membranes				
Mouse liver cells	46	54	2–4	0.85
Human erythrocyte	49	43	8	1.1
Amoeba	54	42	4	1.3
HeLa cells	60	40	2.4	1.5
Nuclear membrane of rat liver cells	59	35	2.9	1.6
Mitochondrial outer membrane	52	48	$(2–4)^a$	1.1
Chloroplast lamellae (spinach)	70	30	$(6)^a$	2.3
Mitochondrial inner membrane	76	24	$(1–2)^a$	3.2
Gram-positive bacteria	75	25	$(10)^a$	3.0
Halobacterium purple membrane	75	25		3.0

[a] Deduced from the analyses and present as either glycoprotein or glycolipid. Data collected by Guidotti (90a).

Membrane proteins may be conveniently classified into two broad groups known as peripheral and integral proteins (62, 90a, 105, 270). The former are weakly bound to the membrane and do not seem to interact with membrane lipids, whereas the latter are tightly bound and can only be dissociated by detergents or organic solvents. Most membrane proteins are integral proteins. Integral proteins may contain a higher proportion of nonpolar residues than do soluble proteins, a finding that is presumably related to their need to interact with membrane lipids.

There is great diversity in the gross proportions of proteins and lipids in different membranes, and this is broadly related to their biochemical and functional complexity (Table I). Thus a simple membrane such as nerve myelin, which functions chiefly as an insulator, has a low protein content, whereas membranes exhibiting complex and varied enzyme activities, such as those of mitochondria and chloroplasts, consist of over

70% protein. The wide variability between membrane types within an organism and between the same membrane types in different species must be accounted for in any general model of membrane structure.

B. STRUCTURE

The concept of a lipid bilayer as the basic feature of membrane structure now appears to be firmly established. It was an essential feature of the Danielli–Davson model proposed in 1935 and originated from the somewhat fortuitous calculations of Gorter and Grendel concerning the lipid content and surface area of erythrocytes. The considerable amount of evidence that has accumulated since then in favor of the lipid bilayer has been thoroughly discussed in numerous reviews (e.g., 24, 121, 271) and is only briefly summarized here. Some of the more compelling evidence is as follows.

1. Modern analyses of membrane lipid content and surface area are generally consistent with a bilayer arrangement.

2. Certain physical properties of synthetic lipid bilayers (e.g., water permeability, transverse dimensions, and electrical capacitance) are remarkably similar to biological membranes.

3. X-ray diffraction and polarization optical studies of nerve myelin and other membranes suggest that they consist largely of a lipid bilayer with nonlipid materials associated with both surfaces.

4. Electron microscopy allows membranes to be visualized and shows a characteristic appearance, after heavy-metal staining, of two electron-dense lines separated by an unstained space (Fig. 5), which is considered to be consistent with the bilayer model. Artificial bilayers have a similar "rail track" appearance after similar treatments.

5. The appearance of membranes after freeze-etching (freeze-fracturing) has also been explained in terms of a bilayer. The hydrophobic interior contains few water molecules and so, on freezing, a line of weakness is formed between the two-lipid layers along which cleavage can occur (see Fig. 4a). The replica so produced reveals the smooth surface of the hydrophobic interior and any particles anchored in this matrix (Fig. 6).

The characteristic appearance of cellular membranes after thin sectioning and heavy-metal staining led Robertson (248) to propose a modified version of the Danielli–Davson model known as the unit membrane, which was considered to represent a universal structure for all biological membranes. The model consisted of a lipid bilayer covered on either

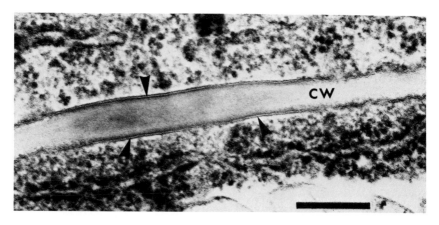

FIG. 5. Electron micrograph of part of two endodermal cells in the root of barley (*Hordeum vulgare*) showing the typical "rail track" appearance of the plasma membrane (arrows) on either side of the cell wall (cw). The plasma membrane is particularly well defined in the region of the Casparian band, where it is closely adpressed to the wall. Scale bar = 0.25 μm. Micrograph provided by A. W. Robards, University of York.

side by proteins in the extended β form. Allowance was made for possible asymmetry, such as glycoprotein content, between the inner and outer surfaces. This model was widely accepted for a few years until serious objections to its universality became apparent. The great variation in thickness, composition, and biological activity of different membranes is not compatible with so standardized a model. Globular units within the plane of the membrane were observed by conventional thin-sectioning techniques and were later clearly demonstrated by freeze-etching (Fig. 6). These particles are embedded in the hydrophobic interior of the membrane, and their density appears to be correlated with the biochemical activity of the membrane (Table II). It is generally assumed that they are predominantly protein in nature. Doubts were also raised as to the nature of heavy-metal staining, and studies on inner mitochondrial membranes revealed a rail track appearance even after 95% of the lipid had been removed. The unit membrane model assumes that proteins react predominantly with the hydrophilic groups of lipids, whereas the rather drastic conditions needed to disrupt membranes suggest that hydrophobic binding is more important. Furthermore, elegant labeling and degradation experiments have shown that proteins may penetrate and, in some cases, traverse the hydrophobic layer. It has also been demonstrated that lateral mobility of proteins in the plane of the

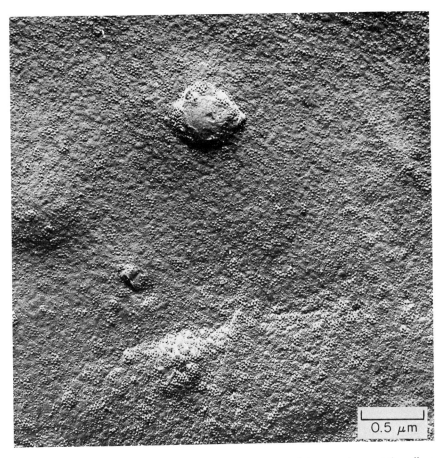

FIG. 6. Electron micrograph of the plasma membrane of *Avena sativa* root tip cell as revealed by freeze-etching. The particles are on fracture face E (i.e., the outer surface of the membrane abutting the cell wall). The large "lump" on the micrograph may represent the fusion of a cytoplasmic vesicle with the plasma membrane. Micrograph provided by B. A. Fineran, University of Canterbury, New Zealand.

membrane may occur; the proteins are considered as "floating" in the lipid bilayer.

Many models alternative to the unit membrane were proposed in the 1960s, and many of these ideas were brought together by Singer and Nicolson (271). Their fluid mosaic model (Fig. 4) is now widely quoted as the most generally acceptable explanation of all available information. Functional membranes are envisaged as a two-dimensional arrangement

TABLE II

DENSITY OF PARTICLES ON FREEZE-FRACTURED MEMBRANE FACES[a]

Type of membrane	Number of particles (μm^{-2})		Membrane area covered with particles (%)
	Densely populated face	Thinly populated face	
Nerve myelin	0	0	0
Plasma membrane (cambial cells of willow)	1608	282	16
Plasma membrane (root tip of onion)	2030	550	15
Tonoplast (cambial cells of willow)	695	140	7
Tonoplast (root tip of onion)	3300	2480	32
Nuclear envelope (root tip of onion)	1790	420	12
Endoplasmic reticulum (root tip of onion)	1700	380	12
Mitochondria			
Outer membrane	2806	770	23
Inner membrane	4208	2120	40
Chloroplast lamellae	3860	1800	80

[a] Data collected by Clarkson (44a).

of globular integral protein dispersed in a fluid lipid matrix. The proteins are amphipathic molecules with their hydrophilic ends protruding into the aqueous phase and their hydrophobic ends embedded in the nonpolar regions of the membrane. The peripheral proteins are not depicted in this model but are considered to be bound to the exposed hydrophilic ends of specific integral proteins (270).

Thus, although this model appears to satisfy much that is known about membranes, the wide diversity in membrane properties and functions suggests caution regarding its adoption as a generalized membrane model. This is especially so because so much of the evidence relating to membrane structure has come from the study of animal plasma membranes. Detailed examination of other membranes is needed before the universality of such a model can be accepted.

C. Transport through Membranes

Having discussed current concepts of membrane structure, it is useful to consider how these models may relate to the mechanisms of transport, because one of the most fundamental properties of membranes is to act as a selectively permeable barrier. Because the lipid bilayer appears to be established as a feature of all membranes, it is important to ask how the transport of water and other small hydrophilic molecules across this hydrophobic barrier occurs. It is generally considered that membrane proteins are intimately associated with this process.

Apart from specific transport mechanisms, an early suggestion derived from classical studies of membranes involving very careful permeability measurements assumed the presence of very small water-filled pores.[2] These pores have never been observed by electron microscopy and, in fact, their calculated dimensions are below the resolution of present instruments. In addition, the pores may well be transient rather than permanent. In terms of the fluid mosaic membrane model, water-filled pores may be generated by subunit aggregates of integral proteins that span the membrane (270) (Fig. 7).

Proteins are assumed to be involved in the more specific facilitated movements of small hydrophilic molecules across membranes. One suggested mechanism involves specific membrane proteins known as carriers. Numerous models have been proposed (see 62, 104, 105), but all involve essentially three steps: solute recognition at the membrane surface in which the solute is bound to the carrier, translocation of the complex across the membrane, and release of the solute into the cell. In active transport, these steps are coupled to energy utilization. At present we know little concerning the nature of these carriers, although one commonly quoted model, involving the rotation of the protein in the membrane, has been criticized on thermodynamic grounds by Singer (270); the movement of the polar and ionic groups of amphipathic proteins through the hydrophobic interior is considered to be a very unlikely event, and little evidence for the rotation of integral proteins within the membrane has been found.

Singer (270) has suggested an alternative model known as the fixed-pore mechanism, in which a conformational change within a hydrophilic protein-lined pore translocates a bound solute across the membrane (Fig. 7). Peripheral proteins may function as binding proteins involved in the specificity of transport, that is, in the solute recognition step.

[2] For early concepts of membrane permeability in plant cells, see Volume II, Chapter 1 of this treatise. (Ed.)

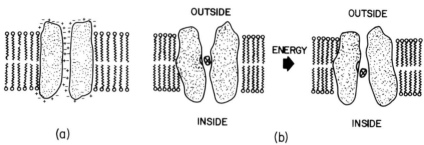

FIG. 7. Diagrammatic representation of possible arrangement of integral proteins to form water-filled pores and a mechanism for the translocation event in active transport. (a) The integral proteins may exist as subunit aggregates that span the membrane to generate water-filled pores. (b) Schematic mechanism for the translocation event in active transport. A specific binding site for a hydrophilic ligand (X) exists on the surface of a pore formed by a particular subunit aggregate. Some energy-yielding process results in a quaternary rearrangement of the subunits, which opens the pore and releases X to the other side of the membrane. Redrawn from Singer (270).

These proteins bind to the subunits of integral proteins that span the membrane. An energy-requiring step results in a conformational change in which the pore opens and the solute is released to the other side of the membrane.

It must be noted that there is no direct evidence for these proposals and that these models have been based very largely on observations of mammalian and bacterial membranes. Their applicability to, say, the higher-plant plasma membrane is not clear. Certainly much more information is needed concerning the properties of various membrane types from higher plants. The evidence for specific binding sites for ions on the plant plasma membrane and the nature of energy coupling to active transport are discussed in Chapter 2 of this volume.

D. MEMBRANE DYNAMICS

Under the electron microscope, cellular membranes necessarily appear as static structures that separate the cell from its environment and subdivide it internally. Yet it is clear from other observations that membranes are not static, and a brief consideration of the more dynamic nature of membranes must be given before considering the details of electron micrographs. Examination of cells such as *Amoeba* or phagocytes by phase contrast microscopy shows that changes in cell shape, implying considerable mobility of the surface membrane, continually occur. Plant protoplasts that lack a cell wall (see 47) will fuse with other protoplasts and show marked volume changes in response to changes in the osmotic potential of the surrounding medium. Examination of the cytoplasmic organelles in living plant cells suggests that there is considerable flexibility in the shape of chloroplasts and mitochondria and that fusion between similar organelles and even between chloroplasts and mitochondria may occur (315). There is evidence of membrane flexibility in the transport of materials into cells by pinocytosis and phagocytosis (see 9), and reversed pinocytosis of vesicles derived from the Golgi bodies at the plasma membrane is a critical step in the synthesis of cell wall material (see Section IV,B). In fact, there is considerable evidence for membrane flow involving the transfer of material from one membrane to another with the endoplasmic reticulum functioning as the major site of membrane synthesis; this is discussed more fully later in this chapter (see Section IV,D).

In addition, radioactive tracer studies have clearly demonstrated that membrane components undergo a regular turnover, being synthesized and broken down many times during the life of a cell. In fact the various components of a particular membrane may show very different rates of turnover, although how the various components are assembled and replaced is not fully understood. Siekevitz (268) has likened membranes to an elastic, deformable brick wall composed of bricks of many colors; each brick is constantly displaced by a similar brick, but these replacements are not in harmony. Thus the membrane remains but its constituent molecules are continuously replaced. If a particular molecule is replaced by one of different structure, chemical modification of the membrane will occur; this may be the mechanism by which membranes differentiate and respond to changes in environment.

III. The Cell Surface

All living cells are delimited from their environment by a membrane, the plasma membrane or plasmalemma, and in most cases they possess some further extracellular structure outside this membrane. In many animal cells a layer of protein and polysaccharide may form a distinct but flexible coat, revealed as a "fuzz" in electron micrographs. Bacteria usually possess a clear, moderately rigid cell wall that provides considerable protection to the cell. Very simply, the wall consists of a framework of covalently linked polysaccharide and peptide chains with various accessory components attached; in Gram-negative bacteria the wall contains an additional lipid-rich coating at the outer surface. In plant tissues the cell wall is the most obvious structure seen under the light microscope, and traditionally it has been considered the major feature that distinguishes plant from animal cells. This carbohydrate-rich, highly organized structure is usually quite thick and tough, giving plant cells and tissues their characteristic form and rigidity and prohibiting sudden, reversible changes in cell shape that may be exhibited by various microorganisms and certain cells in animals. Thus the physiology of cells that possess a cell wall is very different from those that lack a wall with respect to such features as water and ionic relations, the availability and utilization of foodstuffs, and surface–surface interactions. Therefore, although the major structural difference between plants and animals is extracellular, the possession or lack of a cell wall influences the whole functioning of the cell.

The plant cell wall and the plasma membrane may be considered, as they have been historically, two distinct and independent structures—one predominantly carbohydrate in nature, the other composed chiefly of protein and lipid. They may be clearly distinguished in electron micrographs, and, although normally closely associated as a result of turgor pressure, cell wall and plasma membrane can be separated by plasmolysis or the cell wall may be removed completely by treatments with enzymes as in the production of isolated protoplasts (47). However, although structurally the relationship between the cell wall and surface membrane may not be a continuous one, the two are now known to be functionally very closely related in many respects. For example, certain components of the cell wall matrix are synthesized intracellularly and transferred to the wall in vesicles that must penetrate, or fuse with, the plasma membrane. Recent evidence suggests that the plasma membrane may be the site of cellulose synthesis and thus must play a crucial role in cell wall development and differentiation. A striking example of this association between cell wall and cytoplasm is seen in the resynthesis of a new cell wall by isolated wall-less protoplasts (see 246). Conversely, the loss of certain permeability properties exhibited by some protoplasts may be related to the removal of crucial cell wall factors (283). Another proposed association may operate in auxin-induced extension growth of the wall. It has been proposed that auxin stimulates a proton pump in the plasma membrane and that the resulting increase in acidity leads to an increase in wall plasticity (see 3). Electron microscopy has revealed the presence of a variety of vesicular and membraneous structures (paramural bodies) between the cell wall and plasma membrane of different plant cells (252), although their significance is not yet clear. In relation to absorption processes, the cell wall may function as an initial barrier to penetrating substances, both as a physical network and as a charged surface, and so it may influence the range of solutes and particles approaching the plasma membrane. If enzymes are in the cell wall, they may modify substances passing through and so again influence the availability of materials for uptake (e.g., hydrolysis of sucrose and phosphate esters by invertase and acid phosphatase, respectively).

These and other examples of cell wall–plasma membrane interactions are discussed more fully later in this chapter. Although frequently treated for convenience as separate entities, current research suggests that in many situations the cell wall and cell membrane should be considered as a functional association that, in turn, may be closely related to certain cytoplasmic components.

A. THE PLASMA MEMBRANE

1. Properties

Many of the prior observations (see Section II,A,B) on membrane structure and composition also relate to plasma membranes, although this observation is based on data derived mostly from animal systems. Plant plasma membranes appear as typical unit membranes in fixed and stained sections (Fig. 5), whereas freeze-etching reveals particles in the size range of 8 to 16 nm (Fig. 6). In addition, staining suggests that a thin layer of predominantly polysaccharide material may exist at the outer face of the plasma membrane, perhaps similar to the glycocalyx found in animal cells. Such material would mark a very important difference from the other cellular membranes (252, 254). The chemical composition of plant plasma membranes is difficult to assess because the isolation of purified fractions still presents considerable difficulties despite recent advances in preparative techniques (62, 104, 109). Therefore, the detailed chemistry of specific plant plasma membranes is still largely unknown.

The plasma membrane of animal cells is clearly the site for a variety of fundamental cellular processes. These include permeability and solute transport, cell recognition and adhesiveness, immunological responses, and binding of hormones, drugs, toxins, and viruses (306). Much less is known about the role of the plant plasma membrane apart from its obvious involvement in permeability and transport properties and, presumably, in cell wall synthesis and perhaps as a major site of hormone action.

The plasma membrane is usually regarded as a more dynamic structure than is apparent from electron microscopy, which, at present, is the most widely used method of examining its properties. Necessarily, electron micrographs present a static picture of the surface membrane, although, like other membranes, the plasma membrane is presumed to undergo turnover and to be capable of considerable flexibility. For example, a freeze-etch study of the plasma membrane during the life cycle of the unicellular alga *Cyanidium caldarium* showed that the membrane was not uniform but possessed humps and ridges of particles (277) in patterns that were not constant but varied during development.

Further evidence comes from investigations of the processes of pinocytosis and phagocytosis (collectively known as endocytosis), which may be observed by light microscopy in certain living animal cells. Electron micrographs of plant cells frequently reveal invaginations of the plasma membrane although the static nature the invaginations is equivocal. However, more clear-cut evidence has been obtained from the study of the uptake of electron-dense heavy metals and, in particular, by following the uptake of large molecules such as ferritin and of polystyrene beads into isolated protoplasts (9, 176, 235). Another striking example is seen in the development of nitrogen-fixing soybean nodules cells (82). Bacteria move into the cells from a bulge on an infection thread by a process of endocytosis and remain enclosed in an infection vacuole bounded by a membrane originating from the host cell. In muscle cells exhibiting pinocytosis, the density of membrane particles seen by freeze-etching increases in areas where this process is occurring, although the functional significance of this observation is not clear (204).

In addition to these inward movements of the plasma membrane, its flexibility is also demonstrated by processes involving movement in the opposite direction. An example is the development of the pollen grain cell wall, briefly described in Section III,B,4. The intine frequently contains protein with a lamellated or tubular appearance, that is formed by separation of fragments of the plasma membrane in a variety of ways (126, 127). For example, tubular evaginations of the surface membrane may penetrate the intine to be cut off later by the establishment of a continuous plasma membrane and further deposition of polysaccharide material. In the development of the exine, sporopollenin is deposited extracellularly on lamellae derived from the plasma membrane. This process is presumably controlled by intracellular factors, which therefore impose the characteristic patterning on the developing exine.

2. Cell Wall Synthesis

In addition to these rather specific and unusual roles in aspects of cell wall development, it is clear that the plasma membrane is generally involved in a number of ways in primary and secondary wall synthesis. It is now well known (see Section IV,B,2) that cell wall matrix polysaccharides are synthesized in the Golgi bodies and move in vesicles to the cell surface where they fuse with the plasma membrane and add their contents to the wall. This process, of course, also results in addition of membranous material to the plasma membrane, and there is evidence that the membranes of these vesicles undergo differentiation before reaching the surface, presumably differentiating to a structure more closely resembling the plasma membrane than the membrane of origin (see Section IV,D).

A second, more direct involvement of the plasma membrane in cell wall development probably occurs in relation to cellulose synthesis. Biochemical and autoradiographic studies on the incorporation of radioactive glucose into cell wall materials indicate that the cytoplasmic organelles are not involved in cellulose synthesis (44). More direct evidence has come from freeze-etch studies using both algae and higher plants. In the unicellular alga *Oocystis*, bands of granules are observed over the surface of the plasma membrane with an arrangement that corresponds to one of the major microfibril directions (249). It was considered that the bands might be concerned directly with cellulose synthesis, although this interpretation has been questioned in a later study on the same species in which additional linear arrays of particles were identified in association with the ends of growing microfibrils; these particles have been interpreted as cellulose-synthesizing complexes, and the granule bands are thought to be involved in the control of microfibril orientation (31). Similar complexes have also been identified at the end of nascent microfibrils in cells of *Zea mays* (188). If these ideas are correct, cellulose synthesis would occur by the addition of glucose units to the ends of growing microfibrils rather than by the insertion or addition of preformed cellulose molecules.

3. Cell Wall–Plasma Membrane Associations

A number of more complex structural associations between cell surface membranes and cell walls must now be considered. These are the paramural bodies, the plasmodesmata, and the walls of transfer cells.

a. Paramural Bodies. Various membranous structures have been observed between the plasma membrane and the cell wall in electron micrographs from a range of plant species; they appear to be particularly abundant in fungi (166). They have been variously described as para-mural bodies, lomasomes, plasmalemmasomes, multivesicular bodies, and boundary formations. The suggestion that they arise as artifacts is unlikely because they have been observed with a variety of fixation pro-cedures. Marchant and Robards (166) have suggested that all such bodies should be classified as paramural bodies and that they should be subdivided into two classes depending on their origin when it is clear: lomasomes, which are derived from cytoplasmic membranes, and plas-malemmasomes, which are formed entirely from the plasma membrane. The function of these bodies is largely unknown, although there are speculations that involve them in the formation of primary walls, in the transport of materials for secondary wall synthesis, or as sites of enzymes involved in wall modifications.

   b. Plasmodesmata.   Plasmodesmata consist of narrow, tubular strands of cytoplasm that extend through cell walls and connect the protoplasts of adjacent cells. They are bounded by the plasma membrane, which is thus continuous from one cell to another. The resulting cytoplasmic continuum is known as the symplast, and it clearly has considerable potential as a pathway for the movement of materials and signals be-tween two cells. However, there are long-standing and great technical difficulties in demonstrating such a translocatory function, and unequiv-ocal evidence of it is still lacking.

   Plasmodesmata are found in higher plants, pteridophytes, bryo-phytes, many algae, and some fungi (244, 245). In higher plants they appear to be present in all young cells but may be lost or occluded with maturity. Their frequency in the wall varies widely between different cells and tissues, usually being high in walls across which high transloca-tory fluxes occur (244, 245). They are rarely arranged symmetrically around the cell. For example, they are grouped together in primary pit fields, later pits (see Section III,B,4), and, in axial structures such as roots, the transverse walls usually show a higher frequency than the longitudinal walls due to the greater stretching of the latter in cell ex-pansion (137). The development of plasmodesmata from their origin in the cell plate has been the subject of a very thorough study by Gunning (92) using the root of *Azolla* in which all cell types are derived ultimately from a single tetrahedral apical cell. There was no evidence of secondary augmentation of numbers of plasmodesmata after the cell plate stage of development, although plasmodesmata were lost from walls of sieve and xylem elements during their differentiation. The relative high density of plasmodesmata in transverse walls appears to be due to the relatively small expansion that these walls undergo. The root of *Azolla* exhibits determinate growth with the apical cell dividing about 55 times. It is interesting to note that plasmodesmatal frequency falls off after about the 35th division, presumably leading to a gradual symplastic isolation of the apex, a process that could account for its limited lifespan.

The detailed structure of plasmodesmata is difficult to study because it is at the limit of resolution of current microscopic techniques. However, uncertainties concerning the structure of plasmodesmata must be resolved if their function in intercellular communication is to be fully understood. In electron micrographs, they appear as tubular canals in longitudinal section and concentric rings viewed on end (Fig. 8). The canals are usually about 30–80 nm in diameter and are lined by the plasma membrane, which is thus continuous from cell to cell. Many plasmodesmata contain a central strand or core about 20 nm in diameter, known as the desmotubule, that runs along the complete axis of the canal. There is evidence that this strand is derived from the ER, which becomes modified when it is trapped in the canal (133, 243–245). The desmotubule may contain a further electron-opaque central rod about 4 nm in diameter, although this may well be an artifact of negative staining (243–245). In some plasmodesmata the cell wall pore narrows at the two end faces to form a distinct collar that is in close contact with the desmotubule. These observations have been combined to give a tentative model for the structure of plasmodesmata (243, 245), which is shown in Fig. 9.

However, it should be recognized that plasmodesmata show enough structural variability to make a generalized model difficult to formulate. They may contain an enlarged median cavity with a corresponding median knot in the desmotuble, or they may show anastomosing arms that meet in a central cavity. They may lack a desmotubule or be incompletely developed.

The formation of these major structural features is more readily understood in relation to the development of plasmodesmata. It is generally believed that plasmodesmata arise as a normal part of the process of cytokinesis at sites where the fusion of cell plate vesicles is prevented by strands of ER (133; Section III,B,1). The ER is often observed to traverse the cell plate, and the desmotubule may be formed as a result of the forces exerted on the membrane trapped within the plasmodesmatal canal. Although this is probably the primary method of formation, plasmodesmata may also occur in more mature walls not directly formed at cell division. Such examples are to be seen in the development of connections between parasites and hosts and also in graft hybrids, although secondary plasmodesmata may also form between like cells, perhaps after sliding growth, or as in the nematode-induced giant cells in higher

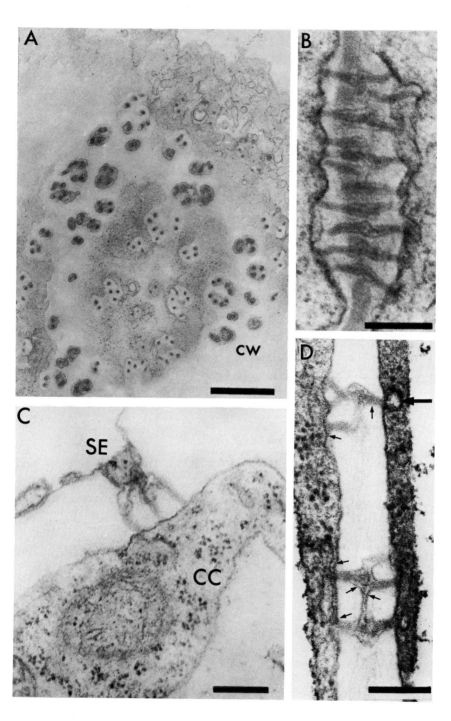

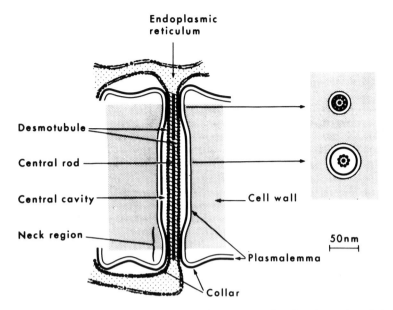

Endoplasmic reticulum

Desmotubule

Central rod

Central cavity

Cell wall

Neck region

50nm

Plasmalemma

Collar

FIG. 9. An interpretation of plasmodesma ultrastructure showing the various features seen in electron micrographs. The endoplasmic reticulum is shown to be continuous with the desmotubule, but in modified form. The diagram is not meant to imply any general uniformity of structure or of specific features. Reproduced from Robards (245).

plants (133). The mechanism of formation of secondary plasmodesmata is uncertain because few cases have been clearly described.

Much that is still uncertain about plasmodesmatal structure has an important bearing on their function (97). For example, little is known about the nature of the plasma membrane that bounds the canal, a membrane that may be highly specialized; this consideration also applies

FIG. 8. Electron micrographs showing aspects of plasmodesma ultrastructure. (A) Glancing section through cell wall (cw) of red beet (*Beta vulgaris*) root cell showing end views of many plasmodesmata arranged in small groups. Scale bar = 1 μm. (B) Numerous plasmodesmata traversing the wall between adjacent cells in wheat (*Triticum aestivum*) coleoptiles. The cell wall is swollen as is often found at sites of compound plasmodesmata. Scale bar = 0.25 μm. (C) A typical branched plasmodesma funneling from a companion cell (CC) to a sieve element (SE) in stem of *Ricinus communis*. Scale bar = 0.25 μm. (D) Compound plasmodesmata in stem of *Suaeda maritima* in which several canals meet in the interior of the cell wall and are connected by a side passage. Elements of endoplasmic reticulum are closely associated with the plasmodesmata (large arrow), whereas the "rail track" appearance of the plasma membrane is seen in places (small arrows). Scale bar = 0.25 μm.

to the adjacent region of the cell wall. There is considerable doubt as to the nature and organization of the desmotubule and central rod and the tightness of the collar of the plasmodesmata on the desmotubule. These features may have important implications for function. For example, bidirectional bulk flow could occur in a single plasmodesma if both the desmotubule (ER) pathway and the cytoplasmic sleeve surrounding the desmotubule were used. Again, the neck constrictions could play the role of valves under certain conditions because there is some evidence for the presence of specialized neck regions that have been called sphincters (see 210). In simple plasmodesmata in leaves of *Salsola* and roots of *Epilobium*, these regions are made particularly visible by staining with tannic acid and heavy metals and are localized just outside the region of the plasma membrane that surrounds the entrance of each plasmodesma (201). It has been proposed that the sphincters may function as valves with the ability to control the rates of symplastic transport through the plasmodesmata. For a fuller discussion of all structural and physiological aspects of plasmodesmata, a volume devoted to this problem should be consulted (98).

c. *Transfer Cells.* The term *transfer cells* has been applied to cells possessing irregular projecting wall ingrowths that result in a greatly increased surface area of the plasma membrane (see 95, 217). Such cells seem to be widespread among herbaceous dicotyledons, rarer in woody dicotyledons, and largely absent in monocotyledons (95, 216, 217). The ingrowths may be distinguished by light microscopy, particularly after staining with the periodic acid–Schiff reaction or with toluidine blue, but they are more easily examined by electron microscopy (Fig. 10). The wall ingrowths are continuous with the secondary walls and are regarded as a specialized form with randomly arranged microfibrils embedded in the matrix. In addition to the characteristic structure of the wall, transfer cells usually contain numerous mitochondria with well developed cristae. There may also be abundant ER that lies close to the plasma membrane.

Transfer cells are assumed to be a means of increasing the total flux across the plasma membrane, although there is little direct evidence for this. Apart from the increased surface area, it is possible that the plasma membrane has special transport properties relating to absorptive or secretory processes (96). Transfer cells are found in plants in a wide variety of tissues; these tissues include glands, absorptive epidermal cells, and the parenchyma surrounding xylem and phloem elements. All these cells may be specialized for short-distance loading or unloading (96, 217). In minor veins, four types of transfer cell are recognized: two in

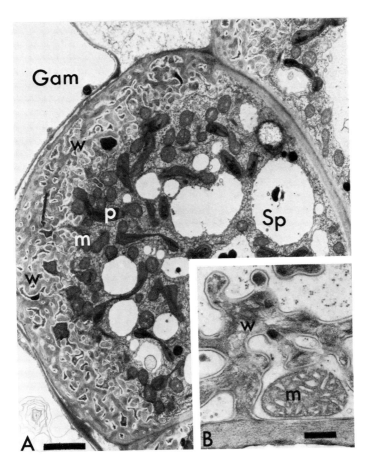

Fig. 10. (A) Longitudinal section of transfer cells at the tip of the sporophyte of *Funaria hygrometrica* prepared by conventional procedures. The sporophyte cells (Sp) are pressed against empty gametophyte cells (Gam). The ingrowths of the wall labyrinth (w) are surrounded by a narrow, clear zone and numerous mitochondria (m) and plastids (p) are present. Scale bar = 2 μm. (B) Freeze-substituted preparation showing the compact nature of the labyrinth (w) and a mitochondria (m) with well-developed cristae. The clear zone is very narrow or absent and may be an artifact of conventional fixation procedures. Scale bar = 0.25 μm. Micrographs provided by A. J. Browning. See Browning and Gunning (32) for further details of methods and interpretation.

the phloem (so-called A and B types) and two in the xylem (C and D types) (96, 216). The A type is the most common, is associated specifically with the sieve elements, and has ingrowths right around the periphery of the cells; it may be anatomically equivalent to a modified phloem companion cell. Type B is more widely scattered throughout the phloem and has ingrowths oriented toward sieve elements or associated cells. Types C and D are much less common and are found in the xylem parenchyma and bundle sheath, respectively, with ingrowths only on walls close to vessels or tracheids. Clearly, much more physiological information, particularly concerning rates of transport of materials and metabolic activity, is needed before the relationship between structure and function in these cells is fully understood.

B. THE CELL WALL

The cell wall is perhaps the most fundamental cytological structure that distinguishes plant from other cells, and its properties are manifested in many of the characteristic features of plant life. In particular, the cell wall is closely associated with the growth and development of plant cells, plant–water relations, and the response of plant cells to pathogens. From being considered solely as a somewhat inert framework surrounding the more active protoplast, the cell wall is now often regarded as a highly organized, dynamic structure that exhibits a clear developmental sequence and pronounced and permanent responses to differentiation. However, the changes that occur in the wall are a consequence and not the cause of differentiation. The commitment to cell types in differentiation is made before any of the familiar wall features emerge at maturity.

Many cell types are easily identified by their characteristic wall structure, and it is clear that the functions of many cell types are closely related to the properties of their cell walls. Although they are outside the plasma membrane, cell walls normally constitute a considerable part of the total dry weight of cells. It follows therefore that a considerable proportion of the economy of cells must be directed toward wall synthesis. Because many of the wall components are synthesized intracellularly, this again implies carefully controlled direction from inside the cell (see Sections IV,B,2; IV,C; and IX,A,2,b). For the present we are concerned with current views of cell wall organization and development, principally as determined by light and electron microscopy.

1. Definitions

A new cell wall is formed at cell division in the equatorial zone between the two daughter nuclei. The first sign is given by the appearance at about anaphase of an organized zone in the cytoplasm known as the phragmoplast. This zone consists of a barrel-shaped region containing branched ER, microtubules, and numerous ribosomes and vesicles. The cell plate, which provides the base for the new cell wall, is formed within the phragmoplast from vesicles that appear to flow into the equatorial region along the path of the microtubules (Fig. 11). The vesicles are thought to originate from the Golgi bodies and also perhaps from the ER (311). The cell plate grows outward by the addition of more vesicles and eventually fuses with the longitudinal cell wall to complete cell division. The membranes of the vesicles may thus contribute to the new plasma membrane. Cytoplasmic continuity between the two cells is maintained by plasmodesmata, which are now generally believed to exist where cytoplasmic continuity persists, at sites where fusion of cell plate vesicles is prevented by the presence of strands of ER (133; Fig. 12).

The zone that persists as a distinct shared region between the two daughter cells is classically known as the middle lamella, and it is thought to function mainly as a cementing layer. The young walls continue to grow and undergo a period of synthesis and consolidation in which two gross phases are usually recognized. The primary cell wall is that formed during all stages of development that accompany increases in cell volume; the wall usually remains thin and relatively undifferentiated during this period. After cell expansion stops, further layers may be added to the cell wall to form the secondary wall. This wall may ultimately be very thick, and it differs both physically and chemically from the primary wall. There are some exceptions to this general description. Collenchyma and some algal cells show increases in wall thickness during growth and the walls of these cells may not be true primary structures. Preston (237) prefers to use the terms "growing" and "nongrowing" walls as being more practical than "primary" and "secondary."

2. Composition

The structure and composition of cell walls, particularly of secondary cell walls, varies widely with cell differentiation, among different tissues, and among similar tissues in different species. However, certain generalizations can be made. The major chemical components of cell walls are cellulose, hemicellulose, pectic polymers, lignin, and structural protein. In addition, walls contain minerals and, in some cases, encrusting or adcrusting substances such as suberin, cutin, or wax. The polysaccharides invariably make up the greater part of the dry weight of the wall although the wall may contain up to 90% water on a fresh weight basis. In general, primary walls are largely composed of cellulose, pectins, and hemicellulose with up to 10% protein and usually little or no

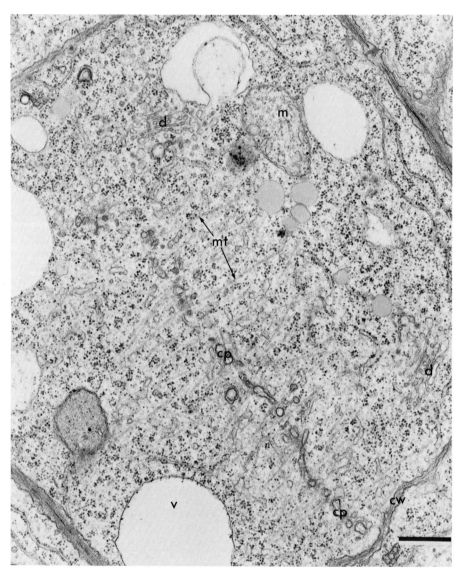

FIG. 11. Dividing cell in the root tip of bean (*Phaseolus vulgaris*) showing the developing cell plate (cp). Numerous microtubules (mt) are present, particularly in the lower half of the micrograph where cell plate formation is more advanced. Scale bar = 0.5 μm. For abbreviations, see Fig. 3. Micrograph by B. A. Palavitz, University of Georgia; courtesy of E. H. Newcomb, University of Wisconsin.

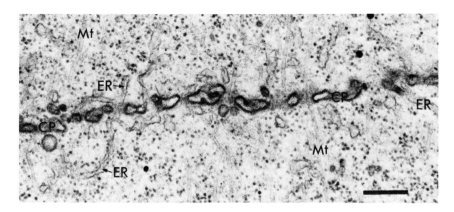

Fig. 12. Early stages in the formation of plasmodesmata in the root tip of *Phaseolus vulgaris*. Section normal to the plane of a developing cell plate (CP). Elements of the endoplasmic reticulum (ER) traverse the fusing vesicular material. Microtubules (Mt) are also present. Scale bar = 0.25 μm. At a later stage the elements of the ER become constricted by the fusing vesicular aggregates. Micrograph by P. K. Hepler and E. H. Newcomb; reproduced from Jones (133).

lignin. The cellulose content is usually higher in secondary walls than in primary walls. In addition, secondary walls show an increase in lignin content to 15–35% in many supporting tissues whereas the pectin content falls to a very low level.

In structural terms, the arrangement of these components to form cell walls is often likened to that of modern composite building materials such as reinforced concrete or glass fiber-reinforced plastic consisting of a dispersed phase of microfibrils within a complex continuous matrix (196); the microfibrils are normally composed of cellulose, and the matrix of combinations of hemicellulose, pectic substances, and lignin. Although such an analogy may often be useful, the dynamic nature of the cell wall, which responds to environmental and developmental changes, must always be remembered; the proportions and interactions of the wall components may change to alter the properties of the wall and ultimately the function of the cell.

It is beyond the scope of this chapter to consider the detailed chemistry of the wall components. These are thoroughly described elsewhere (3, 196, 236, 237), and only certain basic features are outlined here.

a. Cellulose. Crystalline microfibrils are an essential part of the cell walls of all green plants, and, with a few exceptions, these microfibrils are composed of cellulose, probably the most abundant organic macro-molecule on earth. The exceptions are found among the algae, where some species contain a linked xylan or linked mannan as the only struc-tural polysaccharide (236). Cellulose is a linear, unbranched chain of β-1 → 4-linked glucose units that may contain 8000–12,000 residues in sec-ondary walls. The chains are usually shorter in primary walls, with per-haps 2500 residues. These linear molecules are held together by hydro-gen bonds to form aggregates, the microfibrils, which are readily visible under the electron microscope, particularly by shadowing after the ma-trix materials have been extracted (see 236). In some regions of the microfibrils, the cellulose chains are highly ordered to form crystalline regions that have a distinct X-ray pattern and give positive birefringence under the light microscope. Elsewhere the chains are more loosely ar-ranged in paracrystalline regions. There are a number of uncertainties concerning microfibrillar structure. It is not clear whether adjacent cel-lulose chains lie in a parallel or antiparallel (in opposite directions) ar-rangement, although the latter is generally favored. There is some dis-pute over the exact diameter of the microfibrils, which may vary between plants. For higher plants, however, each microfibril appears to contain a crystalline core about 4 nm in diameter and surrounded by a less or-dered cortex giving a total diameter of about 10 nm (236). There is also evidence that noncellulosic polysaccharides are associated with microfi-brils, particularly with the paracrystalline cortex. These polysaccharides may be particularly rich at the surface of the cortex and important in the interaction between the microfibrils and the matrix (196, 236).

   b. Matrix Polysaccharides.   Two broad groups of noncellulose polysac-charides, the pectins and hemicellulose, have been defined on an empiri-cal basis in terms of their solubility in various inorganic solvents. Unlike cellulose, which is chemically identical from plant to plant and cell to cell, the nature of these matrix compounds varies widely and involves a com-plex group of linear and branched polymers. These polysaccharides have proved to be difficult to study because the extraction methods are not wholly specific, rarely provide homogeneous polymers, and are fur-ther complicated by degradation products produced during extraction. Perhaps the most clearly defined study has been that of cultured cells of sycamore (*Acer pseudoplatanus*), which have been analyzed using a spe-cific wall-degrading enzyme isolated from a fungal pathogen (3, 282). This study is discussed in Section III,B,3 in relation to the molecular structure of the primary wall.

   Pectic substances are relatively soluble and may be extracted by hot water, dilute acid, or EDTA. The major constituents of the pectic fraction are various acidic (galacturonorhamnans) and neutral (arabinoga-lactans) polysaccharides (196). In sycamore cell walls, it has been sug-gested that a branched arabinan and linear galactan occur as side chains to the galacturonorhamnan, and that the wall hemicellulose (a xyloglu-can) is covalently linked to the galactan chains, thus forming a link between the hemicellulose and pectic components (282). Isolated pectins form viscous solutions and reversible gels with water, and this phenome-non is presumably related to their role as filler substances within the matrix, particularly in relation to water distribution (148, 196).

The hemicelluloses are extracted after the pectins by solution in alkali. Again, this fraction constitutes a complex group of polymers, including xylans (which form the bulk of the hemicellulose fraction in angio-sperms), arabinoxylans, mannans, glucoglucomannans, and galactoglu-comannans (which form the bulk of the hemicellulose fraction in gym-nosperms) (196). These polysaccharides are able to form paracrystalline structures, although they lack the degree of stabilization found in cellu-lose. Xylan molecules appear to be oriented parallel to and closely ap-plied to the surface of the microfibrils where they may function as sub-stances linking the microfibrils with the matrix (196). In sycamore cell walls, the principal hemicellulose is a xyloglucan that appears to be cova-lently linked to the pectic polysaccharides and noncovalently bound to the microfibrils (12).

c. Protein and/or Glycopeptides. The primary cell walls of almost all green plants examined to data appear to contain a hydroxyproline-rich glycoprotein. Although only traces of this amino acid are found in the cytoplasm, it has been stated that up to 30% of the amino acid residues of the protein are hydroxyproline, and that the protein may account for 2–10% of the cell wall preparations (146, 147). The glycoprotein ap-pears to be a protein–arabinogalactan complex (147, 148). When bulk cell wall preparations have been fractionated into their components, the bulk of the hydroxyproline remains with the cellulose fraction (118). However, combined hydroxyproline complexes are also found in situa-tions where the walls are not cellulosic (178). In sycamore (*Acer pseudo-platanus*), it has been suggested that the pectic galacturonorhammans are linked to the hydroxyproline-rich protein through an arabinogalactan (140).

It must be remembered, however, that the chemical identity of these glycopeptides was established initially upon examination of isolates from bulk biochemical preparations, and careful consideration is required before relating them unequivocally to "cytological" cell walls. Attempts to localize bound hydroxyproline in intact cells by high resolution autoradiography have produced contradictory findings. For example, when rapidly growing carrot cultures were supplied with [^{14}C]proline or [^{3}H]proline and high resolution autoradiography was used to study the distribution of radioactivity fixed as proline and hydroxyproline, the visible cytological cell wall was free of radioactivity, which nevertheless showed a nonrandom distribution within the cells (131a). Furthermore, sporelings of *Valonia ventricosa*, which in culture develop a cellulose cell wall that has been investigated in detail, accumulate combined proline and hydroxyproline from a proline source without incorporation into the cellulose walls (280a). In contrast, the application of a similar technique to sycamore suspension cells reveals heavy labeling from the proline source over the cell walls (247a). These differences may well be a reflection of the state of development of these cells, because a further study of sycamore suggested that actively growing cells show most radioactivity in the nucleus and cytoplasm, whereas cells with thickened walls accumulate radioactivity in these walls (280b). If this distribution is a general property of plant cell walls, then the need for caution advised previously and the value of complementary cytological observations are fully justified.

d. Lignin. Lignin constitutes the chief nonpolysaccharide component of cell walls and is laid down very largely during secondary thickening. However, not all secondary walls contain an appreciable proportion of lignin. It is insoluble, very resistant to chemical degradation, and variable from plant to plant, and so it has proved difficult to characterize. Lignin appears to be a high molecular weight, aromatic polymer derived from the enzymatic dehydrogenization of coumaryl, coniferyl, and sinapyl alcohols (77). Lignin acts as a hydrophobic filler material that gradually replaces the water in the cell wall to encrust the microfibrils and matrix substances (196), giving great rigidity and chemical resistance to the wall.

e. Suberin, Cutin, and Wax. The outer walls of leaf epidermal cells and the walls of cork cells are examples of adcrusting with cutins and suberins, although the methods of deposition may be very different. Certain walls may also contain additional impregnations with wax. These adcrustations greatly restrict the diffusion of water and solutes and, in general, form a protective covering. Cutin and suberin are complex mixtures of polymeric cross-esterified fatty acids with hydroxy fatty acids predominating (148, 196). Waxes are a heterogeneous group of compounds containing long-chain alkanes, alcohols, ketones, and fatty acids (148). A cuticle usually consists of an outer layer of wax covering a skin of cutin, the lipid material of the cutin penetrating the outer polysaccharides of the wall (148, 168).

3. Structure and Growth of the Primary Wall

The most thoroughly characterized cell wall in terms of molecular composition is that of cultured cells of sycamore (see 3, 12, 140, 282), and information derived from these studies has led to the formulation of a tentative model for the macromolecular structure of these primary walls (3, 140). The primary cell wall has been considered as a single macromolecule with the galacturonorhamnan, arabinogalactan, xyloglucan, and, probably, the protein interconnected by covalent bonds, whereas the xyloglucan and cellulose are strongly associated by numerous hydrogen bonds. Galactan functions as a link between the xyloglucan–cellulose complex and the pectic galacturonorhamnan. A coherent model may also be constructed using only the linkages between cellulose and xyloglucan and between xyloglucan and the pectic polysaccharides (Fig. 13). A single pectic polymer is attached through xyloglucans to more than one microfibril, and a single microfibril is connected in the same way to more than one pectic compound. This model does not

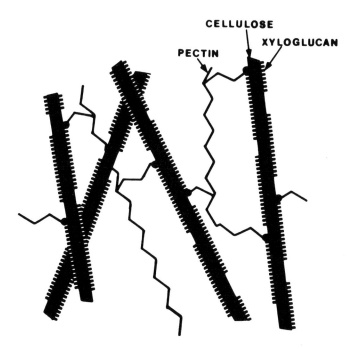

FIG. 13. Highly simplified model of the primary cell wall of a dicotyledon. See text for explanation. Reproduced from Albersheim (3).

involve the cross-linking role of protein, although both of these types of interconnections may well be present in many cell walls.

These models are, of course, constructed from data obtained from cultured cells whose walls may differ significantly from the walls of intact plants. Certainly, detailed information is now required from a range of plant types. Nevertheless, Albersheim (3), in considering the results obtained from cultured sycamore cells and from cell wall preparations from certain whole plants, suggests that the models may serve as a basis for an architectural plan of the primary wall of other dicotyledenous and also monocotyledenous plants. For example, although xyloglucan is not the dominant hemicellulose in many plants, both dicots and monocots contain other hemicelluloses that bind tightly to cellulose and so may replace xyloglucan in the proposed structure.

Thus the work on sycamore cell cultures provides a working model for further studies on the structure, synthesis, and extension of primary cell walls. We shall see later that much is now understood concerning the sites of synthesis of the molecules that make up the wall, particularly the matrix components. However, the mechanisms by which they are assembled to give the macromolecular structures described above are still largely unknown.

A cell may often increase some 10–50-fold in length (far more in the case of many fibers) as it expands from its meristematic to mature state, and this increase involves an enormous increase in the area of the cell wall. This growth requires both the deformation (stretching) of existing material and the deposition of new material. Thus wall thickness remains approximately the same at all stages of expansion in most cases. The addition of material to the wall is believed to occur either by the deposition of layer upon layer (appositional growth) or by the insertion of new material within the existing structure (intussusceptional growth). Electron microscopic autoradiography indicates that both processes occur and that their relative contributions may be influenced by auxin (239). Appositional growth is involved in "passive" wall extension driven by the tensile stress induced by turgor pressure and is thought to account for the greater part of new material incorporation (236). Deposition by intussusception may set up "growth pressures" and so cause cell extension; this process is probably more important with respect to the incorporation of noncellulosic components and plays a relatively minor, modifying role in cell wall extension.

The basic structural components of the wall are the cellulose microfibrils, and, as we shall see later, these appear to be deposited largely by apposition to form specific patterns that have been the subject of many investigations and considerable speculation. The hypotheses that have been proposed to account for microfibrillar arrangements have been fully and critically discussed elsewhere (236, 251, 255, 256). The most widely accepted theory was proposed by Roelofsen and Houwink (see 251) and is known as the multinet growth theory. In the innermost lamellae of the wall, adjacent to the plasma membrane, the microfibrils are laid down more or less transversely to the direction of growth. During cell elongation they become passively stretched and reoriented roughly parallel to the axis of growth, while at the same time new fibrils are being laid down at right angles on the inner face (Fig. 14). Thus a fully extended wall contains a gradation in microfibrillar orientation with the outermost lamellae reflecting the growth axis of the cell. This theory was developed from observations on tip growth by cotton hairs and has been confirmed for certain other higher and lower plant cells. In general, microfibrillar patterns have been examined by shadowing after most of the wall matrix has been removed. However, the multinet theory cannot explain all the variety of wall patterns that have been described. For example, corner thickening in collenchyma appears to require localized microfibrillar synthesis, which conflicts with the uniform distribution predicted by the multinet hypothesis. Other cell walls have been observed to contain longitudinal ribs of microfibrils that are present throughout the growth of the wall. These and other observations suggest that many of the patterns developed during growth are too ordered and complex to be determined solely by mechanical extension. In addition, the drastic nature of the shadowing techniques widely used in these studies has been criticized because they may well produce displacement and scattering of the wall components (255, 256).

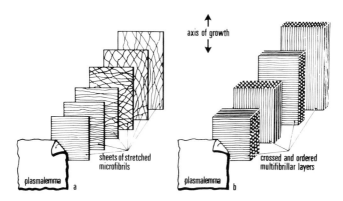

FIG. 14. Comparison between the multinet growth hypothesis (a) and the ordered-fibril hypothesis (b). See text for explanation. Provided by J.-C. Roland; after Roland *et al.* (255).

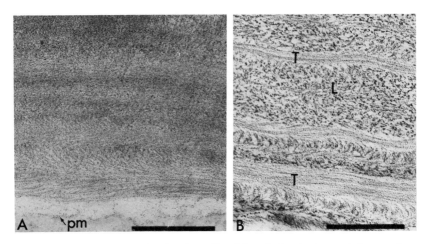

Fig. 15. Electron micrographs showing cell wall lamellation in the cortical parenchyma of mung bean [*Vigna radiata* (*Phaseolus aureus*)] hypocotyl (A) and the collenchyma of celery (*Apium graveolens*) (B). Resin-embedded sections were stained for polysaccharides using the periodic acid–thiocarbohydrazide–silver proteinate procedure. T,L, tranverse and longitudinal polysaccharides; pm, plasma membrane. Scale bar = 0.5 μm. Reproduced from Roland *et al.* (255).

An alternative concept, known as the ordered fibril hypothesis, is based on observations using less drastic techniques such as ultracryotomy with negative staining (255). This concept proposes that the primary wall is made up of superimposed lamellae, each having its own characteristic ordered organization and its own response to growth regulators and stress (Figs. 14, 15). Thus the wall would have a more important role in the control of cell growth, the potential for growth of a given cell being determined essentially by the structure of the primary cell wall. Both longitudinal and transverse layers of fibrils are found throughout the wall. The transverse elements are envisaged as providing resistance to cell enlargement, whereas cell elongation is allowed by sliding of fibrils or fibrillar lamellae against each other. This elongation process implies a linkage of these elements by labile bonds, their breakage or formation controlling the rate and direction of expansion. The question of how the orientation of the lamellae is determined remains unanswered. Similarly, the mechanisms involved in initiating and controlling wall loosening and cell expansion, which include consideration of the mechanism of auxin action, are the subject of considerable debate though of little real understanding. This topic falls outside the present discussion but is thoroughly analyzed elsewhere (3, 110). One interesting possibility is that a transglycosylase in the wall could temporarily effect connections between cellulose fibers by making or breaking bonds between interconnecting polysaccharides (3). The enzyme is envisaged as having a low pH optimum and thus as more active when the wall pH is lowered by an auxin-activated proton pump. With connections loosened, the fibers could move past each other in response to turgor pressure.

4. Secondary Walls

Secondary walls are formed largely after cell expansion has stopped and differ both chemically and structurally from primary walls, although, of course, not all cells produce secondary walls. The major chemical changes generally associated with secondary wall formation are an increased amount of cellulose, a wider range of hemicelluloses, and the addition of a significant proportion of lignin. Wax, cutin, and suberin may also be added at this stage. However, it must be remembered that there is an enormous diversity in the thickness and composition of secondary walls. Some may contain up to 95% cellulose; at the other extreme, the wall may be dominated by the matrix polysaccharides forming masses of mucilaginous materials. This diversity reflects the wide range of different cell types (e.g., vessels, tracheids, fibers, and phloem elements) whose functions are closely related to secondary wall structure.

The cellulose microfibrils found in the secondary walls frequently show a densely packed and highly ordered arrangement because they are not subjected to the stretching effects of cell enlargement and generally lie more parallel to the axis of growth than in primary walls. In many fibers they are closely parallel to the growth axis whereas in others they exhibit a helical arrangement. However, in the vessels of angiosperms, the fibrils lie more or less at right angles to the direction of growth in an annular arrangement. Many secondary walls, particularly those of tracheids and fibers, show a division into three distinct, superimposed layers that have been designated S_1, S_2, and S_3, with S_1 the outer layer next to the primary wall (Fig. 16). These layers are distinguished by the orientation of the microfibrils, usually helically arranged, and each may consist of a number of separate lamellae. The layers may also differ chemically from each other, both in cellulose content and in the proportions of the various hemicelluloses (236). How these patterns of deposition are directed and controlled is not clear, although microtubules probably play an important role in the orientation of the microfibrils (see Section IX,A,2,b).

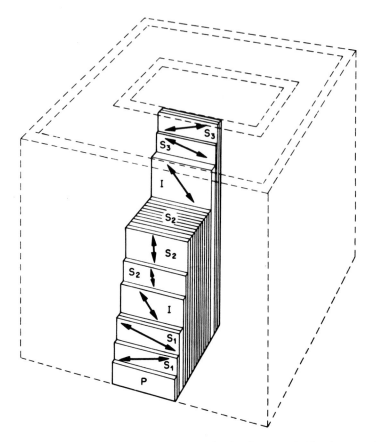

FIG. 16. Diagrammatic representation of a tracheid wall as revealed by electron micros-copy. The dotted lines depict a segment through a tracheid. Double-headed arrows show the general orientation of the microfibrils in the various layers (for details see text). The intermediate microfibrils (I), which represent a third microfibrillar lamella in the S_1 and S_2 layers, are not always present, and S_3 is absent altogether in some species. P, primary wall. Reproduced from Preston (236).

Secondary wall formation also involves the addition of various en-crusting or adcrusting substances. The most important encrustation is lignin, which penetrates the wall from the outside (primary wall) inward, gradually replacing water and so allowing the formation of strong hy-drogen bonding between the matrix polysaccharides and cellulose and between the various matrix components. Covalent bonds may also be formed between the carbohydrates and lignin. The wall then resembles

FIG. 17. Scanning electron micrograph showing bordered pit in the side walls of xylem tracheids in pine (*Pinus*). The pit "membrane" consists of a lens-shaped disc of lignified wall material [the torus (T)] suspended in the cavity between the pair of borders by bundles of microfibrils (MF). The diameter of the torus is somewhat greater than the aperture of the borders, and it may act as a valve by moving on its slings of microfibrils. Scale bar = 2 μm. Reproduced from Gunning and Steer (99).

a synthetic glass fiber composite, with great tensile strength due to the microfibrils and a rigidity derived from the lignified matrix (196). Cutin and suberin may also be deposited on the wall to form further protective layers that help to bind the cells together. These materials are usually found on outer surfaces, such as the epidermis and periderm, but also occur internally in the endodermis. Perhaps the most striking example of the protective function of these secondary wall deposits is seen in pollen grains, which possess walls of great structural complexity (126, 127). The pollen grain cell wall, known as the sporoderm, consists of an inner layer, the intine, and an outer layer, the exine. The intine is similar in structure to normal primary walls and often contains proteinaceous material, such as lamellae or tubules, which is derived from the cell surface and incorporated by a variety of mechanisms (see Section III,A). The exine forms the very resistant outer coating, which often shows characteristic sculpturing that provides a basis for pollen identification. It contains carbohydrates, lignin, and a material known as sporopollenin. The latter is a polymer of carotenoid and carotenoid derivatives and is extremely resistant to both chemical and biological degradation. Sporopollenin appears to be deposited on special lamellae formed at the plasma membrane, with deposition occurring as soon as the lamellae are extruded extracellularly (99, 126, 127).

Secondary wall thickening does not normally occur over the whole surface of a cell because this would result in the loss of cytoplasmic continuity between cells through the plasmodesmata (see Section III,A,3,b). Thickening does not occur over small localized regions called pits, which are usually formed over regions of the primary wall that contain numerous plasmodesmata and that are known as primary pit fields. These perforations in the secondary wall may be straight-sided (simple pits) or may show an overarching of the primary pit field to form a raised border (bordered pits) (Fig. 17). The primary wall, known as the pit "membrane," may remain unmodified or, as in many conifers, may develop a thickened central region known as the torus, which is formed by the deposition of microfibrils in a circular arrangement. The surrounding marginal zone, the margo, shows a radial arrangement of thinly dispersed microfibrils. The possible functions of these pits and the torus in controlling flow between cells is fully discussed by Preston (236).

IV. The Endoplasmic Reticulum and Golgi Apparatus

Although it was possible to see the endoplasmic reticulum and Golgi bodies in certain cells by light microscopy, the widespread occurrence of these membrane systems in plant and animal cells was established only after development of electron microscopy. Evidence now strongly suggests that the ER and Golgi apparatus comprise a structurally and functionally integrated membrane system that is particularly involved in the synthesis, transport, and secretion of proteins, glycoproteins, and polysaccharides (44, 186, 207). Perhaps the best described system is that concerned with protein secretion by the pancreatic exocrine cells (207). In these cells proteins are synthesized on the polyribosomes attached to membranes of the rough ER (see following discussion). The secretory proteins are segregated into the cisternal space of the ER and transported via the transitional elements of the system to the Golgi complex. The proteins are concentrated and stored temporarily in certain large vacuoles before being discharged extracellularly by fusion of secretory granules with the plasma membrane in a process known as exocytosis (or reversed pinocytosis). In addition to this continuity between the ER and the Golgi apparatus, there is good evidence that the ER is also closely associated with the nuclear envelope (see Section IV,D).

A. THE ENDOPLASMIC RETICULUM (ER)

1. Structure and Range of the ER

In plants, the ER usually appears in thin sections as sheet-like cisternae (lines of double unit membranes), often in parallel array, that permeate the cytoplasm. Alternatively, it may also appear as short, randomly distributed tubules. The former type usually contains ribosomes attached to the outer (cytoplasmic) surface of the membrane and is known as rough or granular ER (Fig. 18). The density of ribosomes varies widely in different situations. ER that lacks ribosomes is known as smooth ER and is more variable in its form; it has been described as having the form of lamellae or as composed of sac-like, tubular, or vesicular structures. This general view of ER structure has been confirmed by freeze-etching, which also demonstrates the presence of the characteristic membrane particles (68, 69). The extent of the ER varies widely in different cell types; it may be very sparse in some cells, whereas others (e.g., certain gland cells) show massive parallel arrays of cisternal or extensive tubular elements. Calculations of the surface area provided by these membranes confirm this variability (99). Pancreatic cells and certain plant glands provide a surface area of 15–20 μm^2 per μm^3 of "average" cytoplasm. Liver cells and tapetal cells contain 6–12 μm^2 per μm^3 of cytoplasm, and root tip cells 1 μm^2 per μm^3 or less. In root cap cells, the ER increases approximately in proportion to cell size (46), but this relationship is not always so simple. The ER may also change mark-

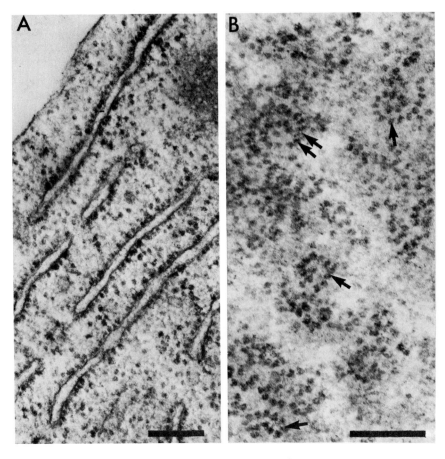

FIG. 18. Electron micrograph of a portion of an epidermal cell in wheat (*Triticum aestivum*) coleoptile showing (A) appearance of the rough endoplasmic reticulum and (B) free ribosomes and polysomes (arrows). A polysome with about 23 ribosomes is indicated by two arrows. Scale bar = 0.2 μm. Reproduced from Pope *et al.* (229).

edly, sometimes very rapidly, during cell development or in response to external stimuli (46). For example, certain glands may show changes in ER with the onset of secretion (260), and the form of the ER changes in cambial cells with the seasons, as does that of other organelles (41). The form of ER may change dramatically when storage tissues are sliced and washed (103, 296), and it shows a marked increase in the cells of the separation zone during leaf abscission (264). In roots of certain grasses, the ER of the root cap cells may be redistributed in response to gravity when the root is displaced from the vertical, although this response is not universal and may be influenced by the movement of heavier plastids (46, 99, 136). The significance of these variations in the form and extent of the ER is discussed together with the functions of this membrane system in Section IV,A,3.

2. Ribosomes

It is appropriate to discuss here the structure of ribosomes because they are often considered in association with the ER, although they also occur separately in the cytoplasm and within certain organelles.

Ribosomes are small subcellular particles found in almost all living cells. They function as sites of protein synthesis where amino acids are assembled in a specific sequence to produce polypeptide chains. They are roughly spherical, about 17–23 nm in diameter, and so are invisible by light microscopy. Ribosomes are osmiophilic and were first observed in plants in electron micrographs of isolated fractions by Robinson and Brown (250). Since then they have been seen in micrographs of almost all living plant cells with the exception of mature sieve elements, which lose their ribosomes, nuclei, and mature Golgi bodies.

Ribosomes occur both free in the cytoplasm and attached to the ER (Fig. 18). They frequently occur as strings or clumps with as many as 20 ribosomes joined by a thin thread about 1 nm wide. The thread is believed to be messenger RNA because it is destroyed by the enzyme ribonuclease. These groups—called polyribosomes or polysomes—are presumed to be ribosomes passing along a strand of messenger RNA, each at a different stage in the assembly of the same polypeptide chain. The number of ribosomes in the chain thus gives an indication of the length of the messenger RNA and hence the size of the protein molecule.

Ribosomes are also found within the nucleus, where they are at least partly assembled, and in chloroplasts and mitochondria. Ribosomes isolated from chloroplasts and mitochondria are distinguishable from cytoplasmic ribosomes and resemble bacterial ribosomes in many respects (see 107, 281). Based on their sedimentation coefficients in the ultracentrifuge, eukaryotic cytoplasmic ribosomes are frequently classified as 80 S, and those from bacteria, chloroplasts, and mitochondria as 70 S. However, these classes also differ in relation to a number of biochemical characteristics, and there appears to be a wide size range within the 70 S class; mitochondrial ribosomes, in particular, have been reported to have much lower S values in some cases.

The detailed ultrastructure of ribosomes is more readily seen after negative staining, which clearly reveals that each ribosome is made up of two subunits, one larger than the other (192, 194). The subunits dissociate if magnesium ions are removed, and they may then be separated by ultracentrifugation. Each consists of 30–50% protein, and the rest is RNA. In animal and bacterial ribosomes (and so presumably in plant ribosomes), the small subunit appears to sit on the large subunit, forming a tunnel that runs between the two particles. The tunnel is thought to be occupied by the messenger RNA and other factors involved in protein synthesis.

3. Functions of the Endoplasmic Reticulum

The ER presents two important structural features to be considered in relation to cell function. First, it greatly increases the surface area of membrane available both for the ordered arrangement of enzymic reactions and for the spatial separation of different metabolic systems. Second, the space enclosed by these membranes is available for storage and as a channel for the transport of materials, both intracellularly and, possibly, intercellularly via plasmodesmata (see Section III,A,3,a). In addition, the involvement of the ER–Golgi apparatus complex in the origin of cellular membranes and subcellular structures suggests that considerable biochemical heterogeneity exists within these membranes and that this heterogeneity may vary with time. Although this functional differentiation is normally not detectable morphologically (except for such rather gross differences as those between rough and smooth ER), the concept of a dynamic, changeable membrane system (see Section II,D) is clearly very apt.

a. Protein Synthesis, Secretion, and Storage. The structural association between the ER and ribosomes and their ,roles in the synthesis and secretion of proteins in animal cells have already been mentioned. Polysomes occur both free in the cytoplasm and bound to the ER, and there is evidence, especially from animal cells, that these two types are responsible for the synthesis of different classes of protein (see 20). In particular, protein for secretion seems to be synthesized on polysomes associated with the ER. There is some evidence that similar systems operate in plants, as seen from correlations between extensive stacks of ER and synthesis of proteins that are exported or, at least, transported intracellularly. For example, in certain insectivorous plants, the gland cells that secrete digestive enzymes may show an increase in the rough ER with the onset of secretion (128). An autoradiographic study of the green alga *Enteromorpha*, which secretes glycoprotein-containing adhesive vesicles, showed that protein is synthesized in the ER and passes through the Golgi bodies before adhesive-containing vesicles discharge their contents to the exterior (38). Another autoradiographic experiment indicated that a similar process might be involved in the secretion of proteins by the petiolar glands of *Mercurialis annua* (64).

Examples of extracellular protein secretion in plants are few, but other interesting examples relating ER development to protein synthesis do occur. For the mobilization of reserves during the germination of cereal grains, hydrolytic enzymes are synthesized in the aleurone cells and secreted into the endosperm. This activity is stimulated by gibberellic acid and is accompanied by an extensive proliferation of ER and attached polysomes, which are thought to be involved in the synthesis and export of the enzyme proteins (61, 134,135).

The ER also appears to be involved in the synthesis of reserve protein. For example, when storage discs are sliced and washed, the ER usually increases in extent and proteinaceous crystals may accumulate in the cisternal space (103, 132). A more striking demonstration is found in the synthesis of storage proteins in developing cotyledons. Just before synthesis commences, there is a proliferation of ER and an increase in the proportion of membrane-bound ribosomes (7, 20, 26, 202). The protein is accumulated in the vacuoles, which are converted into structures known as protein bodies or aleurone grains. Labeling experiments, including autoradiography, suggest that the protein is first synthesized in the ER and later transferred to the vacuoles (7). It has been postulated that the newly synthesized protein is released into the ER cisternae and then gradually transported to the vacuoles; the Golgi bodies may not be involved (7, 174). Later, as the seed matures and is dehydrated, there is a loss of membrane-bound ribosomes and an increase in free ribosomes.

b. Polysaccharide Synthesis and Secretion. Ultrastructural studies of membrane distribution indicate that the ER may be associated with cell wall formation in certain situations. The possible involvement of the ER in cell plate formation has already been mentioned (see Section III,B,1). During maturation of xylem vessels in wheat (*Triticum*) coleoptiles, the ER is arranged in a definite way so that profiles of sheets of the ER are seen at right angles to the wall between each thickening when longitudinal sections are examined (195). A similar morphological association is found in the development of sieve plates, where elements of the ER appear to denote the positions of pore formation and the probable sites of deposition of callose (195). However, as O'Brien (198) has indicated, particularly in reference to xylem differentiation, the relationship between static electron micrographs and the dynamic reality is at best uncertain. Certain biochemical and cytochemical evidence suggests that the ER plays only a minor role in the formation of wall polysaccharides (44). However, the ER could contribute by providing precursors and enzymes to the plasma membrane for cellulose synthesis, as is suggested in a study of wall regeneration by isolated protoplasts (246). Cultured protoplasts show an increase in cell organelles including the ER, which also exhibits close connections with the plasma membrane. It is suggested that the rough ER synthesizes precursors and enzymes for cellulose synthesis, which are then transferred to the plasma membrane. Whether this hypothesis applies to wall development in "normal" cells remains to be seen.

c. Lipid Biosynthesis. Biochemical studies have shown that microsomal fractions isolated from both plant and animal tissues contain most of the enzymes required for the synthesis of a range of lipids and sterols. In some cases there is evidence that these fractions are rich in smooth ER (184). Mitochondria and chloroplasts may also synthesize certain lipids. Furthermore, pulse chase labeling experiments indicate that precursors of membrane lipids usually are incorporated first into microsomal fractions rich in ER (184). However, isolation methods for specific membrane fractions are not yet at the stage where problems due to contamination may be discounted. There is some fine structural evidence to support these biochemical observations. In particular, gland cells that produce lipophilic substances are often very rich in smooth ER (260), and both rough and smooth ER are prominent in the tapetal cells of anthers, which manufacture the lipid cores of orbicules and the precursors of sporopollenin (see Section III,B,4; and 57, 99).

B. The Golgi Apparatus

The structure that we now call the Golgi apparatus is generally considered to have been first described in the nervous tissue of the owl by Camillo Golgi in 1898. He used silver nitrate staining after fixation to reveal a network of fine filaments known as the internal reticular apparatus. The observation aroused much controversy concerning the nature of the structure and this controversy persisted for over fifty years because there were suspicions that the findings were due to artifacts of the fixation and staining methods then employed. However, the use of the electron microscope and rapid improvements in fixation and preparative techniques in the 1950s established the Golgi apparatus as a real and almost universal organelle present in living and fully competent eukaryotic cells. (It is absent from mature sieve elements.) Interesting accounts of the history of the discovery of the Golgi apparatus and the elucidation of its functions are given by Beams and Kessel (13) and, in more detail, by Whaley (311). From these and other accounts the diversity in the terminology used to describe this structure is apparent, although electron microscopy has demonstrated that the basic ultrastructure is fairly consistent in a wide variety of plants and animals. The network appearance observed by Golgi is by no means universal; in many cells heavy-metal staining suggested that the Golgi apparatus was divided into a number of smaller and apparently independent units, which were called dictyosomes. This term and the term Golgi body are now widely used to describe these bodies, particularly in plant cells. It has been suggested (180) that the individual dictyosomes, and there may be several hundred per cell, are interconnected by membranous elements to form the Golgi apparatus. However, in many electron micrographs, the dictyosomes appear as discrete units with no obvious interconnections. Alternatively, the whole population of dictyosomes within a cell, whether they are structurally connected or not, may be considered to constitute the Golgi apparatus. In many unicellular algae this distinction does not arise because there is only one dictyosome per cell. In this chapter, the term dictyosome refers to individual units seen as stacks of membranous cisternae by electron microscopy, and the term Golgi apparatus describes the properties and function of the whole population of dictyosomes.

1. Ultrastructure

Each dictyosome usually consists of a stack of membrane-bounded, flattened sacs or cisternae, usually about 1–3 μm in diameter, with which are associated various vesicles (Figs. 19 and 20). The number of cisternae per dictyosome varies widely. There are commonly 4–7 in higher plants, but certain algal and animal cells may contain between 20 and 30. Both the degree of vesiculation and the number of cisternae associated with the dictyosomes vary widely both between different cell types and with the stage of development and activity of a particular cell. The cisternae are not uniform within a stack. They may differ both in diameter and in staining properties (see following discussion), although most are dilated at the periphery. In addition, the edges of the cisternae may show considerable reticulation to form a network of interconnecting tubules (Fig. 19). This network is seen most clearly in negatively stained preparations of isolated dictyosomes, although its significance is not yet understood. In some dictyosomes, intercisternal fibers or elements have been dem-

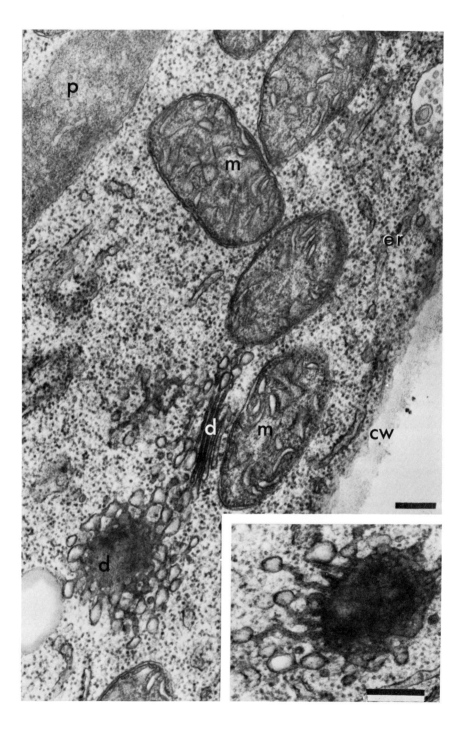

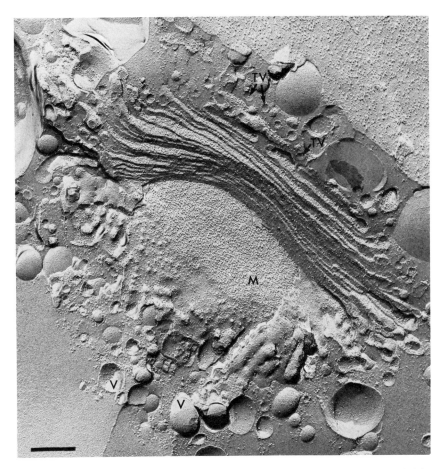

Fig. 20. Electron micrograph of a dictyosome in the alga *Micrasterias* as revealed by freeze-etching. The small vesicles at the upper right are transition vesicles (TV) at the forming face of the dictyosome. The larger vesicles (V) at the bottom represent the membrane-bounded products formed at the maturing face. The extensive area M probably represents the interior of a membrane of a mature cisterna. Scale bar = 0.25 μm. Reproduced from Staehelin and Kiermayer (278).

Fig. 19. Electron micrograph showing dictyosomes and mitochondria in a thin section of pea (*Pisum sativum*) root cells. The dictyosomes are seen in both cross-sectional and face views. The inset shows the reticulate network of tubules that are sometimes observed at the edge of the cisternae. Scale bar = 0.25 μm. For abbreviations, see Fig. 3.

onstrated. These rod-shaped structures are 3–5 nm in diameter and are usually found in the space between the flattened portions of adjacent cisternae. They may hold the central regions of the cisternae together and perhaps restrict the production of vesicles to the marginal regions (144). However, in maize (*Zea mays*) root cap cells, they have been observed in association with forming secretory vesicles, which suggests that they may aid in the organizing and/or shaping of these unusual elongate or kidney-shaped vesicles (181). In addition, some dictyosomes show a dense material between the cisternae (see 311). Its composition is unknown, but it is presumed to be a cementing material that holds the cisternae together.

The cisternae within a single Golgi stack are not uniform and there is evidence suggesting that dictyosomes are polar structures that contain a gradation of activities from one face to the other. A variety of ultrastructural observations support this concept. These include an increase in membrane thickness across the stack and gradations in heavy-metal poststaining (187), metal impregnation, staining for polysaccharides (224), and differences in certain enzyme activities (54, 318). By freeze-etching it has been shown that there is a gradual increase in the density of membrane particles (see Section II,B) on successive cisternae and from one face to another of the dictyosomes of the alga *Micrasterias*; a drop in density occurs at the peripheral regions where packaging of secretory products should take place (278).

The close proximity of the nuclear envelope or an element of the ER and its associated vesicles to one face of the dictyosomes, which has been observed in some cells, has led to the view that the Golgi cisternae are formed as these vesicles coalesce (99, 170, 185). The polarity of the dictyosomes would thus be due to the presence of a forming face and, opposite it, a maturing face where cisternae are lost by the production of secretory vesicles. This hypothesis also relates to membrane synthesis and flow and to the endomembrane concept (see Section IV,D).

2. Functions

Electron microscopy and associated techniques represent the major approach to the study of the Golgi apparatus, although cell fractionation procedures for the isolation of preparations enriched in dictyosomes are now in widespread use (see 21, 63, 219, 234). However, as stressed earlier, there are still outstanding problems relating to the presumed dynamic properties of a system that is observed in a series of static micrographs. For example, if vesicles both enter and leave the dictyosomes, it is clearly difficult to determine the direction of their movements. However, the use of autoradiography combined with pulse chase experiments and chemical analysis of the labeled products has helped overcome these problems to a considerable extent (see 58, 218, 223).

Fine structure studies of a variety of plant and animal cells have shown that the Golgi apparatus and associated vesicles are particularly well developed in cells that are actively involved in secretion. The process of secretion appears to involve first the accumulation of the secretory product in distended cisternae or in vesicles that are still attached to the cisternae by tubules. Vesicles are then detached and move across the cytoplasm to the plasm membrane where their contents are discharged to the extracellular space by reversed pinocytosis. In this process the vesicle membrane must fuse with the plasma membrane, and the significance of this membrane flow is discussed further in relation to the endomembrane concept (see Section IV,D).

The role of the Golgi apparatus in relation to secretory processes in animal cells has been well described in a number of reviews and monographs (13, 311). The number of relevant papers concerning plant cells is vast. For example, the Golgi apparatus appears to be involved in the secretion of scales and wall materials in various algae, in water secretion in other algae, in the secretion of various slimes and mucilages, in the production of noncellulosic wall materials in higher plants, and in the secretion of glycoproteins and various digestive enzymes (for detailed references, see 44, 99, 186). There is one example of the synthesis and export of a cellulosic polymer in an algal system (30, 125), and there is evidence from higher plants that cellulose synthetase may be moved to the plasma membrane by the Golgi apparatus (240, 266, 267, 294). However, the ER may also contribute to many of these activities (see Section IV,A,3). In addition, the Golgi apparatus may also be involved in the intracellular movement of materials, as for example, in cell plate formation and in the development of the cell vacuole. To illustrate these varied roles of the Golgi apparatus in plant cells, a few well defined examples are now discussed.

a. Scale Production in Algae.   Among the visually striking examples of the function of the Golgi apparatus is scale production by various unicellular flagellate algae; studies of these systems were initiated by Manton and coworkers (162–164). Flagellates such as *Mesostigma*, *Chrysochromulina*, and *Platymonas* possess distinctive scales or particles at the cell surface. The scales of these algae are predominantly carbohydrate in nature, but scales of other algae may contain considerable formations of calcite and silica. The scales frequently have a distinctive appearance and are easily recognizable in electron micrographs, both at the cell surface and, at various stages of development, within the cells. The scales are formed within the cisternae of the Golgi apparatus, successively more mature scales being found in successive cisternae moving from the forming face. Cisternae bearing mature scales then separate from the stack and fuse with the plasma membrane, discharging their contents to the exterior. Thus the whole process takes place within a single cisterna of the Golgi apparatus. Some algae may produce several distinct types of scale, and these may be seen in successive cisternae of the Golgi apparatus. In such cases, an individual cisterna may produce a particular type of scale.

A chemically and structurally well characterized example of scale production is that in the alga *Pleurochrysis scherffelii*, studied by Brown and coworkers (30, 125). This work has additional important implications because it shows that, in these cells, cellulose may be synthesized intracellularly within the Golgi apparatus. In higher plants, cellulose is thought to be synthesized extracellularly at the outer surface of the plasma membrane (see Section III,A). The scales of this alga are composed of three subunits: radial microfibrils composed of noncellulosic polysaccharides, concentrically or spirally arranged cellulosic microfibrils linked to peptides, and an amorphous coating deposited on the fibrillar network. The stepwise assembly of these scale components has been followed using a silver–methenamine–periodic acid staining procedure for carbohydrate and using standard electron microscopical techniques. The forming face of the Golgi apparatus is produced by the coalescence of vesicles from the ER, and it contains the precursor pools and enzymes for scale formation. The radial microfibrils are assembled first in a bilayer of parallel microfibrils. These unfold, allowing the deposition of the spiral bands of cellulosic microfibrils and, later, the addition of the amorphous material (Fig. 21). The cisternal membranes seem to possess distinct recognition sites that function in scale assembly. Finally, the fully assembled scale is secreted, the whole process requiring only about 2 min. Thus, in a single cisterna, a range of polysaccharides and proteins are synthesized and assembled in a process that must require highly specialized control mechanisms. These may include control of entry of precursors into the cisternae, the size and shape of the cisternae, the arrangement of membrane-bound synthetases, and the initial arrangement of the radial microfibrils that form the first visible template for later deposits. Similar control processes may operate in the dictyosomes of higher plants. The role of the Golgi apparatus in cellulose synthesis in higher plants is contrasted with this algal system in Section IV,B,2,c.

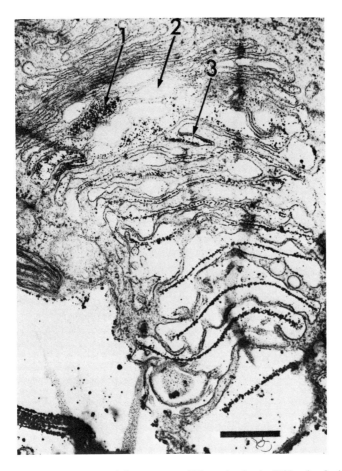

Fig. 21. Overall view of the Golgi apparatus of *Pleurochrysis scherffelii* stained with silver methenamine for carbohydrate. The radial precursor pools are dilated and react heavily with silver (arrow 1), whereas the cellulose precursor pools are electron transparent (arrow 2). A stage of folded radial microfibrils is indicated by arrow 3. The distal scales react more heavily with silver, indicating the addition of amorphous polysaccharide. The cell wall is in the lower left corner. Scale bar = 0.5 μm. Reproduced from Brown *et al.* (30).

b. Secretion of Root Cap Slime. The outer root cap cells of a number of higher plants secrete a polysaccharide slime, which is presumed to protect and to aid the movement of the root through the soil. Maize roots have been particularly well studied, and examination of their root cap cells shows that the dictyosomes undergo a marked change in form in the succession from the cap initials to the outer detached cells (see 46, 54, 198). The cap initials show only a few small vesicles associated with the dictyosomes. However, toward the outer cap there is a steady increase in secretory vesicle production with the peripheral cells exhibiting a typical hypertrophied form of dictyosome with large, irregularly shaped secretory vesicles. In the outer detached cells, where secretion has presumably ceased, the production of these large vesicles stops and the dictyosomes again possess only small spherical vesicles. The slime may well be derived from the concomitant breakdown of starch in the amyloplasts. The pattern of discharge of this secretion is not random, because it accumulates only against the wall that is, or will be, the outer tangential one (138). This and other directional movements of Golgi vesicles are discussed later.

This circumstantial evidence for dictyosome function has received strong support from the application of autoradiographical techniques. Pulse chase experiments with root tips fed with [³H]glucose demonstrated that radioactivity is incorporated into high molecular weight material, first in the dictyosomes and then moving via the Golgi vesicles to the cell wall (197); later experiments using [³H]galactose (53) and [³H]fucose (218, 219) have shown similar patterns of labeling. The use of [³H]fucose with maize roots provides specific labeling, because this compound is a major component of the slime in this species, but it is found elsewhere only in trace amounts in the polysaccharides of maize seedlings (115, 116, 317). Further support for the involvement of dictyosomes comes from sucrose density gradient centrifugation of homogenates of [³H]fucose-labeled root cap tissue; these show that the peak of radioactivity in nondialyzable material corresponds with a fraction rich in dictyosomes, and autoradiography of these fractions shows that the radioactivity is associated almost exclusively with dictyosomes (218, 219). Detailed chemical analysis of maize root slime indicates that it consists of a central short β-1→4-linked glucan rendered soluble by a coating of hydrophilic polysaccharides (317). Because the glucan component is probably synthesized within the dictyosome, this site is different from the generally accepted site of cellulose synthesis (see Section III,A). These studies and work on algal scales (IV,B,2,a) suggest that in some cases glucan synthesis can be initiated within the Golgi apparatus.

Other examples of the involvement of the Golgi apparatus in the secretion of slimes and mucilages are provided by various glands (156a, 261) and the secretions of certain algae (59, 60).

   c. Cell Wall Formation and Development.   The evidence presented suggests that the Golgi apparatus secretes materials that contribute to the extracellular polysaccharides and, it is widely believed, to the formation of primary and secondary walls in higher plants. However, the cell wall is a complex structure containing a number of different components, most of which cannot be seen directly in the electron microscope. Unequivocal evidence for the participation of the Golgi apparatus is lacking, although there are indications that it functions in the formation and secretion of the matrix polysaccharides (see Section III,B,2,b; 195, 196).

It is widely believed that during cell division vesicles derived from the dictyosomes and ER invade the spindle region, perhaps by some association with the spindle fibers, and that they fuse to form the cell plate (see Section III,B,1). However, as O'Brien (198) has indicated, this explanation, though plausible, is without clear proof. The events involved in the formation of the phragmoplast and cell plate are very complex and may vary between species and with conditions. The problems of interpreting these dynamic events from a series of static micrographs are considerable. Autoradiographic studies have not clarified the situation. Although there is obvious incorporation of labeled glucose and galactose into the cell plate region of dividing cells, labeling of dictyosomes is not consistently observed, and analysis is made difficult by the small size of the Golgi stacks and associated vesicles (53, 223).

In the case of primary walls there is again no unequivocal evidence that the Golgi apparatus contributes to its formation. However, improved methods for the isolation of preparations enriched in dictyosomes and vesicles have provided important evidence that these organelles are involved in the synthesis of matrix polysaccharides. These fractions have been shown to be enriched in exportable polysaccharides (21, 22, 115, 116, 295) and to contain various glycosyl transferases (234, 294).

Another important finding is that isolated Golgi fractions from various plants have β-glucan synthetase (cellulose synthetase) activity (120, 240, 294). Similar activity has also been reported in fractions thought to be enriched with plasma membranes (see 62, 109, 294); this finding is consistent with the idea that cellulose synthesis in higher plants occurs at the surface of the plasma membrane. Thus the Golgi apparatus may not be involved in the synthesis of cellulose directly (although it may produce hemicellulosic glucans as precursors), but it may have a role in the transfer of the synthetase to the cell surface.

Fine structure studies suggest that the Golgi apparatus of the alga *Micrasterias denticulata* participates in the transfer of cellulose synthetase and a matrix for microfibrillar assembly to the cell wall (142). This mechanism may be contrasted with the intracellular production of cellulose in the alga *Pleurochrysis* (see Section IV,B,2,a) and the situation in higher plants, where, as yet, there is no evidence that a cellulose template occurs within the dictyosomes. The difference between these systems may simply be in the timing of the sequence of events needed for cellulose synthesis.

With regard to secondary wall formations, there is again no unequivocal evidence of Golgi apparatus involvement, although it is quite widely believed that Golgi vesicles are associated with, for example, the bands of secondary thickening in the xylem. Although dictyosomes are abundant in developing vascular tissues and may show a hypertrophied appearance, there is little more direct evidence of their participation in secondary thickening (198, 199). There is some evidence from studies of secondary thickening that microtubules play a part in controlling cell wall deposition; this is discussed fully in Section IX,A,2,b.

Although circumstantial evidence is often presented in this and previous sections, a generalized scheme of polysaccharide synthesis in higher plants is shown in Fig. 22. This represents a current view of the synthesis, transport, and deposition of cell wall polysaccharides. Nevertheless, many of the steps represented in this scheme are tentative; direct evidence is often lacking, largely as a result of limitations in present day techniques.

C. MECHANISMS OF SECRETION

Secretion of a particular product by the ER or Golgi apparatus is clearly a highly coordinated process involving, for example, a supply of energy, precursor molecules, and enzymes, the occurrence of specific changes in the form and movement of membrane materials, and control over the site and rate of secretion. The mechanisms involved in these processes are still very obscure.

There is much uncertainty concerning the supply of energy; although low temperatures and metabolic inhibitors may reduce secretion (44, 99), it has not been possible to determine which (or how many) stages in the sequence are affected. The critical inhibition may be either that of product synthesis or that of steps in the movement or transformation of membranes. Protein synthesis is apparently not required for the secretion of those macromolecules that are present in the cytoplasm and ready for export, although continued secretion is affected by the presence of protein synthesis inhibitors (44). Adverse conditions often produce the apparently contradictory finding of increased numbers of cisternae or vesicles. Presumably the rate of synthesis of new membranes exceeds that of the processes occurring at the maturing face of the

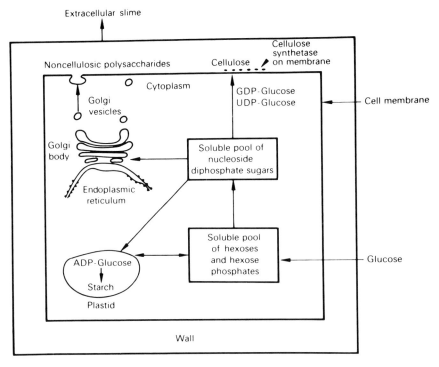

FIG. 22. Diagram to show possible sites of polysaccharide synthesis and transport in a growing plant cell. Redrawn from Northcote (195); reproduced from Hall *et al.* (107).

dictyosomes or of the movement of vesicles for export. The rates of secretion may be quite rapid; 10–15 min has been estimated as the elapsed time from synthesis of macromolecules to their arrival in the wall of higher plants (44). Assuming that a dictyosome contains four or five cisternae, it has been calculated that a cisterna matures every 1–2 min. Similar rates have been measured for scale production in the alga *Pleurochrysis* (see Section IV,B,2,a).

The guidance of secretion products to sites of deposition presents other problems. Diffusion and cytoplasmic streaming may be involved in the delivery of substances or vesicles to the surface, although these apparently random processes do not alone explain the very localized deposition that occurs in some cells. It has been suggested that microtubules and microfilaments may function as cytoplasmic guides for the movement of secretory vesicles; this concept is discussed in Section IX.

There is also the question of whether hormones control certain secre-

tory processes; the best described system here is the secretion of hydrolytic enzymes by barley (*Hordeum vulgare*) aleurone layers, which is stimulated by gibberellic acid (see 44, 135, 229, 300). However, the major
effect of this hormone appears to be on the release of the enzymes from
the cell wall into the surroundings rather than on the secretion of the
enzymes into the wall, and there is no direct evidence in this system for
the involvement of secretory vesicles (6, 44, 135). Ethylene is apparently
involved in the synthesis and secretion of cellulase during leaf and fruit
abscission, although its precise mode of action is uncertain (1, 44). Thus
direct evidence for hormonal action on secretory processes involved with
plant cells is lacking, although there are numerous examples of hormonal control of the release of secretory products in animal cells (141,
262).

D. The Endomembrane Concept

The preceding sections have highlighted the dynamic nature of the
cytoplasmic membrane systems, their interrelationships, and the transfer of membranous materials to the cell surface. The material derived
from the ER and Golgi cisternae is presumably replaced, and this is
thought to occur primarily from outgrowths of the ER. In fact, the ER is
generally considered to be a major site of membrane biogenesis involved
not only in its own formation but in the production of other cellular
membranes (184). The rough ER may be continuous with the outer
nuclear membrane indicating that these two systems are also closely related. Thus these observations give rise to the concept of membrane
transfers between cellular structures that may be associated with membrane transformation. Such ideas are embodied in the concept of the
endomembrane system, which is envisaged as a functional continuum
involving the physical transfer of membrane material from one cellular
structure to another, leading to the generation of the cellular membrane
systems (185–187; Fig. 23). There are two possible routes of "flow" of
vesicles supposed to contain membrane materials, one to the cell surface
and the other to the vacuole. The endomembrane system is thought to
involve the outer membrane of the nuclear envelope, rough and smooth
ER, and the Golgi apparatus and secretory vesicles that give rise to the
plasma membrane and vacuolar membranes. The inner membranes of
the nuclear envelope, chloroplasts, and mitochondria are not considered
as part of the system, although there is some evidence that the outer
membranes of the semiautonomous organelles are involved. The term
"transition element" has been used to describe those membranes that are

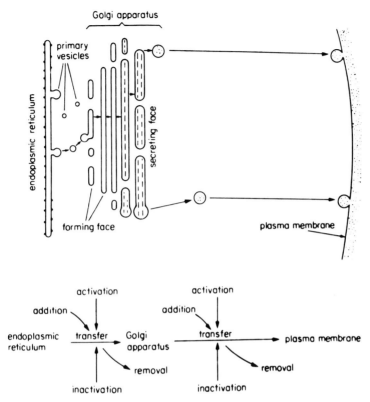

FIG. 23. Diagrammatic representation of the endomembrane concept involving membrane flow and differentiation. New membranes are synthesized at the rough endoplasmic reticulum, from which small primary vesicles arise and move to the forming face of the Golgi apparatus. Secretory vesicles from the secreting face of the Golgi apparatus migrate to and fuse with the plasma membrane. Various types of transformation that might account for the changes in membrane composition occurring during this process are depicted. The alternative route of vesicle movement to the vacuole is not shown. Redrawn from Morré and Mollenhauer (185); reproduced from Hall and Baker (105).

incapable of protein synthesis and possess properties intermediate between those of the rough ER and nuclear envelope at one extreme and those of the plasma membrane and tonoplast at the other; transition elements include the smooth ER, Golgi apparatus, and various kinds of vesicles.

Thus the essential features of this concept are the origin of membranes in the rough ER and outer nuclear membrane and the processes of membrane flow and differentiation to provide a wide variety of cellu-

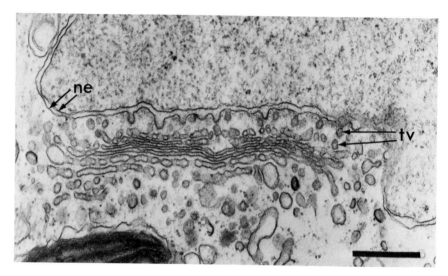

FIG. 24. Dictyosome in the alga *Tribonema* showing transition vesicles (tv) between the nuclear envelope (ne) and the forming face of the dictyosome. The invaginations of the outer membrane of the nuclear envelope are presumed to represent stages in the formation of transition vesicles. Scale bar = 0.5 μm. Reproduced from Massalski and Leedale (170).

lar membranes. The evidence in support of the endomembrane concept is covered elsewhere (185–187) and is discussed only briefly here.

There is considerable evidence from both plants and animals that the rough ER is a major site of synthesis of membrane proteins and lipids. Although free ribosomes, soluble systems, chloroplasts, and mitochondria are involved in the synthesis of certain membrane components, the rough ER appears to contain the most complete array of biosynthetic machinery (184). The origin of membranes in the ER is supported by ultrastructural evidence that shows connections between the nuclear envelope and the ER and the apparent origin of Golgi cisternae from the coalescence of vesicles derived from the ER or nuclear envelope (185; Fig. 24). However, there are problems associated with such a straightforward interpretation of these micrographs (311). For example, the dimensions of the vesicles assumed to be in transit rarely correspond to the dimensions of the cisternae, whereas in a number of cases the secreting face of the dictyosome appears to be in association with the ER or nuclear envelope. Interconnections between cellular membranes are, however, readily seen, particularly in thick sections that have been impregnated with heavy metals (40); numerous connections between the nuclear membrane, ER, and dictyosomes are often visible.

There are many cytochemical and biochemical observations that point to the process of membrane differentiation. The polarity of the dictyosomes has already been discussed (see Section IV,B,1). There appears to be a progressive increase in membrane thickness in the sequence: nuclear envelope to ER to dictyosome to plasma membrane. However, staining with a phosphotungstic acid–chromic acid procedure shows staining of the plasma membrane and of mature secretory vesicles but not of immature vesicles or ER (252, 301). Further evidence of progressive differentiation is provided by biochemical analysis of isolated membrane fractions. Protein and lipid composition and enzyme activities of rat liver fractions often show progressive changes in the sequence from ER to plasma membrane; enzyme activities characteristic of the plasma membrane may be added or increased, while those characteristic of the ER may be reduced or lost (185, 186).

As we have seen, the passage of membrane material to the cell surface has been extensively studied, but there is also some evidence of a relationship between the ER and other cellular membranes. For example, a close structural association between the ER and the vacuoles has been reported (68), and a possible origin of the vacuoles from the ER remains to be discussed. A possible relation of endomembranes with the vacuole concerns the incorporation of endocytotic vesicles into the vacuolar apparatus. The concept of membrane flow, which was originally applied to endocytosis (15), is yet to be developed (see Section V). There is also evidence of connections between the ER and the outer mitochondrial membrane, which would allow transfer of materials from one structure to another (e.g., mitochondrial proteins may be synthesized outside the mitochondria) and yet still maintain the compartmentalization of the luminal contents (23).

V. Vacuoles

The presence of a vacuole system (the vacuome) is a characteristic feature of differentiated, living plant cells, and it may often consist of a single large vacuole that occupies over over 90% of the total cell volume. The vacuole is surrounded by a semipermeable membrane known as the tonoplast, the reality of which was demonstrated by permeability and microdissection experiments long before it was visualized in thin sections by electron microscopy. The turgor pressure resulting from the accumulation of osmotically active substances within the vacuole normally pushes the cytoplasm to a thin layer against the cell wall and gives rise to the general rigidity of plant tissues and organs. The swelling of the vacuole and its interaction with the cell wall are also critical factors in the extension growth of young cells. However, the concept of the vacuole as a single, large water-filled space containing dissolved solutes describes only one aspect of vacuolar structure and function. The plant vacuolar apparatus varies greatly, both in its morphology and in the nature of its contents. In addition, plant vacuoles are now considered to play a more active role in plant metabolism, in particular in relation to molecular turnover and other digestive processes.

A. Range, Structure, and Origin

The size, shape, and number of vacuoles varies widely among different plant cells. A few cells (e.g., tapetal cells) contain no vacuoles, whereas certain meristematic cells may possess only a few small vacuoles or their precursors. However, in the development of the latter into parenchymatous cells, small vesicles appear, fuse, and expand to form the vacuoles observable by light and electron microscopy. This process often leads to a single large vacuole with a shape similar to that of the cell. There are two common fallacies involving vacuoles in meristematic cells. The first is that these cells never contain large vacuoles; this is generally true for apical meristems, although cambial cells may be highly vacuolated (41). Second, it has sometimes been suggested in physiological studies that cells of root tips are essentially nonvacuolated. Yet these cells may contain numerous, very small vacuoles with a tonoplast surface area greater perhaps than that of a single, large vacuole (9). Apart from the increase in a central vacuolar volume associated with cell growth, the number and size of vacuoles may show marked changes during the life of a cell. For example, in cambial cells the vacuolar apparatus may undergo marked seasonal variations, changing from a central vacuole to a more diffuse irregular network and then to a great number of smaller vacuoles in the autumn before reverting back to a central vacuole in the spring[3] (41). In the development of extrafloral nectaries in *Ricinus communis*, the development of the epidermis into a secretory zone is accompanied by the transformation of the single vacuole in each cell into a

[3] Compare the account of vacuoles and the vacuome in Volume II, pp. 285–287, where these aspects are described and illustrated. (Ed.)

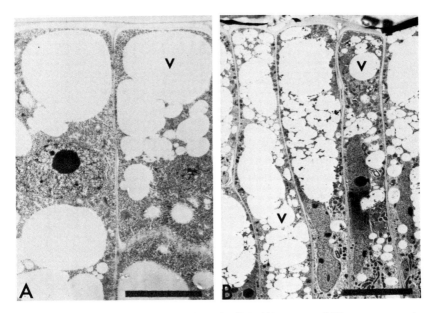

FIG. 25. Electron micrographs of epidermal cells in (A) young and (B) mature, secreting nectaries of *Ricinus communis*. Numerous vacuoles (v) and mitochondria are observed in the secreting cells. Scale bars: (A), 5 μm; (B), 10 μm. Reproduced from Baker *et al.* (10).

number of smaller vacuoles, the extent of the vacuolation increasing with the development of the secretory process (10; Fig. 25). Thus, because there may be more than one vacuole within a single cell and the degree of vacuolation may exhibit developmental, functional, or seasonal changes, the classical term vacuome (see 91, 280) may still be a useful one to describe the whole vacuole system in a plant cell. This word has also been used by de Duve (56) to describe all those membrane systems involved in lytic processes in animal cells, although the term lysosome still appears to be more widely used for this purpose. It is somewhat fortuitous that plant vacuoles are now believed to have a lysosome-like function (see Section V,B,3). The term vacuome has also been applied more broadly in plants to include the contents of the main vacuole and of the ER cisternae, the spaces enclosed by the Golgi membranes, nuclear envelope and microbody membranes, the inner and outer membranes of mitochondria and chloroplasts, and the thylakoid compartment of chloroplasts (174). This definition may well be premature and lead to confusion, because there is certainly no unequivocal evidence that *all* of these compartments are either structurally or ontogenetically related or possess lytic properties. Until such relationships have been firmly established, the classical definition should be retained.

The appearance of vacuoles in thin sections of material fixed and stained by standard procedures shows a great diversity of shapes, the greatest irregularity usually occurring after fixation in potassium permanganate (see 66). The narrow extensions of the vacuole often observed in such preparations have been interpreted by microscopists as various stages of vacuolar development or as connections with the ER. However, the vacuoles of freeze-etched roots are mainly spherical, and it has been suggested that this is a more accurate representation of vacuolar shape and that irregular forms result perhaps from shrinkage during chemical fixation (66).

Vacuolar contents are commonly considered to consist of optically clear, watery fluids containing a wide variety of solutes (see Section V,B) and perhaps crystals of more insoluble substances (e.g., calcium oxalate). However, vacuoles from all species studied may also contain structures recognizable in electron micrographs, which have been called inclusion bodies (see 67 and Section V,B,3). In the early stages of their formation these appear to be surrounded by a single membrane, which is presumably a result of their origin as invaginations of the tonoplast. Inclusion bodies may consist of simple vesicles or may contain recognizable cytoplasmic material such as ribosomes and mitochondria. After their incorporation into the vacuole, these structures normally degenerate, although the simple vesicular type may increase in size and accumulate electron-dense deposits. The significance of these inclusions in relation to the autophagic role of vacuoles remains to be discussed.

Clearly, many of the important properties of vacuoles are closely related to the vacuolar membrane, the tonoplast. This usually appears to be about 10 nm wide in thin sections and, as with the plasma membrane, may show some asymmetry after fixation in osmium tetroxide, the vacuolar face being more heavily stained than the cytoplasmic side. Freeze-etch studies support this observation; the density of membrane particles may be very different on opposite faces of the membrane (65, 69). Cytochemical studies show the presence in the tonoplast of enzymes such as peroxidase, ATPase, and acid phosphatase (103, 106, 108, 211, 233), although there is very little additional information available concerning tonoplast composition and properties. However, recently developed techniques for the large-scale isolation of intact vacuoles should improve this situation (154, 305). Using these techniques, vacuolar membranes have been shown to possess ion-stimulated ATPase activity that may be involved in the regulation of ionic balance (155).

The old question of the origin of vacuoles is still an area of uncer-

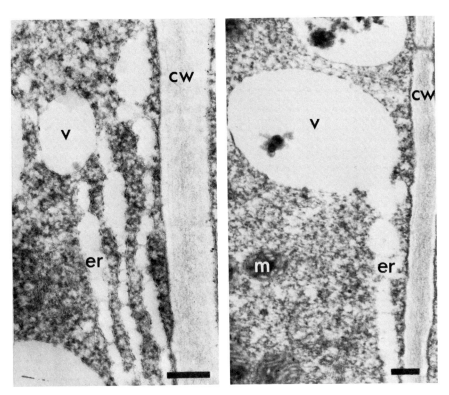

FIG. 26. Electron micrographs of epidermal cells in a secreting nectary of *Ricinus communis* showing small vacuoles apparently arising from swellings and vesiculations of the endoplasmic reticulum. Scale bars = 0.25 μm. For abbreviations, see Fig. 3. Reproduced from Baker *et al.* (10).

tainty. The cells of apical meristems appear to contain very few vacuoles, if any, but, as these cells increase in volume, small vesicles or provacuoles appear that expand and fuse to produce a number of larger vacuoles and, eventually, the single main vacuole. Many authors subscribe to the view that the provacuoles originate from the ER, the tonoplast consisting of a differentiated portion of that membrane system (35, 37, 46, 172–176, 185). Others suggest that this is unlikely because of the profound differences between the two membranes (99). The two systems may show different staining properties (e.g., for peroxidase; see 108 and 211), although this may simply be an indication of membrane differentiation. Perhaps we know too little about the properties of these membranes to make such a distinction. In relation to the endomembrane concept (see Section IV,D), it could be argued that most cellular mem-

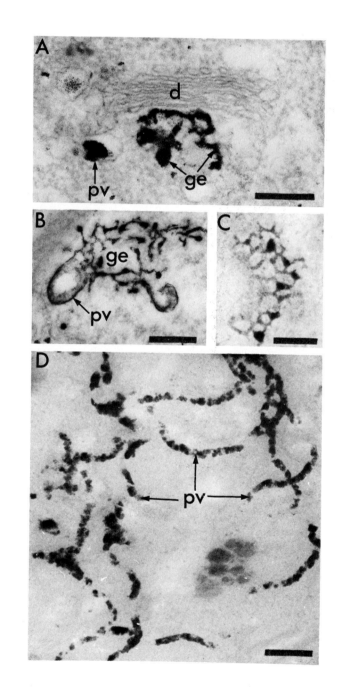

branes originate in the ER. Certainly a number of fine structure studies suggest that provacuoles may arise by dilation and vesiculation of the ER (10, 17, 177; Fig. 26), although again such static micrographs should be interpreted with caution. More convincing evidence for an origin from the ER has come from Marty (169a) using various cytochemical methods with *Euphorbia* roots. These include techniques for the localization of acid phosphatase and esterase activities, which are believed to be marker enzymes for lysosomes (see Section V,B,3), and incubation in a mixture of zinc oxide and osmium tetroxide, which is thought to stain the endo-membrane system specifically. The latter technique was also used in conjunction with high voltage electron microscopy, which allows thicker sections to be examined and provides insight into the three-dimensional arrangement of cell structures. This study suggests that the provacuoles arise from the so-called GERL (Golgi-associated ER, from which lyso-somes form), as has been postulated for the origin of lysosomes in ani-mal cells. The GERL consists of a network of anastomosing tubules situated close to the maturing face of a Golgi stack, and the provacuoles appear to be budded from the junctions of these tubules (Fig. 27a–c). In more differentiated cells, provacuoles become elongated to form tubes (Fig. 27d), which may become wrapped around portions of the cyto-plasm to produce autophagic vacuoles (see Section V,B,3). Lysosomal enzymes appear to be concentrated and packaged in the provacuoles, which swell and fuse to give larger vacuoles; these continue to collect the GERL-derived vesicles throughout the life of the cell.

A number of other suggestions have been made for vacuolar origin. The production of vesicles by the Golgi apparatus is one possibility, although these vesicles may enter the vacuole by invagination of the tonoplast and may not be involved in membrane fusion (172). Alterna-tively, it has been proposed that vacuoles may arise from degenerating organelles or, independently of other organelles, from provacuoles that persist and multipy by division, or *de novo* by the association of hydro-philic substances and water, followed by the attraction of proteins and lipids to form a membrane.

FIG. 27. Electron micrographs of portions of meristematic cells from the root tip of *Euphorbia characias*. (A),(C),(D) Incubated for acid phosphatase activity and showing in-tense staining in the GERL (see text) and provacuole channels. (B) Incubated with zinc oxide and osmium tetroxide and showing staining of the GERL and a nascent provacuole. (C),(D) Thick sections examined by high-voltage electron microscopy. Scale bars: (A)–(C), 0.25 μm; (D), 0.5 μm. d, dictyosome; ge, GERL; pv, provacuole. Reproduced from Marty (169a).

B. Functions of Vacuoles

Plant vacuoles contain a great variety of chemical compounds. Dissolved substances include mineral ions, sugars, organic acids, amides and amino acids, gums, tannins, lipids, pigments, and proteins (173, 174).[4] The protein content includes both storage proteins and a variety of enzymes, particularly hydrolases such as acid phosphatase, proteinase, and RNase (173, 174). In addition, vacuoles may contain crystalline materials and structural elements (inclusion bodies) that have already been described. Although the vacuolar content may vary widely among different plants and among different cells in the same plant, this general description gives a clear indication of the range of vacuolar function. Vacuoles clearly have a role as storage organelles for a variety of compounds. The turgor pressure produced by osmotically active substances in the vacuolar sap plays a critical role in plant osmoregulation and in cell growth. The presence of hydrolases and degenerating cellular structures suggests a further digestive function. Although these properties may be closely related (e.g., in the mobilization of reserves), they are considered for convenience under three broad headings, namely, storage, osmoregulation, and hydrolytic activity. Because these topics are covered in other chapters in this treatise, the emphasis here is on structural observations.

1. Storage

From the structural point of view, the very many low molecular weight solutes found in cell vacuoles are, of course, of little interest. However, the storage of macromolecules may lead to interesting changes in vacuolar structure. A particularly well studied example is provided by aleurone grains (protein bodies), which are vacuoles specialized for the storage of protein in seeds (see 220). These are thought to arise by the accumulation of protein in vacuoles, possibly by transport through the ER cisternae (see Section IV,A,3,a). The reserve proteins legumin and vicilin have been clearly demonstrated in aleurone grains of bean (*Vicia faba*) cotyledons by the use of fluorescent antibodies (85). In addition to protein, aleurone grains may contain other reserves in the form of phytin, which consists of the insoluble salts of inositol hexaphosphoric acid. Under the electron microscope, aleurone grains appear bounded by a single membrane that contains a proteinaceous matrix. Embedded in

---

[4] Also see Volume 2, pp. 255 ff. (Ed.)

this matrix, aleurone grains may show a protein crystal and globular bodies of phytin. During the course of germination, the reserve protein is mobilized, and the aleurone grains inflate and coalesce to form vacuole-like structures. The mobilization of these reserves is presumably accomplished by the proteinases and other hydrolases known to be present in these particles (171, 173).

Spherosomes provide another example of storage organelles, although in them the reserve is lipid in the form of triglycerides. Spherosomes have been defined as spherical particles with high fat content that are bounded by a single membrane (46). They may be considered as a modified vacuole (172), although not all authors classify them in this way (99). The structure of the bounding membrane is somewhat anomalous because it does not show a complete unit membrane appearance; only one electron-dense layer is observed, which measures 2–3.5 nm in width (2, 320). Peanut spherosomes contain membrane structural protein, and the microscopic observations are interpreted as representing a half unit membrane, the polar surfaces of which face the cytoplasm with the nonpolar lipoidal surface in contact with the internal storage lipid (319).

2. Osmoregulation

The role of cell vacuoles in osmoregulation is very crucial to the life of plants. In its interaction with the cell wall, the vacuole is essential for the maintenance of turgidity and so of general rigidity of plant tissues and for the growth in volume of plant cells. These functions are not generally reflected in major fine structure changes in the vacuole, except in vacuolar development as discussed earlier, and so they are not discussed in any detail here. However, there are a number of interesting exceptions. In guard cells, the vacuome may change from a series of small vacuoles in the dark, inflating and fusing with the induction of opening to form a single large vacuole. This cycle may be related to a reversible deposition of osmotically active solutes in the vacuole (100). These changes are similar to those occurring in the budding cycle of yeast, where inflation and shrinkage of the vacuome appears to be related to the deposition and withdrawal of soluble reserves (173).

The use of a silver precipitation technique in conjunction with X-ray analysis to localize chloride ions has demonstrated the presence of two types of vacuole in the mesophyll cells of mangrove (*Aegiceras corniculatum*) (297). The predominant type contains large amounts of osmiophilic organic solute and little chloride, whereas the other type shows the reverse distribution. It has been suggested that the former type may utilize organic solutes for osmotic adjustment, whereas the latter is concerned with the control of chloride movement from the conducting elements to the salt glands via the symplast.

Changes in solute absorption patterns may also be reflected in the appearance of the tonoplast. Increased rates of solute absorption in washed *Zea mays* roots result in increased tonoplast electron density in fixed and stained tissue (112); this increase is presumably related to changes in the molecular makeup of the membrane that occur in response to washing.

3. Hydrolytic Activity

The concept of plant vacuoles as lysosome-like organelles has developed from fine-structural and cytochemical observations and biochemical studies on isolated vacuolar fractions. The term lysosome, which means a lytic or digestive body, was originally used by de Duve to describe single membrane-bound organelles that occur in animal cells and contain a variety of acid hydrolases. This term now covers a variety of different forms that are heterogeneous with regard to their enzyme content. However, they all form part of an intracellular digestive system that is involved in both normal and pathological metabolism. The material undergoing breakdown by lysosomal action may be of endogenous or exogenous origin; these processes are known as autophagy or heterophagy, respectively. Autophagy includes the digestion of macromolecules in the normal processes of cellular turnover, the controlled breakdown of cytoplasm in the degradative stages of differentiation and metamorphosis, and the clearance of dead cells. Heterophagy generally involves the uptake of particles by phagocytosis for nutritional or defensive purposes. Alternatively, lysosomal enzymes may be secreted to participate in extracellular digestion. Full details of the range of structure, properties, and functions of animal lysosomes can be found in a number of comprehensive reviews (e.g. 56, 56a, 227).

Therefore, in general terms, similar digestive activity is required in the metabolism, differentiation, and senescence of plants. As in all living cells, plant cells show turnover of their constituent molecules and organelles, and there are numerous examples where specific digestive activity is required (e.g., mobilization of reserves, differentiation of xylem elements, and abscission). Hence plant cells may also show compartmentation of hydrolase activity, and there is now considerable evidence that the vacuole is the plant equivalent of the lysosome.

Under the electron microscope, vacuoles frequently show inclusion bodies that appear to have originated from the cytoplasm by a process analogous to phagocytosis (Fig. 28a,b; and see Section V,A). It is tempting to conclude that these bodies represent stages in the autophagy of cellular material. This is supported by numerous cytochemical studies that show the localization of acid phosphatase and other hydrolytic enzymes within the vacuolar cavity and associated with the inner face of the tonoplast (106, 171, 233, 265; Fig. 28c). Autophagic vacuoles can apparently also be formed by the enclosure of portions of the cytoplasm in continuous ER envelopes (169a, 172). These observations are supported by biochemical analysis of vacuolar fractions, aleurone grains, and spherosomes, which show the presence of a wide range of hydrolases (see 172, 173).

This descriptive evidence is now supported by a number of examples of lysosome-like function in plant development. The role of hydrolase activity in the mobilization of reserves in aleurone grains has already been described. In studies of development of protoxylem elements in roots, staining for acid phosphatase in light microscopy reveals a change from a particulate to a diffuse cytoplasmic appearance, which is assumed to be related to the production of dead xylem elements (79). Similar changes occur in the development and death of root cap cells (265). In the fading corolla of *Ipomoea purpurea*, the fall in levels of certain macromolecules and increase in activity of some hydrolases (e.g., DNase) are accompanied by apparent autophagic activity of the large vacuole, when observed by electron microscopy (175). Similar autophagic activity is evident in the development of laticifers in *Euphorbia characias* (169) and in seeds of *Fraxinus* subjected to prolonged dormancy (304). Lysosomal activity may also be involved in plant disease such as the infection of potato (*Solanum tuberosum*) tubers by *Phytophthora erythroseptica*. This results in a swelling and disruption of lysosome-like particles (227, 228), whereas infection of potato leaves by *P. infestans* is accompanied by a decrease in activity of particulate acid phosphatase and an increase in that of the soluble phase (227). It has been suggested that such responses may be involved in the hypersensitive reaction of plants to infection, in which necrotic areas are formed at sites of infection and the pathogen is inactivated or killed (227).

VI. Mitochondria

Mitochondria were probably first reported around 1890, and the term mitochondrion which describes the granular and/or thread-like appearance under the light microscope, was used by Benda a short time later.

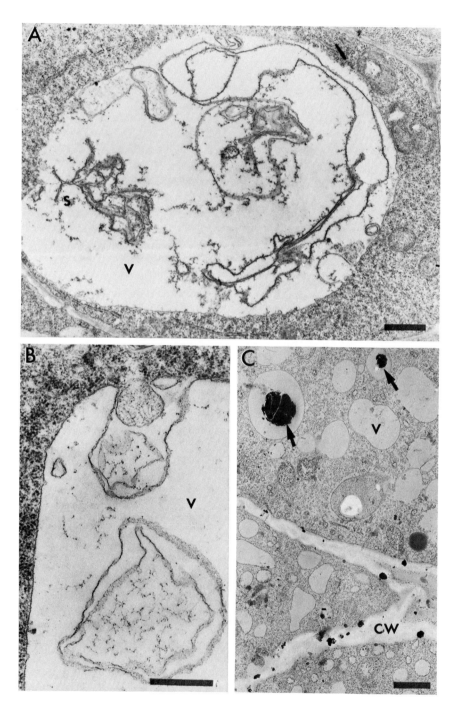

These structures were also known as chondriosomes, although the term finds little use today. Mitochondria occur in all aerobic eukaryotes and are sites of sequential events within the cell involved in the processes of respiration and other aspects of metabolism (see accounts in Chapters 2 and 4). Mitochondria contain the enzymes of the tricarboxylic acid cycle (TCA or Krebs cycle), in which pyruvate from glycolysis is oxidized to CO_2 and water, and they contain the reactions of the respiratory chain and oxidative phosphorylation that result in the formation of ATP and the consumption of molecular oxygen. The first electron micrographs of plant mitochondria that had been isolated and dried were published by Buchholz in 1947; the basic structural features common to all mitochondria, first revealed by thin sectioning of animal cells, were described independently by Palade and Sjöstrand some six years later. Since then, cell fractionation and electron microscopy have been successfully combined to provide a good understanding of the relationship between structure and function in mitochondria. Much of this work has been carried out with animal systems and so is referred to whenever information on plants is lacking. The discovery that mitochondria (like chloroplasts) contain DNA and ribosomes opened a whole new area of interest in these organelles, and studies concerned with mitochondrial genetics, the degree of genetic autonomy of mitochondria, and their interactions with nucleus and cytoplasm followed.

A. GENERAL FEATURES AND RANGE OF MITOCHONDRIA

Under the light microscope, only the overall shape of mitochondria can be seen, although observations on living cells reveal that mitochondria move, fuse, branch and divide, and may show rapid changes in shape. These dynamic properties are not so evident in electron micrographs, in which mitochondria usually appear as cylinders (about 3–6 μm long and 0.5–1.0 μm in diameter) or as spheres (about 0.5–1.5 μm in diameter) (Fig. 19). They may also show other, less regular forms (see 99, 190), although these forms are perhaps less common than might be predicted from phase contrast observations, indicating that some

---

FIG. 28. Autophagic activity of vacuoles. (A),(B) Note cellular fragments within the vacuole (v) and the apparent phagocytotic activity of the tonoplast. (C) Incubated for acid phosphatase activity and showing staining associated with the cell wall (cw) and staining within the vacuolar cavity (arrows). (A),(C) *Pisum* root, (b) *Suaeda* root. Scale bars: (A),(C), 1 μm; (B), 0.5 μm. (A),(B) Reproduced from Hall *et al.* (107); (C) from Sexton *et al.* (265).

fragmentation and "rounding off" of mitochondria may occur during processing for electron microscopy.

The real elucidation of mitochondrial structure came with the application of electron microscopy and constitutes one of the earliest success stories of this technology. Although, as we have seen, they show considerable variability in gross morphology, all mitochondria possess certain essential characteristics. They are bounded by an envelope consisting of two membranes, each about 5 nm wide, with the inner membrane invaginated to form projections called cristae that extend into the central matrix (Figs. 19, 29). The space between the inner and outer membranes, normally about 8.5 nm wide, is known as the perimitochondrial space, and the region within the invaginations is known as the intercristal space. Again, as we shall see, these features show some variation among species, in response to fixation, and in response to environmental and developmental changes.

The numbers of mitochondria per cell also show considerable diversity (46, 99, 190). Some unicellular algae contain only one, although it may be a complex, branched structure, whereas the cells of higher plants may contain up to several thousand mitochondria. The meristematic cells of maize (*Zea mays*) root tips have been estimated to contain about 200, whereas the much larger outer root cap cells may possess over 2000 (46). These numbers are of the same order as those found in animal cells such as those of rat liver. In some specialized cells, such as certain spermatozoa and erythrocytes, the mitochondria are lost during maturation. The critical factors to consider, however, may not be numbers but the volume of cytoplasm occupied by these organelles and the surface area of their inner membranes.

Mitochondria may appear to be randomly distributed throughout the cytoplasm or may show a marked localization in one region of the cell. There are numerous examples of the latter in animal cells, including the association between mitochondria and muscle fibers in striated muscle and the close application of mitochondria to infoldings of the plasma membrane in a variety of cells engaged in active solute transport (190). These nonrandom distributions are presumably related to localized high energy requirements. There are only a few reports of analogous situations in plant cells. For example, transfer cells (see Section III,A,3,c) usually contain abundant mitochondria with well developed cristae (96); the mitochondria may lie close to the wall ingrowths (316), although this arrangement does not appear to be a consistent feature of these cells. Various unicellular flagellates may show mitochondria associated with the base of the flagellum, whereas in the epidermal cells of the extrafloral nectaries of *Ricinus communis* the mitochondria are grouped at the

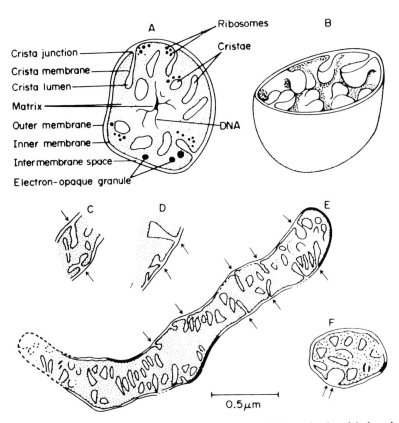

FIG. 29. Schematic representation of the morphology of plant mitochondria based on chemical fixation and thin-sectioning techniques. The intermembrane space and crista lumen may be exaggerated. (A) Cross section. (B) Three-dimensional reconstruction. (C)–(F) Actual tracings of electron micrographs to illustrate the form of the junctions of cristae and the inner membrane. The matrix is stippled; indistinct (obliquely sectioned) membranes are dotted. Figs. C–E show the common situation in which the crista junction (arrow) appears narrow. A wide junction (F; double arrows) is rarely found. If plate-like cristae had openings that extended the whole length of the fold, wide junctions would be seen more frequently. From Opik (203), copyright 1974 McGraw-Hill Book Co. (U.K.) Ltd. Reproduced by permission of the publisher.

base of the cells, although the significance of this polarity is not clear (10). The association of mitochondria with other organelles has also been reported occasionally. Connections between the ER and outer mitochondrial membrane are mentioned in Section IV,D. An interesting interaction with chloroplasts has been proposed from observations of living cells by phase contrast microscopy (315; and see discussion in Chapter

3, Section II,E,3). Mitochondria have been observed to become stationary within the concavity of chloroplasts, and this association appears to be related to the origin of starch grains within the chloroplasts; the mechanism of such an interaction is purely speculative at the present time. In many cells, mitochondria are moved around by cytoplasmic streaming (129, 314), which presumably aids the distribution of the products of mitochondrial metabolism to all parts of the cell; there is also evidence that their distribution is controlled by microtubules (118a; and see Section IX,A,2).

B. Fine Structure of Mitochondria

1. Membranes

It is now generally agreed that the inner and outer mitochondrial membranes differ in composition and function and in their appearance under the electron microscope. Both membranes show the typical unit membrane appearance (see Section II,B) after fixation in OsO_4 or $KMnO_4$. The outer membrane generally appears smoother and more weakly stained, although it has also been reported to be pitted in negatively stained preparations (214); this feature is not widely found, and it may be an indication of damage during preparation. There is also some doubt about the perimitochondrial space. It usually appears to be empty, and its exact width varies with the cell and tissue employed. Material prepared by rapid freezing shows the two membranes to be closely adpressed, suggesting that this outer compartment may not exist in the living cell (203, 279).

In most mitochondria the surface area of the inner membrane is increased by invaginations known as cristae, which show considerable diversity in form in different systems. Cristae have been broadly classified into three types, lamellar, tubular (villous), and tubulosaccular, with the lamellar form being the most common, particularly in animal cells (190). In general, plant cells do not show the regular stacked arrangement of cristae found in animal mitochondria. The range is wide, and clear classification is difficult as mitochondrion morphology appears to be very dependent on fixation (203). Furthermore, the considerable changes in the appearance of cristae that can be induced in isolated mitochondria suggest that the cristae may not have a constant shape in living cells.

Another uncertainty concerns the connections between the intracristal space and the perimitochondrial space. Clear openings from one compartment to another are not commonly seen, and it is generally assumed that the junctions are very narrow (99, 190, 203). In some cells, however, numerous connections may be observed (e.g., in blowfly flight muscle and *Neurospora*) (190). A study of mouse liver cells by serial sections led Daems and Wisse (52a) to suggest that cristae are attached to the inner membrane by means of one or more narrow tubes known as *pediculi cristae*; this conclusion has been supported by observations of a range of other cells (see 190). The wide range of forms of cristae in animal and plant cells are described thoroughly by Munn (190).

A controversial aspect of mitochondrial membrane structure is the presence of stalked particles found regularly spaced along the matrix side of the inner membrane of negatively stained membrane preparations (see 46, 107, 190, 203; Fig. 30). The particles are somewhat standard in size and consist of a head about 8–10 nm in diameter borne on a stalk about 4–6 nm long and 4 nm in diameter; they are usually present at a density of 2000–4000 particles per μm^2. Their true nature is, however, debatable. They are not normally seen in sections of intact mitochondria, nor by freeze-etching, and their density may vary with the preparation and staining procedure. It has been suggested that they are artifacts produced by the extrusion of proteins and/or lipid material, either as a result of membrane disruption or by the drying process in negative staining. However, there are some observations that suggest that they are real structures. They have been seen occasionally in thin sections of intact mitochondria, and it now seems that the freeze-fracture procedure reveals the inner face of the membrane and not the surface (see Section II,B). In addition, the particles have been observed in membranes that have been fixed prior to staining in preparations using a wide variety of negative stains with widely different properties. Finally, the particles have been isolated and identified as the enzyme system responsible for the synthesis of ATP during oxidative phosphorylation (190, 203, 209, 313). Unstalked particles are now called coupling factors and under normal conditions they may not project from the membrane surface. Recent evidence shows that the stalks are real and may serve to link electron transport in the basal membrane to phosphorylation in the headpiece.[5]

2. Matrix

The mitochondrial matrix shows considerable variation in electron density from the almost transparent to the very dense. Such differences

[5] For a discussion of the biochemical interpretation of these structures, see Chapter 2, Sections III,B and IV,C. (Ed.)

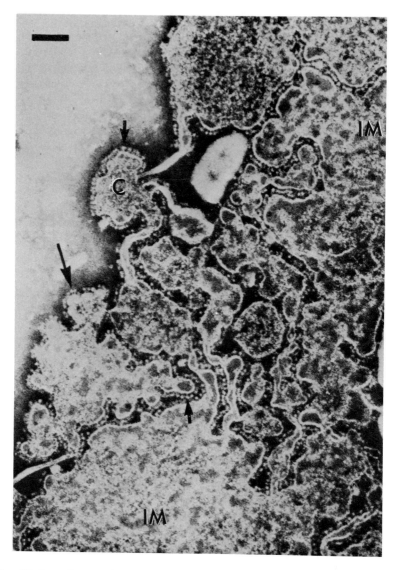

FIG. 30. Part of a negatively stained mitochondrion from summer squash (*Cucurbita pepo*). The large area (IM) is presumed to be inner membrane, and cristae (C) appear rounded. The inner membrane subunits (arrows) each have a head that is about 10 nm in diameter. Scale bar = 0.1 μm. Reproduced from Parsons *et al.* (214), by permission of the National Research Council of Canada.

may occur between adjacent mitochondria in the same fixed cell or even between different regions of the same organelle. The matrix is thought to have a high protein content and also to contain lipid. In sectioned material, a variety of mitochondrial inclusions have been observed including DNA filaments, ribosomes, and calcium-containing granules.

The mitochondrial matrix contains a number of fibrous inclusions, some of which are DNA and which may be identified by their removal with DNAase treatment. The DNA fibrils are often found in the electron-transparent regions of the matrix (107, 190, 279) and in some cases are connected to the inner membrane or the cristae. The DNA may appear clumped or finely filamentous depending very much on the conditions of fixation. There may be one or several strands per mitochondrion, the number depending to some extent on the size of the organelle. All functional mitochondria appear to contain DNA that has many similarities to bacterial DNA (20, 279, 313); the role of mitochondrial DNA is discussed later.

Mitochondria contain ribosomes, which may be demonstrated by treatment with RNase. Mitochondrial ribosomes are generally smaller than their cytoplasmic counterparts and have many similarities with prokaryotic ribosomes (20, 279, 313; and see Section IV,A,2).

Finally, the mitochondrial matrix contains various granules and crystalline inclusions. The granules, which are usually 20–40 nm in diameter, consist of various insoluble salts, particularly calcium phosphate. Crystalline inclusions, probably protein in nature, have also been found in mitochondria. They appear to be more frequently observed in animal cells (see 190) but have also been found in plant mitochondria (203). Their role is uncertain. They have been reported to develop in response to cytochalasin B treatment in coleoptiles (39), although this effect seems more likely due to the presence of dimethyl sulphoxide (229).

C. STRUCTURE AND FUNCTION IN MITOCHONDRIA

Mitochondria are sites of chemical reactions of respiration, where substrates are oxidized and coenzymes are reversibly oxidized and reduced so that the released energy may be conserved as ATP (see Chapter 2 of this volume). Very simply, pyruvate produced largely from glycolysis enters the mitochondrion and is oxidized via the TCA cycle to CO_2 and water. Acetyl-CoA from fatty acid oxidation and the carbon skeletons of amino acids may also enter the TCA cycle. The reduced coenzymes produced by these oxidations feed electrons into the respiratory chain where electron flow to molecular oxygen is coupled to the phosphorylation of ADP to ATP in the process known as oxidative phosphorylation. In addition to this crucial role at the center of metabolism, mitochondria carry out a variety of other enzymic reactions that are linked to various major metabolic pathways (107, 209, 313). They contain transaminases, which produce amino acids from TCA cycle acids and hence can give rise to a variety of nitrogenous compounds; they also are the sites of the enzymes associated with the β-oxidation of fatty acids, although in some tissues this pathway occurs in the glyoxysomes (see Section VIII,B). As already mentioned, mitochondria contain DNA, ribosomes, and related macromolecules, and, presumably, they carry out the processes of transcription and translation (20, 203, 313). In addition, these many functions clearly require the transport of organic and inorganic solutes between the mitochondria and cytoplasm, and thus mitochondria possess specific transport mechanisms for the movement of these metabolites (111, 119). The integration of this wide variety of functions is closely related to the structure of the organelle. Much is now known of the spatial organization of the various enzymes within mitochondria, particularly mammalian mitochondria, although only the essential features of their intramitochondrial distribution can be described here.

With the exception of succinic dehydrogenase, the enzymes of the TCA cycle are found in the mitochondrial matrix (107, 119, 209). Succinic dehydrogenase remains bound to the inner membrane when mitochondrial are disrupted, and cytochemical staining shows high activity associated with the cristae (see 263). However, the TCA cycle enzymes may not be randomly distributed within the matrix but organized into a functional complex (210) perhaps loosely associated with the inner membrane.

The inner and outer membranes of mitochondria are very different. Apart from certain distinct fine-structural features noted earlier, they show marked differences in chemical composition and enzymic activity (62, 119, 209). The inner membrane is richer in protein, and there are marked differences in lipid content; for example, the inner membrane has a much higher proportion of unsaturated fatty acids, which may contribute to its greater flexibility in swelling and contraction (111, 183). The outer membrane is permeable to most solutes of less than 10,000 Daltons and is generally ignored as a solute barrier; functionally, it may be regarded as part of the extra mitochondrial space (111, 119).

The inner membrane has a dual role. It is osmotically responsive and is impermeable to many solutes; it is the site of transport of many metabolites and is assumed to possess specific exchange carriers or porters (111, 119, 313). It also contains succinic dehydrogenase and the respiratory chain and the enzyme system involved in ATP synthesis. The latter is believed to be located in the stalked particles or coupling factors that have already been described. In addition, it has been possible to isolate four separate complexes made of components of the respiratory chain and lipid material, and it has been suggested that these complexes are normally structurally integrated within the intact membrane (203, 209, 313). The considerable biochemical activity associated with the inner mitochondrial membrane is reflected in the high protein content of the membrane and the high density of membrane particles when compared with many other cellular membranes (see Section II,B).

Thus there is good evidence that various functions of mitochondria are spatially separated and organized within the organelle. This relationship between structure and function may also be reflected in the gross morphology of mitochondria. Isolated mitochondria may swell or contract under different metabolic conditions, and this change can be detected by the light scattering properties of mitochondria. Mammalian mitochondria may show two reversible structural states. In the presence of ATP they have an orthodox appearance similar to those fixed *in situ*, whereas in the presence of ADP (when they are actively synthesizing ATP), they appear condensed with a greatly reduced matrix volume (101, 107, 209). The outer membrane appears to play little part in these structural changes. The significance of these findings is not fully understood. One suggestion is that the structural changes reflect conformational changes in the inner membrane associated with energy conservation. However, it must be remembered that conformational changes in proteins may take only milliseconds, whereas gross changes can take several minutes, too long to represent a mechanism of ATP synthesis (190, 203). Furthermore, there are numerous inconsistencies in the literature. For example, one report of mitochondria isolated from plants found no structural changes in response to the respiratory state; all mitochondria showed the condensed form (11). However, Malone *et al.* (158) have described three morphological types (both *in vivo* and *in vitro*) in different tissues of etiolated maize (*Zea mays*) shoots; again the meaning of these changes is unclear. Perhaps they are part of a mechanism of metabolic control (190). The size of the matrix in relation to the surface area of the inner membrane could have a marked effect on the passage of metabolites into and out of the matrix.

Correlations are seen between the gross structure of mitochondria in thin sections and the physiological or developmental state of the tissue. The mitochondria of meristematic cells often have poorly developed cristae, which increase in number with cell differentiation (46, 190, 203). The cristae are particularly numerous in mitochondria of highly active cells such as companion cells, secretory cells, and transfer cells. The most striking development of cristae is seen in certain animal mitochondria, such as those in dipteran flight muscle and heart tissue (190) where there are great demands for energy release. In yeasts and multicellular organisms growing at low oxygen concentration, the number of cristae per mitochondrion is generally low. However, no significant differences are found in the ultrastructure of the mitochondria in rice (*Oryza sativa*) seedlings grown under aerobic or anaerobic conditions, although this may well be a special case (203).

D. Origin and Autonomy of Mitochondria

It has been mentioned that mitochondria contain DNA, ribosomes, and a variety of enzymes and that they are similar to bacteria in a number of respects. These characteristics and the bounding of mitochondria by two membranes have been cited in support of the hypothesis that mitochondria originated as intracellular symbiotic microorganisms (167, 279, 313); similar arguments may also be applied to chloroplasts. This concept is not encountered in relation to the single-membrane-bounded organelles, and it raises the question of the degree of genetic autonomy of mitochondria (and chloroplasts) and the nature of their interaction with the nuclear genes and the cytoplasm in mitochondrial development and function.

Mitochondria appear to be present in all eukaryotic cells, although some unicellular algae possess only one, and in other cells the mitochondria may have a rudimentary structure. For example, in yeast grown under anaerobic conditions, the mitochondria may be poorly developed, and mitochondria may be difficult to see in dry, dormant seeds due to fixation difficulties and the presence of only a few cristae (203). There is still not complete agreement concerning the cellular origin of mitochondria. The degree of genetic autonomy conveyed by the presence of DNA and related macromolecules suggests the possibility of mitochondrial continuity from cell to cell and from generation to generation, and this hypothesis is supported by a number of interesting pieces of evidence indicating that mitochondria arise from preexisting mitochondria. Many phase contrast and electron microscopic observations have been interpreted in terms of mitochondrial division or budding. The interpretation of these images is, of course, somewhat limited—the former by resolution and the latter by the aforementioned problem of static micrographs. However, a clear-cut example is provided by certain unicellular algae that possess a single mitochondrion that divides at each cell division (160). Division may be initiated when a critical size or DNA content is reached, and it involves the partitioning of the matrix into two compartments by part of the inner membrane, followed by constriction and eventual separation (99). Further, more dynamic evidence has come from labeling experiments combined with autoradiography. A choline-requiring mutant of *Neurospora* fed with the precursor[³H]choline incorporated radioactivity into the mitochondrial membranes (156). The fungus was then transferred to a medium containing unlabeled choline, and the distribution of isotope was followed by autoradiography of isolated mitochondria through three cycles of mass doubling. After each cycle the mitochondria were found to be evenly labeled with a decrease in label per mitochondrion, all of which is consistent with the idea that the organelles arise by division of preexisting mitochondria. A similar conclusion followed when the DNA of *Tetrahymena* mitochondria was labeled with [³H]thymidine (215). The limited evidence available suggests that the mitochondrial multiplication, which needs to keep pace with cell division, occurs during interphase (5). An alternative proposal for the origin of mitochondria is that they arise by evagination of the nuclear membrane (14), but the evidence for this is not widely supported (203).

Finally, brief mention should be made of current views of mitochondrial autonomy. Mitochondrial DNA is both functional and essential. It can be replicated and transcribed, and mitochondria contain the machinery for the translation of messenger RNA into proteins (20, 313). The question of what mitochondrial proteins are synthesized within the organelle may be tackled by the use of mutants and specific inhibitors and by studies of isolated mitochondria. It has been demonstrated that mitochondria are only partially autonomous in terms of proteins coded and synthesized within the organelle; for example, the DNA of animal mitochondria is capable of coding for no more than 30 small proteins, far fewer than the total mitochondrial protein complement (313). One interesting finding is that some of the subunits that make up certain proteins are synthesized within the mitochondria, whereas others are synthesized in the cytoplasm. Similar considerations also apply to chloroplasts (see Section VII,F), and they raise the fascinating, but as yet ill-understood question of how these spatially separated protein-synthesizing systems are controlled and integrated.

VII. Plastids

The term plastid describes a fairly diverse group of double-membrane-bounded organelles found in plant cells, which show a developmental relationship to each other or to a common precursor. They are usually relatively large organelles and may be pigmented or possess highly refractive contents, making them readily visible by light microscopy. Plastids, or at least chloroplasts, were first described by van Leeuwenhoek some 300 years ago, and the term plastid was used by Schimper in the 1880s. Some forms of plastid are now believed to be present in all eukaryotic plant cells with the exceptions of a few specialized algal cells and certain male gametes (see 99), although plastids are absent from the cells of animals, fungi, and prokaryotes. Plastids may thus be regarded as the most constant feature distinguishing eukaryotic plant cells from other living cells.

Plastids form, both functionally and structurally, a quite diverse group of organelles, although they have a number of features in common. They are all surrounded by an envelope consisting of two membranes, and, in addition, all possess an internal membrane system, although in some cases this is poorly developed. Chloroplasts and certain other plastids are known to possess a genetic system consisting of DNA, RNA, and ribosomes; this system is generally considered to be a characteristic feature of all plastids, although it has perhaps not been unequivocally demonstrated in all types. In addition, most plastids contain small droplets known as plastoglobuli. Thus, in their possession of a double-membraned envelope and a genetic apparatus, plastids resemble mitochondria, and, as with mitochondria, the degree of autonomy of plastids and their relationship with nucleus and cytoplasm are active areas of research.

The various types of plastid possessing these common features may be found within the different tissues of a single plant, and plastids are classified in terms of color and contents, features that are of course closely related to function. Thus there are two major types of colored plastid: chloroplasts, which are usually green and contain chlorophyll and other photosynthetic pigments, and chromoplasts, which are usually yellow or orange and photosynthetically inactive. The colorless plastids, collectively known as leucoplasts, include those that store starch (amyloplasts), protein (proteinoplasts), or lipid (elaioplasts). In addition, there are two other important types of colorless plastid: proplastids, which are small precursors of other plastids and are particularly common in non-pigmented structures such as roots, and etioplasts, which are found in dark-grown leaves and form a transitory stage between proplastids and chloroplasts. The structure of these individual types will now be described; their origin, development, and interrelations are considered later. The various structural components of plastids (e.g., envelope, lamellae, ribosomes, etc.) are discussed in detail in the following subsections in relation to chloroplasts because, for obvious reasons, the chloroplast has been by far the most thoroughly studied plastid.

A. PROPLASTIDS

Proplastids are the simplest recognizable form of plastid. They are variable in shape, usually somewhat larger than mitochondria, and possess a double-membrane envelope that often shows little development of the internal membrane system. They contain ribosomes, plastoglobuli, often dense aggregates of phytoferritin particles, and sometimes starch (Fig. 31a; 99, 103, 287, 293). In some cells, proplastids may show a more extensive development of the internal membranes, distinguishable perhaps into two types known as the peripheral and central systems (Fig. 31a; 103, 293). The former membranes are normally situated at the ends of the plastids and may be continuous with the inner bounding membrane whereas the latter membranes are usually arranged in centrally situated loops.

Plastoglobuli appear to be present in all types of plastid. They are small, usually osmiophilic droplets that are 10–500 nm in diameter and are found in the matrix or stroma of the plastid. They are not bounded by a membrane; they are similar in appearance to lipid droplets. They contain much of the total plastoquinone of plastids, which is an essential component of chloroplast membranes associated with electron transport and phosphorylation (see Section VII,B,2), together with other, lipidic and lipophilic compounds. The abundance of plastoglobuli often appears to be correlated with the stage of development of the internal chloroplast membranes or thylakoids (99, 286). They are usually small in proplastids but larger and more common in chloroplasts and some chromoplasts, particularly when breakdown of thylakoids is taking place (e.g., during senescence or when chloroplasts are changing to chromoplasts). In developing etioplasts showing membrane proliferation, the number of plastoglobuli may fall, presumably as their lipid reserves are used in membrane synthesis. The composition of plastoglobuli can thus change during development, and this change may be reflected in osmiophilic properties seen by electron microscopy.

Phytoferritin consists of an iron–protein complex, which is similar to the ferritin found in animal cells. Ferritin molecules are large (up to 10 nm in diameter) and are visible in the electron microscope without staining due to the electron-dense iron. In plants, phytoferritin is found mostly in the plastids, and up to 80% of the total iron content of leaves may be found in chloroplasts; phytoferritin is particularly common in

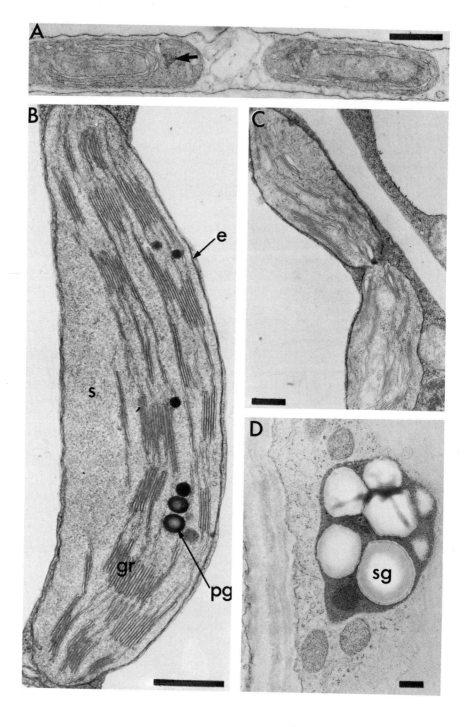

proplastids, senescing chloroplasts, and developing chromoplasts (99, 286). Phytoferritin may be randomly scattered or may appear aggregated in certain regions of the stroma. It may represent a store of iron that can be used to produce iron-containing proteins, such as ferredoxin, during chloroplast development.

Thus proplastids are presumed to act as precursors of other, more specialized plastids, and the presence of some form of plastid in almost all living plant cells guarantees the continuation of this group of organelles. However, little is known concerning the metabolic role of proplastids, although it is clear that they can make starch and certain other storage products. Yet proplastids can occupy up to 10% of the cytoplasm of meristematic cells (99), and so it would be surprising if they were not shown to be more metabolically active in such cells.

B. CHLOROPLASTS

All photosynthetic organisms possess a system of membranes or lamellae that contain the photosynthetic pigments. In photosynthetic prokaryotes, these systems may be dispersed throughout the cytoplasm or grouped to form distinct chromatophores. In higher plants, however, the lamellae are compartmentalized within double-membrane-bounded organelles known as chloroplasts, which also perform various photosynthetic reactions not directly associated with membranes (see Chapter 3).

Although algal chloroplasts show a wide range of shapes and sizes, those from higher plants are usually ellipsoidal or discoidal in shape, measuring 5–10 μm long and with a thickness of 2–3 μm (Fig. 31c). Figures quoted for chloroplast numbers vary widely from one in some algal cells to several hundred in some cells of higher plants; it appears that there may be 20–500 chloroplasts per photosynthetic cell in a leaf, and they may occupy 40–90% of the cytoplasmic volume.

As might be expected, major advances in the elucidation of chloro-

FIG. 31. Electron micrographs illustrating the variability in plastid fine structure. (A) Proplastids in red beet (*Beta vulgaris*) roots showing the double-membrane envelope, internal membranes, and phytoferritin-like particles (arrow). (B) Chloroplast from the leaf of *Suaeda maritima* showing the two membranes of the chloroplast envelope (e), side views of grana (gr), plastoglubuli (pg), and the granular stroma (s). (C) Chloroplast in the stem of *Suaeda maritima* that appears to be in the process of division. (D) Amyloplasts showing prominent starch grains (sg) in beet tissue that was excised and washed for 72 hr. Amyloplasts are presumed to develop from the proplastids shown in Fig. A. Scale bars = 0.5 μm. (D) Reproduced from Hall (103).

plast structure came with the development of electron microscopy, and some of the most striking early electron micrographs were made of the chloroplast lamellar system after shadowing isolated fragments (49, 78, 86). Since then, electron microscopy of thin sections has demonstrated that chloroplasts show three major structural regions. These are the two outer membranes that make up the envelope, the internal lamellar system, which includes the grana seen by light microscopy, and the amorphous matrix or stroma. These regions, which show distinct functional specializations, are considered separately for convenience, although they are undoubtedly highly interrelated *in vivo*.

Finally, before considering the detailed aspects of chloroplast fine structure, the static nature of electron micrographs must be emphasized once again. Cinephotomicrography of living cells demonstrates the remarkably pliant nature of chloroplast shape (129, 314). The outermost layer or mobile phase of chloroplasts constantly changes shape, giving rise to protuberances that may retract or, in some cases, apparently segment to give rise to free streaming structures. Protuberances have been observed in electron micrographs (286), although these are usually less striking; presumably many of these transient structures are lost during fixation. Another apparently dynamic interaction of the mobile phase, that involving the fusion of chloroplasts with mitochondria, has already been mentioned[6]. Both light and electron microscopy have also shown that chloroplast volume changes *in vivo* in response to changes in light or in the ionic environment (193). Thus these dynamic aspects of chloroplast structure must not be overlooked when interpreting structure in terms of electron microscopy.

1. *The Chloroplast Envelope*

The chloroplast envelope consists of two membranes, each variously reported to be 6–10 nm thick and separated by an electron-translucent space about 1–20 nm in width (49, 286). The envelope is clearly one of the most important bounding membranes in living cells and because it controls the movement of photosynthetic metabolites into and out of the chloroplasts. Yet, surprisingly, there have been relatively few studies of its ultrastructure. Chloroplasts are among the easiest organelles to isolate and purify, and a number of attempts have been made to separate and characterize the envelope (see 62). The lipid composition of the envelope is different from that of the internal lamellae; the envelope has a

---

[6] See Section VI,A and discussion in Chapter 3 Section II,E. (Ed.)

lower proportion of galactolipid and higher levels of phosphatidylcholine. The envelope also lacks chlorophyll and has a very different carotenoid content. The two membranes of the envelope also differ in their permeability properties. The outer membrane appears to be nonspecifically permeable to sucrose and some other small molecular weight compounds, whereas the inner membrane is the site of a variety of specific transport systems for solutes such as phosphate, amino acids, and adenine nucleotides (62). The gap between the inner and outer membranes is referred to as the sucrose-permeable space.

Relatively few comparative studies have been made of the ultrastructure of the two membranes of the chloroplast envelope. The outer membrane generally has a smooth appearance, whereas the inner membrane is frequently invaginated with finger-like protusions into the stroma. These protrusions are more pronounced in proplastids, and it has been suggested that they are involved in the formation of the internal membrane system, although the chemical composition of the envelope and lamellae are very different.

A feature of the choroplast envelope seen in many C_4 plants is the invagination of the envelope membrane to form a complex arrangement of anastomosing tubules known as the peripheral reticulum (49, 144a, 274, 286; Fig. 32). Although it is widely used as an ultrastructural marker for chloroplasts of C_4 plants, it should be noted that the reticulum is sometimes poorly developed and that it is also found in certain C_3 species (see 274). This proliferation of the envelope clearly increases its surface area and presumably results in a rapid exchange of metabolites across the membrane.

2. The Internal Lamellar or Grana–Fretwork System

The internal lamellar system of chloroplasts contains the chlorphyll and forms the structural framework for the primary reactions of photosynthesis. The most obvious structural features of this system, the grana, are visible by light microscopy as small green discs and were described by Meyer as early as 1883. The chloroplast was considered to be divided into two regions, the grana and stoma, although the presence of the grana was not fully established until the 1930s when they were extensively studied by light microscopy. The grana were shown to be roughly cylindrical, usually about 0.2–0.4 μm in diameter, with generally about 40–60 grana per chloroplast in a typical photosynthetic cell. Electron microscopy conclusively demonstrated the presence of grana and showed that they consisted of stacks of flattened lamella-bounded sacs known as thylakoids

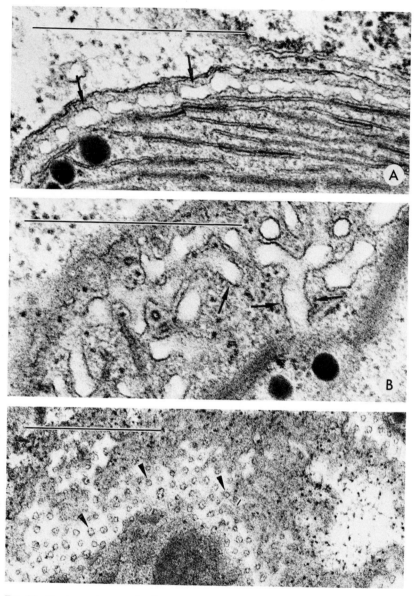

Fig. 32. Electron micrographs illustrating different aspects of the peripheral reticulum (arrows) in chloroplasts of mesophyll cells of *Portulaca oleracea*. Both vesicular (A) and tubular (B) appearances are shown. Scale bar = 1 μm. Reproduced from Sprey and Laetsch (274).

(Figs. 31c,37). In addition, the grana were seen to be linked by inter-granal lamellae (stromal thylakoids), called frets. The whole internal membrane structure is thus known as the grana–fretwork system. It has an extremely complex three–dimensional organization, the structure of which has been slowly unraveled from studies of the surface of chloro-plast fragments, by serial sections through the lamellar system, and by the use of freeze-fracture techniques (Fig. 33). The grana–fretwork sys-tem is now regarded as a single membrane continuum with structurally differentiated regions that probably permit considerable functional spe-cialization.

The structure of the grana–fretwork system has been the subject of a very large number of papers, and no attempt will be made to give a comprehensive coverage in this short survey. Further details and view-points can be obtained from a number of useful reviews (49, 99, 189, 213, 286); the terminology employed here is that developed by Weier and co-workers (e.g., 307,308), and it seems to have gained the most widespread use.

Each granum consists of a stack of thylakoids, which may be precisely aligned to form a compartmentalized cylinder. The simplest granum consists of only two thylakoids, but typically there may be 10–100 thyla-koids in a single stack. Although selected micrographs and diagram-matic representations often show cylindrical grana, irregular forms are frequently observed with thylakoids that vary in diameter and that may extend from granum to granum to give a branched appearance. Again, it should be noted that the often presented micrograph showing grana with thykaloids running parallel to the long axis of the chloroplast may not be fully representative. Other orientations are observed, and it is possible that grana can move in relation to light (286) as well as can the chloroplasts.

The various regions of grana have been given specific names (Fig. 34). The region between two adjacent thylakoids is a very characteristic fea-ture of grana and has been termed the partition; its precise structure has been difficult to resolve because its appearance in thin sections de-pends very much on the method of preparation and staining. The re-gion may appear to be filled with heavily stained material; alternatively, with methods that retain lipid, the region may be resolved into two electron-dense layers separated by an electron-translucent gap. The lat-ter has been termed the A space, although this does not mean that it is empty. The internal translucent regions enclosed by the thylakoids are termed loculi, and these may be continuous with spaces enclosed by the fret lamellae. The ends of the thylakoids in contact with the stroma are

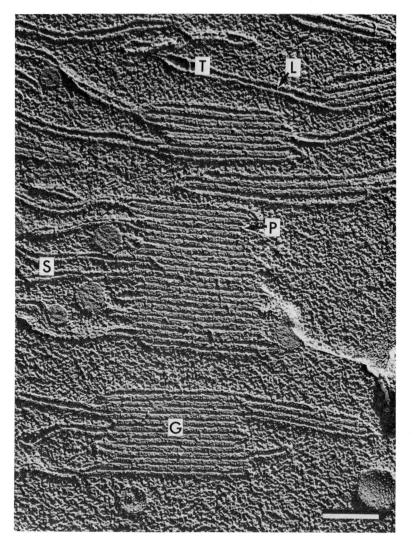

FIG. 33. Cross-sectional view of a spinach (*Spinacia oleracea*) chloroplast as revealed by freeze-etching and showing the arrangement of grana and stromal thylakoids. T, thylakoid; L, lumen; P, partition; G, grana; S, stroma; Scale bar = 0.2 μm. Reproduced, after E. Wehrli, from Mühlethaler (189).

called margins, whereas the thylakoids at each end of the stack are termed the end granal membranes.

It is now clear that the fretwork lamellae are not simply flat expanses of membrane but that they may be highly perforated to form anastomos-

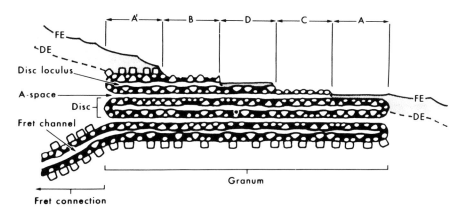

FIG. 34. Diagrammatic representation of thylakoid disc and fret membranes based on observations obtained by freeze- and deep etching and by negative staining. The top surface is imagined to be exposed as in a freeze-etched preparation, yielding images of faces labeled B and C. If additional ice (shaded layer) is removed by deep etching, external faces of membranes (A', D, and A) are also revealed. The large B-face particles and the smaller C-face ones are shown embedded in the membranes. Molecules of coupling factor for phosphorylation are drawn as squares on the A' surface. The solid blank representation of the membrane matrix ignores any spatial separation of lipid and protein components that may exist in the nonparticulate part of the membrane. Reproduced from Gunning and Steer (99).

ing membrane-bound channels linking two or more grana (Fig. 35). Two important features of these connections have been established. First, an individual fret lamella may be connected to a number of thylakoids in a single granum and appear to take the form of a right-handed helix ascending the stack. Second, each thylakoid may be connected to a number of different fret systems that in turn branch and connect with other granal stacks. The maximum amount of integration is indicated in the model of Paolillo (210a). Thus the grana–fretwork system is envisaged as a complex three-dimensional system of anastomosing channels connecting the thylakoids in an individual stack and in adjacent grana. This is essentially the concept of the chloroplast envisaged by Heslop-Harrison (125a) which is that of an organelle consisting of single, enormously complex lamellar system separating the stroma proper from the spaces within the thylakoids (which are probably continuous). The initial concept of the thylakoids as separate structures is no longer valid, although "thylakoids" remains a useful term to describe the sac-like structures seen in thin sections by electron microscopy (213). The model shown in Fig. 35 is an idealized one because the degree of interconnection may very greatly depending on factors such as the number of thylakoids per

Fig. 35. Schematic representation illustrating the development of the concept of membrane stacking in the grana of chloroplasts of higher plants. (A) Early concept showing grana formed of individual discs (small thylakoids) linked by more extensive stromal discs (large thylakoids). (B) Perforated stromal lamellae (frets) linking grana at several levels. (C) Frets connecting with several thylakoids within the same grana. (D) Spiral formation of grana membranes. (E) Spiral fretwork arranged around and interconnecting the individual grana. Reproduced from Coombs and Greenwood (49).

stack, the number of fret connections per thylakoid, and the degree of branching of the fret system.

  a. Substructure of the Thylakoid Membranes.   Current ideas concerning the substructure of biological membranes have been discussed in Section II,B. The structure of the thylakoid lamellae has been the subject of many studies using a variety of techniques, presumably due to the importance of these membranes as sites of the primary reactions of photosynthesis. However, these findings are frequently not considered in generalized models of membrane structure, presumably due to their very special nature resulting from the presence of photosynthetic pigments. Nevertheless, many of the characteristic features of biological membranes in general are found in the thylakoid membranes.

  The high protein content and high density of membrane particles seen by freeze-etching (Table II) suggest that chloroplast membranes are biochemically very active and specialized (see Section II,B). When examined in thin sections, they often show the typical tripartite unit membrane appearance with a thickness of 5–10 nm. Other studies, however, indicate the presence of globular subunits, particularly when $KMnO_4$ is used for fixation or staining. These globular structures are usually 6–7 nm in diameter, and it has been noted that it is not easy to see how such structures can give a simple subunit appearance in sections of about 50 nm thickness. The problems of interpreting these images have been discussed by a number of authors (189, 213, 288). However, further support for the subunit concept has come from the observation of particles in negatively stained and shadowed membranes. On the basis of these latter observations, the term *quantasome* was coined for these particles, because it was thought that they were the morphological expression of the light-harvesting photosynthetic unit predicted by physiological measurements (213). Indeed the photosynthetic unit was thought to contain about 250 chlorophyll molecules, and it was estimated that each quantasome could contain about 230 chlorophyll molecules. However, this relatively straightforward explanation has been compromised by more recent observations, particularly studies attempting to equate quantasomes with particles seen by freeze-etching. For example, large particles are rarely found in the fret lamellae, yet these membranes are known to contain chlorophyll and to be photosynthetically active.

The presence of globular subunits in the thylakoid membranes was confirmed by the freeze-etch procedure, which allows an examination of both the surface and internal structures of membranes (see Section II,B). Freeze-fracturing reveals the internal surface, whereas the outer surfaces may be viewed by the process of deep-etching. The following nomenclature is now widely used to describe the faces revealed by these procedures (Fig. 34).

The exterior surface of a thylakoid next to the stroma on a fret or at the top and bottom of a granum is designated A', and the corresponding face within grana backing the partitions is labeled A. The other exterior face that bounds the loculi is labeled D. The two internal faces exposed by freeze-fracture are not identical and are designated B and C; the former is the exposed face that backs the loculus, whereas the latter is adjacent to the A' or A surfaces.

These various surfaces show characteristic arrangements of particles. The A' face carries particles that are mostly about 11 nm in diameter and can be removed by treatment with EDTA. The corresponding A faces lack these particles and appear smooth. The B surface contains very large particles that are estimated to be 13–17 nm in diameter and 8–11 nm high. Such large particles have been found only in photosynthetic membranes and are assumed to be associated with the primary processes of photosynthesis. They are numerous in the grana but their density decreases or may be zero in the fret lamellae. The C face of both granal and fret thylakoids contains a dense array of smaller particles estimated to be 9–12 nm in diameter, whereas the D surface shows characteristic large bulges, about 18.5 by 15.5 nm, which are thought to be due to the largest B face particles on this side of the membrane.

Thus these observations are consistent with the model of biological membrane structure discussed in Section II. The membrane consists of various particles fully or partially embedded in the lipid matrix, which (in the case of the thylakoids) also contains the photosynthetic pigments. As discussed previously, there is evidence that the lipid molecules are in a fluid state, allowing free movement of particles within the membrane (189). Broadly, there appear to be two classes of particles, which are approximately either 17.5 or 11 nm in diameter. The larger particles occur mainly in the grana, whereas the smaller are found in both granal and fret membranes, clearly suggesting that the two membranes have differences in function. A third, loosely bound particle is attached to membranes in contact with the stroma. The position of the chlorophyll molecules in relation to thylakoid structure is uncertain and is usually not depicted in membrane models. Chlorophyll appears to be conjugated to protein in photosynthetic membranes, although little is known of the three-dimensional organization of these complexes or how they interact with various membrane components (288).

b. The Relationship between Structure and Function within the Grana–Fretwork System. Since the experiments of Englemann in the 1890s, which showed that the reactions of photosynthesis (or at least the processes of light absorption and oxygen evolution) occur within the chloroplasts, there has been a slow unraveling of the complex relationships between structure and function in these organelles. It is now clear that the thylakoid membranes are differentiated not only structurally, as we have just seen, but also functionally to form an organized assembly of subunits, each with its particular place in the three-dimensional framework. It is now possible to assign certain functions to specific regions and particles within the thylakoid membrane.

In broad terms, the process of photosynthesis is divided into two stages—the light reactions and the dark reactions (see Volume I,B). The former involve the absorption of light and the associated electron transport reactions resulting in the phosphorylation of ADP to ATP and the reduction of ferredoxin, which in turn reduces $NADP^+$ to NADPH (see Chapter 2, Section V). The dark reactions utilize the energy and reducing power generated by the light reactions to catalyze the reduction of CO_2 to carbohydrate (see Chapter 3, Section II). Experiments in the 1960s established that the light reactions and associated processes are located in the chloroplast lamellae and that the dark reactions occur in the stroma (see 212)[7]; the latter is discussed in detail later.

The flow of electrons from water to ferredoxin involves two light reactions that are known as photosystems I and II (PSI and PSII). The absorption of light by PSI leads to a reduction of ferredoxin and the oxidation of a chain of electron carriers that link the two photosystems. These carriers are reduced by electrons that come ultimately from the oxidation of water in a process involving PSII. The flow of electrons and the associated energy drop along these carriers is coupled to the phosphorylation of ADP to ATP.

The question of the functional differentiation of thylakoids may be tackled at two levels: first, in terms of the overall three-dimensional arrangement of the grana–fretwork system, and second, with regard to the integration of the various components of the light reactions within the plane of the membrane. The grana system clearly provides a large surface area of membrane, and it has been estimated that the combined surface area of membrane within leaf chloroplasts exceeds by many times the total area of the leaf (99). A more difficult question is whether the grana stack arrangement is the most efficient way of displaying this membrane. Certainly the ratio of grana formation to total membrane varies widely in different plants and under different conditions.

Particularly extensive development of grana is seen in shade plants, and it has been suggested that the grana provide a high pigment concentration per unit area, which allows a more efficient collection of light at low intensity (257). An interesting observation is provided by the bundle sheath cell chloroplasts of malate-transporting C_4 plants, which, unlike the mesophyll chloroplasts, are essentially devoid of grana. In these plants there is a much reduced requirement for NADPH production in bundle sheath chloroplasts, an activity associated with grana membranes. In addition, as we shall see below, there is some evidence that ATP synthesis is associated with the fret membranes and not the grana membranes, suggesting that the more extensive stromal development in

---

[7] For a detailed discussion of the light reactions and electron transport in chloroplasts, see Chapter 2, Sections V and VI; for carbon reactions see Chapter 3. (Ed.)

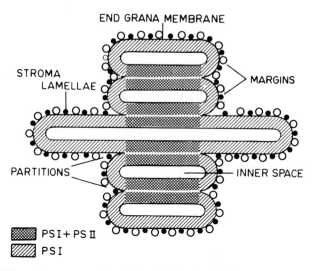

FIG. 36. Schematic representation of the distribution of photosystem I (PSI), photosystem II (PSII), NADP$^+$ reductase (●), and ATPase (○) in the lamellar structure of chloroplasts. After Sane (257).

the bundle sheath chloroplasts could provide the additional ATP needed for certain types of C_4 metabolism (see Chapter 3, Section IV,B,3). The role of grana may be to provide flexibility and allow plants to adjust to the prevailing light conditions (257). However, because the development of grana varies with other physiological conditions, other roles for grana should also be considered.

At the submembrane level, attempts have been made to link structure and function by disruption of membranes using detergents or mechanical means, followed by biochemical and fine-structural characterization of the separated fractions. The interpretations of the distribution found for various biochemical activities has been discussed in detail in a number of articles (189, 213, 257, 286), and only a general summary is given here. The evidence indicates a correlation between the small (11 nm) particles observed by freeze-fracture and PSI activity, suggesting that PSI is associated with both fret and grana membranes. The large (17.5 nm) particles appear to separate with PSII activity, which would thus be restricted to the partitions in the grana thylakoids (Fig. 36). It is not clear whether there are two types of PSI, one physically close to PSII and the other isolated in the fret membranes.

Both photosystems are linked to the synthesis of ATP in the process of photosynthetic phosphorylation, which has many similarities to oxidative phosphorylation in mitochondria.[8] In the latter, ATPase and the coupling factor are believed to be located in the stalked particles seen on the inner membrane. In chloroplasts, the coupling factor appears to be associated with the loosely bound particles attached to the A' surfaces of the fret and grana thylakoids (Fig. 34). Immunological evidence suggests that NADP$^+$ reductase is also found on these surfaces (257). If this is the case, then it is necessary to explain how the processes of electron transport and ATP synthesis are linked. However, if the type of vectorial electron flow predicted by Mitchell's chemiosmotic hypothesis establishes a proton gradient of pH difference between the inner and outer spaces, then an ATPase situated anywhere in the system could use this gradient to drive the synthesis of ATP.

There is considerable uncertainty concerning the morphological equivalent of the photosynthetic unit that was defined by classical physiological measurements (see Section VII,B,2a). The quantasomes observed by negative staining or shadowing are thought to represent the views of the D face of the thylakoids after freeze-etching (i.e., they are thought to be the underlying, large B-face particles). However, these particles are not found to any great extent on the fret membranes, which nevertheless contain chlorophyll and are photosynthetically active. Thus there is no clear evidence for a structural counterpart to the photosynthetic unit.

This brief summary of current thought suggests that both structural and functional differentiation occurs within the fret and grana lamellae. The morphological evidence gained by freeze-fracturing further indicates that the membrane is asymmetric, and it is assumed that multienzyme complexes are inserted into the membrane to form an integrated photosynthetic network. There are now various proposals that locate the components of the two photosystems within the thylakoids; because these schemes do not depend on morphological evidence, they are not detailed here. They may be found in a number of useful reviews (189, 257, 291). A full understanding of the topography of the thylakoid membranes is a major aim of future research.[9]

3. The Stroma

The stroma is the name given to the granular matrix of chloroplasts and other plastids that is enclosed by the plastid envelope and in which

---

[8] See Section VI and discussion in Chapter 2. (Ed.)
[9] See discussion in Chapter 2, Sections V and VI. (Ed.)

the internal lamellar system is embedded. It is considered to have a high protein content due largely to the presence of fraction I protein, which makes up about half of the soluble protein of leaves and has ribulosebisphosphate carboxylase as its major catalytic activity. In addition, the stroma contains an assortment of particles including various crystals and fibrillar aggregates, ribosomes, DNA, starch grains (discussed in Section VII,D), and plastoglobuli and phytoferritin (described under proplastids).

The degree of granularity of the stroma depends very much on the fixation and staining procedures employed. It is assumed that many of the particles observed are fraction I protein, because this protein is a large molecule (MW ~500,000) and represents a large proportion of the total protein. In addition, various crystalline arrays from a number of different plants have been described. For example, Gunning *et al.* (99a) have described a fibrillar spherulite or *stromacenter* that may be aggregates of fraction I in the stroma of oat leaf plastids. Such aggregates are particularly common in the plastids of plants subjected to various types of physiological stress, and their formation may be related to an increased dehydration of the stroma (see 286). For example, they are frequently found in protoplasts isolated from leaf cells that have been subjected to the plasmolyzing conditions of the isolation procedure (see 284). However, the arrangement of the crystalline arrays varies in different species, and it is not certain that the arrays represent the same material in each case.

It should also be noted that crystalline inclusions are found within the lamellae of chloroplasts (see 273, 286; Fig. 37). Fractions enriched in these intrathylakoid inclusions have been isolated from stressed spinach chloroplasts and have been shown to possess ribulosebisphosphate carboxylase activity with some ATPase (275). It has been concluded that the crystals are most likely inactive ribulosebisphosphate carboxylase, possibly mixed with some ATPase. The function and mode of formation of these crystals are not clear, although the presence of similar structures in the stroma suggests that under stress conditions stroma constituents are transported across the thylakoid membrane into the thylakoid lumen (273).

Another characteristic feature of the chloroplast stroma is the presence of ribosomes and polysomes (see Section IV,A,2). Chloroplast ribosomes are about 17 nm in diameter and are slightly smaller than their cytoplasmic counterparts. In many important respects they resemble prokaryotic ribosomes more than the cytoplasmic ones (see 20, 107, 281). Polysomes are frequently observed in close association with the thylakoid membranes, particularly during periods of rapid membrane synthesis, and

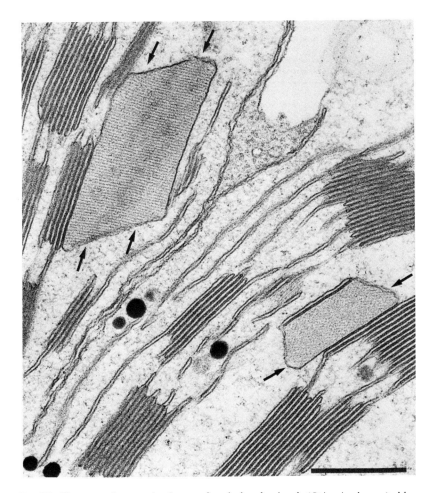

FIG. 37. Electron micrograph of part of an isolated spinach (*Spinacia oleracea*) chloroplast with two membrane-bounded inclusions (arrows). When thin-sectioned perpendicular to the thylakoid plane, the contents are predominantly striated (upper inclusion). Dilation of the grana thylakoid (lower inclusion) has forced the granum apart without altering the parallel orientation of the grana lamellae. Scale bar = 0.5 μm. Reproduced from Sprey (273).

fractionation studies suggest that this is a real attachment rather than a fortuitous close positioning (99).

The stroma of a variety of plastids has been shown to contain DNA, which, as with mitochondria (see Section VI,B,2), usually appears in electron micrographs as a mesh of 2.5-nm fibrils within a clear zone. These regions are often referred to as nucleoids because they resemble the DNA-containing regions of prokaryotes. There may be up to five nucleoids per chloroplast, and, in general, the larger the plastid the more nucleoids it is likely to contain. The DNA differs from nuclear DNA in relation to its base composition, buoyant density, lack of histone protein, and other features. It is extensively reiterated and, because the number of DNA copies exceeds the number of nucleoids, the latter may be assumed to be polyploid.

C. ETIOPLASTS

The term etioplast describes organelles that develop from proplastids in dark-grown (i.e., etiolated) leaves. These plastids contain little chlorophyll and lack a grana–fretwork system. Instead the proplastids develop into etioplasts, which are characterized by an internal membrane system that forms a central paracrystalline lattice known as the prolamellar body (Fig. 38). Otherwise etioplasts show the general features of plastids; they may be up to 5 μm in diameter, possess a double membrane envelope, and contain a few thylakoids, DNA, and ribosomes (99, 151). When etiolated plants are subsequently exposed to light, protochlorophyllide is converted to chlorophyll while, as its name implies, the prolamellar body is dispersed and transformed into the characteristic thylakoids of mature chloroplasts; the term etiochloroplast has been used to describe the developing chloroplasts in etiolated, illuminated leaves (151, 152). Etioplasts have often been considered to be abnormal plastids, although they are useful in the study of chloroplast development. However, prolamellar bodies are commonly found in young monocotyledonous leaves that are shielded from light by the coleoptile or leaf sheath, and so they may represent a normal step in chloroplast development in these plants (151).

The prolamellar body consists of a regular three-dimensional arrangement of membranous tubules that branch and interconnect to form an extensive lattice, which may be several micrometers across. A full description of the structure, variability, and development of etioplasts has been provided by Gunning and Steer (99) and is not repeated here. It should be noted that the lattice is composed of repeating units, the most common consisting of tetrahedrally branched tubes interconnected in various ways. For example, "zinc sulfide" lattices resemble the symmetrical crystal lattices found in deposits of zinc sulfide and consist of tetrahedral units fused to form a mesh of six-sided rings in one plane and connected to other rings above and below. The prolamellar bodies form a reserve of membrane with a very high ratio of surface area to volume. The conformations are assumed to represent stable arrange-

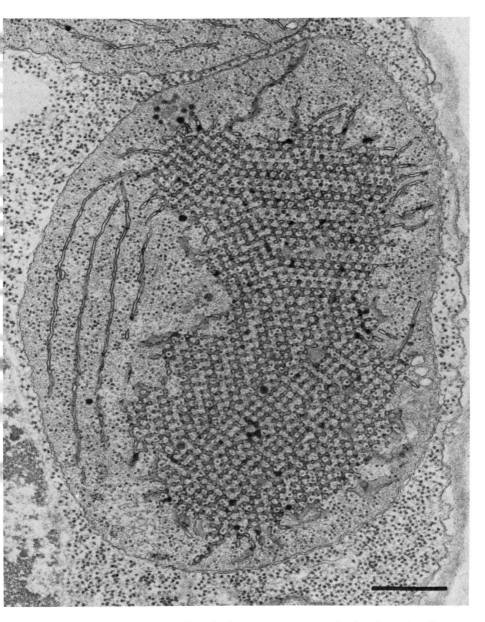

Fig. 38. Electron micrograph of an etioplast from *Avena sativa* showing the prolamellar body. Scale bar = 0.5 μm. Provided by B. E. S. Gunning and M. W. Steer; reproduced from Hall *et al.* (107).

ments, the different regular forms depending on the initial type, which may be formed by chance.

D. AMYLOPLASTS

Amyloplasts are plastids usually found in the nongreen parts of plants that synthesize and store starch to the exclusion of other functions. Although proplastids and chloroplasts may also contain starch grains, they rarely develop to the size found in amyloplasts, where they may distend the plastid envelope and may be seen by light microscopy (Fig. 31d). Amyloplasts contain sparse internal membranes and few ribosomes.

Starch is deposited as granular particles that are built up by apposition of concentric layers around a center or hilum, which may be centrally or eccentrically placed. Starch grains may be simple or compound, the latter consisting of several segments that have fused during growth. The size, shape, and other features of reserve starch grains are often characteristic of a species, although the starch grains found in chloroplasts invariably have a simple lens shape.

One interesting and specialized role of amyloplasts is suggested in relation to the geotropic response. The gravity-sensitive zones of many plants (e.g., the root cap and coleoptile tips) contain certain cells, the statocytes, which are characterized by the presence of starch statoliths (amyloplasts) that sediment under gravity and apparently initiate the preception of gravity through the cytoplasm of the statocyte cell (136). This statolith movement is presumably translated into some form of chemical gradient that results in the geotropic curvature, but although there are many theories, the exact cause of the response is unknown at present.

E. CHROMOPLASTS

The term chromoplast describes colored, nonphotosynthetic plastids that may develop directly from proplastids, chloroplasts, or amyloplasts. Chromoplasts possess the general features of plastids described previously; they are characterized by the accumulation of various storage products (frequently carotenoid pigments) and by little development of internal membranes, although in some there is an extensive thylakoid system. The distinction between chromoplasts and chloroplasts is particularly unclear during certain transition periods. Dedifferentiation of chloroplasts to form chromoplasts involves a progressive loss of the internal membrane system, which may be accompanied by the accumulation of carotenoids; often a thykaloid complex resembling a prolamellar body is retained (312). If tissue regreening occurs, the chromoplasts may form new grana to resemble typical chloroplasts again.

The pigments accumulated in chromoplasts can take several forms. In some plants they appear to be associated almost entirely with the plastoglobuli, whereas in others they form separate filamentous or crystalline structures. Crystalline forms of β-carotene and lycopene are found in carrot roots and tomato fruits, respectively, and the deposition of these crystals appears to commence in special thylakoids (99). Chromoplasts occur in a variety of tissues, although they are most commonly associated with reproductive structures (e.g., petals and fruits) where they are assumed to function in the attraction of animals involved in pollination and seed dispersal.

F. PLASTID DEVELOPMENT AND INTERRELATIONS

The various types of plastid found in plants are generally believed to originate from proplastids. The proplastids are perhaps totipotent for all kinds of plastid; their subsequent development may depend upon the environment in which the plant grows. However, the possibility that there are genetically distinct races of plastids should also be considered (45).

The meristematic cells of the shoot apex contain proplastids but no chloroplasts. The proplastids are normally about 1 μm in diameter and contain only a few thylakoids. As leaves develop in the light, there is a massive proliferation of the internal membranes which produces many flattened thylakoids that become stacked on top of each other to produce grana. Contrary to some earlier beliefs, there is no clear evidence that this membrane is derived from invaginations of the inner membrane. It is assumed that membrane components are synthesized and inserted randomly over the lamellar surface.

Amyloplasts may also develop directly from proplastids. Proplastids often contain small starch grains, and amyloplasts develop by an increase in starch accumulation that in some cases is initiated simply by the presence of the substrates needed for starch synthesis.

The aforementioned are examples of plastid formation directly from proplastids; these examples suggest a unidirectional pathway of development. However, it is also clear that many interconversions can occur and that differentiation can take place. For example, some amyloplasts can produce green thylakoids when exposed to light, whereas both amyloplasts and chromoplasts may develop from chloroplasts under appropriate conditions. In certain colored tissues that undergo regreening, chromoplasts may form new thylakoids and grana (312). Whatley (312) has revived an idea Schimper proposed in the 1800s: that plastid development follows a cyclic—not a unidirectional—pathway. Both plastid dedifferentiation and redifferentiation are thought to occur, with the mature chloroplast as the fundamental plastid structure in which the internal membrane system is continuously degraded and replaced. Alterations in the balance between degradation and synthesis can lead to differentiation or redifferentiation provided that senescence has not begun because senescence may involve a separate, irreversible mechanism that branches from the cyclic pathway.

Closely related to any discussion of plastid development is the question of plastid replication. There are many unknowns; for example, it is not clear whether plastid division occurs only at certain stages of development or whether all plastid types can divide. There is good evidence that both proplastids and chloroplasts can divide, although information is sparse for other plastids. Division of chloroplasts has been filmed in giant algal cells (87), whereas in higher plants division is implicated from both kinetic (152, 231, 232) and morphological observations (152; Fig. 31c). Division has also been observed with isolated chloroplasts (241). In spinach leaves, and probably in other species as well, the vast majority of chloroplasts arises from the fission of grana-containing chloroplasts rather than directly from proplastids. Chloroplast DNA synthesis occurs in association with chloroplast replication, such that the DNA doubles during the division cycle and segregates to give similar amounts to each daughter chloroplast (231). It is likely that plastid division is not an independent process but is under general cellular control. For example, division may be related to cell enlargement; it responds to exogenous kinetin but not to certain other growth substances (152).

Division of higher plant chloroplasts may occur in one of two ways (52). One is by gradual constriction, forming an ever narrowing middle region (Fig. 31c), whereas the other appears to involve an invagination of the inner envelope, which eventually extends across the chloroplast to produce two compartmens. Finally it should also be noted that fusion of chloroplasts has been observed in a number of plant groups, and this presumably provides another means of controlling plastid number.

VIII. Microbodies

Microbodies were discovered by electron microscopy, although their rather indistinct morphology meant that their widespread occurrence and metabolic importance was not recognized for some time. However, it has become clear that microbodies are a distinct and ubiquitous class of subcellular organelles with specific metabolic functions. The term microbody was first used to describe single-membrane-bounded particles of unknown function observed in animal cells; soon similar bodies were found in plants. In addition to the word microbody, other terms have been used to describe these or similar structures, including crystal-containing body, cytosome, lysosome, and spherosome; the last two terms are now used for separate organelles having very distinctive functions (see Sections V,B,1 and V,B,3). The term microbody is now applied to a single-membrane-bounded organelle with a granular matrix that may contain an electron-dense core or crystal and that often shows a close relationship with the ER. Microbodies have also been characterized biochemically and so may be defined more precisely than only their morphology allows. In addition to specific enzyme complements related to their physiological function, they contain flavin-linked oxidases, which generate hydrogen peroxide, and catalase, which destroys the peroxide (107, 259, 290), and they therefore contribute to the overall gas exchange of cells. Microbodies differ from mitochondria structurally and in their lack of cytochromes, and the former are not involved in the conservation of energy as ATP. The presence of catalase has proved to be of great value as a marker enzyme. It may be stained for electron microscopy using 3,3'-diaminobenzidine (DAB) (Fig. 39), so it has been extremely useful in the identification of microbodies in a variety of different cells (74, 89, 269, 302, 303).

Microbodies have been observed in cells of every eukaryotic plant examined in any detail (75, 99, 269, 290). Two major types of microbody, known as peroxisomes and glyoxysomes, have been identified and characterized biochemically. Peroxisomes are involved in the metabolism of glycolate in green leaf cells, whereas glyoxysomes function in the conversion of lipid to sugar during the germination of certain seeds. In addition, microbodies occur in a variety of other cells and tissues, although most have not been examined with any thoroughness. These apparently less specialized organelles are usually grouped together in a third category, although further research may well reveal its diverse and specialized metabolic functions.

A. ULTRASTRUCTURE OF MICROBODIES

Microbodies are usually roughly spherical, ranging from 0.2 to 1.5 μm in diameter, although elongated, dumbbell-shaped, and other, more irregular forms occur. Microbodies consist of a single smooth membrane surrounding a granular matrix of moderate electron density, which may

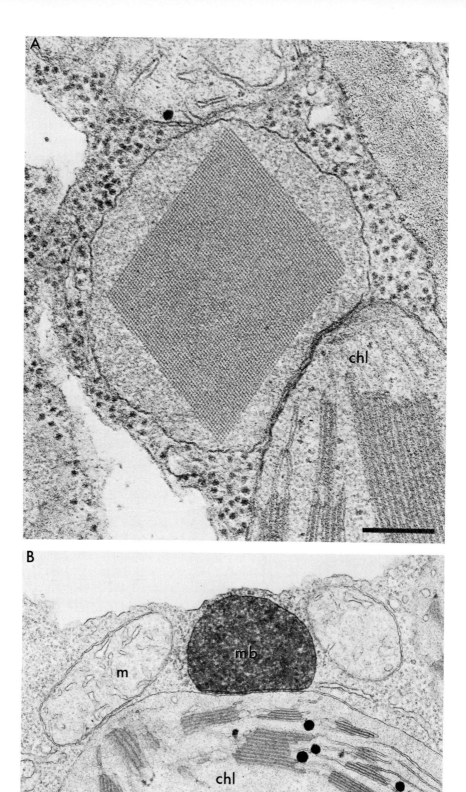

contain fibrillar, amorphous, or crystalline inclusions (73, 75, 76, 269; Fig. 39). Microbodies may show a close association with the ER, and this observation is discussed in more detail in relation to microbody origin. They do not contain DNA, ribosomes, or internal membranes.

The nature and function of the various inclusions that may occur in microbodies are not fully understood. When preparations are stained for catalase, the reaction product is usually distributed through the matrix in those microbodies that lack inclusions. When inclusions are present, they often show more intense staining than the surrounding matrix (see 303). However, isolated crystalline material does not appear to be particularly rich in catalase (130a). These inclusions do not appear to be essential features of microbodies; for example, they are not commonly found in roots (75, 76) or green algae (269). Microbodies lacking inclusions are found predominantly in younger tissues, and the formation of crystalline inclusions seems to represent a normal pattern of development, often starting near the surrounding membrane of the organelle (76). In addition to being closely associated with the ER which has already been mentioned, glyoxysomes may be found closely apposed to lipid bodies and peroxisomes situated next to mitochondria and chloroplasts. The possible significance of these arrangements is discussed in the next section.

B. FUNCTION AND DEVELOPMENT OF MICROBODIES

Two major classes of microbody have been described, and, although they have certain basic structural and biochemical features in common, they perform very different metabolic tasks.

During the germination of certain seeds, there is a massive conversion of stored lipid into sucrose. This involves the breakdown of long-chain fatty acids by the process of β-oxidation, followed by the conversion of acetyl-CoA to succinate (a precursor of sugars) by the glyoxylate cycle. Using castor bean endosperm tissue, it has been demonstrated that the two essential enzymes of the glyoxylate cycle, isocitratase and malate synthetase, are present in an isolated fraction containing microbody-like particles, which were thus called glyoxysomes (25, 50). A later cytochemical study has clearly demonstrated the specific association of malate

---

FIG. 39. Electron micrographs of peroxisomes in tobacco (*Nicotiana*) leaf cells. (A) Peroxisome with a large crystalline inclusion. (B) Peroxisome stained for catalase using 3,3'-diaminobenzidine, which produces an osmiophilic (electron-dense) product. Note the close association between chloroplast (chl) and peroxisome (mb) and the presence of mitochondria (m) nearby. Scale bars = 0.2 μm. Reproduced from Frederick and Newcomb (74).

synthetase with glyoxysomes (34). A further study of these particles revealed that they also contain the enzymes catalyzing the β-oxidation of fatty acids (51). Glyoxysomes clearly play the essential role in the conversion of fats to sugars.

Fine-structural studies lend strong support to this conclusion. Lipid-metabolizing tissues possess numerous microbodies that are often closely adpressed to spherosomes or lipid bodies (292, 302, 303), presumably aiding the efficient transfer of fatty acids to the microbodies. As germination proceeds in castor bean (*Ricinus communis*), spherosomes gradually disappear and the number of glyoxysomes increases (302). In the final stages of germination in this tissue, glyoxysomes are removed from the cytoplasm with the formation of autophagic vacuoles.

Peroxisomes occur in green leaves, where they are frequently observed in close contact with the chloroplast envelope (73, 74, 76, 89, 90; Fig. 39b). Futhermore, they have been shown to occur in greater numbers per cell in grasses with high rates of photorespiration when compared to those with lower rates (74). Both of these observations are consistent with a large body of biochemical evidence that demonstrates that peroxisomes have a major role in the metabolism of glycolate and in the process of photorespiration (107, 259, 290). Photorespiration is the light-dependent evolution of carbon dioxide and uptake of oxygen, which is most readily demostrable in plants with the C_3 pathway of carbon dioxide fixation (see Chapter 3, Section III). The carbon dioxide is released during the metabolism of glycolate (a product of photosynthesis in the chloroplasts), which moves to the peroxisomes where it is oxidized by glycolate oxidase, producing glyoxylate and hydrogen peroxide. The hydrogen peroxide is removed by the characteristic enzyme of microbodies, catalase. The glyoxylate may be returned to the chloroplast where it is reduced to glycolate, thus removing excess NADPH produced during photosynthesis. The major pathway of glyoxylate metabolism, however, is via the glycolate cycle to produce glycine, serine, hydroxypyruvate, and glycerate in succession. The conversion of two molecules of glycine to one of serine produces a molecule of carbon dioxide, which is evolved in photorespiration. Most of the enzymes involved are found in the peroxisomes, although the reaction producing serine from glycine occurs in the mitochondria (259). Thus three organelles are involved in glyoxylate metabolism, and their close proximity is observed in many leaf cells (Fig. 39b). The significance and the mechanism of control of the cooperative processes between the three organelles and the cytoplasm are, however, far from being fully understood. [For a complete discussion of glycolate metabolism and photorespiration, see Chapter 3, Section III. (Ed.)]

The close association often seen in electron micrographs between the

ER and microbodies, an association suggesting direct membrane continuity, has led to the proposal that microbodies are derived from the rough ER, perhaps by budding (76, 90, 302). This proposal is supported by studies using isolated cell fractions, which show that the microbodies and ER have certain biochemical properties in common (130a), and by measurements of membrane turnover, which suggest that the lecithin of the glyoxysomal membrane originates in the ER (139). In leaves the formation of peroxisomes does not appear to require light or products of photosynthesis, because the organelles occur in equal numbers per cell in the green and colorless regions of, for example, variegated leaves (89) and in leaves below ground (90). However, as bean (*Phaseolus vulgaris*) leaves expand and turn green, the microbodies enlarge, and the specific activities of peroxisomal enzymes increase as much as tenfold (90).

An interesting case of microbody development is found in the cotyledons of certain species (e.g., *Cucumis sativus*), in which the microbodies initially function as sites of lipid storage in the seed and later become photosynthetic. This development is accompanied by an initial increase and then a decrease of glyoxysomal activity, the latter corresponding to a rise in peroxisomal activity (292, 303). There are then a number of possibilities regarding the microbodies; either there is a change in the enzyme complement of a single organelle, or both types are present but only one if functional at a given time, or the glyoxysomes are removed and replaced by a new population of peroxisomes. Although glyoxysomes appear to be removed by autophagic vacuoles in castor bean tissue where there is no functional continuation (302), electron microscopy shows no evidence of glyoxysomal breakdown or peroxisomal formation in *Cucumis* cotyledons during the transition period (292). Furthermore, the use of cytochemical staining methods for malate synthetase (a glyoxysomal marker) and glycolate oxidase (a peroxisomal marker) shows that all microbodies contain both enzymes during the transition period, which suggests that only one type of microbody occurs that undergoes a change in enzyme complement during the course of germination (34). These observations are supported by a biochemical study that shows no new synthesis of catalase to suggest *de novo* synthesis of peroxisomes (80).

IX. Microtubules and Microfilaments

Microtubules and microfilaments are slender proteinaceous strands that appear to be present in all eukaryotic cells, and they have been implicated in various aspects of morphogenesis and motility. These structures are assumed to constitute at least part of an organized mi-

croarchitecture or cytoskeleton within the cell. Acting separately or together, they may be responsible for the various elastic properties of cells,
the movement of cytoplasm and organelles, and the motility of single-
cell organisms. Microtubules have now been studied quite extensively in
plants; microfilaments have been much less studied. Much of what we
know of the latter depends on observations on animal cells. The structure and functions of microtubules and microfilaments are described
separately, although the possibility that microtubules and microfilaments interact in certain processes must certainly be considered.

A. MICROTUBULES

1. Structure and Properties

Microtubules were first described in the moss *Sphagnum* by Manton
(159) in 1957, although the name microtubule was not used in that work.
The widespread existence of these structures in eukaryotic cells was
established in the 1960s with the introduction of glutaraldehyde as a
tissue fixative. Microtubules are usually destroyed by permanganate or
osmium fixation, and they also depolymerize at the low temperatures
often used for fixation. Thus fixation by aldehydes (e.g., glutaraldehyde
or glutaraldehyde–formaldehyde mixtures) at room temperature is rec-
ommended for the preservation of microtubules.

Microtubules appear in electron micrographs as long, unbranched
tubular filaments about 22–25 μm in diameter (Fig. 40). In transverse
section, they show an electron-translucent core and a densely stained
sheath or wall. The wall is made up of globular subunits, each about 4–5
nm in diameter, usually arranged in 13 longitudinal rows, known as
protofilaments. The subunits show an axial displacement between protofilaments, producing a pitch relative to the longitudinal axis of the microtubule (Fig. 41). Electron micrographs may also reveal a clear zone
along the sides of the microtubules, which perhaps indicates that some
additional material is present.

More difficult to assess is the length of microtubules. In one of the
earliest descriptions of these structures, they were considered to be arranged like hoops around the cell (149); this concept has since been
reiterated in a number of publications. However, recent careful serial
sectioning of microtubules in a number of developmental stages in plant
cells has shown that, although the microtubules are arranged circumfer-
entially, they are only about 2–4 μm in length (i.e., about one-eighth of
the cell circumference) (94,113,114; Fig. 42). The arrays consist of overlapping microtubules of various lengths, which are linked by cross-

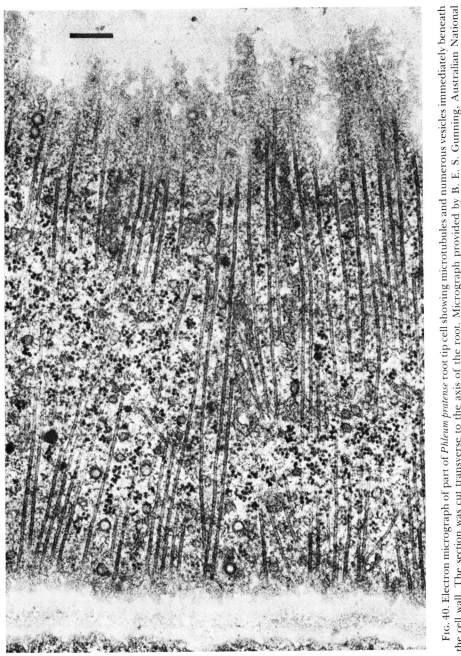

Fig. 40. Electron micrograph of part of *Phleum pratense* root tip cell showing microtubules and numerous vesicles immediately beneath the cell wall. The section was cut transverse to the axis of the root. Micrograph provided by B. E. S. Gunning, Australian National University.

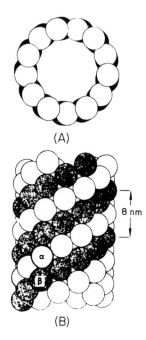

(A)

(B)

FIG. 41. Model of a microtubule showing (A) in transverse section, the 13 globular subunits comprising the wall and (B) in longitudinal view the axial displacement of subunits in adjacent protofilaments and a possible arrangement of the α- and β-monomers (3.5 × 4 nm). Redrawn after Bryan (32a).

bridges. The possibility that microtubules pass around the cell is very unlikely.

As just mentioned, microtubules have been observed to possess electron-dense arms or cross-bridges that project from their outer surface. These are usually 2–5 nm thick and 10–40 nm long. The arms are most easily observed between adjacent tubules that are in highly organized, stable arrays, but they also occur between microtubules and various membranes such as the plasma membrane (117, 122, 124, 289). They are assumed to act as stabilizing agents and perhaps to determine the pattern of microtubular arrangements. They may also function in motility, and this possibility is discussed later.

The subunits of microtubules from a wide variety of sources are remarkably similar, and all are made of a protein known as tubulin. Tubulin was first isolated and identitified using the specific binding of radioactive colchicine, a drug that is known to inhibit mitosis by its disruptive action on microtubules. Tubulin consists of a dimer with a molecular

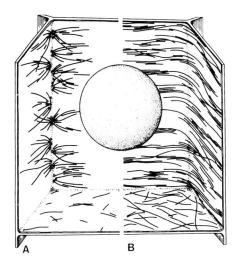

FIG. 42. Diagram illustrating the emergence of microtubules from complexes lying at intervals along a cell edge (A), compared with their position later (B). In the latter, the arrays of microtubules consist of overlapping component microtubules. Reproduced from Gunning *et al.* (94).

weight of 120,000; when denatured, the dimer yields two slightly different monomers, known as α and β tubulin, each with a molecular weight of about 55,000 (117, 122, 124). Colchicine and another antimitotic agent, vinblastine, bind at specific but different sites on the dimer.

Although tubulin from different sources appears chemically similar, it should be remembered that there have been very few studies of tubulin from plants. There is also some evidence that different types of microtubules may exist, because microtubules in different situations may show quite large differences in sensitivity to depolymerizing conditions. For example, a very much higher concentration of colchicine is needed to block division in *Haemanthus katherinae* endosperm cells than is required for HeLa cells (122). Such differences may reflect variation in the degree of cross-bridging or may result from as yet undiscovered biochemical differences.

Depolymerization of microtubules has already been mentioned. Microtubules are difficult to preserve; they can be readily broken down and reassembled. It is believed that cells contain a pool of microtubular subunits that under appropriate conditions polymerize to form visible tubules. Thus there is a dynamic assembly–disassembly equilibrium. Colchicine is thought to produce its effect by binding to and inactivating free subunits thereby disrupting this equilibrium. However, it is not known how this balance is controlled in normal cells, although it is clear that microtubules can appear at specific times during the cell cycle and at specific sites within the cell. It is known that tubulin binds a guanidine nucleotide (GDP) and that both GDP and calcium ions affect polymerization. Cyclic AMP, which is important in the control of calcium levels in animal cells, might also play an important role.

In principle the simplest way of controlling ordered patterns of microtubules is by controlling the initiation of their assembly (289). Once begun, polymerization proceeds by end growth as shown by experiments using radioactive tubulin. The initiation sites thus determine the location and direction of growth of the microtubules; they have been called nucleating centers or, more commonly, microtubule organizing centers. However, although this concept is a very interesting one, clear evidence for the existence and nature of such centers is not easy to obtain. In electron micrographs recognizable zones of microtubule assembly usually appear to contain moderately stained, amorphous, or flocculent material into which the tubules are inserted (see 99, 122, 124). Clear examples are provided by the kinetochores, the sites at which microtubules are attached to chromosomes (see Section IX,A,2a), and by the spindle poles of lower plants. However, there are also other examples where the presence of MTOCs would be expected but where none have been observed. The first evidence for the presence of nucleating regions for cortical (cytoplasmic) microtubules in higher plants has come from careful serial sectioning of polyhedral cells in the root apex of *Azolla* (94). The edges of these cells have been observed to develop transitory regions in the form of complexes of vesicles, microtubules, and an electron-dense matrix surrounding parts of the microtubules (Fig. 42). These complexes appear after cytokinesis has been completed and persist for a period that is not yet clearly defined. They also have been seen among phragmoplast microtubules (see Section III,B,1) and where the preprophase bands of microtubules pass around the cell edge (see Section IX,A,2a). It is thought that these complexes may have vectorial properties that allow them to control the orientation of microtubules on the different faces of the cell. Whether MTOCs are solely responsible for the more highly ordered and oriented arrangement of microtubules found in many situations remains an important topic for future investigations.

2. Functions

The potential versatility and importance of microtubules is suggested by the wide range of functions that have been proposed for them. For convenience a distinction is sometimes made between nuclear and cyto-plasmic (or cortical) microtubules. The former are thought to be in-volved in the processes of cell division, including chromosome move-ment and cell plate formation. Cytoplasmic microtubules are believed to function as a cytoskeleton and are credited with generating cell shape by processes that include the organization of cellulose microfibrils in the cell wall. They have also been implicated in the intracellular transport and location of organelles and other particles, although this function may involve the microfilaments as well. For example, an association between microtubules and mitochondria in cultured animal cells has been dem-onstrated using immunofluorescence techniques at the light microscope level (118a). Mitochondria have been observed to be arranged along the cytoplasmic microtubules, with a fairly dense concentration around nu-clei. When the microtubules are disrupted with colcemid, redistribution of mitochondria occurs, which suggests that the microtubules are in-volved in the regulation of the intracellular distribution of mitochon-dria. Finally, microtubules are involved in cell motility in the form of the major structural components of cilia and flagella.

   a. Mitosis and Cytokinesis. The fibrous nature of the spindle appa-ratus, indicated by its birefringence in the polarizing light microscope, is now widely believed to be due to bundles of microtubules (122, 224a), although this view is not universally held (see 122). The sequential distri-bution of microtubules during mitosis has been described in detail in a number of reviews (see 99, 117, 122, 224a) and is only outlined here. Much of the detailed work on plants has been done with the endosperm cells of *Haemanthus katherinae* (see 122).

   In early prophase, randomly orientated microtubules appear in the clear zone surrounding the nucleus, which is also visible by light micros-copy, and then gradually become oriented parallel to each other and to the longitudinal axis of the spindle. It seems that these microtubules, which form outside the nuclear envelope, produce the continuous spin-dle, but, as the envelope disintegrates during prometaphase, they invade the nucleus and become oriented along the spindle axis. At the same time, microtubules appear attached to the chromosomal kinetochores, although it is not clear whether these microtubules originate from the invading ones or arise *de novo*. The latter origin is more probable be-cause the kinetochores appear to act as MTOCs. Thus the spindle con-tains at least two types of microtubules: those that traverse from pole to pole (the continuous fibers) and those that connect the chromosomes to the poles (the chromosomal or kinetochore fibers).

   During metaphase the chromosomes become aligned at the center of the spindle, and the number of microtubules attached to each kinetochore generally increases. The separation of the chromatids to opposite poles occurs during anaphase at a uniform rate ranging from 0.2 to 4 μm/min. This involves both a shortening of the chromosomal fibers and an elongation of the spindle itself. During telophase the kinetochore microtubules begin to disappear, while at the same time there is a proliferation of fibers in the interzonal region, leading to the formation of the phragmoplast and cell plate (see Section III,B,1).

The events of cell division raise a number of problems that relate to microtubules. These include questions such as how the plane of division is defined, where the microtubules originate, and how the chromosomes move.

In many plant growth regions, cell divisions occur in apparently predetermined planes that are often closely related to the differentiation that is to follow (e.g., root tip meristems and stomatal complexes). Of great interest in regard to the possible mechanisms whereby these planes of division are prescribed was the discovery of a transient band of microtubules that appears in the cytoplasm of many premitotic cells (224a, 225, 226). This "pre-prophase band" generally consists of about 200 closely packed tubules arranged in rows 5 or so deep; the rows encircle the cell at right angles to its longitudinal axis where the future cell plate will become attached to the "parental" walls. Thus the band appears to predict the plane of cell division and to play an important role in determining the position and orientation of the nucleus. For example, in the root apex of *Azolla,* it is possible to predict within a fraction of a micrometer where a new cell wall will be located; the midline of a band 1.5–3 nm wide anticipates this location, and each daughter cell inherits approximately half of the pre-prophase band site (93). However, other observations have raised doubts about the importance of this band. For example, although the band has been observed in many plants, it is by no means universal (93, 122, 224a). Other experiments have shown that the band appears in some cells after general polarization of the cytoplasm and nucleus has occurred, and it may well be a result, rather than a cause, of this polarization. Pickett-Heaps (224a) believes that the band probably represents a pool of structural material that is destined for reutilization during spindle formation. Thus the factors that induce polarization in these cells remain unknown.

Another poorly understood process is that of spindle formation. Many earlier investigators believed that the centrioles were closely associated with the formation of the spindle, although it is now apparent that this is not so. In fact centrioles do not exist in cells of higher plants (99, 124, 224a). In certain algae and fungi, spindle pole structures, characterized by the presence of amorphous flocculent material and associated microtubules, have been widely observed, although they are absent from the

poles of higher plant spindles (124). The continuous fibers of microtubules terminate within a broad region at the poles, but there is no obvious structure that could act as the spindle organizer. It is possible that
the MTOCs in these regions are simply difficult to detect by current
techniques. Alternatively, higher plant cells may lack specific MTOCs at
the spindle poles; microtubule polymerization could be initiated over a
much broader area, with spindle formation depending on the interactions and alignment of the polymerized structures. In other parts of the
mitotic apparatus the MTOCs are more clearly defined. The structure of
kinetochores shows an increasing degree of complexity from lower to
higher organisms (99). In higher plants kinetochores generally consist of
a ball of fibrous or flocculent material attached to the chromatid in which
numerous microtubules terminate. The phragmoplast also contains a
centrally placed MTOC, although it is less well defined than the kinetochore.

A third, very controversial question concerns the mechanism of chromosome movement during mitosis. There are many hypotheses, but none
has found widespread acceptance. One possibility is that the dynamic
interchange between the microtubules and the free subunits can provide
the force necessary for chromosome separation (131). It is envisaged
that the chromosomal microtubules shorten by depolymerization at the
poles, whereas the poles are pushed apart by elongation of the continuous fibers, perhaps utilizing the free subunits produced by the depolymerization. However, there are objections to this proposal. For example,
there is no obvious structure at the poles in higher plant cells to which
the chromosomal microtubules are anchored, and induced dissociation
of the fibers close to the poles by the use of ultraviolet microbeams has
had no effect on chromosome movement (8).

An alternative suggestion is that there is another component to the
microtubules and that a shearing force exists between the tubules and
the matrix (205). This force may be produced by interaction between
actin-like microtubules and a myosin-like matrix. However, although
there is some evidence for actin-like filaments in the spindles (see 117),
there is none for myosin-like proteins.

A third possibility involves interactions among the microtubules themselves causing the sliding of adjacent tubules (179). This is thought to be
mediated through the cross-bridges that bind to specific sites on adjacent
tubules and act as mechanochemical elements. Sliding occurs by breaking of cross-bridges and rebinding to the next site along the adjacent
tubule. Microtubules are considered to be polar structures, sliding occurring only between antiparallel tubules. Thus the chromatids move
initially by interaction between antiparallel chromosomal and continuous fibers.

These and other models have been discussed by Hepler (122), who concluded that although microtubules should play a central role in chromosome movement, the question of how this is achieved remains unanswered.

   b. Cell Wall Formation.   The possible role of microtubules in directing the movement of vesicles to the developing cell plate has been mentioned previously (see Section III,B). Interest in possible microtubule involvement in the development of older walls originated from the observation by Ledbetter and Porter (149) that the disposition of cortical microtubules mirrors the orientation of cellulose microfibrils in the adjacent cell wall. This correlation has since been confirmed in numerous reports using both algae and higher plants and in relation to the growth of both primary and secondary walls (see 122, 124, 195, 224a). Particularly striking examples are seen in relation to the discrete bands of secondary thickening in xylem elements, where the cortical microtubules are specifically located over the developing thickenings and are largely absent from the adjacent primary walls (27, 123, 276). Again, in differentiating guard cells that have characteristic shapes and thickenings, microtubule orientation is directly correlated with microfibril pattern in the thickened wall (208). Further support for the role of microtubules in wall development has come from observations of the effect of colchicine and the herbicide isopropyl-N-phenylcarbamate (IPC) on xylem differentiation. Both of these compounds affect microtubule systems and both cause major disruptions in the patterns of secondary wall thickening (27, 123). Similarly, colchicine application leads to abnormal wall thickening in developing guard cells (208).

   However, not all studies have shown a clear-cut correlation between the alignment of microtubules and wall formations. For example, microtubules are absent in certain algae when, nevertheless, ordered arrays of microfibrils are laid down (see 124), and there are other instances in which microfibrils and microtubules do not appear to be mutually aligned. However, the lack of parallelism between the orientation of these structures as seen in some micrographs of collenchyma cells was attributed to a stage in reorientation prior to the deposition of a new cell wall layer (42). It is also probable that other factors apart from microtubules are involved in the ordered deposition of cell wall material. In developing xylem treated with colchicine, the secondary wall pattern in a young cell is strongly influenced by the pattern laid down in an adjacent mature cell; the thickenings are formed directly opposite those of the neighboring cell (123). The nature of this factor, which presumably resides in the wall itself, is unknown.

Thus, there is general agreement that microtubules play a role in the orientation of cellulose microfibrils and the pattern of cell wall thickening. How this is achieved is a much more difficult question. Certainly microtubules do not appear to be involved in the actual synthesis of cellulose microfibrils. There are a number of examples of cells, such as the tips of pollen tubes, that are active in cell wall formation and yet appear to lack microtubules. Again, although colchicine affects the pattern of cell wall thickening, it does not appear to inhibit the synthesis of cell wall materials (see 122).

There are a number of ways in which microtubules could control the orientation of microfibrils and the more general deposition of cell wall components. One possibility is that the microtubules direct cytoplasmic streaming and/or the flow of Golgi vesicles that bring cell wall precursors or synthetases to the cell surface. However, it has been shown that the direction of streaming does not always correspond with the orientation of microtubules. Streaming is also unaffected by colchicine, not an unexpected result because this process is probably controlled by microfilaments (see IX,B,1). Results also suggest that microtubules are not involved in the movement of secretory vesicles to the cell surface (182).

Another proposal is that control of microfibrillar orientation is exerted through the microtubule cross-bridges that are attached to and produce shearing forces in the plasma membrane, thus generating a directed flow of, perhaps, cellulose synthetase complexes (122–124). However, these possibilities remain mere speculations at present.

Similarly, the role of microtubules in the movement of Golgi vesicles is uncertain. For example, a widely accepted view is that microtubules control wall formation by directing the vesicles along defined channels to the cell surface. However, in the development of primary tracheary elements of cucumber, the size of Golgi vesicles is larger than the distance between microtubules, which therefore appear to act as a barrier (83, 84); it has been proposed that cell wall materials pass through the plasma membrane between the thickenings and are then transported through the primary wall to the sites of thickening. It is also interesting to note that cytochalasin B but not colchicine inhibits the migration of secretory vesicles to the surface in maize (*Zea mays*) root cap cells, suggesting that the microfilaments but not microtubules are involved in this movement (182,229).

This discussion suggests that microtubules could play an important role in determining the shape of cells such as xylem elements and guard cells; they could also act as a cytoskeleton and control the shape of cells that lack a rigid cell wall. The characteristic pear shape of the unicellular alga *Ochromonas* is maintained by two sets of microtubules and is destroyed by colchicine (28). Similarly, the generative cell in *Haemanthus katherinae* endosperm has a spindle shape that is maintained by a sheath of microtubules; this becomes spherical in the presence of colchicine (258).

c. Cilia and Flagella. Microtubules are the major structural components of cilia and flagella. Cilia are much shorter than flagella although their structure is basically similar; neither is found in flowering plants.

Cilia and flagella consist of a cylindrical membrane sheath originating from the plasma membrane and surrounding a group of fibers that arise from a basal body located in the cytoplasm. The overall diameter is about 200 nm. The fibers have a characteristic arrangement with a circle of nine outer filaments surrounding a central pair (the 9 + 2 configuration). Electron microscopy has shown that these filaments are composed of microtubules. Each consists of a pair of tubules—one (the A tubule) is just inside the circle compared with the other (the B tubule). The A tubule has two short arms that project toward the B tubule in the adjacent fiber. Some links are also observed between the outer microtubules and the membrane and between the central and outer doublets.

The arms possess ATPase activity, which is thought to be essential for the functioning of these structures; isolated flagella can be observed to resume beating if ATP is added. It is thought that the hydrolysis of ATP induces conformational changes in the tubules, which, when coordinated along the length of the flagella, produce the characteristic movement. How this is achieved is not yet clear.

The basal body at the base of each flagellum is identical in structure to the centriole of animal or lower plant cells. It consists of nine sets of triplet microtubules surrounded by a membrane. The centrioles are located at the poles of the spindle during mitosis and subsequently move to the surface of flagellated cells to initiate the assembly of the flagellar apparatus.

B. MICROFILAMENTS

1. Structure

Microfilaments are narrower than microtubules, being about 5–8 nm in diameter. They usually occur in roughly parallel bundles that stain more lightly than other cellular structures (Fig. 43), and they are frequently located in the path of streaming cytoplasm and in regions of cells undergoing contraction. They have not been widely observed in higher plant cells, perhaps due to preservation problems, but their appearance seems to be correlated with the direction of cytoplasmic streaming (124).

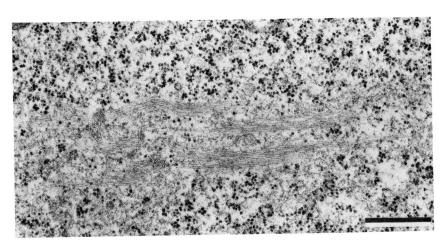

FIG. 43. Electron micrographs showing bundles of microfilaments (each microfilament 2–5 nm in diameter) in a root tip cell of spinach (*Spinacia oleracea*). Scale bar = 0.5 µm. Reproduced from Gunning and Steer (99).

Particular attention has been given to the similarity between microfilaments and the actin filaments found in muscle. Very briefly, actin consists of globular subunits (G-actin), which polymerize under appropriate conditions to form fibrillar F-actin. These filaments are able to bind to the protein meromyosin (the ATPase protein of myosin), and this binding is used as a means of identification. Myosin consists of thicker filaments than actin, and muscle contraction involves the sliding of the actin filaments over the myosin filaments under the control of the myosin ATPase cross-bridges. However, although there is considerable evidence that microfilaments consist of actin-like protein, the presence of myosin has not been widely established in nonmuscle cells (124).

2. Functions

Much of the evidence for the functions of microfilaments comes from the use of the drug cytochalasin B, which is thought to exert its effects through the disruption of these structures; microfilaments are believed to be the contractile machinery of nonmuscle cells, and sensitivity to CB implies the presence of a contractile microfilament system (310). Thus CB is known to produce a rapid cessation of cytoplasmic streaming in many plant cells and to inhibit the movement of cellular structures such as nuclei (124, 310). Cytochalasin B has been shown to inhibit the migration of dictyosome vesicles to the surface of maize (*Zea mays*) root cap cells, a process that is not affected by colchicine (182, 229; Fig. 44). Thus

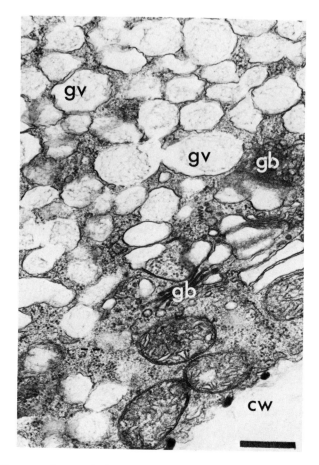

FIG. 44. Electron micrograph showing portion of a maize (*Zea mays*) root cap cell that was treated with cytochalasin B. Note the accumulation of vesicles (gv) around the dictyosomes (gb) and the absence of vesicles near the cell surface. cw, cell wall. Scale bar = 0.5 μm.

microfilaments and not microtubules are implicated in the movement of these secretory vesicles. However, colchicine does affect the deposition of wall thickening in developing xylem elements (see Section IX,A,2), which implies that both these elements may play a role in secretion; microfilaments might be involved in transport and microtubules in specific plasma membrane–wall interactions. Although strong evidence for microtubule–microfilament cooperation is lacking, this is a possibility to be considered in relation to cellular movement (124).

Furthermore, the mode of action of CB is not clearly defined. Although it appears to alter microfilament morphology in many cases, this is not always so (4, 272). There have also been conflicting reports of its effects on several cellular processes (124, 272), and so the site and the mode of its action remain in doubt.

The way in which microfilaments produce movment is also a topic for speculation. Although myosin-like proteins have not been convincingly demonstrated in nonmuscle cells, some action of the type described for muscle remains a possibility. Microfilaments are found anchored to various membranes, particularly the plasma membrane, and such attachments may play a crucial role in the production of movement (124, 272). A suggestion is that the primary site of action of CB is at the plasma membrane; CB disrupts the linkages between the filaments and the membrane and thus blocks the function of the microfilaments (272).

More needs to be known about the mode of action of CB and the sites of attachment of microfilaments within the cell. It is possible that microfilaments are the "muscle" of the cell and that microtubules play a more static, cytoskeletal role.

X. The Nucleus

The nucleus is the largest of the cellular organelles and was first described by Robert Brown in 1831. It contains the major part of the cellular DNA, the hereditary material responsible for the storage, replication, and expression of genetic information. The presence of a descrete membrane-bounded nucleus is a feature distinguishing eukaryotes from prokaryotes, although the former also have a more highly structured cytoplasm. Nuclei are thus found in all living higher plant cells with a few notable exceptions, such as the mature cells of sieve tubes.

In nondividing cells the nucleus is usually roughly spherical and about 10 μm in diameter, although considerable changes in both volume and shape can occur as cells undergo differentiation. For example, in elongating pea root cells the nuclei change from a spherical to a more elongated shape; this change is associated with a rapid increase in both volume and surface area of the nuclei, both of which then decrease slightly in mature cells (43). Changes also occur in the form of the chromatin and nucleolus. In some mature cells, the nuclei may be highly lobed. Such lobing leads to an increase in the relative surface area of the nucleus and may indicate an increased activity between nucleus and

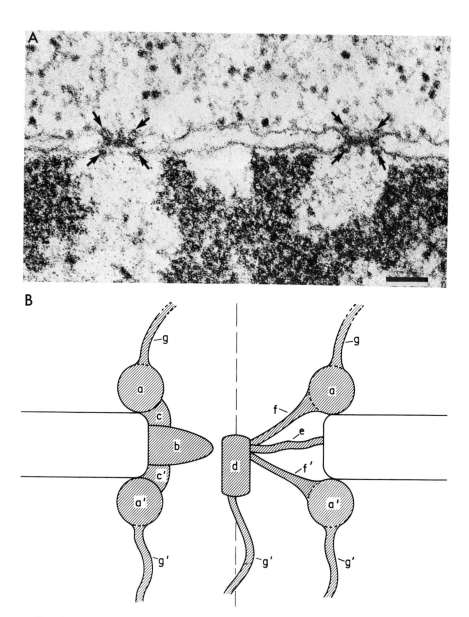

FIG. 45. (A) Electron micrograph of nuclear envelope of onion (*Allium cepa*) root tip cells. In the interpore region the inner membrane is attached to the peripheral condensed chromatin. The arrows denote the annular granules lying in the pore margin. Note the peripheral granules, which project from the pore wall into the lumen, and the centrally placed electron-opaque elements. The outer nuclear membrane bears ribosomes. Scale bar = 0.1 μm. (B) Diagrammatic view of the individual components of the nuclear pore

cytoplasm, although such associations are not easily demonstrated. For example, in the extrafloral nectaries of *Ricinus*, the nucleus is roughly spherical in young epidermal cells and clearly lobed and elongated in the secretory zone (10). Although most cells contain just one nucleus, multinucleate cells are quite common (e.g., the endosperm of certain higher plants). Again, cell development may sometimes be accompanied by an increase (2–16 ×) in DNA content (polyploidy), which is reflected in an increase in either chromosome number or chromosome size and a proportional increase in nuclear volume.

Both light and electron microscopy show that the nucleus is made up of a number of distinct components. The nuclear contents, consisting very largely of nucleic acids and proteins, are separated from the cytoplasm by the nuclear envelope or membrane. Within this envelope, nuclei contain one or more prominent spherical structures (known as nucleoli) and regions that stain heavily with basic dyes (the chromatin). When cell division begins, the nuclear envelope and nucleoli disappear, and the chromatin is organized to form the chromosomes.

A. The Nuclear Envelope

The nucleus is surrounded by an envelope consisting of two unit membranes, each 5–10 nm wide (they appear wider with freeze-etching than after chemical fixation) and separated by a gap known as the perinuclear space. This space varies between 10 and 60 nm in width, although it is fairly constant for a particular cell type. The envelope is pierced by pores which are normally between 10 and 40 nm in width; these pores can be seen in thin sections (Figs. 45, 48) but are most dramatically demonstrated by freeze-etching (Fig. 46). The structural and biochemical properties of the nuclear membrane have been extensively discussed by Franke and co-workers (70, 222) and are only outlined here.

The two membranes of the envelope appear to be quite similar structurally, although they differ in their associated components. The outer membrane is frequently studded with ribosomes and may be continuous with the ER and so is very much a component of the endomembrane system, whereas the inner one is associated with the peripheral chroma-

---

complex. a, annular granules; b, peripheral granules; c, material between annular and peripheral granules; d, central granule; e, equatorial filaments of the pore interior; f, inner pore filaments connecting annular and central granules; g, axial filaments. The prime denotes components of the nucleoplasmic side. Reproduced from Franke (70).

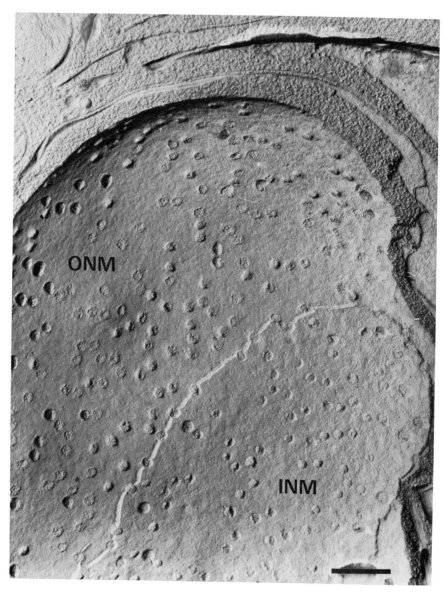

Fig. 46. Freeze-etched preparation of nucleus of pea (*Pisum sativum*) root cell showing nuclear pores. ONM, outer nuclear membrane; INM, inner nuclear membrane. Scale bar = 0.25 μm. Micrograph provided by K. Roberts; reproduced from Hall *et al.* (107).

tin and may function in the control of transcription (70, 185). Isolated nuclear membrane fractions from plants, which presumably contain both membranes, are closely similar to the ER in their biochemical characteristics (222). The distance between the inner and outer membranes is assumed to be maintained by the numerous nuclear pore walls, although, in some nuclei only a few pores exist and membrane-to-membrane cross-bridging threads may be important in this respect (70). Little is known about the contents of the perinuclear space, although it certainly includes proteins that may assume a crystalline form.

The nuclear envelope is perforated by pores that are formed by the fusion of the inner and outer membranes. Their diameter varies greatly in different situations, the average ranging from 45 to 110 nm (285). The density of pores is also greatly variable and ranges from 6 to 135 per μm^2 (285). The pores are not simply holes; they have a very specific structure at the pore boundary (known as the pore complex), which appears to be similar in plants and animals. Although the pores can occupy up to 25% of the total envelope area (8–10% is a more frequent estimate), they probably do not represent regions of free diffusion between nucleus and cytoplasm (see 70). The evidence available is, however, limited and somewhat contradictory. Some observers suggest that ions and many small molecules can penetrate the nucleus quite readily, whereas electrical resistance measurements across the membrane in *Drosophila* salivary gland cells are very much higher than would be expected through open channels. The nuclear membrane acts as a sieve for particles larger than 6 nm and totally excludes particles larger than 15 nm. The latter are considerably smaller than the average pore diameters, suggesting the presence of material in the lumen that reduces the effective pore size. Furthermore, these exchange characteristics seem to vary with the stage of cell development and among different cell types.

The structure of the pore complex has been studied with a variety of microscopic techniques and has been the subject of many controversial interpretations (see 70, 247). However, there is now some general agreement about certain features of nuclear pores. The ring-like material at the pore margin is known as the annulus, and both inner and outer annuli are made up of distinct globular subunits known as the annular granules, each about 18 nm in diameter. There are usually, but not always, eight granules that are attached to the membrane in a symmetrical arrangement, with much of their mass positioned outside the pore perimeter (Fig. 45). The pores may also contain an additional inner series of electron-dense clumps, the peripheral granules, which are attached to the membrane in the equatorial plane. These are more difficult to identify, and it is not clear whether they are ubiquitous or not. In many pores the center is occupied by a central granule of variable size (usually about 15 nm diameter), which may be connected to the pore wall by fibrillar material. The remainder of the pore often appears to be filled with an amorphous material with increased electron density. Franke (70) has emphasized that the various components do not always have a compact globular form but may appear as aggregates of fine filaments. This observation is of interest in relation to the possible mechanism of transport of RNA from the nucleus to the cytoplasm. There is considerable experimental evidence that transport of ribonu-cleoproteins occurs through the nuclear pores, yet their small size (about 15 nm) should exclude this possibility. However, the nuclear pore com-plex is (after the ribosomes and polysomes) the most RNA-rich structure known, and it has been suggested that the pore complex material may consist largely of ribonucleoproteins in various states of coiling (70). One possibility is that ribonucleoprotein in the nuclear sap uncoils when it comes into contact with the nuclear membrane and passes as a thread through the pore complex (70). The central granule could then corre-spond to RNA molecules in the process of passing from nucleus to cyto-plasm, with recoiling and recombination with proteins occurring in the cytoplasm.

Finally mention should be made of the fate of the nuclear envelope during mitosis, a process that differs considerably in different orga-nisms. In lower organisms, the envelope may persist to some considera-ble degree to give the so-called closed spindles. There then appears to be a progressive disintegration of the envelope up the evolutionary scale to produce the open spindles; a completely dispersed envelope is found in higher plants and animals. The retention of the envelope, attachment of chromosomes to the envelope for segregation, and the lack of a well-defined spindle are all considered to be primitive features (153). When the envelope disintegrates in higher plants, it becomes indistinguishable from components of the ER and its fate is unknown. For example, it may or may not contribute to the envelope that is reformed after cell division. The old envelope certainly contributes to the new one in cells with closed spindles, and this process may well persist in higher organisms.

B. Chromatin and Chromosome Structure

The chromosomes are the major DNA-containing structures in cells and so carry most of the hereditary information. In interphase nuclei, the chromatin is dispersed throughout the nuclear matrix and usually two regions can be distinguished: the condensed, heavily stained heterochromatin and the more dispersed, lightly stained euchromatin. Chromatin can thus be considered as the interphase chromosomes. The DNA does not occur naked but is complexed with basic proteins known as histones to form a nucleoprotein complex. Analyses of isolated chromatin from plant sources show that it is composed very largely of DNA and histones in a ratio of about 1 to 1.1 together with smaller amounts of acidic proteins and RNA (19a, 33). A discussion of the role of these proteins in gene activity is beyond the scope of this chapter, but it is well covered elsewhere (19a, 33, 88, 191). The organization of DNA is more readily observed after the formation of the chromosomes during mitosis. The chromosomes are formed largely from the euchromatin, whereas the heterochromatin remains condensed at specific sites along the chromosomes. Genetically, heterochromatin appears to be very much less active than euchromatin. The chromosomes later become divided longitudinally to form two chromatids. Under the light microscope these may appear to be composed of a coiled thread-like structure, the chromonema, which has a number of bead-like structures, the chromomeres, arranged along it. However, the relationship between these structures and the fibrils seen by electron microscopy (discussed later) is by no means clear.

The DNA in chromosomes is very tightly packed so that, if it were uncoiled, it would be many times longer than the chromosome itself. So-called *packing ratios* (length of DNA to length of chromatin fibril) range from 25 to 150 to 1 (191). Fine-structural studies of chromosomes have been very much concerned with how the DNA is folded and coiled to produce the compact chromosome. The high-voltage electron microscope, which permits the examination of these highly ordered structures in relatively thick sections, has proved to be very valuable. Current ideas in this rapidly developing field have been reviewed by Ris (242), and these are summarized here.

Electron microscopy has revealed a globular repeating unit in chromatin that takes the form of a linear arrangement of beads, each about 10 nm in diameter, which have been termed nucleosomes. These are thought to consist of DNA containing about 200 base pairs wound around a histone core. These contain the bulk of the chromatin in eukaryotic cells, although their functional significance is uncertain. A chain of these nucleosomes forms a fibril that is approximately 10 nm in diameter, and this in turn is organized to form the 20-nm fibril, which is considered to be the structural unit of both interphase and mitotic chromosomes. The 20-nm fibril may be produced by helical coiling of the 10-nm fibril or by close packing of groups of nucleosomes (the so-called *superbeads* consisting of 6–10 nucleosomes). Nonhistone proteins may clamp this fibril at intervals into chromomere-sized regions. In metaphase the 20-nm fibril is further condensed, probably by helical coiling, to form a 50-nm fibril, although it is not clear how this is in turn organized to form the chromonema. This formation may involve another supercoil, although the mechanism could prove to be more complex.

Fine-structural studies have also suggested that transcriptionally inactive chromatin and transcribed chromatin are organized differently (72). The inactive chromatin appears to be condensed into nucleosomes, which are grouped to form supranucleosomal structures, whereas in transcribed regions the chromatin appears as extended thin fibrils. However, these observations are at variance with certain biochemical findings, and the resolution must await future research. Furthermore, the mechanisms that produce these transitions and, indeed, the factors that govern the degree of compaction of chromatin are largely unknown.

C. THE NUCLEOLUS

Nucleoli are prominent structures found in almost all eukaryotic nuclei (Figs. 2, 25) They are roughly spherical and there are usually one to four in each nucleus, although in some cases there may be many more. They contain a large proportion of the cellular RNA and are known to be sites of ribosomal RNA synthesis. They also contain some DNA, although over 70% of the nucleolus is composed of protein. Nucleoli are not bounded by a membrane but contain densely packed components, and a number of distinct regions are usually recognized in electron micrographs (81, 145; Fig. 47) There is a granular region that is usually found at the periphery of the nucleolus but that may form irregular patches within its mass. This region consists of densely staining granules, about 15–20 nm in diameter, that are composed of ribonucleoprotein and are thought to be precursors of cytoplasmic ribosomes. The fibrillar regions are composed of fibers about 8 nm in diameter and 20–40 nm in length. The granules and fibers may be intermixed, or the granular region may surround a central fibrillar zone. The fibrils are composed of ribonucleoprotein and may be precursors of the granular region. The nucleolus also contains chromatin, which takes the form of coiled filaments about 2 nm in diameter that are dispersed through the fibrillar and, to a lesser extent, granular material. In addition, empty spaces known as *vacuoles* may also be present.

During mitosis, the nucleolus undergoes a series of structural changes that usually lead to its disappearance and, later, reappearance at the end of mitosis, although there are many deviations from this pattern. The

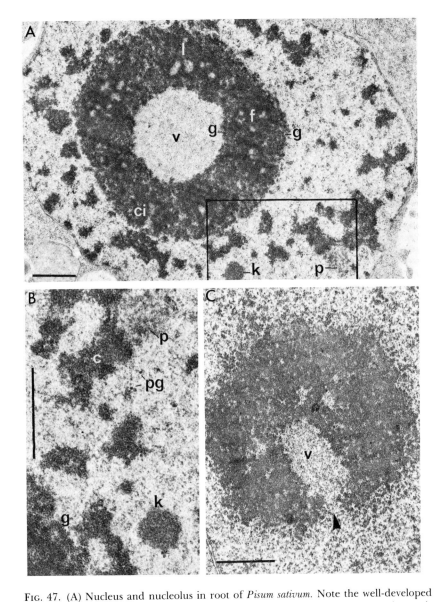

Fig. 47. (A) Nucleus and nucleolus in root of *Pisum sativum*. Note the well-developed nucleolar vacuole (v) surrounded by lacunae (l). A chromatin intrusion (ci) to a lacuna is present. (B) enlargement of boxed area in Fig. A showing details of a puff (p) associated with condensed chromatin (c), a karysome (k), and prechromatin granules (pg). (C) Nucleolus with a vacuole (v) with a channel (arrow) to the exterior that could represent discharge of vacuolar contents. g, granular zone of nucleolus; f, fibrillar zone of nucleolus. Scale bars = 1 μm. Reproduced from Chaley and Setterfield (43), by permission of the National Research Council of Canada.

details of the nucleolar cycle have been described by Lafontaine (145). The reappearance of the nucleolus is intimately associated with a specific region of certain chromosomes that is known as the nucleolar organizer. The organizer appears as a mass of fine fibrils adhering to the decondensing chromosome. The chromatin is probably the only permanent structure of the nucleolus, and the fibrillar material appears to be the first product of its activity. There may be more than one chromosome with a nuclear organizer, but the developing nucleoli often fuse to form a single nucleolus. Fusion often increases as the cells mature.

D. NUCLEOPLASM AND NUCLEAR INCLUSIONS

The results of improved fixation and staining procedures for electron microscopy have indicated that the nucleoplasm surrounding the chromatin is more structured than was previously thought (145). For example, fine fibrillar material from the chromatin can be seen to merge into the adjacent nucleoplasm. In addition, the latter has been shown to

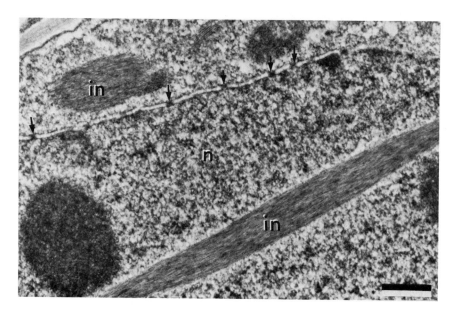

FIG. 48. Electron micrograph of part of an epidermal cell in a secreting nectary of *Ricinus communis* showing the presence of spindle-shaped inclusions (in) in the nucleus (n) and cytoplasm. Nuclear pores are also clearly visible (arrows). Scale bar = 0.5 μm. Reproduced from Baker *et al.* (10).

contain a wide range of granules and to possess more finely particulate texture; chromatin and nucleoplasm may be both physically and structurally integrated.

Both light microscopy and, more recently, electron microscopy have shown that the nuclear inclusions occur widely in both plants and animals (309). These include crystalline, fibrous, and amorphous structures (e.g., Fig. 48). Most appear to be proteinaceous, although lipid-containing bodies also occur. They may be several micrometers long and often appear to be related to the nucleoli. In the fern *Campyloneurum* nuclear inclusions appear as the nucleolus dissociates (309).

XI. Prospects

Our knowledge of cellular structure has expanded greatly although clearly there is still a great deal to be learned. Many aspects of fine structure are not clearly understood, particularly those at the limits of resolution of present microscopes. For example, we have seen that there is no widely accepted view of the detailed aspects of cell wall structure; controversies exist concerning the nature of plasmodesmata, nuclear pores, and micofilaments. The fine details of chromosome structure are uncertain, and the role of various less commonly observed structures, such as lomasomes, is far from clear. Even more uncertain are the interrelationships between subcellular structures and, in particular, how they are integrated in a developmental sequence that, for example, produces a highly specialized xylem vessel or sieve element from a meristematic cell. Many of these purely structural queries may well be answered fairly soon as techniques improve. As in the past, developments in microscopy will be of crucial importance.

Present-day microscopy is limited not only by resolution but, more seriously, by the techniques available for the preparation and analysis of cells. Living cells cannot be examined by electron microscopy, and our view of the cell is based largely on static, two-dimensional images of what in reality are very dynamic structures. The dynamic nature of living cells, clearly revealed by phase contrast light microscopy, may only be deduced at the ultrastructural level by the use of a sequence of photomicrographs or by the technique of autoradiography. Much more must be learned of the chemical nature and properties of cellular components at the level of the electron microscope. For example, we know little of the chemistry and effects of fixation, and the specificity of many conventional staining procedures is even more poorly understood. Again, tech-

niques for the direct analysis of the biochemical nature of cellular structures are very limited. For example, although methods exist for the
localization of certain enzymes, they are not always reliable and represent only a fraction of the enzyme population that has been identified by
in vitro analysis (263). Similarly, techniques for the direct localization of
antibodies, carbohydrates, and ions are still in their infancy (143, 253,
298). Improvements will undoubtedly come from three main directions
as they have in the past. From instrumentation, from specimen preparation, and from analytical procedures (130). Thus the development of
freezing techniques for specimen preparation in conjunction with analytical electron microscopy should produce essential information concerning the intracellular distribution of ions, whereas the wider use of
freezing procedures, which avoid chemical denaturation, will greatly
improve the versatility of many staining methods. The range of enzymes
that may be localized by electron microscopy should also increase, particularly with the development of immunochemical techniques. Regarding
instrumentation, major advances should come from a number of directions, particularly with the wider availability of higher voltage electron
microscopy, which can combine high resolution with the examination of
thicker specimens. This more readily allows a three-dimensional reconstruction of cellular structures (40, 169a, 230, 242), and its potential is
illustrated by its application to the study of the cytoplasmic ground
substance, the matrix, in which the various membranous and filamentous structures described earlier are suspended. This approach suggests
that the ground substance is composed of a finely divided lattice of
slender trabeculae, which supports and contains the larger components
of the cytoplasm and connects these and the cell surface into an organized cytoplast (230, 315a). Developments such as this should add considerably to our knowledge of the organization and integration of cellular structure in the future.

Acknowledgment

I should like to thank Mr. J. R. Thorpe for his help in providing many of the photomicrographs used in this chapter.

References

1. Abeles, E. B., and Leather, G. R. (1971). Abscission: control of cellulase secretion by
ethylene. *Planta* **97**, 87–91.

2. Adams, C. A., and Novellie, L. (1975). Composition and structure of protein bodies and spherosomes isolated from ungerminated seeds of *Sorghum bicolor* (Linn.) Moench. *Plant Physiol.* **55**, 1–6.

3. Albersheim, P. (1976). The primary cell wall. *In* "Plant Biochemistry" (J. Bonner and J. E. Varner, eds.), pp. 225–274. Academic Press, New York.

4. Allison, A. C., and Davies, P. (1974). Mechanisms of endocytosis and exocytosis. *In* "Transport at the Cellular Level" (M. A. Sleigh and D. H. Jennings, eds.) pp. 419–446. Cambridge Univ. Press, London/New York.

5. Anton-Lamprecht, I. (1967). Anzahl und vermehrung der zellorganen im scheiterlmeristem von *Epilobium. Ber. Dt. Bot. Ges.* **80** 747–754.

6. Ashford, A. E., and Jacobsen, J. V. (1974). Cytochemical localization of phosphatase in barley aleurone cells: the pathway of gibberellic acid-induced enzyme release. *Planta* **120**, 81–105.

7. Bailey, C. J., Cobb, A., and Boulter, D. (1970). A cotyledon slice system for the electron autoradiographic study of the synthesis and intracellular transport of seed storage protein of *Vicia Faba. Planta* **95**, 103–118.

8. Bajer, A. S., and Mole-Bajer, J. (1972). Spindle dynamics and chromosome movements. *Int. Rev. Cytol. Supple. 3*.

9. Baker, D. A., and Hall, J. L. (1973). Pinocytosis, ATP-ase and ion transport by plant cells. *New Phytol.* **72**, 1281–1291.

10. Baker, D. A., Hall, J. L., and Thorpe, J. R. (1978). A study of the extrafloral nectaries of *Ricinus communis. New Phytol.* **81**, 129–137.

11. Baker, J. E., Elfvin, L. G., Biale, J. B., and Honda, S. I. (1968). Studies on ultrastructure and purification of isolated plant mitochondria. *Plant Physiol.* **43**, 2001–2022.

12. Bauer, W. D., Talmadge, K. W., Keegstra, K. and Albersheim, P. (1973). The structure of plant cell walls. II. The hemicellulose of the walls of suspension-cultured sycamore cells. *Plant Physiol.* **51**, 174–187.

13. Beams, H. W., and Kessel, R. G. (1968). The Golgi apparatus: structure and function. *Int. Rev. Cytol.* **23**, 209–276.

14. Bell, P. R., and Mühlethaler, K. (1964). The degeneration and reappearance of mitochondrial in the egg cells of a plant. *J. Cell Biol.* **20**, 235–248.

15. Bennett, H. S. (1956). The concepts of membrane flow and membrane vesiculation as mechanisms for active transport and ion pumping. *J. Biophys. Biochem. Cytol.* **2**, 99-103.

16. Benson, A. A., and Jokela, A. T. (1976). Cell membranes. *In* "Plant Biochemistry" (J. Bonner and J. E. Varner, eds.) pp. 65-89. Academic Press, New York.

17. Berjak, P. (1972). Lysosomal compartmentation: ultrastructural aspects of the origin, development, and function of vacuoles in root cells of *Lepidium sativum. Ann. Bot.* **36**, 73–81.

18. Black, C. C. Campbell, W. H., Chen, T. M., and Dittrich, P. (1973). The monocotyledons: their evolution and comparative biology. III. Pathways of carbon metabolism related to net carbon dioxide assimilation by monocotyledons. *Q. Rev. Biol.* **48**, 299–313.

19. Bonner, J. (1976). Cell and subcell. *In* "Plant Biochemistry", (J. Bonner and J. E. Varner, eds.), pp. 3–14. Academic Press, New York.

19a. Bonner, J. (1976). The nucleus. *In* "Plant Biochemistry" (J. Bonner and J. E. Varner, eds.), pp. 37–64. Academic Press, New York

20. Boulter, D., Ellis, R. J., and Yarwood, A. (1972). Biochemistry of protein synthesis in plants. *Biol. Rev.* **47**, 113–175.

21. Bowles, D. J., and Northcote, D. H. (1972). The sites of synthesis and transport of extracellular polysaccharides in the root tissues of maize. *Biochem. J.* **130,** 1133–1145.

22. Bowles, D. J., and Northcote, D. H. (1974). The amounts and rates of export of polysaccharides found within the membrane system of maize root cells. *Biochem. J.* **142,** 139–144.

23. Bracker, C. E., and Grove, S. N., (1971). Continuity between cytoplasmic endomembranes and outer mitochondrial membranes in Fungi. *Protoplasma* **73,** 15–34.

24. Branton, D. (1969). Membrane structure. *Annu. Rev. Plant Physiol.* **20,** 209–238.

25. Breidenbach, R. W., and Beevers, H. (1967). Association of the glyoxylate cycle enzymes in a novel subcellular particle from castor bean endosperm. *Biochem. Bophys. Res. Commun.* **27,** 462–469.

26. Briarty, L. G., Coult, D. A., and Boulter, D. (1969). Protein bodies of developing seeds of *Vicia Faba. J. exp. Bot.* **20,** 358–372.

27. Brower, D. L., and Hepler, P. K. (1976). Microtubules and secondary wall deposition in xylem: the effects of isopropyl *N*-phenyl-carbamate. *Protoplasma* **87,** 91–111.

28. Brown, D. L., and Bouck, G. B. (1973). Microtubule biogenesis and cell shape in *Ochromonas.* The role of nucleating sites in shape development. *J. Cell Biol.* **56,** 360–378.

29. Brown, R. (1960). The plant cell and its inclusions. *In* "Plant Physiology" (F. C. Steward, ed.), Vol. 1A, pp. 3–129. Academic Press, New York.

30. Brown, R. M. Jr., Herth, W., Franke, W. W., and Romanovicz, D. (1973). The role of the Golgi apparatus in the biosynthesis and secretion of a cellulosic glycoprotein in *Pleurochrysis:* a model system for the synthesis of structural polysaccharides. *In* "Biogenesis of Plant Cell Wall Polysaccharides" (F. Loewus, ed.), pp. 207–257. Academic Press, New York.

31. Brown, R. M., Jr., and Montezinos, D. (1976). Cellulose microfibrils: visualization of biosynthetic and orientating complexes in association with the plasma membrane. *Proc. Natl. Acad. Sci. USA* **73,** 143–147.

32. Browning, A. J. and Gunning, B. E. S. (1977). An ultrastructural and cytochemical study of the wall-membrane apparatus of transfer cells using freeze-substitution. *Protoplasma* **93,** 7–26.

32a. Bryan, J. (1974). Biochemical properties of microtubules. *Fed. Proc. Fed. Am. Soc. Exp. Biol.* **33,** 152–157.

33. Bryant, J. A. (1976). Nuclear DNA. *In* "Molecular Aspects of Gene Expression in Plants" (J. A. Bryant, ed.), pp. 1–51. Academic Press, New York/London.

34. Burke, J. J., and Trelease, R. N. (1975). Cytochemical demonstration of malate synthase and glycolate oxidase in microbodies of cucumber cotyledons. *Plant Physiol.* **56,** 710–717.

35. Buvat, R. (1963). Electron microscopy of plant protoplasm. *Int. Rev. Cytol.* **14,** 41–155.

36. Buvat, R. (1969). "Plant Cells." Weindefeld and Nicolson, London.

37. Buvat, R. (1971). Origin and continuity of cell vacuoles. *In* "Origin and Continuity of Cell Organelles" (J. Reinert and H. Upsprung, eds.), pp. 127–157. Springer-Berlin, Heidelberg/New York.

38. Callow, M. E., and Evans, L. V. (1974). Studies on the shipfouling alga *Enteromorpha.* III Cytochemistry and autoradiography of adhesive production. *Protoplasma* **80,** 15–27.

39. Cande, W. Z., Goldsmith, M. H. M., Ray, P. M. (1973). Polar auxin transport and auxin-induced elongation in the absence of cytoplasmic streaming. *Planta* **111,** 279–296.

40. Carasso, N., Favard, P., Mentre, P., and Poux, N. (1974). High voltage transmission electron microscopy of cell compartments in araldite thick sections. *In* "High Voltage Electron Microscopy" (P. R. Swann, C. J. Humphries, and M. J. Gorine, eds.), pp. 414–418. Academic Press, New York.

41. Catesson, A. M. (1974). Cambial cells. *In* "Dynamic Aspects of Plant Ultrastructure" (A. W. Robards, ed.), pp. 358–390. McGraw-Hill, New York.

42. Chafe, S. C., and Wardrop, A. B. (1970). Microfibril orientation in plant cell walls. *Planta* **92**, 13–24.

43. Chaly, N. M., and Setterfield, G. (1975). Organization of the nucleus, nucleolus, and protein-synthesizing apparatus in relation to cell development in roots of *Pisum sativum. Can. J. Bot.* **53**, 200–218.

44. Chrispeels, M. J. (1976). Biosynthesis, intracellular transport, and secretion of extracellular macromolecules. *Annu. Rev. Plant Physiol.* **27**, 19–38.

44a. Clarkson, D. T. (1977). Membrane structure and transport. *In* "The Molecular Biology of Plant Cells" (H. Smith, ed.), pp. 24–63. Blackwell, Oxford.

45. Clowes, F. A. L. (1971). Cell organelles and the differentiation of somatic plant cells. *In* "Origin and Continuity of Cell Organelles" (J. Reinert and H. Ursprung, eds.), pp. 323–342. Springer-Verlag, Berlin.

46. Clowes, F. A. L., and Juniper, B. E. (1968). "Plant Cells." Blackwell, Oxford.

47. Cocking, E. C. (1972). Plant cell protoplasts—isolation and development. *Annu. Rev. Plant Physiol.* **23**, 29–50.

48. Collander, R. (1959). Cell membranes: their resistance to penetration and their capacity for transport. *In* "Plant Physiology. A Treatise" (F. C. Steward, ed.) Vol. II, pp. 3–102. Academic Press, New York.

49. Coombs, J., and Greenwood, A. D. (1976). Compartmentation of the photo synthetic apparatus. *In* "The Intact Chloroplasts" (J. Barber, ed.), pp. 1–51. Elsevier and North Holland Amsterdam.

50. Cooper, T. G., and Beevers, H. (1969). Mitochondria and glyoxysomes from castor bean endosperm. *J. Biol. Chem.* **244**, 3507–3513.

51. Cooper, T. G., and Beevers, H. (1969). B-oxidation in glyoxysomes from castor bean endosperm. *J. Biol. Chem.* **244**, 3514–3520.

52. Cran, D. G., and Possingham, J. V. (1972). Two forms of division profile in spinach chloroplasts. *Nature (London) New Biol.* **235**, 142.

52a. Daems, W. TH, and Wisse, E. (1966). Shape and attachment of the cristae mitochondriales in mouse hepatic cell mitochondria. *J. Ultrastruct. Res.* **16**, 123–140.

53. Dauwalder, M., and Whaley, W. G. (1974). Patterns of incorporation of ³H-galactose by cells of *Zea mays* root tips. *J. Cell Sci* **14**, 11–27.

54. Dauwalder, M., Whaley, W. G., and Kephart, J. E. (1969). Phosphatases and differentiation of the Golgi apparatus. *J. Cell Sci.* **4**, 455-498.

55. de Duve, C. (1971). Tissue Fractionation, past and present. *J. Cell Biol.* **50**, 200–550.

56. de Duve, C, and Wattiaux, R. (1966), Functions of lysosomes. *Annu. Rev. Physiol.* **28**, 435–492.

56a Dingle, J. T. (1977). "Lysosomes. A Laboratory Handbook," 2nd ed. North Holland, Amsterdam.

57. Echlin, P. (1971). The role of the tapetum during microsporogenesis of Angiosperms. *In* "Pollen: Development and Physiology" (J. Heslop-Harrison, ed.), pp. 41–61. Butterworths, London.

58. Evans, L. V., and Callow, M. E. (1978). Autoradiography. *In* "Electron Microscopy and Cytochemistry of Plant Cells" (J. L. Hall, ed.), pp. 235–277. Elsevier and North Holland, Amsterdam.

59. Evans, L. V., Callow, M. E., Percival, E., and Fareed, V. (1974). Studies on the synthesis and composition of extracellular mucilage in the unicellular red alga *Rhodella. J. Cell Sci.* **16**, 1–21.
60. Evans, L. V., Simpson, M., and Callow, M. E. (1973). Sulphated polysaccharide synthesis in brown algae. *Planta* **110**, 237–252.
61. Evins, W. H., and Varner, J. E. (1971). Hormone-controlled synthesis of endoplasmic reticulum in barley aleurone cells. *Proc. Natl. Acad. Sci. USA* **68**, 1631–1633.
62. Falk, R. H., and Stocking, C. R. (1976). Plant membranes. *In* "Encylopedia of Plant Physiology. Transport in Plants. III" (C. R. Stocking and U. Heber, eds.), pp. 3–50. Springer-Verlag, Berlin.
63. Farquhar, M. G., Bergeron, J. J. M., and Palade, G. E. (1974). Cytochemistry of Golgi Fractions prepared from rat liver. *J. Cell Biol.* **60**, 8–25.
64. Figier, J. (1969). Incorporation de glycine ^{3}H chez les glandes pétiolaires de *Mercurialis annua* L. *Planta* **87**, 275–289.
65. Fineran, B. A. (1970). Organization of the tonoplast in frozen-etched root tips. *J. Ultrastruct. Res.*, **33**, 574–586.
66. Fineran, B. A. (1970). An evaluation of the form of vacuoles in thin sections and freeze-etch replicas of root tips. *Protoplasma* **70**, 457–478.
67. Fineran, B. A. (1971). Ultrastructure of vacuolar inclusions in root tips. *Protoplasma* **72**, 1–18.
68. Fineran, B. A. (1973). Association between endoplasmic reticulum and vacuoles in frozen-etched root tips. *J. Ultrastruct. Res.* **48**, 75–87.
69. Fineran, B. A. (1978). Freeze-etching. *In* "Electron Microscopy and Cytochemistry of Plant Cells" (J. L. Hall, ed.), pp. 279–341. Elsevier and North Holland, Amsterdam.
70. Franke, W. W. (1974). Structure, biochemistry and functions of the nuclear envelope. *Int. Rev. Cytol. Suppl.* **4**, 71–236.
71. Franke, W. W., Kartenbeck, J., Zeutgraf, H., Scheer, U., and Falk, H. (1971). Membrane-to-membrane cross-bridges. A means to orientation and interaction of membrane faces. *J. Cell Biol.* **51**, 881–888.
72. Franke, W. W., Zengraf, H., and Scheer, U. (1978). Supranucleosomal and nonnucleosomal chromatin configurations. *Int. Congr. Electron Microsc.* 9th (J. M. Sturgess, ed.), Vol, 3, pp. 573–586. Microscopic Society of Canada, Toronto.
73. Frederick, S. E., and Newcomb, E. H. (1969). Microbody-like organelles in leaf cells. *Science* **163**, 1353–1355.
74. Frederick, S. E., and Newcomb, E. H. (1971). Ultrastructure and distribution of microbodies in leaves of grasses with and without CO_2–photorespiration. *Planta* **96**, 152–174.
75. Frederick, S. E., Gruber, P. J., and Newcomb, E. H. (1975). Plant microbodies. *Protoplasma* **84**, 1–29.
76. Frederick, S. E., Newcomb, E. H., Vigil, E. L., and Wergin, W. P. (1968). Fine-structural characterization of plant microbodies. *Planta* **81**, 228–252.
77. Freudenberg, K. (1968). Constitution and biosynthesis of lignin. *In* "Molecular Biology, Biochemistry and Biophysics" (A. Kleinzeller, ed.), Vol. 2, pp. 45–122. Springer-Verlag, Berlin.
78. Frey-Wyssling, A. (1953). "Submicroscopic Morphology of Protoplasm." Elsevier, Amsterdam.
79. Gahan, P. B., and Maple, A. J. (1966). The behaviour of lysosome-like particles during cell differentiation. *J. Exp. Bot.* **17**, 151—155.
80. Gerhardt, B. (1973). Untersuchungen zur funktionsanderung der microbodies in den kiemblattern von *Helianthus annuus* L. *Planta* **110**, 15–28.

81. Ghosh, S. (1976). The nucleolar structure. *Int. Rev. Cytol.* **44**, 1–28.
82. Goodchild, D. J., and Bergersen, F. J. (1966). Electron microscopy of the infection and subsequent development of soybean nodule cells. *J. Bact.* **92**, 204-213.
83. Goosen-de Roo, L. (1973). The relationship between cell organelles and cell wall thickenings in primary tracheary elements of the cucumber. I. Morphological aspects. *Acta Bot. Neerl.* **22**, 279–300.
84. Goosen-de Roo, L. (1973). The relationship between cell organelles and cell wall thickenings in primary tracheary elements of the cucumber. II. Quantitative aspects. *Acta. Bot. Neerl.* **22**, 301–320.
85. Graham, T. A., and Gunning, B. E. S. (1970). The localization of legumin and vicilin in bean cotyledon cells using fluorescent antibodies. *Nature (London)* **228**, 81–82.
86. Granick, S., and Porter, K. R. (1947). The structure of the spinach chloroplast as interpreted with the electron microscope. *Am. J. Bot.* **34**, 545–550.
87. Green, P. B. (1964). Cinematic observations on the growth and division of chloroplasts in *Nitella*. *Am. J. Bot.* **51**, 34–342.
88. Grierson, D. (1977). The nucleus and the organization and transcription of nuclear DNA. *In* "The Molecular Biology of Plant Cells" (H. Smith, ed.), pp. 210–255. Blackwell, Oxford.
89. Gruber, P. J., Becker, W. M., and Newcomb, E. H. (1972). The occurrence of microbodies and peroxisomal enzymes in achlorophyllous leaves. *Planta* **105**, 114–138.
90. Gruber, P. J., Becker, W. M., and Newcomb, E. H. (1973). The development of microbodies and peroxisomal enzymes in greening bean leaves. *J. Cell Biol.* **56**, 500–518.
90a. Guidotti, G. (1972). Membrane proteins. *Annu. Rev. Biochem.* **41**, 731–752.
91. Guilliermond, A. (1941). "The Cytoplasm of the Plant Cell." Chronica Botanica
92. Gunning, B. E. S. (1978). Age-related and origin-related control of numbers of plasmodesmata in cell walls of developing *Azolla* roots. *Planta* **143**, 181–190.
93. Gunning, B. E. S., Hardham, A. R., and Hughes, J. E. (1978). Pre-prophase bands of microtubules in all categories of formative and proliferative cell division in *Azolla* roots. *Planta* **143**, 145–160.
94. Gunning, B. E. S., Hardham, A. R., and Hughes, J. E. (1978). Evidence for the initiation of microtubules in discrete regions of the cell cortex in *Azolla* root-tip cells, and an hypothesis on the development of cortical assays of microtubules. *Planta* **143**, 161–179.
95. Gunning, B. E. S., and Pate, J. S. (1969). Transfer cells. Plant cells with wall ingrowths, specialized in relation to short distance transport of solutes—their occurence, structure and development. *Protoplasma* **68**, 107–133.
96. Gunning, B. E. S., and Pate, J. S. (1974). Transfer cells. *In* "Dynamic Aspects of Plant Ultrastructure" (A. W. Robards, ed.), pp. 441–480. McGraw-Hill, London.
97. Gunning, B. E. S., and Robards, A. W. (1976a). Plasmodesmata; current knowledge and outstanding problems. *In* "Intercellular Communication in Plants: Studies on Plasmodesmata" (B. E. S. Gunning and A. W. Robards, eds.), pp. 297–311. Springer-Verlag, Berlin.
98. Gunning, B. E. S., and Robards, A. W. (eds.) (1976). "Intercellualar Communication in Plants: Studies on Plasmodesmata." Springer-Verlag, Berlin.
99. Gunning, B. E. S., and Steer, M. W. (1975). "Ultrastructure and the Biology of Plant Cells." Arnold, London.
99a. Gunning, B. E. S., Steer, M. W., and Cochrane, M. P. (1968). Occurrence, molecular structure, and induced formation of the stromacentre in plastids. *J. Cell Sci.* **3**, 445-456.

100. Guyot, M., and Humbert, C. (1970). Les modifications du vacuome des cellules stomatiques d'*Anemia rotundifolia*. *C. R. Acad. Sci. Paris* **270**, 2787–2790.

101. Hackenbrock, G. R. (1966). Ultrastructural basis for metabolically linked mechanical activity in mitochondria. I. Reversible ultrastructural changes with change in metabolic state in isolated liver mitochondria. *J. Cell Biol.* **30**, 269–297.

102. Hall, J. L. (1971). Cytochemical localization of ATP-ase activity in plant root cells. *J. Microsc.* **93**, 219–225.

103. Hall, J. L. (1977). Fine structure and cytochemical changes occurring in beet discs in response to washing. *New Phytol.* **79**, 559–566.

104. Hall, J. L., and Baker, D. A. (1975). Cell membranes. *In* "Ion Transport in Plant Cells and Tissues" (D. A. Baker and J. L. Hall, eds.), pp. 39–77. North Holland, Amsterdam.

105. Hall, J. L., and Baker, D. A. (1977). "Cell Membranes and Ion Transport." Longman, London.

106. Hall, J. L., and Davie, C. A. M. (1971). Localization of acid hydrolase activity in *Zea mays* L. root tips. *Ann. Bot.* **35**, 849–855.

107. Hall, J. L., Flowers, T. J., and Roberts, R. M. (1974). "Plant Cell Structure and Metabolism." Longman, London.

108. Hall, J. L., and Sexton, R. (1972). Cytochemical localization of peroxidase activity in root cells. *Planta* **108**, 103–120.

109. Hall, J. L., and Taylor, A. R. D. (1979). Isolation of the plasma membrane from higher plant cells. *In* "Methodological Surveys in Biochemistry: Plant Organelles" (E. Reid, ed.), pp. Ellis Horwood, Chichester.

110. Hall, M. A. (1976). The cell wall. *In* "Plant structure, Function and Adaptation" (M. A. Hall, ed.), pp. 49–90. MacMillan, New York.

111. Hanson, J. B., and Koeppe, D. E. (1975). Mitochondria. *In* "Ion Transport in Plant Cells and Tissues" (D. A. Baker and J. L. Hall, eds.), pp. 79–99. North-Holland, Amsterdam.

112. Hanson, J. B., Leonard, R. T., and Mollenhauer, H. H. (1973). Increased electron density of tonoplast membranes in washed corn root tissues. *Plant Physiol.* **52**, 298–300.

113. Hardham, A. R., and Gunning, B. E. S. (1977). The length and deposition of cortical microtubules in plant tissues fixed in glutaraldehyde–osmium. *Planta* **134**, 201–203.

114. Hardham, A. R., and Gunning, B. E. S. (1978). Structure of cortical microtubule arrays in plant cells. *J. Cell Biol.* **77**, 14—34.

115. Harris, P. J., and Northcote, D. H., (1970). Patterns of polysaccharide biosynthesis in differentiating cells of maize root tips. *Biochem. J.* **120**, 479–491.

116. Harris, P. J., and Northcote, D. H. (1961). Polysaccharide formation in plant Golgi bodies. *Biochim. Biophys. Acta* **237**, 56–64.

117. Hart, J. W., and Sabnis, D. D. (1977). Microtubules. *In* "The Molecular Biology of Plant Cells" (H. Smith, ed.), pp. 160–181. Blackwell, Oxford.

118. Heath, M. F., and Northcote, D. H. (1971). Glycoprotein of the wall of sycamore tissue cultured cells. *Biochem. J.* **125**, 953–961.

118a. Heggeness, M. H., Simon, M., and Singer, S. J. (1978). Association of mitochondria with microtubules in cultured cells. *Proc. Natl. Acad. Sci. USA* **75**, 3863–3866.

119. Heldt, H. W. (1976). Transport of metabolities between cytoplasm and the mitochondrial matrix. *In* "Encyclopedia of Plant Physiology. Transport in Plants. III" (C. R. Stocking an U. Heber, eds.), pp. 235–254. Springer-Verlag, Berlin.

120. Helsper, J. P. F. G., Veerkamp, J. H., and Sassen, M. M. A. (1977). B-glucan synthetase activity in Golgi vesicles of *Petunia hybrida*. *Planta* **133**, 303–308.

121. Hendler, R. W. (1971). Biological membrane ultrastructure. *Physiol. Rev.* **51**, 66–97.
122. Hepler, P. K. (1976). Plant microtubules. *In* "Plant Biochemistry" (J. Bonner and J. E. Varner, eds.), pp. 147–187. Academic Press, New York.
123. Hepler, P. K., and Fosket, D. E. (1971). The role of microtubules in vessel member differentiation in *Coleus. Protoplasma* **72**, 213–236.
124. Hepler, P. K., and Palevitz, B. A. (1974). Microtubules and microfilaments. *Annu. Rev. Plant Physiol.* **25**, 309–362.
125. Herth, W., Franke, W. W., Stadler, J., Bittiger, J., Keilich, G., and Brown, R. M. Jr. (1972). Further characterization of the alkalistable material from the scales of *Pleurochrysis scherffelii:* a cellulosic glycoprotein. *Planta* **105**, 79–92.
125a. Heslop-Harrison, J. (1963) Structure and morphogenesis of lamellae systems in grana-containing chloroplasts. I. Membrane structure and lamellar architecture. *Planta* **60**, 243–260.
126. Heslop-Harrison, J. (1971). The pollen wall: structure and development. *In* "Pollen: Development and Physiology" (J. Heslop-Harrison, ed.), pp. 75–98. Butterworths, London.
127. Heslop-Harrison, J. (1975). The physiology of the pollen grain surface. *Proc. R. Soc. London Ser. B* **190**, 275–299.
128. Heslop-Harrison, Y. (1975). Enzyme release in carnivorous plants. *In* "Lysosomes in Biology and Pathology" (J. T. Dingle and R. T. Dean, eds.), pp. 525–578. North-Holland, Amsterdam.
129. Hongladarom, T., Honda, S. I., and Wildman, S. G. (1965). Appearance and behaviour of organelles in living plant cells. 16mm cinephotomicrographic film, Extension Media Center, Univ. of California, Berkeley.
130. Horne, R. W. (1976). The development of the electron microscope and its application of biology in the past decade. *Proc. R. Microsc. Soc.* **II**, 117–130.
130a. Huang, A. H. C., and Beevers, H. (1973) Localization of enzymes within microbodies. *J. Cell Biol.* **58**, 379–389.
131. Inoueé, S., and Sato, H. (1967). Cell motility by labile association of molecules. The nature of mitotic spindle fibres and their role in chromosome movement. *J. Gen. Physiol. Suppl.* **50**, 259–292.
131a. Israel, H. W., Salpeter, M. M., and Steward, F. C. (1968). The incorporation of radioactive proline into cultured cells. *J. Cell. Biol.* **39**, 698–715.
132. Jackman, M. E., and Van Steveninck, R. F. M. (1967). Changes in the endoplasmic reticulum of beetroot slices during aging. *Aust. J. Biol. Sci.* **20**, 1063–1068.
133. Jones, M. G. K. (1976). The origin and development of plasmodesmata. *In* "Intercellular Communication in Plants: Studies on Plasmodesmata" (B. E. S. Gunning and A. W. Robards, eds.), pp. 81–105. Springer-Verlag, Berlin.
134. Jones, R. L. (1969). Gibberellic acid and the fine structure of barley aleurone cells. I. Changes during the lag phase of α-amylase synthesis. *Planta* **87**, 119–133.
135. Jones, R. L. (1973). Gibberellins: their physiological role. *Annu. Rev. Plant Physiol.* **24**, 571–598.
136. Juniper, B. E. (1976). Geotropism. *Annu. Rev. Plant Physiol.* **27**, 385–406.
137. Juniper, B. E. (1977). Some speculations on the possible roles of the plasmodesmata in the control of differentiation. *J. Theor. Biol.* **66**, 583–592.
138. Juniper, B. E., and Pask, G. (1973). Directional secretion by the Golgi bodies in maize root cap cells. *Planta* **109**, 225–231.
139. Kagawa, T., Lord, J. M., and Beevers, H. (1973). The origin and turnover of organelle membranes in castor bean endosperm. *Plant Physiol.* **51**, 61–65.
140. Keegstra, K., Talmadge, K. W., Bauer, W. D., and Albersheim, P. (1973). The structure of plant cell walls. III. A model of the walls of suspension-cultured syca-

more cells based on the interconnections of the macromolecular components. *Plant Physiol.* **51**, 188–196.

141. Kerbey, A. L., McPherson, M., Sheterline, P., and Schofield, J. G. (1974). Co-factor requirements for the stimulation of ox growth hormone release in vitro. *In* "Transport at the Cellular Level" (M. A. Sleigh and D. H. Jennings, eds.), S. E. B. Symposia, Vol. 28, pp, 375–397. Cambridge Univ. Press, London/New York.

142. Kiermayer, O., and Dobberstein, B. (1973). Dictyosome derived membrane complexes as templates for the extraplasmic synthesis and orientation of microfibrils. *Protoplasma* **77**, 437–451.

143. Knox, R. B., and Clarke, A. E. (1978). Localization of proteins and glycoproteins by binding to labelled antibodies and lectins. *In* "Electron Microscopy and Cytochemistry of Plant Cells" (J. L. Hall ed.), pp. 149–185. North-Holand, Amsterdam.

144. Kristen, U. (1978). Ultrastructure and possible function of the intercisternal elements in dictyosomes. *Planta* **138**, 29–33.

144a. Laetsch, W. M. (1974). The C_4 syndrome: a structural analysis. *Annu. Rev. Plant Physiol.* **25**, 27–52.

145. LaFontaine, J. G. (1974). The nucleus. *In* "Dynamic Aspects of Plant Ultrastructure" (A. W. Robards, ed.), pp. 1–51. McGraw-Hill, London.

146. Lamport, D. T. A. (1965). The protein component of primary cell walls. *Adv. Bot. Res.* **2**, 151–218.

147. Lamport, D. T. A. (1970). Cell wall metabolism. *Annu. Rev. Plant Physiol.* **21**, 235–270.

148. Lauchli, A. (1976). Apoplastic transport in tissues. *In* "Transport in Plants II. Part B. Tissues and Organs" (U. Lüttge and M. G. Pitman, eds.), pp. 3–34, Encyclopedia of Plant Physiology. Springer-Verlag, Berlin.

149. Ledbetter, M. C., and Porter, K. R. (1963). A 'microtubule' in plant cell fine structure. *J. Cell Biol.* **19**, 239–250.

150. Ledbetter, M. C., and Porter, K. R. (1970). "Introduction to the Fine Structure of Plant Cells." Springer-Verlag, Berlin.

151. Leech, R. M. (1976). Plastid development in isolated etiochloroplasts and isolated etioplasts. *In* "Perspectives in Experimental Biology" (N. Sunderland ed.), Vol. 2, Botany, pp. 145–162. Pergamon, Oxford New York.

152. Leech, R. M. (1976). The replication of plastids in higher plants. In "Cell Division in Higher Plants" (M. M. Yeoman, ed.), pp. 135–159. Academic Press, New York.

153. Leedale, G. F. (1970). Phylogenetic aspects of nuclear cytology in the algae. *Ann. N. Y. Acad. Sci.* **175**, 429–453.

154. Leigh, R. A., and Branton, D. (1976). Isolation of vacuoles from root storage tissue of *Beta vulgaris* L. *Plant Physiol.* **58**, 656–662.

155. Linn, W., Wagner, G. J., Siegelman, H. W., and Hind, G. (1977). Membrane-bound ATP-ase of intact vacuoles and tonoplasts isolated from mature plant tissue. *Biochim. Biophys. Acta* **465**, 110–117.

156. Luck, D. J. L. (1963). Genesis of mitochondria in *Neurospora crassa. Proc. Natl. Acad. Sci. USA* **49**, 233–240.

156a. Lüttge, U., and Schnepf, E. (1976). Organic substances. *In* "Encylopedia of Plant Physiology. Transport in Plants II B" (U. Lüttge, and M. G. Pitman, eds.), pp. 244–277. Springer-Verlag, Berlin.

157. Lwoff, A. (1962). "Biological Order." MIT Press, Cambridge, Massachusetts.

158. Malone, C. P., Koeppe, D. E., and Miller, R. J. (1974). Corn mitochondrial swelling and contraction—an alternative interpretation. *Plant Physiol.* **53**, 918–927.

159. Manton, I. (1957). Observations with the electron microscope on the cell structure of the antheridium and spermatozoid of *Spagnum. J. Exp. Bot.* **8**, 382–400.

160. Manton, I. (1959). Electron microscopical observations on a very small flagellate: the problem of *Chromulina pusilla* Butcher. *J. Mar. Biol. Assoc. U. K.* **38**, 319–333.

161. Manton, I. (1975). Microscopy for fun. *J. Exp. Bot.* **26**, 639–655.

162. Manton, I. (1966). Observations on scale production in *Pyramimonas amylifera* Conrad. *J. Cell Sci.* **1**, 429–438.

163. Manton, I. (1967). Further observations on the fine structure of *Chrysochromulina chiton* with special reference to the haptonema, 'peculiar' Golgi structure and scale production. *J. Cell Sci.* **2**, 265–272.

164. Manton, I., Rayns, D. G., Ettl, H., and Parke, M. (1965). Further observations on green flagellates with scaly flagella: the genus *Heteromastix Korshikov. J. Mar. Biol. Assoc., U. K.* **45**, 241–255.

165. Marchant, H. J. (1976) Plasmodesmata in algae and Fungi. In "Intercellular Communication in Plants: Studies on Plasmodesmata" (B. E. S. Gunning, ed.). Springer-Verlag, Berlin.

166. Marchant, R., and Robards, A. W. (1968). Membrane systems associated with the plasmalemma of plant cells. *Ann. Bot.,* **32**, 457–71.

167. Margulis, L. (1970). "Origin of Eukaryotic Cells." Yale Univ. Press, New Haven, Connecticut.

168. Martin, J. T., and Juniper, B. E. (1970). "The Cuticles of Plants." Arnold, London.

169. Marty, F. (1970). Rôle du système membranaire vacuolaire dans la differenciation des lacticifères d'*Euphorbia characias* L. *C. R. Acad. Sci. Paris* **271**, 2301–2304.

169a. Marty, J. (1978). Cytochemical studies on GERL, provacuoles and vacuoles in root meristematic cells of *Euphorbia. Proc. Natl. Acad. Sci., USA* **75**, 852–856

170. Massalski, A., and Leedale, G. F. (1969). Cytology and ultrastructure of the Xanthophyceae. I. Comparative morphology of the zoospores of *Bumilleria sicula* Borzi and *Tribonema vulgare* Pascher. *Br. Phycol. J.* **4**, 159–180.

171. Matile, Ph. (1974). Lysosomes. In "Dynamic Aspects of Plant Ultrastructure" (A. W. Robards, ed.), pp. 178–218. McGraw-HIll, London.

172. Matile, Ph. (1975). "The Lytic Compartment of Plant Cells." Springer-Verlag, Vienna/New York.

173. Matile, Ph. (1976). Vacuoles. In "Plant Biochemistry' (J. Bonner and J. E. Varner, eds.), pp. 189–224. Academic Press, New York.

174. Matile, Ph., and Wiemken, A. (1976). Interactions between cytoplasm and vacuole. In "Encyclopedia of Plant Physiology. Transport in Plants. III" (C. R. Stocking and U. Heber, eds.), pp. 255–287. Springer-Verlag, Berlin.

175. Matile, Ph., and Winkenbach, F. (1971). Function of lysosomes and lysosomal enzymes in the senescing corolla of the morning glory (*Ipomoea purpurea*). *J. Exp. Bot.* **22**, 759–771.

176. May, M. A., and Cocking, E. C. (1969). Pinocytotic uptake of polystyrene latex particles by isolated tomato fruit protoplasts. *Protoplasma* **68**, 223–230.

177. Mesquita, J. F. (1969). Electron microscope study of the origin and development of vacuoles in root-tip cells of *Lupinus albus* L. *J. Ultrastruct. Res.* **26**, 242–250.

178. Miller, D. H., Mellman, I. S., Lamport, D. T. A., and Miller, M. (1974). The chemical composition of the cell wall of *Chlamydomonas gymnogama* and the concept of a plant cell wall protein. *J. Cell Biol.* **63**, 420–429.

179. McIntosh, J. R., Hepler, P. K., and van Wie, G. (1969). Model for mitosis. *Nature (London)* **224**, 659–663.

180. Mollenhauer, H. H., and Morré, D. J. (1966). Golgi apparatus and plant secretion. *Annu. Rev. Plant Physiol.* **17**, 27–46.

181. Mollenhauer, H. H., and Morré, D. J. (1975). A possible role for intercisternal

elements in the formation of secretory vesicles in plant Golgi apparatus. *J. Cell Sci.* **19,** 231–237.

182. Mollenhauer, H. H., and Morré, D. J. (1976). Cytochalasin B, but not colchicine, inhibits migration of secretory vesicles in root tips of maize. *Protoplasma* **87,** 39–48.

183. Moreau, F., Dupont, J., and Lance, D. (1974). Phospholipid and fatty acid composition of outer and inner membranes of plant mitochondria. *Biochim. Biophys. Acta* **345,** 294–304.

184. Morré, D. J. (1975). Membrane biogenesis. *Annu. Rev. Plant Physiol.* **26,** 441–481.

185. Morré, D. J., and Mollenhauer, H. H. (1974). The endomembrane concept: a functional integration of endoplasmic reticulum and Golgi apparatus. *In* "Dynamic Aspects of Plant Ultrastructure" (A. W. Robards, ed.), pp. 84–137. McGraw-Hill, London.

186. Morré, D. J., and Mollenhauer, H. H. (1976). Interactions among cytoplasm, endomembranes and the cell surface. *In* "Encyclopedia of Plant Physiology. Transport in Plants III" (C. R. Stocking and U. Heber, eds.), pp. 288–344. Springer-Verlag, Berlin.

187. Morré, D. J., Mollenhauer, H. H., and Bracker, C. E. (1971). Origin and continuity of the Golgi apparatus. *In* "Origin and Continuity of Cell Organelles" (J. Reinert and H. Upspring, eds.), pp. 82–126. Springer, Berlin/Heidelberg/New York.

188. Mueller, S. C., Brown, R. M. Jr., and Scott, T. K. (1976). Cellulose microfibrils: nascent stages of synthesis in a higher plant cell. *Science* **194,** 949–951.

189. Mühlethaler, K. (1977). Introduction to the structure and function of the photosynthetic apparatus. *In* "Encyclopedia of Plant Physiology. Photosynthesis I" (A. Trebst and M. Avron, eds.), pp. 503–521. Springer-Verlag, Berlin.

190. Munn, E. A. (1974). "The Structure of Mitochondria." Academic Press, New York/London.

191. Nagl, W. (1976). Nuclear organization. *Annu. Rev. Plant Physiol.* **27,** 39–69.

192. Nanninga, N. (1973). Structural aspects of ribosomes. *Int. Rev. Cytol.* **35,** 185–188.

193. Nobel, P. S. (1975). Chloroplasts. *In* "Ion Transport in Plant Cells and Tissues" (D. A. Baker and J. L. Hall, eds.), pp. 101–124. North-Holland Amsterdam.

194. Nonomura, Y., Blobel, G., and Sabatini, D. (1971). Structure of liver ribosomes studied by negative staining. *J. Mol. Biol.* **60,** 303–323.

195. Northcote, D. H. (1968). The organization of the endoplasmic reticulum, the Golgi bodies and microtubules during cell division and subsequent growth. *In* "Plant Cell Organelles" (J. B. Pridham, ed.), pp. 179–197. Academic Press, New York/London.

196. Northcote, D. H. (1972). Chemistry of the plant cell wall. *Annu. Rev. Plant Physiol* **23,** 113–132.

197. Northcote, D. H., and Pickett-Heaps, J. D. (1966). A function of the Golgi apparatus in polysaccharide synthesis and transport in the root-cap cells of wheat. *Biochem. J.* **98,** 159–167.

198. O'Brien, T. P. (1972). The cytology of cell-wall formation in some eukaryotic cells. *Bot. Rev.* **38,** 87–118.

199. O'Brien, T. P. (1974). Primary vascular tissues. *In* "Dynamic Aspects of Plant Ultrastructure" (A. W. Robards, ed.), pp. 414–440. McGraw-Hill, London.

200. O'Brien, T. P., and McCully, M. E. (1969). "Plant Structure and Development." MacMillan, Toronto.

201. Olesen, P. (1979). The neck constriction of plasmodesmata. Evidence for a peripheral sphincter-like structure revealed by fixation with tannic acid. *Planta* **144,** 349–358.

202. Opik, H. (1968). Development of cotyledon cell structure in ripening *Phaseolus vulgaris* seeds. *J. Exp. Bot.* **19,** 64–67.

203. Opik, H. (1974). Mitochondria. *In* "Dynamic Aspects of Plant Ultrastructure" (A. W. Robards, ed.), pp. 52–83. McGraw-Hill, London.

204. Orci, L., and Perrelet, A. (1973). Membrane-associated particles: Increase at sites of pinocytosis by freeze-etching. *Science* **181**, 868–869.

205. Ostergren, G., Mole-Bajer, J., and Bajer, A. (1960). An interpretation of transport phenomena at mitosis. *Ann. N. Y. Acad. Sci.* **90**, 381–408.

206. Palade, G. E. (1971). Albert Claude and the beginnings of biological electron microscopy. *J. Cell Biol.* **50**, 5D–19D.

207. Palade, G. (1975). Intracellular aspects of the process of protein secretion. *Science* **189**, 347–358.

208. Palevitz, B. A., and Hepler, P. K. (1976). Cellulose microfibril orientation and cell shaping in developing guard cells of *Allium:* the role of microtubules and ion accumulation. *Planta* **132**, 71–93.

209. Palmer, J. M. (1976). Structures associated with catabolism. *In* "Plant structure, Function and Adaptation" (M. A. Hall, ed.), pp. 91–124. MacMillan, London.

210. Palmer, J. M. (1976). The organization and regulation of electron transport in plant mitochondria. *Annu. Rev. Plant Physiol.* **27**, 133–157.

210a. Paolillo, D. J. Jr. (1970). The three-dimensional arrangement of intergranal lamellae in chloroplasts. *J. Cell Sci.* **6**, 243–255.

211. Parish, R. W. (1975). The lysosome-concept in plants. I. Peroxidases associated with subcellular and wall fractions of maize root tips: implications for vacuole development. *Planta* **123**, 1–13.

212. Park, R. B. (1965). The Chloroplast. *In* "Plant Biochemistry" (J. Bonner and J. E. Varner, eds.), pp. 124–150. Academic Press, New York.

213. Park, R. B., and Sane, P. V. (1971). Distribution of function and structure in chloroplast lamellae. *Annu. Rev. Plant Physiol.* **22**, 395–430.

214. Parsons, D. F., Bonner, W. D., and Verbon, J. G. (1965). Electron microscopy of isolated plant mitochondria and plastids using both the thin-section and negative-staining technique. *Can. J. Bot.* **43**, 647–655.

215. Parsons, J. A., and Rustad, R. C. (1968). The distribution of DNA among dividing mitochondria of *Tetrahymena pyriformis. J. Cell Biol.* **37**, 683–693.

216. Pate, J. S., and Gunning, B. E. S. (1969). Vascular transfer cells in angiosperm leaves. A taxonomic and morphological survey. *Protoplasma* **68**, 135–156.

217. Pate, J. S., and Gunning, B. E. S. (1972). Transfer cells. *Annu. Rev. Plant Physiol.* **23**, 173–196.

218. Paull, R. E., and Jones, R. L. (1975). Studies on the secretion of maize root-cap slime. III. Histochemical and autoradiographic localization of incorporated fucose. *Planta* **127**, 97–110.

219. Paull, R. E., and Jones, R. L. (1976). Studies on the secretion of maize root cap slime. IV. Evidence for the involvement of dictyosomes. *Plant Physiol.* **57**, 249–256.

220. Pernollet, J-C. (1978). Protein bodies of seeds: Ultrastructure, biochemistry, biosynthesis and degradation. *Phytochemistry* **17**, 1473–1480.

221. Peters, R. (1969). The problem of cytoplasmic integration. *Proc. R. Soc. London Ser. B* **173**, 11–19.

222. Philipp, E.-I., Franke, W. W., Keenan, T. W., Stadler, J., and Jarasch, E.-D. (1976). Characterization of nuclear membranes and endoplasmic reticulum isolated from plant tissues. *J. Cell Biol.* **68**, 11–29.

223. Pickett-Heaps, J. D. (1967). Further observations on the Golgi apparatus and its functions in cells of the wheat seedling. *J. Ultrastruct. Res.* **18**, 287–303.

224. Pickett-Heaps, J. D. (1978). Further ultrastructural observations on polysaccharide localization in plant cells. *J. Cell Sci.* **3**, 55–64.

224a. Pickett-Heaps, J. D. (1974). Plant microtubules. *In* "Dynamic Aspects of Plant Ultra-structure" (A. W. Robards, ed.), pp. 219–255. McGraw-Hill, London.

225. Pickett-Heaps, J. D., and Northcote, D. H. (1966). Organization of microtubules and endoplasmic reticulum during mitosis and cytokinesis in wheat meristems. *J. Cell Sci.* **1**, 109–120.

226. Picket-Heaps, J. D., and Nothcote, D. H. (1966b). Cell division in the formation of the stomatal complex of the young leaves of wheat. *J. Cell Sci.* **1**, 121–128.

227. Pitt, D. (1975). "Lysosomes and Cell Function." Longman, London.

228. Pitt, D., and Coombes, C. (1968). The disruption of lysosome-like particles of *Solanum tuberosum* cells during infection by *Phytophthora erythroseptica* Pethybr. *J. Gen. Microbiol.* **58**, 197–204.

229. Pope, D. G., Thorpe, J. R., Al-Azzawi, M. J., and Hall, J. L. (1979). The effect of cytochalasin B on the rate of growth and ultrastructure of wheat coleoptiles and maize roots. *Planta* **144**, 373–383.

230. Porter, K. R. (1978). Organisation in the structure of the cytoplasmic ground substance. *Int. Cong. Electron. Microsc. 9th* (J. M. Sturgess, ed.), Vol 3, pp. 627–639. Microscopic Society of Canada, Toronto.

231. Possingham, J. V., and Rose, R. J. (1976). Chloroplast replication and chloroplast DNA synthesis in spinach leaves. *Proc. R. Soc. London Ser. B* **193**, 295–305.

232. Possingham, J. V., and Saurer, W. (1969). Changes in chloroplast number per cell during leaf development in spinach. *Planta* **86**, 186–194.

233. Poux, N. (1973). Localization d'activities enzymatiques en microscopie electronique application aux cellules vegetales. *Ann. Univ. ARERS* **11**, 81–94.

234. Powell, J. T., and Brew, K. (1974). Glycosyltransferases in the Golgi membranes of onion stem. *Biochem. J.* **142**, 203–209.

235. Power, J. B., and Cocking, E. C. (1970). Isolation of leaf protoplasts: macromolecule uptake and growth substance response. *J. Exp. Bot.* **21**, 64–70.

236. Preston, R. D. (1974). "The Physical Biology of Plant Cell Walls." Chapman and Hall, London.

237. Preston, R. D. (1974). Plant cell walls. *In* "Dynamics Aspects of Plant Ultrastructure" (A. W. Robards, ed.), pp. 256–309. McGraw-Hill, London.

237a. Preston, R. D. (1979). Polysaccharide conformation and cell wall function. *Annu. Rev. Plant Physiol.* **30**, 55–78.

238. Quinn, P. J. (1976). "The Molecular Biology of Cell Membranes." MacMillan, London.

239. Ray, P. M. (1967). Radioautographic study of cell wall deposition in growing plant cells. *J. Cell Biol.* **35**, 659–674.

240. Ray, P. M., Shininger, T. L., and Ray, M. M. (1969). Isolation of β-glucan synthetase particles from plant cells and identification with Golgi membranes. *Proc. Natl. Acad. Sci. USA* **64**, 605–612.

241. Ridley, S. M., and Leech, R. M. (1970). Division of chloroplasts in an artificial environment. *Nature (London)* **227**, 463–465.

242. Ris, H. (1978). Higher order structures in chromosomes. *Int. Cong. Electron. Microsc., 9th* (J. M. Sturgess, ed.), Vol 3. pp. 545–556. Microscopic Society of Canada, Toronto.

243. Robards, A. W. (1971). The ultrastructure of plasmodesmata. *Protoplasma* **72**, 315–323.

244. Robards, A. W. (1975). Plasmodesmata. *Annu. Rev. Plant Physiol* **26**, 13–29.

245. Robards, A. W. (1976). Plasmodesmata in higher plants. *In* "Intercellular Communication in Plants: Studies on Plasmodesmata" (B. E. S. Gunning and A. W. Robards, eds.), pp. 15–57. Springer-Verlag, Berlin.

246. Robenek, H., and Paveling, E. (1977). Ultrastructure of the cell wall regeneration of isolated protoplasts of *Skimmia japonica* Thunb. *Planta* **136**, 135–145.

247. Roberts, K., and Northcote, D. H. (1970). Structure of the nuclear pore in higher plants. *Nature (London)* **228**, 385–386.

247a. Roberts, K., and Northcote, D. H. (1972) Hydroxyproline: observations on its chemical and autoradiographical localization in plant cell wall protein. *Planta* **107**, 43–51.

248. Robertson, J. D. (1959). The ultrastructure of cell membranes and their derivatives. *Biochem. Soc. Symp.* **16**, 3–43.

249. Robinson, D. G., and Preston, R. D. (1972). Plasmalemma structure in relation to microfibril biosynthesis in *Oocystis*. *Planta* **104**, 234–246.

250. Robinson, E., and Brown, R. (1953). Cytoplasmic particles in bean root cells. *Nature (London)* **171**, 313.

251. Roelofsen, P. A. (1965). Ultrastructure of the wall in growing cells and its relation to the direction of growth. *Adv. Bot. Res.* **2**, 69–148.

252. Roland, J.-C. (1973). The relationship between the plasmalemma and the plant cell wall. *Int. Rev. Cytol.* **36**, 45–92.

253. Roland, J.-C. (1978). General preparation and staining of thin sections. In "Electron Microscopy and Cytochemistry of Plant Cells" (J. L. Hall, ed.), pp. 1–62. Elsevier/North Holland, Amsterdam.

254. Roland, J.-C, and Vian, B. (1971). Réactivité du plasmalemme végétal. Etude cytochimique. *Protoplasma* **73**, 121–137.

255. Roland, J.-C., Vian, B., and Reis, D. (1975). Observations with cytochemistry and ultracryotomy on the fine structure of the expending walls in actively elongating plant cells. *J. Cell Sci.* **19**, 239–259.

256. Roland, J. C., and Vian, B., and Reis, D. (1977). Further observations on cell wall morphogenesis and polysaccharide arrangement during plant growth. *Protoplasma* **91**, 125–141.

257. Sane, P. V. (1977). The topography of the thylakoid membrane of the chloroplast. In "Encyclopedia of Plant Physiology. Photosynthesis I" (A. Trebst and M. Avron, eds.), pp. 522–542. Springer-Verlag, Berlin.

258. Sanger, J. M., and Jackson, W. T. (1971). Fine structure study of pollen development in *Haemanthus katherinae* Baker. Microtubules and elongation of the generative cells. *J. Cell Sci.* **8**, 303–315.

259. Schnarrenberger, C., and Fock, H. (1976). Interaction among organelles involved in photorespiration. In "Encyclopedia of Plant Physiology. Transport in Plants. III" (C. R. Stocking and U. Heber, eds.), pp. 3–50. Springer-Verlag, Berlin.

260. Schnepf, E. (1974). Gland cells. In "Dynamic Aspects of Plant Ultrastructure" (A. W. Robards, ed.), pp. 331–357. McGraw-Hill, London.

261. Schnepf, E. and Busch, J. (1976). Morphology and kinetics of slime secretion in glands of *Mimulus tilingii*. *Z. Pflanzenphysiol.* **79**, 62–71.

262. Schramm, M., and Selinger, Z. (1975). The functions of cyclic AMP and calcium as alternative second messengers in parotid gland and pancreas. *J. Cyclic Nucl. Res.* **1**, 181–192.

263. Sexton, R., and Hall, J. L. (1978). Enzyme cytochemistry. In "Electron Microscopy and Cytochemistry of Plant Cells" (J. L. Hall, ed.), pp. 63–147. North Holland, Amsterdam.

264. Sexton, R., and Hall, J. L. (1974). Fine structure and cytochemistry of the abscission zone cells of *Phaseolus* leaves. I. Ultrastructural changes occurring during abscission. *Ann. Bot.* **38**, 849–854.

265. Sexton, R., Cronshaw, J., and Hall, J. L. (1971). A study of the biochemistry and

cytochemical localization of β-glycerophosphatase activity in root tips of maize and pea. *Protoplasma* **73**, 417–441.

266. Shore, G., and Maclachlan, G. A. (1975). The site of cellulose synthesis. Hormone treatment alters the intracellular localization of alkali-insoluble β-1, 4-glucan (cellulose) synthetase activities. *J. Cell. Biol.* **64**, 557–571.

267. Shore, G., Raymond, Y., and Maclachlan, G. A. (1975). The site of cellulose synthesis. Cell surface and intracellular β-1, 4-glucan (cellulose) synthetase activities in relation to the stage and direction of cell growth. *Plant Physiol.* **56**, 34–38.

268. Siekevitz, P. (1972). Biological membranes: the dynamics of their organization. *Annu. Rev. Physiol.* **34**, 117–129.

269. Silverberg, B. A. (1975). An ultrastructural and cytochemical characterization of microbodies in the green algae. *Protoplasma* **83**, 269–295.

270. Singer, S. J. (1974). The molecular organization of membranes. *Annu. Rev. Biochem.* **43**, 805–833.

271. Singer, S. J., and Nicolson, G. (1972). The fluid mosaic model of the structure of cell membranes. *Science* **175**, 720–731.

272. Spooner, B. S. (1973). Cytochalasin B: toward an understanding of its mode of action. *Dev. Biol.* **35**, F13–F18.

273. Sprey, B. (1977). Lamellae-bound inclusions in isolated spinach chloroplasts. I. Ultrastructure and isolation. *Z. Pflanzenphysiol.* **83**, 159–179.

274. Sprey, B., and Laetsch, W. M. (1978). Structural studies on peripheral reticulum in C₄ plant chloroplasts of *Portulaca oleracea* L. *Z. Pflanzenphysiol.* **87**, 37–53.

275. Sprey, B., and Lambert, C. (1977). Lamella-bound inclusions in isolated spinach chloroplasts. II. Identification and composition. *Z. Pflanzenphysiol.* **83**, 222–248.

276. Srivastava, L. M., and Singh, A. P. (1972). Certain aspects of xylem differentiation in corn. *Can. J. Bot.* **50**, 1795–1804.

277. Staehelin, L. A. (1968). Ultrastructural changes of the plasmalemma and the cell wall during the life cycle of *Cyanidium caldarium*. *Proc. R. Soc. London Ser. B* **171**, 249–259.

278. Staehelin, L. A., and Kiermayer, O. (1970). Membrane differentiation in the Golgi complex of *Micrasterias denticulata* Breb. visualized by freeze-etching. *J. Cell Sci.* **7**, 787–792.

279. Steinert, M. (1969). The ultrastructure of mitochondria. *Proc. R. Soc. B* **173**, 63–70.

280. Steward, F. C. and Sutcliffe, J. F. (1959). Plants in relation to inorganic salts. *In* "Plant Physiology. A Treatise" (F. C. Steward, ed.), Vol. II, pp. 253–478. Academic Press, New York.

280a. Steward, F. C., Mott, R. L., Israel, H. W., and Ludford, P. M. (1970). Proline in the vesicles and sporelings of *Valonia ventricosa* and the concept of cell wall protein. *Nature (London)* **225**, 760–762.

280b. Steward, F. C., Israel, H. W., and Salpeter, M. M. (1974). The labeling of cultured cells of *Acer* with [¹⁴C]proline and its significance. *J. Cell Biol.* **60**, 695–701.

281. Stutz, E. (1976). Ribosomes. *In* "Plant Biochemistry" (J. Bonner and J. E. Varner, eds.). Academic Press, New York.

282. Talmadge, K. W., Keegstra, K., Bauer, W. D., and Albersheim, P. (1973). The structure of plant cell walls. I. The macromolecular components of the walls of suspension-cultured sycamore cells with a detailed analysis of the pectic polysaccharides. *Plant Physiol.* **51**, 158–173.

283. Taylor, A. R. D., and Hall, J. L. (1976). Physiological properties of protoplasts isolated from maize and tobacco. *J. Exp. Bot.* **27**, 383–391.

284. Taylor, A. R. D., and Hall, J. L. (1978). Fine structure and cytochemical properties

of tobacco leaf protoplasts and comparison with the source tissue. *Protoplasma* **96**, 113–126.

285. Thair, B.W., and Wardrop, A. B. (1971). The structure and arrangement of nuclear pores in plant cells. *Planta* **100**, 1–17.

286. Thomson, W. W. (1974). Ultrastructure of mature chloroplasts. *In* "Dynamic Aspects of Plant Ultrastructure" (A. W. Robards, ed.), pp. 138–177. McGraw-HIll, London.

287. Thomson, W. W., Foster, P., and Leech, R. M. (1972). The isolation of proplastids from roots of *Vicia Faba*. *Plant Physiol.* **49**, 270–272.

288. Thornber, J. P., and Alberte, R. S. (1977). The organization of chlorophyll in vivo. *In* "Encyclopedia of Plant Physiology. Photosynthesis I" (A. Trebst and M. Avron, eds.), pp. 574–582. Springer-Verlag, Berlin.

289. Tilney, L. G. (1971). Origin and continuity of microtubules. *In* "Origin and Continuity of Cell Organelles" (J. Reinert and H. Ursprung, eds.), pp. 222–260. Springer-Verlag, Berlin.

290. Tolbert, N. E. (1971). Microbodies—peroxisomes and glyoxysomes. *Annu. Rev. Plant Physiol.* **22**, 45–74.

291. Trebst, A. (1974). Energy conservation in photosynthetic electron transport of chloroplasts. *Annu. Rev. Plant Physiol.* **25**, 423–458.

292. Trelease, R. N., Becker, W. M., Gruber, P. J., and Newcomb, E. H. (1971). Microbodies (glyoxysomes and peroxisomes) in cucumber cotyledons. *Plant Physiol.* **48**, 461–475.

293. Tulett, A. J., Bagshaw, V., and Yeoman, M. M. (1969). Arrangement and structure of plastids in dormant and cultured tissue from artichoke tubers. *Ann. Bot.* **33**, 217–226.

294. Van der Woude, W. J., Lembi, C. A., Morré, D. J., Kindinger, J. I., and Ordin, L. (1974). β-glucan synthetases of plasma membrane and Golgi apparatus from onion stem. *Plant Physiol.* **54**, 333–340.

295. Van der Woude, W. J., Morré, D. J., and Bracker, C. E. (1971). Isolation and characterization of secretory vesicles in germinated pollen of *Lilium longiflorum*. *J. Cell Sci.* **8**, 331–351.

296. Van Steveninck, R. F. M. (1975). The "washing" or "aging" phenomenon in plant tissues. *Annu. Rev. Plant Physiol.* **26**, 237–258.

297. Van Steveninck, R. F. M., Armstrong, W. D., Peters, P. D., and Hall, T. A. (1976). Ultrastructural localization of ions. III. Distribution of chloride in mesophyll cells of mangrove (*Aegiceras corniculatum* Blanco). *Aust. J. Plant Physiol.* **3**, 367–376.

298. Van Steveninck, R. F. M. and Van Steveninck, M. E. (1978). Ion localization. *In* "Electron Microscopy and Cytochemistry of Plant Cells" (J. L. Hall, ed.), pp. 187–234. Elsevier/North-Holland, Amsterdam.

299. Varner, J. E., and Ho, D. T.-H (1976). Hormones *In* "Plant Biochemistry" (J. Bonner and J. E. Varner, eds.), pp. 713–770. Academic Press, New York.

300. Varner, J. E., and Ho, D. T.-H., (1977). Hormonal control of enzyme activity in higher plants. *In* "Regulation of Enzyme Synthesis and Activity in Higher Plants" (H. Smith, ed.), pp. 83–92. Academic Press, New York/London.

301. Vian, B., and Roland, J.-C. (1972). Différenciation des cytomembranes et renouvellement du plasmalemme dans les phénomènes de sécrétions végétales. *J. Microsc.* **13**, 119–136.

302. Vigil, E. L. (1970). Cytochemical and developmental changes in microbodies (glyoxysomes) and related organelles of castor bean endosperm. *J. Cell Biol.* **46**, 435–454.

303. Vigil, E. L. (1973). Plant microbodies. *J. Histochem, Cytochem.* **21,** 958–962.
304. Villiers, T. A. (1972). Cytological studies in dormancy. II. Pathological ageing changes during prolonged dormancy and recovery from dormancy release. *New Phytol.* **71,** 145–152.
305. Wagner, G. J., and Siegelman, H. W. (1975). Large-scale isolation of intact vacuoles and isolation of chloroplasts from protoplasts of mature plant tissues. *Science* **190,** 1298–1299.
306. Wallach, D. F. H. (1972). "The Plasma Membrane: Dynamic Perspectives, Genetics and Pathology." English Univ. Press, London.
307. Weier, T. E. (1961). The ultramicrostructure of the starch free chloroplasts of fully expanded leaves of *Nicotiana rustica. Am. J. Bot.* **48,** 615–630.
308. Weier, T. E., Stocking, C. R., Thomson, W. W., and Drever, H. (1963). The grana as structural units in chloroplasts of mesophyll of *Nicotiana rustica* and *Phaseolus vulgaris. J. Ultrastruct. Res.* **8,** 122–143.
309. Wergin, W. P., Gruber, P. J., and Newcomb, E. H. (1970). Fine structural investigation of nuclear inclusions in plants. *J. Ultrastruct. Res.* **30,** 533–557.
310. Wessels, N. K., Spooner, B. S., Ash, J. F., Bradley, M. O., Luduena, M. A., Taylor, E. L., Wrenn, J. T., Yamada, K. M. (1971). Microfilaments in cellular and developmental processes. *Science* **171,** 135–143.
311. Whaley, W. G. (1975). "The Golgi Apparatus," Cell Biology Monographs, Vol. 2. Springer-Verlag, Vienna.
312. Whatley, J. M. (1978). A suggested cycle of plastid developmental interrelationships. *New Phytol.* **80,** 489–502.
313. Whittaker, P. A., and Danks, S. M. (1978). "Mitochondria: Structure, Function and Assembly." Longman, London.
314. Wildman, S. G., Hongladarom, T., and Honda, S. I. (1962). Chloroplasts and mitochondria in living plant cells: cine photomicrographic studies. *Science* **138,** 434–436.
315. Wildman, S. G., Jope, C., and Atchison, B. A., (1974). Role of mitochondria in the origin of chloroplast starch grains. *Plant Physiol.* **54,** 231–237.
315a. Wolosewick, J. J., and Porter, K. R. (1979) Microtrabecular lattice of the cytoplasmic ground substance. *J. Cell Biol.* **82,** 114–139.
316. Wooding, F. B. P. (1969). Absorptive cells in protoxylem: association between mitochondria and the plasmalemma. *Planta* **84,** 235–238.
317. Wright, K., and Northcote, D. H. (1976). Identification of β1–4 glucan chains as part of a fraction of slime synthesised within the dictyosomes of maize root caps. *Protoplasma* **88,** 225–239.
318. Zaar, K., and Schnepf, E. (1969). Membranfluss and nucleosiddi-phosphatase-reaktion in wurzelhaaren von *Lepidium sativum. Planta* **88,** 224–232.
319. Yatsu, L. Y., and Jacks, T. J. (1972). Spherosome membranes. Half unit membranes. *Plant Physiol.* **49,** 937–943.

PREAMBLE TO CHAPTERS TWO, THREE, AND FOUR

In life organisms produce and maintain, from the raw materials they receive, their unique structures and forms and, concomitantly, they acquire the chemical composition that sets them apart from their environments. Only in dissolution and death is true equilibrium with the environment ultimately restored. Conversely, therefore, life is a continuing uphill battle, demanding the energy needed to build highly improbable systems and to function in nonequilibruim steady states.

An introduction to Volume III, written in 1963, dealt historically with the emerging concepts of metabolism (i.e., the ways in which organisms perform their chemical functions). In the inanimate world combustion and heat production are the familiar means that drive the engines that do work. Organisms, however, operate in more subtle ways. In their cells they operate as molecular machines in biochemical systems that draw on energy-rich complex organic reserves, breaking them down by many small steps. By these means, molecular breakdown, energy release, and utilization are negotiated without a "molecular conflagration." But for all this to occur smoothly and continuously, cells not only use an array of interlocking, catalyzed biochemical reactions; they also provide the morphological settings in which these reactions can operate in an orderly manner.

Logically, perhaps, concepts of bioenergetics should have dealt first with photosynthesis in autotrophic green plants and then have moved to the cells and organisms that are ultimately dependent upon photosynthesis for their energy supply. The green cells of autotrophic plants in the light have the long-recognized ability to draw upon light energy from the sun and fix it in excess of their immediate needs in biologically usable forms. In darkness these cells or other, nongreen cells can draw upon the energy so fixed to perform useful work; thus heterotrophic plants, animals, and bacteria are ultimately powered through the energy derived from green plants in the light. Historically, however, the means by which the energy of the elaborated photosynthate is released, as in the intermediate steps of respiratory metabolism, proved to be easier to unravel, as recounted in previous volumes of this treatise. By that time (the mid 1960s), the broad principles of energy fixation by plants (as in photosynthesis) and of the complex events of energy release and use (as in respiration and metabolism) seemed to be known satisfactorily.

Older analogies between respiration and combustion have long given way to the recognition of a large number of sequential, coordinated, reversible enzyme- and coenzyme-mediated steps that degrade the carbon of sugar ultimately to CO_2. In so doing, the energy content of sugar is also passed, as it were, "downhill" over a number of small oxidation–reduction steps that eventually terminate in the final act of combination with molecular oxygen. However, the oxidation reactions by which this is finally achieved are only compatible with organic acid substrates. To render the carbon of sugar compatible with the tricarboxylic acid cycle, it must perforce be first split and anaerobically worked into the C_3 acid form in which it may enter the oxidative cycle. Over the years, therefore, there have emerged the now familiar concepts of glycolysis (i.e., the splitting of sugar without the participation of molecular oxygen) and also the sequential roles of glycolysis and the tricarboxylic acid cycle. But neither the measurable emergence of carbon dioxide, nor its often equivalent absorption of oxygen (thus reversing the gas exchange of photosynthesis), nor even any excess heat lost to the environment were to be finally regarded as the primary energetic events. The recognition of this came from the understanding of the role, throughout, of the phosphorylation intermediates and the progressive formation of the "high energy" phosphorylated compounds. Through the pioneer concepts of Lipmann these compounds were, and still are, regarded as the energy currency by which organisms pay, as it were, for the energy they expend in useful—chemical, physical, or morphological—work.

Therefore, cellular respiration came to be seen in terms of the synthesis, breakdown, and resynthesis of those intermediary high energy compounds that at their synthesis incorporate some energy ultimately drawn from sugar and at their breakdown release it again at the point of further use. That the simple combination with inorganic phosphate and its subsequent release is the means usually adopted to achieve this in a given unit step becomes all the more remarkable when it is realized how specific and complex are the molecules that are, as it were, reversibly charged and discharged with this energy. Equally specific are the enzyme systems through which the chemical transactions are negotiated. Thus, cellular respiration came to be seen in terms of the ways by which cells degrade the reduced carbon compounds, which store energy, down to carbon dioxide and water. Concomitantly, the ways that their energy is dissipated were understood in terms of the removal and transfer of electrons through the "electron transport chain."

Since so much had seemingly been accomplished and had already been generally understood, even so long ago as the 1950s and 1960s, one may well ask why such tremendous later activity, both experimental and theoretical, seemed necessary to understand how cells negotiate all the stages of their bioenergetics. There was activity, so much so that Ephraim Racker fortified the 175 pages of compact text of the "New Look at Mechanisms in Bioenergetics" (1976) with over 300 references drawn mostly from the preceding decade and also supplied a glossary of over 50 abbreviations of terms (acronyms), almost all of which would have been unintelligible a generation ago! And Chapter 2 in this volume has its own bibliography of 550 references and a long list of abbreviations. Moreover, both of these works and other, similar ones are essentially unconcerned with organs or organisms (cf. Chapter 4), that is, with the levels in which bioenergetic mechanisms are integrated.

The modern bioenergetic trend, therefore, builds upon the former knowledge of glycolysis and the oxidative cycle, on the paths of carbon from and to carbon dioxide, and on the passage of hydrogen from and to water; but it now focuses intensively on how energy is passed from one form into another. The operative term here is "transduction," literally "the act of leading, or the bringing across" of energy from one form or situation to another. In other words, the preoccupation is now with the actual transfer of energy from sunlight to photosynthate and thence from sugar into energy rich compounds in such form that it may be linked to useful work. Although nature seems to achieve these objectives with deceptive ease, the unraveling of the achievement is itself a task of very great complexity; this is especially so in the complexity of the systems involved in photosynthesis, as Chapter 2 shows.

But this preoccupation with energy flow per se presupposes developments in other fields, notably the now greater knowledge of the morphological settings in which the biochemical events occur. This knowledge derives from the ultrastructure of organelles (mitochondria, chloroplasts, membranes, ribosomes, endoplasmic reticulum, etc.; see Chapter 1) that furnish the physical settings conducive to the enzymatically catalyzed events that occur *in situ*. Knowledge has also flowed from the recognition that the actual catalytic systems that negotiate *in vivo* many of the unit steps also have an essential morphology of their own, a morphology that if disrupted in extraction may be difficult to reconstitute *in vitro*. Finally, the bioenergetic trend has also flowed from the now familiar experience that, as in the case of the cytochromes, nature has variously elaborated far more numerous catalytic systems that achieve major ends by many small intermediate steps than a first overall view implied. Consequently, enzymes and their enzymology have proliferated almost beyond recognition.

The mitochondria, the organelles that convert the organic food of cells into usable fuel in the form of ATP, have long been recognized as veritable chemical power houses. In them morphology and biochemistry seemed to be, and are, wedded to produce the unique high energy chemical end product. But another and different step is now interposed among the biochemical acts of energy transduction. This step is inherent in the membrane of mitochondria, where a physicochemical step is now interposed between the oxidations per se and the final synthesis of ATP. This event occurs when hydrogen ions (protons), driven by the oxidations, move across a mitochondrial membrane that has selective properties and there create both an excess of ions and an electric potential. Thus, according to the "chemiosmotic view," the synthesis of ATP is now seen to be powered by the flow of hydrogen ions down an electrochemical gradient. Moreover, similar intervention of a physicochemical event is conceived to occur in chloroplasts, where the energy source is light and the energy is again built into phosphorylated compounds. But this is not all.

The catalytic systems that bring about all the steps of respiratory breakdown and that negotiate energy transfers, complex as they are *in vivo*, together with the morpological settings in which they work, embody in their orderly structural array an energy moiety. This moiety, genetically directed, appears as the negative-entropy component of the developed structures that function or work. How genetics, information, and energy achieve all this is an even more far-reaching problem.

When all these developments are faced, as in the ensuing chapters, the dilemma is to keep the narrative clear and comprehensible as to its essentials while recognizing that interpretations must necessarily derive from a command of the often complicated detail. It would be tedious and repetitious to attempt here, by an epitome of Chapter 2, to chart a course through the complicated events that it presents. However, each topic is introduced by a short paragraph that describes its essential features. This is followed by a detailed account of the evidence on which the conclusions are based. To retain sufficient detail while emphasizing the overall perspective, the device has been adopted in Chapter 2 of setting the more detailed material in distinctive type. The hope is that this will contribute to the ease with which those who are less familiar with the biochemical and biophysical details may nevertheless comprehend the salient developments that have occurred and anticipate their significance.

The ultimate dilemma remains to account in the living world for the harnessing of energy to complexity, to nonrandomness, to reduced entropy in a physical environment in which the overall energy flow is toward increased entropy and to the progressive unavailability of energy. Some, like Morowitz ["Energy Flow in Biological Systems" (1968)], have concerned themselves with energy flow per se. The view is that intervening compartments along this line of energy flow may nevertheless support events involving reduced entropy and greater order. The emphasis here is not on specific discrete acts of energy transfer but on the *flow* of energy through systems that behave as nonequilibrium steady states. Furthermore, the Nobel Laureate Ptigogyne (1977) has developed the concept of *dissipative structures,* which, as open systems, exchange energy with their environment. Living organisms are in this view to be regarded as such dissipative structures, and they maintain their structure by the continual flow of available energy through their system. The more complicated the structure, the more integrated and organized it is, and more energy "flow-through" is presumably required to maintain it.

Thus the concepts of bioenergetics extend in their scope even beyond their implications for metabolism, for they impinge upon topics that will be the concern of later volumes in the Treatise.

F.C.S.

CHAPTER TWO

Energy Metabolism in Plants[1]

D. T. Dennis

[1] Abbreviations used in this chapter: ADP, adenosine diphosphate; ATP, adenosine triphosphate; CCCP, carbonyl cyanide m-chlorophenylhydrazone; CF_0, thylakoid-membrane-embedded part of the chloroplast coupling factor complex (CF_0 has the proton channel of the complex); CF_1, part of the chloroplast coupling factor complex that is attached to CF_0 and that protudes into the chloroplast stroma (CF_1 has the ATPase activity of the complex); mCLAM, m-chlorobenzohydroxamic acid; DAD, 2,3,5,6-tetramethyl-p-phenylenediamine; DBMIB, 2,5-dibromo-3-methyl-6-isopropyl-p-benzoquinone; DCCD, N,N'-dicyclohexylcarbodiimide; DCMU, 3-(3,4-dichlorophenyl)-1,1-dimethylurea; DEAE, diethylaminoethyl; DNP, 2,4-dinitrophenol; EDTA, ethylenediaminetetraacetic acid; EGTA, 1,2-di(2-aminoethoxy)ethane-N,N,N',N'-tetraacetic acid; EPR, electron paramagnetic resonance; F_0, mitochondrial inner-membrane-embedded part of the mitochondrial coupling factor complex (F_0 has the proton channel of the complex); F_1, part of the mitochondrial coupling factor complex attached to F_0 and protruding into the mitochondrial matrix (F_1 has the ATPase activity of the complex); FAD, flavin adenine dinucleotide; FCCP, carbonyl cyanide-p-trifluoromethoxyphenylhydrazone; Fd, ferredoxin; FMN, flavin mononucleotide; GTP, guanosine triphosphate; HiPIP, high-potential iron protein; ITP, inosine triphosphate; NAD^+, NADH, nicotinamide adenine dinucleotide, oxidized or reduced; $NADP^+$, NADPH, nicotinamide adenine dinucleotide phosphate, oxidized or reduced; NEM, N-ethylmaleimide; NMR, nuclear magnetic resonance; OSCP, oligomycin-sensitivity-conferring protein; P_i, inorganic phosphate; PMS, N-methylphenazonium methosulfate; PQ, plastoquinone; Q cycle, protonmotive ubiquinone cycle; SDS, sodium dodecylsulfate; SH, sulfhydryl; SHAM, salicylhydroxamic acid; TMPD, tetramethyl-p-phenylenediamine; Tris, tris(hydroxymethyl)aminomethane; UQ, UQH_2, ubiquinone, oxidized and reduced.

Plant Physiology
A Treatise
Vol. VII: Energy and Carbon Metabolism

I. Introduction

Since Lipmann (290) first formalized the concepts of cell energetics, our knowledge of this field has expanded rapidly, and the rate of production of papers on this subject is now so great that few can keep pace with them. Yet our understanding of the detailed mechanisms by which cells transduce energy is still incomplete; there are still contrasting, incompatible theories. It would appear that as we learn more about energy transduction the explanation of the phenomenon becomes more elusive.

Why should this be so? There is a fundamental difference between this area of biochemistry and the related field of enzymology. The enzymologist can, with moderate effort in most cases, isolate an enzyme and purify it to homogeneity. The enzyme can then be treated as a complex chemical catalyst and its mode of action determined by various types of kinetic analysis. The active site on the enzyme surface can be probed by specific reagents; the structure of the enzyme can be determined, and the relationship between structure and function elucidated. By contrast, systems involved in cell energetics require intact structures, or at least a minimum level of structural integrity, and once that structure is disrupted the system can no longer synthesize ATP. It is possible, as shown by the elegant experiments of workers such as Racker, to synthesize ATP in reconstituted systems (284), but the key to the problem is the word reconstitution. The individual parts will not perform the overall reactions of energy transduction; they have to be reconstituted to a minimum level of structure before the system functions. The researcher is always at some distance from the subject studied and has to make assumptions, extrapolations, and correlations, some of which may not be valid. There is often no way of knowing which step in the process is rate limiting, and the components examined are sometimes just absorption peaks or EPR signals that disappear when the system is disrupted.

Moreover, there is a critical aspect of cell energetics that is not usually fully appreciated and yet is an important factor in living systems. Within a single cell there may be several thousand enzymes and metabolic intermediates. A limiting factor for metabolic activity in the cell is the very small volume available, which limits the capacity of the cell to dissolve all these components. This concept of the effect of limited solvent capacity in cell metabolism has been developed by Atkinson (16). In order to maintain the cellular and structural components such as nucleic acids and proteins at sufficiently high levels, the concentrations of intermediates must be kept extremely low. Much of the energy utilized by a cell is, therefore, used to displace the equilibrium of intermediary reactions so that these low concentrations are maintained. The energy is not wasted, but is used to maintain the solvent capacity of the cell. As a corollary, enzymes must be highly efficient so that they need be present only in low concentrations. Associated with this concept of solvent capacity is that the reactions concerned with energy transduction occur in a very limited space and in areas where free water is limited or nonexistent. It is under these conditions that energy transduction takes place, and such conditions may be impossible to replicate in the test tube.

In spite of these problems, energy transduction is one of the most exciting and challenging areas of biochemical endeavor. It is not intended in this chapter to discuss all the elementary aspects of cell energetics, which have been covered in recent biochemical textbooks,[2] but to give an overview of energetics in plant cells. The aim is to present an introduction to the various aspects of energetics that are of interest to plant biologists who have some background in biochemistry and plant physiology. The references are by no means complete, but it is hoped that a sufficient number are included for a start to be made into the literature. On the whole, only those papers are cited that have been published in biochemical or plant physiology journals rather than conference proceedings, which are not always readily available.

---

[2] For example, see A. L. Lehninger, *Biochemistry*, 2nd edition, Worth Publishers, Inc. New York, 1975.

Certain areas of plant energetics such as the primary reactions of photosynthesis are highly specialized and are only briefly alluded to. On the other hand, some topics are included that may seem out of place. For example, an outline of the animal mitochondrial energy-transducing system is presented because it has been developed more fully than the equivalent plant system. Therefore, in order to have an idea of how this system operates, it is necessary to discuss the animal system. It also provides an opportunity to present the differences between plant and animal mitochondria.

II. Theories of Energy Transduction

In the present state of knowledge it is not possible to give a clear or complete account of energy transduction in mitochondria or chloroplasts. However, this section describes the various lines of approach to the problems that have been adopted and the ideas that have emerged. There are subsections on the chemical coupling (Section II,A) and chemiosmotic (Section II,B) theories of energy transduction, on the concept of proton mobility and transfer (Section II,C), and on the ways in which electrochemical gradients across mitochondrial membranes may be used to drive ATP synthesis (Section II,D). For the sake of clarity in the presentation of the required experimental detail, Sections II,A–D are presented in distinctive type. It will be evident that the various concepts need to be integrated with each other and with the subcellular organization in which the phenomena occur, as summarized in Section II,E. Necessarily, the detail is intricate and involved, but it is nevertheless presented here as a summation of the current state of knowledge, incomplete though it may be.

A. THE CHEMICAL COUPLING THEORY

It was initially thought that energy coupling could be achieved through the actual transfer of a phosphate or other group from one molecule to another. The reactions of what is termed substrate-level phosphorylation, that is, the oxidation of a substrate with the concurrent phosphorylation of ADP by P_i, have been used, by analogy, as model systems for oxidative phosphorylation. Of the two relevant examples, the oxidation of glyceraldehyde 3-phosphate to 3-phosphoglyceric acid by the enzymes glyceraldehyde 3-phosphate dehydrogenase and 3-phosphoglycerate kinase and the oxidation of succinyl-CoA to succinate by succinate thiokinase, the former initially provided the model for oxidative phosphorylation. Although there is now little evidence for this viewpoint, the following account

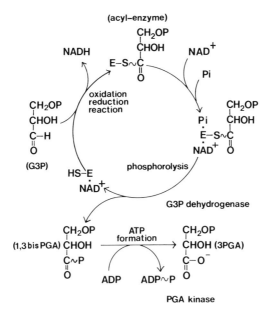

FIG. 1. Substrate-level phosphorylation by glyceraldehyde 3-phosphate dehydrogenase and 3-phosphoglycerate kinase. Not all intermediates are shown. The $\sim$ represents a bond with a high free-energy of hydrolysis; G3P, glyceraldehyde 3-phosphate; 3PGA, 3-phosphoglyceric acid; 1,3bisPGA, 1,3-bisphosphoglyceric acid. Modified from Duggleby and Dennis (127).

is presented to show the historical development of the topic. Nevertheless, the actual catalysis that occurs on the surface of the enzyme catalyzing ATP synthesis, the ATPase, has not yet been elucidated, and the possibility that chemical intermediates occur at the active site of the enzyme has not been ruled out. Some transitory enzyme-linked intermediates similar to those described later may still, therefore, be found.

A reaction mechanism for glyceraldehyde 3-phosphate dehydrogenase was postulated in 1953 by Segal and Boyer (436). This mechanism has been confirmed by kinetic analysis of the enzyme from both plants (127) and animals (322, 496; Fig. 1). In this scheme glyceraldehyde 3-phosphate is oxidized with the concomitant reduction of NAD. The product of the reaction, 3-phosphoglyceric acid, is not released but remains bound to the enzyme as a thioester, so that although most of the energy released during this oxidation is conserved as NADH, some is also conserved by retention of the thioester bond instead of its hydrolytic cleavage. The thioester bond is instead cleaved by a phosphorolysis reaction to form 1,3-bisphosphoglyceric acid rather than 3-phosphoglyceric acid, and the energy is conserved as a mixed acid anhydride bond between 3-phosphoglyceric acid and P_i. In the subsequent reaction, catalyzed by 3-phosphoglyceric acid kinase, P_i is transferred to ADP with the formation of ATP and 3-phosphoglyceric acid.

In the original concept of chemical coupling, Slater (453) suggested that during the oxidation of a component of the electron transport chain the energy released is trapped by

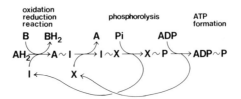

FIG. 2. Possible mechanism of oxidative phosphorylation as postulated by the chemical coupling theory. The ~ represents a bond with a high standard-free-energy of hydrolysis; A, B, components of the electron transport chain; I, X, postulated intermediates. Based on work by Slater and co-workers (453).

the formation of a bond between that component and a molecule that is not a part of the chain. This bond is analogous to the acyl thioester in substrate-level phosphorylation. To show the special nature of the bond so formed, it is written as ~ in various schemes. To accommodate the effects of known inhibitors and uncouplers, a series of reactions was proposed in which the energy is retained as ~ bonds until finally ADP is phosphorylated by a postulated high-energy intermediate, X ~ P, as shown in Fig. 2. In this diagram, A and B are respiratory chain intermediates, and I and X are unknown components that are not part of the chain. A detailed account of the development of this model has been given by Mahler and Cordes (302). Uncouplers were thought to act by hydrolyzing one of the intermediates, which was postulated to have a high free energy of hydrolysis (453).

The formulation of this scheme for chemical coupling led to an intensive search for phosphorylated intermediates that have a high free energy of hydrolysis. Although a number of compounds were suggested (e.g., phosphohistidine and protein acyl phosphates), none of them proved to have the properties required for an intermediate in oxidative phosphorylation (70). A candidate is oleoylphosphate, which has been postulated to be formed by a series of reactions involving lipoic acid termed the oleoyl cycle (171). The exact manner in which the oxidation–reduction reactions of the electron transport chain may be coupled to this cycle is not clear. Support for this scheme has been provided by the observation that isolated ATPase will synthesize ATP when supplied with dihydrolipoate and oleoyl-CoA or oleic acid in a reaction that is sensitive to uncouplers and oligomycin (170). ATPase has also been reported to catalyze the synthesis of ATP from oleoyl phosphate and ADP (238). It has been suggested that even in the purple membrane from *Halobacterium halobium*, which has been used extensively for reconstitution experiments in oxidative phosphorylation (see Section II,B,3), lipoic acid is present in the membrane and is involved in the phosphorylation reactions (172). However, a role for lipoic acid and unsaturated fatty acids in energy transduction is not generally accepted, and there is some evidence that, at least in *Escherichia coli*, they are not involved (451, 452).

Although the chemical coupling theory is no longer accepted by the majority of workers, other theories do not usually describe in the same detail the actual reaction mechanism at the active site. ADP and P must bind at the active site of ATPase and it is possible that a transitory phosphoryl–enzyme intermediate is formed that is too unstable to be isolated. Racker (392) suggested that a proton gradient or conformational change is involved in the reaction of ADP with a phosphoenzyme followed by the release of ATP. Although a phosphoenzyme has not been isolated in the case of mitochondrial ATPase, a phosphoenzyme has been demonstrated for the calcium-transporting ATPase from sarcoplasmic reticulum, which generates ATP on the addition of Ca^{2+} (391). Until the actual reaction mechanism at the active site of the proton-translocating ATPase has been elucidated there will continue to be controversy about the mechanism of energy transduction.

B. THE CHEMIOSMOTIC THEORY

The problem with chemical coupling theories is that they overlook the organization of the system in which energy transduction takes place. This is prominently so with respect to the role of membranes. The chemiosmotic theory, elaborated by Mitchell in 1961 (327) and more fully in 1966 (328), bridges this gap by its emphasis on the importance of the membrane.

The fundamental features of Mitchell's scheme is the separation of the actual phosphorylation process from the oxidation–reduction reactions of the electron transport chain. Phosphorylation has been postulated to occur on a "membrane-located, reversible, proton-pumping ATPase," a process later shown to be possible in the complete absence of electron transport (252). The ATPase can catalyze a hydrolysis of ATP to form ADP and P_i, but *in vivo* it catalyzes the reverse reaction, that is, the dehydration of ADP and P_i to form ATP and water. In addition to these activities, the ATPase will also pump protons during hydrolytic activity, and protons must pass through the ATPase during synthetic activity. The energy liberated by the flux of electrons through the electron transport chain was thought to develop a gradient of protons across the inner mitochondrial membrane, which is impermeable to protons. The discharge of this gradient through the ATPase releases the energy required for ATP synthesis. Therefore, the two components of the system are coupled, but only indirectly, through the proton gradient (Fig. 3).

The original concept of chemiosmosis (327) depends completely upon a gradient of protons or pH gradient (ΔpH) across the membrane. When this gradient builds up to sufficient magnitude the pressure of the gradient on the proton translocating ATPase becomes great enough to reverse the ATPase reaction and generate ATP. Conversely, when the ratio of ATP to ADP ratio is increased, the proton gradient required to generate more ATP must also increase, until electron transport is inhibited. In this way ATP synthesis is indirectly coupled to electron transport. The pH gradient calculated to reverse the ATPase reaction is larger than that measured in mitochondria. The theory was, therefore, modified to include not only a pH gradient but also an electrochemical gradient resulting from other ions (328), to which the membrane must also be impermeable.

1. Generation of a Proton Gradient

From these considerations Mitchell developed the concept of proticity (329, 333). Proticity is the transport of protons through aqueous media, in which they are highly mobile, and is analogous to electricity, the conduction of electrons in metals.

Because the inner mitochondrial membrane is generally impermeable to ions, the ionic composition inside mitochondria may differ considerably from that in the external medium. The protonmotive potential difference across the membrane is the sum of the electrical potential difference (Ψ) and the proton concentration difference ($-\Delta$pH). Thus the protonmotive potential (Δp) is equal to

$$\Delta p = \Psi - Z\,\Delta\text{pH},$$

where $Z = 2.303\,RT/F$, where F is the Faraday constant. The protonmotive force is, therefore, made up of these two components.

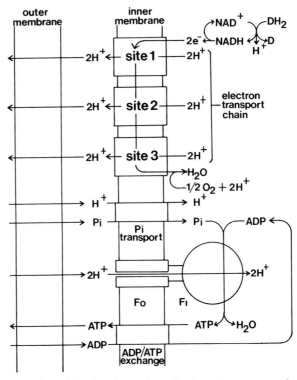

FIG. 3. Representation of the chemiosmotic mechanism of energy transduction in mitochondria. Three sites of proton translocation associated with the electron transport chain
are shown. The coupling factor complex (ATPase) is shown with a stoichiometry of two
protons for each molecule of ATP synthesized. The stoichiometries of proton translocation and the ratio of H$^+$ to ATP have been questioned in later work (see Section II,B,2).
Also shown are the phosphate translocator and the ADP–ATP exchange translocator.
Based on work by Mitchell (327, 328).

The actual synthesis of ATP is postulated to occur on an ATPase, which is located on the
inner mitochondrial membrane. The structure of this enzyme complex is described in
more detail in Section III,B. It consists of two parts: F_1, which has the ATPase activity, and a
membrane-located portion, F_0. There is a channel through F_0 that specifically translocates
protons (333). It has been suggested by Mitchell that this "proton well" in the ATPase
converts the electric membrane potential into a pH gradient (335), so that ultimately it is a
pH gradient that generates ATP.

The chemiosmotic theory requires that the mitochondrial membrane be impermeable to
protons. Under such circumstances there should be a strict stoichiometry between H$^+$
transported through the ATPase and ADP phosphorylated (P). That is, the ATPase
should have a strict quotient $\rightarrow$H$^+$/P, where $\rightarrow$H$^+$ represents the vectorial transport of
protons (i.e., transfer in one direction only). Similarly, the electron transport chain should
have a quotient $\rightarrow$H$^+$/2e$^-$ and the overall process of oxidative phosphorylation a quotient

P/$2e^-$, so that

$$\frac{P}{2e^-} = \frac{\rightarrow H^+/2e^-}{\rightarrow H^+/P}$$

The quotient $\rightarrow H^+/2e^-$ has been measured by Mitchell and Moyle (338, 340) and shown to be 2 for each coupling site (although values higher than this have been reported; subsequently; see Section II,B,2), which would require a $\Delta p = 250$ mV across each coupling region under mitochondrial conditions (333).

One of the most forceful arguments supporting the chemiosmotic hypothesis is that it offers a simple explanation for the mechanism by which phosphorylation can be uncoupled from electron transport. Compounds that are capable of translocating protons across the membrane dissipate the protonmotive force developed by the electron transport chain. These compounds are soluble in the membrane and must be able either to carry protons or provide a channel for proton flow. In the presence of an uncoupler the electron transport chain continues to transport electrons and translocate protons, but the uncoupler allows the protons to travel back across the membrane so rapidly that a protonmotive force cannot be built up; hence, ATP is not synthesized.

a. Redox loops. An important concept developed by Mitchell is that of vectorial processes in bioenergetics and in many other mechanisms as well [reviewed by Mitchell (337)]. All enzymes have distinct domains for binding substrates and products. When the enzyme (or enzyme complex) is incorporated into a membrane, the reaction can become vectorial or specifically directed. Thus the binding and release of some substrates and products may occur on opposite sides of the membrane resulting in a gradient of metabolites across the membrane. The components of the electron transport chain in mitochondria and chloroplasts are arranged across the inner membrane so that some reactions take place on the inside of the membrane and others on the outside (115). Because the membrane is relatively impermeable to most ions, a gradient can develop across the membrane due to electron transport.

Some of the components of the electron transport chain are electron carriers, whereas others are hydrogen carriers (328, 333). The hydrogen carriers in the mitochondrial chain include FMN and UQ and, in the photosynthetic chain, PQ. The remaining carriers are electron carriers. When a hydrogen carrier is reduced by an electron carrier it takes up protons, and, similarly, when it is oxidized by an electron carrier it releases protons. If these two reactions are vectorially arranged so that the reduction and oxidation reactions are placed on opposite sides of the membrane, there will be a net transport of protons across the membrane. Furthermore, any reaction that consumes or releases protons specifically on one side of the membrane will contribute to the proton gradient developed by electron transport.

Mitchell first proposed that the electron transport chain is organized in a series of loops in such a manner that the hydrogen carriers transport protons across the mitochondrial membrane (Fig. 4; 327, 328). This scheme envisaged two proton-translocating sites in the mitochondrion that are mediated by the hydrogen carriers FMN and UQ. The terminal oxidase, cytochrome oxidase, would utilize a proton for every electron passed on to oxygen and, hence, also would contribute to the proton gradient. A problem with the scheme in Fig. 4 is that there are only two proton-pumping loops in the chain, whereas experimental data indicate that there are three coupling sites. The missing loop is between cytochrome c and cytochrome oxidase. The stoichiometry of the scheme also indicates that fewer protons are translocated than are required for the known amount of ATP synthesized. In addition, the chain contains two b cytochromes, the function of which is not accounted for.

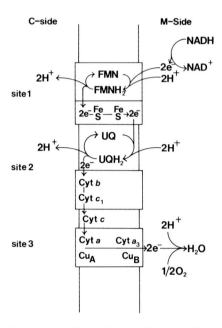

FIG. 4. Proton translocation associated with possible redox loops in the electron transport chain of the mitochondrion. C- M-side, cytosolic and matrix side of the mitochondrial membrane. Based on work by Mitchell (328, 333, 337).

To overcome these problems, the concept of the protonmotive ubiquinone cycle, or Q cycle, was introduced (331, 332); this cycle has been discussed in detail (334). An example of a Q cycle is shown in Fig. 5, where UQ is reduced by one electron from an iron–sulfur center and by a second electron from one of the b cytochromes. This reduction is presumed to occur on the matrix side (M side) of the membrane, and to complete the reaction two protons are absorbed from the matrix. The reduced ubiquinone, carrying two hydrogens, diffuses across the membrane to the cytosolic side, (C side), where it is oxidized by passing one of the two electrons to cytochrome c_1. This electron ultimately goes to oxygen via cytochrome c and cytochrome oxidase. The second electron is transferred to a different b cytochrome and finally back to oxidized UQ on the inside of the membrane via a second cytochrome b molecule. Because the oxidation of UQ occurs on the outside of the inner membrane, two protons are released to the outside of the mitochondrion. The UQ then diffuses back to the matrix side of the membrane to complete the cycle. During the operation of the protonmotive Q cycle, one electron is transported from an iron–sulfur center to cytochrome oxidase, the second electron cycles within the cytochrome $b-c_1$ complex, and in the process two protons are translocated across the inner mitochondrial membrane.

The protonmotive Q cycle provides the stoichiometry required for the amount of ATP synthesized as two electrons pass down the electron transport chain; it also completes a third site of proton transport and provides a function for the two b cytochromes. A similar

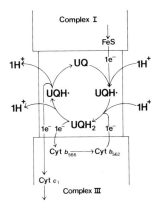

FIG. 5. Possible protonmotive Q cycle associated with ubiquinone. Modified from Mitchell (334).

protonmotive Q cycle has been suggested for the electron transport chain of chloroplasts, employing PQ in place of UQ (337).

Although the system of redox loops incorporating the protonmotive Q cycle is an ingenious suggestion for transporting protons at the required stoichiometry, there is no direct evidence for its presence in mitochondria, and there is increasing evidence that it either does not operate (e.g., 526) or is only one of a number of proton-translocating mechanisms. Topological studies of the mitochondrial membrane indicate that the arrangement of some electron transport components across the membrane does not agree with the proposed redox loops (115, 135). It is well established that cytochrome oxidase transports electrons from cytochrome c on the outside of the inner mitochondrial membrane to oxygen on the inside, but no hydrogen carrier is associated with this region of the chain. In the Mitchell scheme, cytochrome oxidase carries electrons from one loop of the Q cycle to oxygen. However, there is evidence that it may act as a proton transporter as well (525), although this view has been contested by Mitchell (346).

In Mitchell's redox loop scheme a stoichiometry of two protons per two electrons at each coupling site is required, and a ratio of two has been found by Mitchell and Moyle (see Section II,B,2). However, more recent measurements have suggested that the ratio may be as high as four protons per coupling site. Such high ratios are not in accord with a redox loop scheme and have led to the development of the concept of proton pumps as an alternative.

b. Proton Pumps. The concept of the proton pump was developed by Papa (373; reviewed in 144, 529). In this model it is suggested that protons are moved across the membrane by a vectorial Bohr mechanism analogous to the Bohr effect in hemoglobin. On oxygenation of hemoglobin, the pK_a values of the N-terminal valine of the α subunit and of a histidine in the β subunit become lower (321). At pH 7.0, protons dissociate from hemoglobin on oxygenation and reassociate on deoxygenation. This effect is important in the transport of both oxygen and carbon dioxide in the blood. The effect is mediated through a conformational change in the hemoglobin tetramer. It was suggested by Papa (373) that reduction of an electron transport component on the matrix side of the inner mitochondrial membrane might increase the pK_a of an acidic group, which would result in proton uptake. Upon oxidation the pK_a is lowered and a proton is released. These changes in the pK_a value may be due to a conformational change in the redox component. If the oxidation and reduction sites are separated by the membrane, there may be rotational movement of the redox component such that uptake and release could occur on opposite sides of the membrane.

Alternatively, a proton channel lined with acidic amino acids may traverse the membrane so that conduction through the membrane may occur by the transfer of protons via these amino acids. Again, the energy for this transfer may be derived from some type of Bohr effect. In addition to a proton channel across the membrane, there must also be a gate to prevent the back flow of protons. Such a gate is present in the proton channel of mitochondrial ATPase (see Section III,B). The transmembrane proton channel is well documented in bacteriorhodopsin (473), in which light absorption in the absence of redox reactions provides the energy to pump protons across the membrane (473). Bacteriorhodopsin has been used as a proton pump in reconstitution experiments (see Section II,B,3).

The evidence is now in favor of a direct proton-pumping mechanism being involved in the generation of a proton gradient. Redox titrations of the components of the electron transport chain show that the redox potential of some of the components changes as the pH is altered, indicating some interdependence between ionizable groups and the redox properties of the components that is similar to a Bohr effect (375). Oxidation of an electron carrier at pH values in the region of its pK_a could result in release of protons. Conversely, its reduction could cause proton uptake (375). Such uptake and release of protons has been demonstrated for cytochromes on reduction and oxidation (375).

2. The Role of a Proton Gradient in Energy Transduction

Jagendorf and co-workers demonstrated in elegant experiments performed with thylakoid membranes that a proton gradient is an obligatory intermediate in energy transduction. They first separated the light and dark processes of photophosphorylation by illuminating chloroplasts with a brief flash of light in the absence of ADP and P$_i$ (216). For short periods of time after the light flash, ATP can be generated if ADP and P$_i$ are added subsequent to the flash. The high-energy state of the chloroplast, generated in the light and termed X$_E$ is labile with a half-life of 1–2 sec. The ratio of ATP formed in the dark to chlorophyll content can be as high as 8 to 1, suggesting that X$_E$ is not identical to any of the known constituents of chloroplasts, such as cytochromes, for these are present only in much lower concentrations. At first there was no indication of the nature of X$_E$, but with the development of the Mitchell theory it was recognized that X$_E$ might be a proton gradient.

A major advance came when the link between a proton gradient and X$_E$ was, in fact, demonstrated. On illumination of chloroplasts, the pH of the external solution increases (354), a change that is reversed in the dark. The magnitude of the pH change is increased in the presence of the redox carrier pycocyanine, which implicates the electron transport chain in the pH change. Compounds that act as uncouplers of oxidative phosphorylation prevent the pH change, thereby correlating it with photophosphorylation. Also, an intact membrane is required, because the detergent Triton X-100 prevents the effect. These data demonstrate that a proton gradient is formed in light and that this gradient is correlated with the high-energy intermediate X$_E$. The pH shift during the light can be as high as one proton per chlorophyll molecule; this is much higher than the concentration of any known electron transport component in the chloroplast. The relationship between the pH shift, X_E, and photophosphorylation was developed further by Jagendorf and co-workers (217, 235), although interpretation is difficult because X_E is unstable and decays rapidly in the dark. The dark-decay reaction competes with ADP and P_i for X_E, so that under experimental conditions the phosphorylation step is only about 60% efficient (234). A further complicating factor is that the initial rate of X_E formation is slower than that of ATP formation in photophosphorylation (234).

An extension of this work, demonstrating ATP formation by chloroplasts entirely in the dark, shows that X_E is indeed a proton gradient. The inner space of the grana can be loaded by placing chloroplasts in an acid medium. Subsequent transfer of the chloroplasts to an alkaline medium generates a pH gradient capable of generating ATP in the dark (236). To ensure that light is not involved, the broken chloroplasts are first preincubated for 1–2 hr in complete darkness. To eliminate electron transport, DCMU is added as an inhibitor of noncyclic electron flow (see Section V,B,2), and no artificial redox dye is added to minimize cyclic electron transport. These precautions ensure that no ATP synthesis by conventional electron transport occurs. The chloroplasts are also incapable of photophosphorylation. The chloroplasts are then incubated at pH 4.0 for 60 sec and then rapidly transferred to buffer at pH 8.0 containing ADP and P_i. Under the alkaline conditions ATP is indeed synthesized. The yield of ATP can be greatly increased by including succinic acid in the acidic (pH 4) medium; the succinic acid that enters the thylakoids acts as a source of protons. A number of other organic acids may be used in the same way (499, 500), indicating that the effect is not due to their metabolism. The formation of ATP by this method is inhibited by uncouplers of photophosphorylation, and the decay of the high-energy state induced by the acid–base transition is the same as that of X_E formed by illumination. Also, the chloroplast membranes must be intact, with sufficient internal volume for acid storage (501). The highest yields obtained have been one ATP per four chlorophyll molecules. These experiments show that the high-energy state produced by acid–base transition, X_E, and photophosphorylation is driven by the same mechanism, namely, a proton gradient across a membrane. This constitutes one of the most effective demonstrations of the chemiosmotic coupling mechanism.

The chloroplast system was, for a time, the only one in which ATP synthesis could be driven by an artificially imposed pH gradient. However, in 1965 Mitchell and Moyle introduced the oxygen pulse technique for the measurement of proton transport in mitochondria (338). In this technique rat liver mitochondria are incubated anaerobically with an electron transport substrate such as hydroxybutyrate. Electron transport is initiated by the injection of small, measured amounts of oxygen, and the pH of the external medium is measured by a very sensitive glass electrode. Mitchell and Moyle (338) found that when oxygen is injected, an almost instantaneous drop in the external pH occurs, followed by a much slower return to the original pH. This pH drop is inhibited by an uncoupler and is prevented completely by Triton X-100, which destroys the integrity of the membrane. The number of protons transported per oxygen atom (i.e., the H/O quotient) is four for succinate and six for hydroxybutyrate. This corresponds to two protons per coupling site as required by the chemiosmotic theory. A reversal of this reaction can be driven by ATP hydrolysis catalyzed by the mitochondrial ATPase. In this case, using the system just described, ATP is injected anaerobically into the mitochondrial suspension in the presence of EDTA, and two protons are transported per molecule of ATP hydrolyzed. These data are in accord with the known P/O ratios of two and three for succinate and hydroxybutyrate, respectively. All this provides strong support for the chemiosmotic coupling mechanism.

That sonicated mitochondria remain active in oxidative phosphorylation was difficult to explain in terms of the chemiosmotic mechanism. It was shown, however, that sonicated mitochondrial preparations still consist of vesicles, but that these vesicles have been inverted during sonication. Such vesicles transport protons, but the polarity is reversed, and protons are transported into the vesicle with the same stoichiometry as they are transported outward in intact mitochondria (339). Later work has confirmed these findings for both electron transport (340) and ATP-mediated transport (341). The ratio of two H^+ per hydrolyzed ATP has been confirmed for inverted mitochondrial vesicles by other workers (488).

However, there is still disagreement about this ratio because proton transport is also associated with the uptake and exchange of ions and P_i, which affects the measured proton transport. For example, when protons are translocated from mitochondria, their electrical neutrality is maintained by the uptake of Ca^{2+} and efflux of Cl^-. Potassium can also perform this role in the presence of valinomycin. Brand et al. (76) repeated the original experiments by Mitchell and Moyle and demonstrated that when phosphate is present in the medium surrounding mitochondria, two protons are transported at each coupling site per two electrons passing through the site. However, when the uptake of phosphate into mitochondria is inhibited by NEM, the ratio per site increases to three, indicating that one proton is being transported inward with each inorganic phosphate ion. Part of the proton gradient developed by electron transport is, therefore, linked to phosphate transport into mitochondria. When mitochondria are maintained aerobically and electron transport is initiated by a pulse of reductant (e.g., succinate), a ratio of H^+ to site as high as four is found. Therefore, the ratio of two H^+ per site may not be correct because higher values are possible. This may reflect adversely on the concept of redox loops put forward by Mitchell.

Measurements by Lehninger and co-workers of proton efflux by mitochondria from various sources have confirmed that for every two electrons passing down the electron transport chain, four protons are translocated at each coupling site (509 and references cited). There is a simultaneous uptake of four K^+ or two Ca^{2+} charge-compensating cations per site (509). The measurement of proton extrusion at coupling sites 1, 2, and 3 (see Fig. 3) also yields a ratio of H^+ to $2e^-$ of four at each site, including the cytochrome oxidase site (21, 285, 389), but this measurement is not universally accepted (374). The passage of two electrons generates one ATP at each site (21, 389). A model for proton transport with a stoichiometry of four for coupling site 2 has been proposed that involves a cytochrome b dimer and a vectorial Bohr effect (511).

The use of an acid–base transition to generate ATP, as shown in chloroplasts, was eventually demonstrated with mitochondrial preparations (489). Inverted submitochondrial particles were used that had the same acid–base polarity as chloroplasts; that is, the proton concentration was higher inside the vesicles. In contrast to chloroplasts, efficient ATP synthesis requires valinomycin in the acid phase and K^+ in the base phase (489). The results indicate that a membrane potential as well as a pH gradient is required for high rates of ATP synthesis because a pH gradient alone produces 25% and a K^+ gradient alone 15% as much ATP as the two together (489). In this system uncouplers of oxidative phosphorylation and inhibitors of ATP synthesis effectively block ATP formation, but inhibitors of electron transport have no effect.

The kinetics of ATP synthesis during acid–base transition and during respiration-driven oxidative phosphorylation reveal that the rate of ATP synthesis is at least as fast during the acid–base transition as during normal oxidative phosphorylation (490). However, there is a lag in respiration-driven phosphorylation, presumably due to the time required to establish the electrochemical proton gradient, a lag that is increased by valinomycin. It has been suggested that the initial phase of ATP synthesis may be due to an electrochemical gradient rather than a pH gradient because valinomycin collapses the potential formed by ion movement but not the pH gradient (490). The initial ATP synthesis during photophosphorylation appears also to be driven by an electrical gradient that is established prior to a pH gradient (249) because the initial, but not steady-state, photophosphorylation is uncoupled by valinomycin (249). Similar results have been reported by Ort and Dilley (368), who suggest that initial photophosphorylation is dependent on an electrical as well as a proton gradient, but in the steady state only the proton gradient predominates.

3. Reconstitution Experiments

Proof of a correlation between ion transport and energy transduction has been difficult to obtain because energy-transducing systems require intact structures for ion movement and ATP synthesis. Much effort has been expended to develop an artificial membrane system that is based on a lipid bilayer and that has one or a minimal number of components so that each component may be characterized. The problem has two aspects—first, the demonstration that isolated single complexes of the electron transport chain can transport ions, particularly protons; second, that when a gradient is developed across a membrane, a specified component can discharge the gradient and generate ATP.

A remarkable breakthrough occurred in the development of artificial energy-transducing systems with the discovery of purple membranes in *Halobacterium halobium* (473). This bacterium is an extreme halophile and develops purple patches on the membrane when the cells are starved. The purple membrane contains a protein, bacteriorhodopsin, that is oriented across the bacterial plasma membrane. Upon light absorption, rapid transitory bleaching is observed, followed by a slightly slower recovery. The bleaching and recovery are associated with proton uptake and release, respectively. The orientation of the bacteriorhodopsin across the bacterial membrane generates a proton gradient by transport of protons out of the cell upon illumination. The discharge of this gradient through a membrane ATPase provides the energy to generate ATP by a chemiosmotic mechanism (356).

Starting with pure purple membranes of *Halobacterium halobium* (357), phospholipids from soybean (253), and the complete ATPase complex from beef heart consisting of F_1, F_0, and OSCP, which binds F_1 to F_0, (253), Racker and Stoeckenius (394) were able to compile a system that could phosphorylate ADP in the light.[3] Lipid bilayers can be formed from the phospholipids by dissolving them in a 2% solution of sodium cholate and then dialyzing the detergent away. When purple membranes are added to the solution before dialysis, they are incorporated into the bilayers. In the light these reconstituted vesicles cause an increase in the pH of the medium; the pH returns to its original level when the light is turned off, demonstrating that a proton gradient has been generated across the membrane. The hydrophobic protein F_0 from beef heart mitochondria is also incorporated into the reconstituted vesicles when it is added to the phospholipid solution. Then OSCP and ATPase (F_1) can be added to the vesicles to reconstitute the ATPase attached to the membrane. When such vesicles are illuminated in the presence of ADP and P_i, ATP is formed. Both the formation of the pH gradient and ATP generation are inhibited by uncouplers. These experiments were the first to demonstrate ATP synthesis in a system constituted entirely from other sources (for reviews, see 251, 395).

[3] For a discussion of the morphology of mitochondrial membranes and the structure of F_1, F_0, and OSCP, see Chapter 1 and Section III,B of this chapter. (Ed.)

C. The Membrane-Localized Proton Theory

Coincidental with the publication of Mitchell's theory of chemiosmosis, Williams put forward a theory of energy transduction that in some respects was similar to Mitchell's, but in other respects was quite different. Like Mitchell, Williams (528) suggested that the electron transport chain transported protons, but that they remain within the membrane. In chemiosmosis these protons are used to form a proton gradient across the membrane; that is, there is an osmotic component to energy storage. In Williams' theory the protons remain localized in the membranes and react directly with ATPase, and ATP is generated on the active site of the enzyme where the membrane-localized protons produce a pH low enough to reverse the normal ATPase reaction. In this case a proton gradient across the membrane is not essential to energy transduction, although it may be used to store energy (529), and artificially imposed gradients may be used to generate ATP via membrane-localized protons (529). In some respects the effect of localized protons on ATPase has much in common with the conformational coupling theory (see following discussion). In this theory, localized effects in membranes are thought to bring about a conformational change in the ATPase molecule. This could quite possibly occur through localized protons interacting with specific ionizable groups on the enzyme. The development of the theory of membrane-localized protons has recently been reviewed by Williams (530, 531), and the differences between chemiosmosis and the localized-proton theory of Williams has been debated in a number of reviews (336, 344, 532).

D. The Conformational Coupling Theory

The conformational coupling theory is based on the concept that an enzyme undergoes a conformational change when a substrate is bound or when a product is released. A conformational change may alter the active site of the enzyme in such a way that the reaction is promoted. An energy input (e.g., from an electron transport chain, a proton gradient, or by ATP hydrolysis) could also cause a conformational change in an enzyme, which would be detected as a change in the affinity of an enzyme for its substrate or product (i.e., a change in the tightness of their binding). The conformational coupling theory suggests that energy is required for a conformational change in the ATPase molecule, which allows substrate binding and product release in ATP synthesis.

The standard free energy of hydrolysis of the acid anhydride bond in ATP is large and negative; it has been tacitly assumed that the formation of the acid anhydride from P_i and ADP would require an energy input equivalent to that released on bond hydrolysis. In the overall process of ATP formation, that is, the formation of ATP in solution from ADP and P_i in solution, this statement must be thermodynamically correct. However, in the ATPase-catalyzed formation of ATP from ADP and P_i, a number of steps must be considered—substrate binding, bond formation, and product release. The energy input into the overall process could occur at any of these steps and not necessarily just at bond formation. (The term "bond formation" is used loosely here to describe the overall process of formation of ATP from ADP and P_i and addresses the rearrangement of a number of bonds in these molecules. The actual formation of a single bond must, of course, release energy.) It was a major breakthrough in thought to suggest, as did Boyer's (73) and Slater's (187, 188) groups in 1973, that the energy-requiring step in both mitochondria and chloroplasts is not the formation of the acid anhydride bond but the release of ATP from ATPase.

Mitochondria and submitochondrial particles, in the absence of complete oxidative

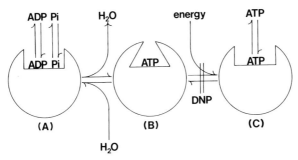

FIG. 6. Diagrammatic representation of conformational changes that might occur in coupling factors, explaining the data from exchange reactions. Based on work by Boyer *et al.* (73).

phosphorylation, will catalyze three exchange reactions that can be followed by isotope incorporation. The three reactions are

$$H—O—H \rightleftharpoons P_i \qquad (1)$$
$$H—O—H \rightleftharpoons ATP \qquad (2)$$
$$P_i \rightleftharpoons ATP \qquad (3)$$

Reactions (1) and (2) can be followed by the incorporation of ^{18}O from water into P_i and ATP, respectively, and Reaction (3) can be followed by measuring $[^{32}P]P_i$ incorporation into ATP. The reaction mechanism illustrated in Fig. 6 for ATP synthesis has been proposed to account for these exchange reactions. In this mechanism P_i in the medium can exchange with P_i on the enzyme surface, as can ADP (Fig. 6A). At the active site the P_i and ADP can react to form water and ATP. The water readily exchanges with the water in the medium, whereas the ATP that is formed is tightly bound (Fig. 6B). It was postulated that energy is required to change the conformation of the coupling factor to a form in which the ATP is loosely bound (Fig. 6C).

To support this model Boyer and co-workers showed that Reaction (1) has an absolute requirement for ADP, the K_m for ADP in this reaction being the same as that in oxidative phosphorylation. This exchange reaction is inhibited by oligomycin, a specific inhibitor of oxidative phosphorylation (244). Reaction (1) is much more rapid than the other two exchange reactions, suggesting that the rate-limiting step is the release of bound ATP (417). Most important, Reactions (2) and (3), but not Reaction (1), are inhibited by uncouplers of oxidative phosphorylation (e.g., DNP), which inhibit at the point of energy input; this indicates that energy is required for ATP release but not for its synthesis.

A more detailed analysis of Reaction (1) has shown that the scheme just described is oversimplified. By labeling either H_2O or P_i with ^{18}O, Boyer and co-workers were able to show that Reaction (1) has two components. One, occurring on the surface of the enzyme, is a fraction of the total reaction and is not sensitive to uncouplers. A second component involves P_i binding to the enzyme and is sensitive to uncouplers, indicating that energy is required for P_i binding as well as for ATP release (256, 417). The energy input for P_i binding in these exchange reactions is provided by ATP hydrolysis (417). As has been predicted from these results, the K_m for P_i binding increases considerably when an uncoupler is present, indicating that under conditions of energy input P_i is more tightly bound

(255). However, the K_m for ADP binding is also increased by an uncoupler, suggesting an energy requirement for the tight binding of both substrates (255).

From these and other data a concept of energy transduction by conformational change was developed that incorporates two active sites on the ATPase complex. In this concept a conformational change of the whole complex, brought about by an energy input either through ATP hydrolysis or electron transport, causes a tight binding at one site and loose binding at a second site (255). The sequence of events begins with the binding of ADP and P_i to a loose binding site; energy input causes a conformational change in the ATPase complex that results in this site becoming a tight binding site. ATP is formed but is not released until a second conformational change causes the site to become a loose binding site. Parallel with this sequence are alternating reactions at a second site [reviewed by Boyer (70, 71)].

A similar concept to that of Boyer was developed simultaneously by Slater and co-workers using a completely different approach. They demonstrated that isolated mitochondrial ATPase (F_1) contains tightly bound ATP and ADP that cannot be removed by repeated washing and precipitation (187). The bound nucleotides exchange only slowly with nucleotides in solution, but they can be released by cold inactivation of the enzyme. From their analysis of the data Slater and co-workers suggested that the equilibrium position of the ATPase reaction (ATP + $H_2O \leftrightharpoons$ ADP + P_i) can be altered by tight binding. Similarly the ATPase from chloroplasts (CF_1) has 2 mol ATP and 1 mol ADP per mole of enzyme, all of which are tightly bound and cannot be exchanged with external nucleotides (188). However, exchange of these bound nucleotides occurs readily when the thylakoid membranes are illuminated, suggesting that energy is required to change the conformation of the ATPase, which in turn changes the affinity of the ATPase for nucleotides (454).

The concept that the release of tightly bound, newly synthesized ATP is the principal energy-requiring step in photophosphorylation has been further explored by Boyer's group (461). They developed a rapid mixing and quenching technique that takes advantage of the acid–base transition method for ATP synthesis described by Jagendorf and Uribe (236) and demonstrated that after an initial lag of 3–7 msec [^{32}P]P_i incorporation into [^{32}P]ATP proceeds linearly. There appears to be no phosphorylated intermediate in this process, although a slow side reaction, in which newly formed [^{32}P]ATP donates the ^{32}P to AMP, gives a small amount of labeled ADP.

In a further refinement of this technique, Smith and Boyer (460) combined an acid–base transition with rapid filtration to show that a small amount of rapidly labeled [^{32}P]ATP is tightly bound to thylakoid ATPase. As in the mitochondrial system, the ADP and P_i that are bound in the ATPase are committed to ATP synthesis. ADP bound to the thylakoid membrane can be exchanged with the medium when the membrane is energized, which supports the concept of alternating sites.

An actual change in conformation of chloroplast ATPase during energy conservation has been detected by the incorporation of tritium from ^3H_2O into the ATPase (422, 423). The incorporation requires energization of the membrane by ATP, light, or an acid–base transition. It is prevented by inhibitors that destroy the high-energy state. A correlation is found between the extent of the conformational change and the high-energy state. These conformational changes have also been detected by the release of tightly bound nucleotides from the ATPase either by an acid–base transition (228) or by subjecting the ATPase to a transmembrane electric field (165). In the latter work it was found that very similar results can be obtained by continuous light, saturating light pulses of 30 msec, or external voltage pulses of 30 msec. The transmembrane potential in all cases is the same, but in continuous light the principal component of this potential is a pH gradient, whereas in the voltage pulse the pH gradient is zero and the potential is due entirely to an electrical potential difference across the membrane. The light pulse is intermediate between the two.

These data indicate that a conformational change can be caused solely by an electrical field across the membrane, and an electrical field can be developed across a membrane by an electron transport chain. It is possible that the electrical field causes a localized pH gradient, and Boyer (69) has suggested a model by which a conformational change of the ATPase could be caused by a local pH gradient. Development of theories of conformational coupling have been reviewed (72, 74).

E. GENERAL COMMENTS AND SUMMARY

The preceding pages have described a number of theories and mechanisms by which energy transduction might take place. Some of these overlap, and the true situation may involve aspects of several or all of these mechanisms. What follows describes some of the ways in which the operation of these mechanisms may be integrated with each other and with the organization of the physical systems in which they operate.

An aspect of both oxidative and photophosphorylation that needs to be emphasized is that ATP synthesis occurs in the specialized environment of the mitochondrial inner membrane or the chloroplast thylakoid membrane, and not in free solution. The value of the free-energy change required for ATP synthesis is usually calculated for the components in solution at neutral pH, and it has been estimated to be as high as 12 kcal/mol in the mitochondrial matrix. Under normal cellular conditions, therefore, ATP hydrolysis is strongly favored. However, under different conditions such as a hydrophobic environment and low pH, the free-energy change might be quite different and could actually favor ATP synthesis. Such conditions may exist at the active site of the ATPase when it is attached to the membrane. No more would be required for the phosphorylation of ADP than the correct orientation of the substrates and the availability of surface groups to catalyze a concerted reaction. Boyer's concept that no energy is required for this step in ATP synthesis is therefore reasonable.

A conformational change would, however, be required to change the active site of the ATPase so that the newly synthesized ATP could be liberated. The change must alter the active site of the ATPase from an enclosed hydrophobic environment that favors synthesis (i.e., dehydration) to one that is in contact with the normal cellular environment, which favors ATP hydrolysis. Energy input must take place at this point to cause the conformational change that results in a shift from tight to loose binding of ATP, and this is the phenomenon that is measured. The energy so supplied must, of course, be equivalent to the free energy of the hydrolysis of ATP under cellular conditions. This energy is not used directly in the synthesis of ATP, but to change the environment of the active site of the ATPase from one that favors synthesis to one that favors hydrolysis.

Another point requiring consideration is that the internal space of a mitochondrion is so small that the normal colligative properties of solutions may not apply. This may be important when considering the role of protons in energy transduction. Assuming that the volume of a mitochondrion is approximately 1 μm^3, one can calculate that at pH 7.5 the internal space contains approximately 20 hydrogen ions. On the further assumption that only 25–50% of the internal space is matrix, the number of free protons in the matrix will be only 5 to 10. The total pool of protons in the mitochondrial matrix available for proton transport is, of course, much larger due to the buffering capacity of the organic acids. However, electron micrographs of the inner membrane suggest that many more ATPase molecules and, presumably, proton-translocating complexes are present than are free protons. The situation is, therefore, very different from that of a normal *in vitro* enzyme reaction, in which the enzyme is free in a larger volume of solution, at a concentration usually at least 10^3-fold lower than its substrate. Thermodynamic calculations of energy derived from gradients are valid for large volumes of dilute solutions at equilibrium. However, in mitochondria or chloroplasts, which have small volumes, surface effects, and localized high concentrations must be taken into account.

In summary, it may be assumed that the synthesis of ATP takes place approximately as follows. Conformational changes that result in the formation of ATP may reasonably occur as the result of the passage of protons through ATPase. In this context, the concept of proticity developed by Mitchell is useful to explain the behavior of a highly mobile charge in the confined space of a mitochondrion. The transport of protons through ATPase must be driven by a potential difference between the inside and the outside of the membrane that is equivalent to the free energy required for ATP synthesis. This potential difference is presumably maintained by the proton-pumping action of the electron transport chain.

The exact details of ATP synthesis will not be fully known until the mechanism at the active site of the ATPase has been determined (see Section III,B). This will require a detailed knowledge of the structure and interaction of the subunits of the ATPase complex, knowledge that is now being rapidly accumulated. There is little doubt that the discharge of a proton gradient through ATPase is the energetic input and that a conformational change is involved. But how the passage of protons through the complex achieves a conformational change, the nature of the conformational change, and its affect on the active site remain to be elucidated.

III. Mitochondrial Electron Transport and ATP Synthesis: Background

A. THE ELECTRON TRANSPORT CHAIN

1. Introduction

Much of the recent research on mitochondrial energetics has been done with systems other than plants. Probable reasons for this are the larger number of scientists, greater research pressures for animal and bacterial studies, and the greater difficulty of working with plant material. Whatever the reason, it seems appropriate in the following sections to give first, as background, an account of our present understanding of the electron transport complexes based on animal and microbial studies (Section III) and to follow this with a description of the situation in plants insofar as the data permit and in the light of the information from nonplant studies (Section IV). Because the basic organization and operation of plant and animal mitochondria are probably similar, it is hoped that the following analysis will provide a useful basis for the understanding of plant mitochondrial energetics.

One of the most fascinating and puzzling aspects of both mitochondrial and photosynthetic electron transport chains is their complexity. The oxidation of NADH by oxygen involves a standard redox potential change of 1.14 V, equivalent to a standard free-energy change of 52.7 kcal/mol of NADH oxidized. The standard free energy of hydrolysis of ATP is 7.3 kcal/mol. However, when one considers the ratios of the concentrations of NAD^+ to NADH and ADP to ATP found in the mitochondrion, the actual free energy of oxidation of NADH can be calculated to be of the order of 45–50 kcal/mol, and the free energy of hydrolysis of ATP approximately 12 kcal/mol. It has been well established that three molecules of ATP are synthesized for every two electrons transferred from NADH to oxygen, so that three redox potential changes in the chain, each approximately 250 mV or 15 kcal/mol, are required for ATP synthesis. There seems to be no evident reason why the chain could not operate with three redox components, each one corresponding to one of the coupling sites and catalyzing an electron transfer reaction that releases enough energy for ATP synthesis.

In fact the mitochondrial electron transport chain is much more complex and contains at least 21 components, as shown in Fig. 7. This degree of complexity may be partly due to use of energy in the mitochondrion not only for synthesis of ATP but also for transport. In addition, the actual mechanism of ATP synthesis is complex, requiring several distinct processes including proton translocation into and out of the mitochon-

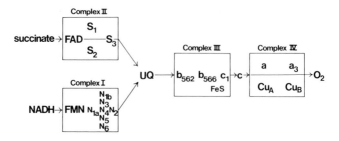

Fig. 7. Components of the electron transport chain in animal mitochondria and their arrangement into the four complexes. See text for detailed explanation.

drion (Section II). The interaction between the energy-tranducing components may also require intermediates because not all redox carriers can interact readily and transfer electrons.

One concept to account for the large number of components in the chain has been suggested by Chance (86, 87). The redox components of the mitochondrial electron transport chain can be divided into four groups depending on their midpoint potentials; the components of each group have similar potentials. The redox span between each group is of magnitude sufficient to allow the synthesis of an ATP molecule for every two electrons passing between the groups. Within each group, the rate of transfer of electrons between the components of that group or pool is rapid, much more rapid than that between the pools. Each group contains four to five components that can be said to be isopotential or equipotential.

In the electron transport chain the individual redox components are aligned so that they can transfer electrons only between designated molecules. A concept such as midpoint potential (i.e., the potential when half of the molecules are reduced when measured in free solution) has, therefore, only a limited meaning. The midpoint potential does indicate the affinity of a single component for electrons, but in the chain it is either fully oxidized or fully reduced. If, however, there are redox pools of several components with electrons in rapid equilibrium between them, then the addition or removal of an electron from the pool will not significantly affect its potential, and defined gaps required for ATP synthesis can be maintained. These isopotential pools have been referred to as redox ballast by Chance (87). Within the electron transport chain there must be high specificity in the electron transfer reactions to prevent short-circuiting of the gaps between the isopotential groups and to inhibit transfer of electrons from one group to the next without ATP formation (87). The concept of redox ballast can explain the number of redox components but not their variety. That is probably associated with the mechanism of energy transduction.

A second approach to the study of the electron transport chain is more biochemical. The inner and outer membranes of the mitochondrion can be separated, and the inner membrane further resolved into a series of complexes that catalyze segments of the electron transport chain. This fragmentation involves the use of mild detergents or chaotropic agents (agents that disrupt the structure of water and destabilize hydrophobic bonds). The first separations of the components of the chain were achieved by Green and Hatefi (168, 169, 194), and early achievements in fractionation of the chain have been reviewed by Hatefi (189). The inner mitochondrial membrane can be distinguished into five complexes, four of which (I–IV) contain components of the electron transport chain; a fifth complex, still occasionally designated complex V, is the coupling factor complex (ATPase).

The electron transport chain complexes catalyze the following reactions.

Complex I. NADH : ubiquinone oxidoreductase

$$NADH + UQ + H^+ \rightleftharpoons NAD^+ + UQH_2$$

Complex II. Succinate : ubiquinone oxidoreductase

$$Succinate + UQ \rightleftharpoons fumarate + UQH_2$$

Complex III. Dihydroubiquinone : cytochrome c oxidoreductase

$$UQH_2 + 2 \text{ ferricytochrome } c \rightleftharpoons UQ + 2 \text{ ferrocytochrome } c + 2 H^+$$

Complex IV. Cytochrome c : oxygen oxidoreductase (Cytochrome oxidase)

$$2 \text{ Ferrocytochrome } c + 2 H^+ + \tfrac{1}{2} O_2 \rightleftharpoons 2 \text{ ferricytochrome } c + H_2O$$

Complexes I–IV occur in a $1:1:1:1$ ratio, and the arrangement of electron transfer components in these complexes is shown in Fig. 7. Although the complexes are normally drawn in a specific order, they are in fact embedded in the lipid bilayer of the inner membrane and are regarded as mobile units in the fluid mosaic model of the membrane. Ubiquinone and cytochrome c act as mobile carriers between the complexes.

The isolated complexes will catalyze the individual reactions of the chain, and when two or more complexes are added together they will reconstitute larger segments of the chain. For example, NADH oxidation with the concomitant reduction of cytochrome c can be achieved by combining complexes I and III (189, 193). The highest efficiency is

obtained when the complexes are present in equimolar ratios. Methods
have been developed for the routine purification of the four electron
transport complexes from a number of sources (190), but not from
plants yet. Methods have also been described for the reconstitution of
the entire chain (193).

Examination of the components of the individual complexes shows
that they bear little relationship to the equipotential pools of redox ballast described previously, which might be expected. All the complexes
except complex II are sites of energy transduction; that is, they are sites
involved in the process of ATP synthesis. The complexes must, therefore,
contain within them the redox gap between isopotential pools. Each
isopotential pool must therefore have components in more than one
complex. The larger number of compounds found in each complex is
probably a result of its function in energy transduction; it may be acting
as a proton transporter. Proton transport in mitochondria has been
reviewed (144). The details of complexes I–IV are presented in distinctive type in Sections III,A,2–5, and the summary and conclusions are
given in Section III,A,6.

2. Structure and Properties of Complex I

Complex I (NADH : ubiquinone oxidoreductase) catalyzes the transfer of electrons from
NADH to UQ. The structure and properties of this complex have been reviewed (195,
398) and methods described for its purification (191). The structure, size, and mechanism
of action of the complex have been difficult to determine because its components are
highly insoluble and detergents and chaotropic agents tend to denature it (398). The
molecular weight of the complex is between 670,000 and 890,000, and it is made up of at
least 16 polypeptides (397, 398). Methods for the resolution of the complex into active
fractions are being developed (462). Some of the polypeptides are embedded in the membrane and some exposed on the surface (154). Associated with a single FMN molecule in
the complex are two polypeptides with molecular weights of 53,000 and 26,000. Also
associated with the complex are 20 nonheme iron atoms and a similar number of labile
sulfur atoms that make up the iron–sulfur centers of the complex (195, 398). Each iron–
sulfur center may contain 2–4 iron and sulfur atoms. On isolation, complex I will not
readily catalyze the transfer of electrons from NADH to UQ, but this activity can be
restored with phospholipids (400).

The nature and number of the iron–sulfur centers that form the bulk of the complex
have been determined by EPR spectroscopy, the EPR signal due to an unpaired electron in
the reduced form of the center. When the complex is titrated at low temperature ($<25°K$)
against a series of redox buffers with differing potentials, a signal is obtained from a center
when the potential of the buffer is sufficiently negative to reduce it. This technique has
been used by Ohnishi (358) and other workers (367) to distinguish the various iron–sulfur
centers in the complex.

The complex may contain at least six iron–sulfur centers, although there is disagreement on the exact number and their nature. The centers are numbered N-1 to N-6, where
N indicates that the center is in complex I (i.e., it is associated with NADH oxidation).
When complex I is titrated with redox buffers, center N-1 can be resolved into two entities
with redox potentials of -231 mV (center N-1b) and -396 mV (center N-1a). The most

positive center is N-2 with a midpoint potential of -20 mV; it is the only one that can be reduced by succinate. Centers N-3, N-5, and N-6 have midpoint potentials of approximately -250 mV. Center N-4 is somewhat anomalous because in intact mitochondria it has a potential of -245 mV, whereas in submitochondrial particles it has a potential of -410 mV. This suggests that the potential is very dependent upon the environment in which the center is embedded. As there are numerous iron–sulfur centers, and because of their relationships, it has been suggested that two forms of complex I may be present (1).

It has long been known that the segment of the electron transport chain represented by complex I contains one site for ATP synthesis, but the mechanism by which the energy released from electron transfer through the complex is made available for ADP phosphorylation has not been elucidated. It has been proposed by Mitchell that one redox loop for the transfer of protons operates in this segment of the chain (Section II,B,1,a), and the arrangement of subunits of complex I is consistent with the presence of a redox loop (462).

Measurement of proton transport associated with this complex (i.e., site 1) has indicated that the ratio of H^+ to $2e^-$ is four (285, 389), which is the same as the ratio measured for coupling sites 2 and 3. This indicates that some form of proton pump is operating in addition to a redox loop.

Purified complex I incorporated into phospholipid vesicles will oxidize NADH and reduce UQ (399), a process coupled to vectorial proton transport with a ratio of H^+ to $2e^-$ approaching two. This proton transport is sensitive to uncouplers. Complex I can therefore act as a proton-transporting system, confirming *in vivo* results. It has been suggested that a very hydrophobic polypeptide with a molecular weight of 33,000 may act as a proton channel (397). If a vectorial Bohr mechanism is operating in this complex, proton transport may also be mediated by protolytic equilibria in the iron–sulfur centers (115, 373). The complex is globular and large enough to span the membrane, so a proton-transporting channel is quite possible.

In contrast to the proton-transporting schemes for ATP synthesis associated with complex I, Ohnishi has proposed a direct involvement of the iron–sulfur centers in phosphorylation (359). Measurements of the midpoint potentials of the iron–sulfur centers by EPR spectroscopy have shown that two of them, N-1a and N-2, change their potentials in the presence of ATP; center N-2 becomes more positive by 125 mV and center N-1a more negative by 60 mV. The other iron–sulfur centers are unaffected by ATP. These results suggest that center N-2 may exist in liganded and nonliganded forms and that the addition and removal of the ligand from the oxidized and reduced form of N-2 is directly linked to ATP synthesis. However, no direct evidence for a liganded form of N-2 is available. A theoretical analysis of the EPR data that supports this scheme has been given by DeVault (118).

It is clear that complex I has many components, and it seems that each must have a specific function. Simple schemes for energy conservation at this site would not require this degree of complexity, especially the multiplicity of iron–sulfur centers. The complex is undoubtedly able to transport protons, and it seems probable that this is a function of the iron–sulfur centers acting as a proton pump or more directly in the phosphorylation process as suggested by Ohnishi (359). Details of the actual mechanism of energy transduction in complex I will not be clear until its structure and function are resolved.

3. Structure and Properties of Complex II

Complex II (succinate : ubiquinone oxidoreductase) catalyzes the transfer of electrons from succinate to ubiquinone with the formation of fumarate. The structure and properties of this complex have been reviewed (195, 450), and methods are available for its purification (196) and resolution into functional units (192). The molecular weight of the complete complex (i.e., the functional unit that can react with succinate and transfer electrons to UQ in the electron transport chain) has been reported to be 200,000 for a number of preparations. Purified complex II will recombine with complexes III and IV to allow electron transfer from succinate to oxygen. Succinate dehydrogenase, the enzyme component of the complex that oxidizes succinate, can be solubilized and separated by chaotropic agents and comprises 50% of the total protein of the complex (192). This component is composed of two subunits with a total molecular weight of approximately 97,000 (112). The large subunit contains nonheme iron and is a flavoprotein with a molecular weight of 70,000. The other is an iron–sulfur protein with a molecular weight of 27,000. The flavin in the large subunit is FAD, which is located close to a thiol group at the active site of the enzyme. The FAD is covalently bound to the succinate dehydrogenase through a histidine residue to give an 8-α-histidyl–FAD link at the active center of the enzyme (262, 513). There is a second thiol on the small subunit that is not part of the active site (84).

In addition to the two subunits of the succinate dehydrogenase, two small subunits with molecular weights of 13,500 and 7,000 are associated with complex II (548). That complex II, but not purified succinate dehydrogenase, can interact with complex III implicates one or both of these smaller subunits in binding the complex to the membrane (548). The larger of the small polypeptides may be necessary for the reaction with UQ or for the transfer of electrons to complex III (320). Complex II spans the mitochondrial membrane with the active site, containing FAD, facing the mitochondrial matrix (266, 513). The larger of the nonenzymatic subunits is associated with the cytoplasmic side of the membrane, and it is probable that these subunits form the major part of the bilayer-intercalated portion of complex II and hence are available for interaction with the lipophilic UQ molecules (266).

Complex II also contains three iron–sulfur centers that have been termed S-1, S-2, and S-3. S-1 and S-2 are present in amounts equal to FAD and exhibit EPR absorbance in the reduced state (361), with midpoint potentials of 0 and -260 mV, respectively. There are indications that these two centers are closely associated (361). Center S-1 can be reduced by succinate directly. This implicates it in the reaction mechanism of the complex. S-2 can only be reduced by dithionite, so its role is as yet unknown (361).

The third center, S-3, is a special type of iron–sulfur center in which the EPR signal is obtained from the oxidized, not the reduced, form. Centers such as S-3 have a high midpoint potential and are termed HiPIP-type centers (369). Unlike other iron–sulfur centers that contain two labile sulfur and two nonheme iron atoms, HiPIP centers contain four and sometimes eight of each of these atoms. It is thought that the HiPIP center in complex II is made up of four labile sulfur and four iron atoms. Iron–sulfur center S-3 is highly labile, so it is difficult to determine its concentration; estimates range from 1 per FAD (360) to less than 0.5 (32). The midpoint potential has been estimated to be 60 mV (360). In inactive complexes, the EPR signal of HiPIP is not observed, suggesting that it plays a role in electron transfer from succinate to UQ.

The location of the iron–sulfur centers in the subunits of the complex has not been definitely determined. It has been suggested that S-3 is located on the small subunit of the succinate dehydrogenase and is involved in the transfer of electrons from FAD to UQ (32, 360). The two centers S-1 and S-2 may be located on the large subunit. In complex II, one is again faced with a multiplicity of components; four subunits and four redox centers are involved in the electron transfer process. Unlike the other complexes, complex II is not involved in energy transduction, so the components are not needed for ATP synthesis or proton pumping. It would seem reasonable to assume that the components are required for the transfer of electrons from the aqueous phase of the mitochondrial matrix to the lipid phase of the membrane where the interaction with UQ takes place.

Membrane-bound and soluble succinate dehydrogenases undergo reversible activation and deactivation (192). The enzyme is activated by a number of compounds or conditions, including succinate, phosphate, ATP, UQH_2, succinyl-CoA, chaotropic agents, acidic pH, and phospholipids (195, 450). The activation process appears to involve the removal of tightly bound oxaloacetic acid, a powerful inhibitor of the enzyme. The activation and deactivation of the enzyme and the mechanism of these processes have been extensively reviewed (195, 450), as has the physiological significance of this effect (450).

4. Structure and Properties of Complex III

Complex III (ubiquinone : cytochrome c oxidoreductase) catalyzes the transfer of electrons between UQ and cytochrome c. The structure and properties of this complex have been reviewed (415). It has a molecular weight of approximately 250,000 in beef heart mitochondria, and negatively stained preparations show the complex to be roughly spherical with a diameter of 75–90 Å. The complex contains cytochrome c_1, cytochrome b, nonheme iron, iron–sulfur protein, antimycin-binding protein, and core protein in the ratio 1 : 2 : 2 : 1 : 1 : 2, respectively (415). The proteins of the complex have been isolated and purified.

It is well known from spectral studies that there are at least two and possibly three b cytochromes in complex III because there are three α-band maxima at 558, 562, and 566 nm (415, 515, 522). The absorbance maxima at 558 and 566 nm may be due, however, to a single cytochrome b molecule. Two cytochrome b species have been isolated from beef heart mitochondria and have molecular weights of 37,000 and 17,000 (547), but the smaller protein may be a breakdown product. These proteins are very hydrophobic and are synthesized within the mitochondrion (515) and coded by the mitochondrial genome. The exact number and nature of the b cytochromes are not yet known, although the data are most consistent with two molecules placed asymmetrically in the membrane with cytochrome b_{566} on the intercristal surface of the membrane and cytochrome b_{562} on the matrix side. In this arrangement cytochrome b_{562} would accept electrons from UQ and pass them to cytochrome b_{566}.

Cytochrome c_1, unlike cytochrome c, is an integral part of complex III and is released only by anionic detergents and chaotropic agents (497). Electrons are transferred directly from cytochrome c_1 to cytochrome c, and there may be protein–protein interactions between these molecules. Cytochrome c_1 has been purified from beef heart mitochondria and shown to have a molecular weight of 30,600. It has been proposed that cytochrome c_1 is present in the membrane in an orientation that exposes acidic residues to the surface. These residues may be involved in an ionic interaction with cytochrome c. Purified cytochrome c_1 is susceptible to trypsin digestion, but the membrane-bound molecule is not. This shows that it is an integral part of the complex (497). The midpoint potential of cytochrome c_1 is 230 mV.

The iron–sulfur center in complex III was first investigated by Rieske (415); it can be removed much more readily from the membrane than can the cytochromes, and it may be peripheral to the main chain. It appears to be in rapid electronic equilibrium with cytochrome c_1, indicating that it is at the cytochrome-c side of the complex. Various values for the midpoint potential have been proposed, from 180 to 250 mV.

Antimycin A is a powerful inhibitor of the action of complex III, inhibiting the transfer of electrons between cytochrome b and c_1 as well as that between cytochrome b and UQ. The binding of antimycin is stoichiometric with cytochrome c_1; that is, there is one binding site in the complex, but the nature of this site is not known. Rieske has reported the isolation of a protein with antimycin-A-binding properties (415). It has also been suggested that antimycin A binds to cytochrome b or to the iron–sulfur protein. The dissociation of the complex by bile salts is inhibited by antimycin A, suggesting that it affects conformation.

Complex III is the second site for energy transduction, but the mechanism by which this occurs is unknown. Four mechanisms have been proposed, namely, protonmotive coupling, energization of cytochrome b_{566}, coupling through the antimycin-A site, and conformational coupling. Submitochondrial particles, on the addition of UQH_2, will transport electrons to cytochrome c, and in the process protons are transported into the particle with a ratio of H^+ to $2e^-$ of four (283, 376). Similarly, isolated complex III will also transport protons when incorporated into phospholipid vesicles (287). In order to account for the observed stoichiometry, Mitchell modified his earlier proposal for the second coupling site and introduced the protonmotive Q cycle for this complex (Section II,B,1). This model takes account of the topography of cytochromes b_{566} and c_1, which are located on the intercristal side of the membrane, and cytochrome b_{562}, which is on the matrix side. In this case the proton translocator is UQ, which is reduced on the matrix side of the membrane. The mechanism of action of UQ as a proton translocator and its function in interconnecting the various complexes has been reviewed (177, 496a).

In contrast, Papa has suggested a vectorial Bohr mechanism for this complex (376, and see Section II,B,1,b). In this mechanism a metalloprotein in the process of transferring electrons also transports protons across the membrane. Such a metalloprotein has not been isolated from the complex, and the other components appear not to have the properties required to act in this manner. However, the concept of specific proton transporters appears to be gaining acceptance to account for the observed ratios of H^+ to $2e^-$ in all parts of the chain (76).

The stoichiometry and nature of proton transport at coupling site 2 have been explored more fully by Lehninger and co-workers (2, 510). They have presented strong evidence for a ratio of H^+ to $2e^-$ of four as electrons flow from succinate to ferricyanide (21, 510 and references cited therein). In ascites tumor mitochondria calcium ions are taken into the mitochondria with a ratio of Ca^{2+} to $2e^-$ of one; the calcium ions act to balance the charge developed across the membrane (510). A detailed examination of proton transport in rat liver mitochondria at site 2 has shown that the ratio of H^+ to $2e^-$ is four whether glycerol phosphate or succinate is the substrate (2). The dehydrogenases for these two substrates are located on opposite sides of the membrane, succinate dehydrogenase on the matrix side and glycerol phosphate dehydrogenase on the cytosolic side. It has been suggested by Alexandre et al. (2) that both substrates deliver two hydrogens ($2e^- + 2H^+$) to UQ (possibly attached to a protein) and that on oxidation UQH_2 releases two electrons to the cytochrome $b-c_1$ complex and two protons to the medium. In order to account for a ratio of H^+ to $2e^-$ of four for electrons passing from succinate to cytochrome c, a site for proton-pumping with a ratio of H^+ to $2e^-$ of two has been suggested for complex III, the complex possibly employing cytochrome b as a proton pump (511). Coupling site 2 would therefore consist of two subsites, 2A and 2B, associated with UQ and complex III, respectively. In this scheme the need for a Q cycle would be eliminated.

In 1970 Wilson and co-workers reported that the midpoint potential of cytochrome b_{566} was changed by the addition of ATP from -30 to 245 mV (90, 429, 534). It was postulated that cytochrome b_{566} is directly involved in energy transduction, and this cytochrome was therefore designated cytochrome b_T (T indicating transduction). The other b cytochrome (b_{562}), with a midpoint potential of 30 mV, was not affected by ATP and was designated cytochrome b_K (in honor of Keilin, who first isolated intact mitochondria). Cytochrome b_K is, therefore, considered to be a simple electron carrier between UQ and cytochrome b_T. Passage of an electron through cytochrome b_T or the presence of ATP causes this cytochrome to be energized, but a mechanism that uses the energized form of cytochrome b_T to form the acid anhydride bond in ATP has not been formulated. The experiments on cytochrome b_T were done by applying a redox buffer to mitochondria or particles and measuring the percentage reduction of the individual cytochromes by using a spectrophotometric technique. It has been suggested that the system is not at equilibrium under the conditions used by Wilson and co-workers and that reversed electron flow is potentiated by ATP so that cytochrome b_T becomes more reduced in the presence of ATP, giving the appearance of a change to a more positive potential. Conformational changes in the complex on the addition of ATP could also change the redox potential of cytochrome b_T (415). Thus the involvement of cytochrome b_T in energy transduction is not certain.

The antimycin-A site has also been implicated as the site of energy transduction (415), because blocking of electron transport by antimycin A to some extent resembles the blocking of electron transport by the high-energy state of mitochondria in the presence of ATP. Finally, it is known that complex III can undergo conformational changes (415), and these may be important in energy transduction as suggested by Boyer and co-workers (Section II,D).

The mechanism by which energy transduction is accomplished in complex III is still, therefore, an open question. It is clear that the complex can transport protons, that it can undergo conformational changes, and that components change their spectral properties when energized. How these functions are interrelated and the manner in which they are used to synthesize the anhydride bond in ATP are still uncertain.

5. Structure and Properties of Complex IV

Complex IV, cytochrome c oxidase (ferrocytochrome c: oxygen oxidoreductase), is the terminal oxidase that binds and reduces oxygen with the concomitant oxidation of cytochrome c, a reaction that is associated with an energy transduction site in oxidative phosphorylation. It is found in all eukaryotic and some prokaryotic cells and has been extensively studied; the author of a recent review commented that at least 1000 papers have been written on it (315).

Cytochrome oxidase in animals and yeast has a molecular weight of approximately 140,000 and consists of at least seven subunits (315) that are all present in a 1:1 ratio. Three of the subunits are coded by mitochondrial DNA and are synthesized in mitochondria; four are synthesized in the cytosol (388). The molecular weights of the polypeptides range from 44,600 for subunit 1 to 4,300 for subunit 7. Many of the polypeptides are strongly hydrophobic in nature, which makes them insoluble and difficult to resolve. Cytochrome oxidase spans the membrane, so some components are exposed to the matrix and some to the cytosol in the inner mitochondrial membrane (115, 433). All subunits, except 1 and 2, are accessible at the surface. Subunit 2 is partly inaccessible, and 1 completely inaccessible (142). One of the smaller subunits may be involved in the binding of cytochrome c to the complex.

The presence of four electron-transferring components in cytochrome oxidase, two molecules of cytochrome a and two copper atoms, has been known for a long time (286, 314), but the detailed function of these components is still in question. After extraction the two cytochrome molecules appear to be identical. They can, however, be distinguished in the intact enzyme because only one of them binds cyanide, carbon monoxide, and azide (286). This cytochrome has been designated cytochrome a_3 and is the electron carrier that reacts with oxygen. Carbon monoxide binds to cytochrome a_3 only when the cytochrome is reduced, and it has little affinity for the oxidized form (314, 535).

The cytochrome molecules have different midpoint potentials (314, 535), possibly due to heme–heme interactions. It has been suggested that the redox potential of cytochrome a is a function of the redox state of cytochrome a_3 (355). Cytochrome a is now thought to contribute 70% of the absorption at 605 nm, whereas previously cytochromes a and a_3 were thought to contribute equally to this band (355). Later it was suggested that the band at 605 nm may be due entirely to cytochrome a (524).

After reduction of 50% of all of the cytochrome oxidase cytochromes a and a_3 appear to be in rapid equilibrium between their oxidized and reduced forms, so they must have very similar midpoint potentials in the transition from the oxidized to the half-reduced complex (524). Previously, it had been proposed that the midpoint potentials of the two cytochromes were different, 220 mV for a, and 375 mV for a_3 (289), but more recent evidence suggests that both cytochromes in the uncoupled state initially have quite positive and similar values, (around 362 mV) (524). Reduction of either heme induces a decrease of approximately 124 mV in the other by heme–heme interactions. The sequential addition of two electrons shows negative cooperativity; that is, the half oxidation of the fully reduced cytochrome oxidase is more favorable than the further oxidation of the half-reduced enzyme (524).

Energizing the membrane in coupled mitochondria has also been reported to change the redox potential of cytochrome a_3 to more negative values. This implicates this cytochrome in energy transduction in a manner similar to that proposed for cytochrome b_T (130, 289). A reexamination of the effect of ATP on the midpoint potential of cytochromes a and a_3 by Chance and co-workers (524) demonstrated that in energized mitochondria the midpoint potentials of both cytochromes are reduced by 104 and 133 mV for cytochromes a and a_3, respectively. This means that cytochrome a_3 becomes a better reductant than a, and the equilibrium at 50% reduction favors the latter in the reduced state. It was suggested that in the energized state the oxygenated form of the enzyme is stabilized, which would shift part of the cytochrome oxidase molecules into a form not involved in the catalytic cycle and hence inhibit activity, giving respiratory control (524).

It has been well established that cytochrome oxidase contains equimolar amounts of cytochrome a and copper and that the minimal functional unit contains two molecules of heme and two of copper (371). The complex requires four electrons for complete reduction, indicating that both hemes and copper atoms are involved in the redox reaction (298). One copper atom, Cu_A, appears to be associated with cytochrome a and the other, Cu_B, with cytochrome a_3. Cu_A and cytochrome a titrate with the same redox potential (220 mV), and the Cu_B–cytochrome-a_3 pair also have the same potential (350 mV) (298). The oxidase, therefore, accepts four electrons from the electron transfer chain before transferring them to a molecule of oxygen in a coordinated reaction to form two molecules of water. Oxygen binds to the cytochrome-a_3 region of the oxidase and is reduced by the addition of two electrons from the heme-a_3–Cu_B electron-donating pair to form a peroxy compound (89). The heme-a–Cu_A pair may act as an electron reservoir and rapidly donate two electrons to the peroxy compound via heme a_3–Cu_B to form water, thus maintaining a minimum concentration of reactive intermediates (88, 387). A model of the reactive center of cytochrome oxidase that accounts for many of the properties of the complex has been presented (371).

Various models for the structure of the oxidase have been proposed (88, 315). Cytochrome c is on the cytosolic side of the membrane, and there are data to indicate that cytochrome a is located on the same side and can interact with cytochrome c (115). Cytochrome a_3 is located either on the matrix side of the membrane or within it, and that the link between cytochrome a and cytochrome a_3 crosses the membrane.

In the Mitchell scheme of chemiosmotic coupling, the electron transport chain is envisaged as a series of redox loops across the inner mitochondrial membrane (Section II,B,1,a). In this scheme, cytochrome oxidase functions as one arm of the final redox loop to transport electrons vectorially from cytochrome c on the outside of the membrane to oxygen on the inside, where protons are taken up to form water. No direct transport of protons across the membrane occurs at this site in this scheme. The stoichiometry of this electron transfer would result in a charge separation of two per two electrons.

This simple mechanism has been challenged by Wilkström and co-workers, who have evidence for direct transfer of protons across the cytochrome oxidase (523, 527; reviewed in 525). In this mechanism four protons are consumed in the matrix by the reduction of an oxygen molecule by four electrons, but, in addition, four protons are pumped across the membrane from inside the mitochondrion. There is, therefore, a charge separation of four for every two electrons transferred across the membrane, and this ratio has been measured (278, 447, 525). In order to observe this direct transport of protons it is necessary to inhibit the phosphate transporter (525), and there are also problems in interpreting the results with different electron donors. As a result, the proton-pumping activity of cytochrome oxidase is not accepted by Moyle and Mitchell, who have stated that "cytochrome oxidase is not a proton pump [346]."

Further evidence for proton pumping by cytochrome c oxidase was provided by the study of proton transport using purified cytochrome oxidase (277). When the purified enzyme is incorporated into phospholipid vesicles, protons are transported with a ratio of H^+ to $2e^-$ close to 1.0, both in reduced-cytochrome-c and oxygen-pulse experiments. It has been suggested that during the passage of electrons through the oxidase there is a conformational change (525). This can be detected by a change in the heme a_3 from a high-spin to low-spin state. There is also a change in the spectrum of cytochrome a_3 as the membrane is energized. The relaxation of the strained conformation may be coupled to electrogenic proton translocation across the membrane. It is assumed that one or more of the subunits in cytochrome oxidase forms a proton channel and that protons are pumped by a mechanism similar to the Bohr effect in hemoglobin (525).

6. Conclusions: Proton Pumping and the Electron Transport Chain in Animal Mitochondria

Although there is still controversy about the number of protons translocated at each coupling site in the electron transport chain, there is increasing evidence for a ratio of H^+ to $2e^-$ at each site of four (e.g., 2,510 and references cited therein). Present ideas on proton transport tend to favor either vectorial proton pumps or some form of redox loop. Evidence in favor of proton pumps is accumulating, and pumps for many ions are known in biological systems, including proton-pumping activity of mitochondrial and chloroplast coupling factors. On the other hand, a vectorial arrangement of redox carriers in mitochondria has been clearly established, so redox loops are also quite probable. It would seem feasible to construct a model for proton transport in the electron transport chain that would involve both mechanisms. Such a model is shown in Fig. 8, where a proton pump is shown associated with complexes I, III, and IV, and redox loops with complex I and UQ. A similar system for coupling site 2 has already been proposed (2).

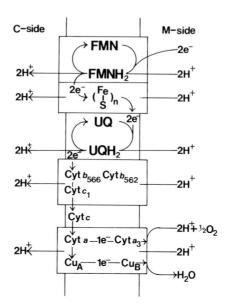

Fig. 8. Possible arrangement of redox loops and proton pumps that would give a stoichiometry of four protons per coupling site. For Complex IV (lower box) a charge separation of four has been assumed; two protons are transported. In other models it has been suggested that four protons are transported. C- and M-side, cytosolic and matrix side of mitochondrial membrane. Based, in part, on work by Mitchell (328), Alexandre *et al.* (2), and Wikström and Krab (525).

B. The Coupling Factor: ATP Synthesis

The mitochondrial coupling factor complex has been termed complex V, but this name is rarely used now and the complex is more commonly called mitochondrial ATPase or, more appropriately, ATP synthase. After isolation, the complex catalyzes the hydrolysis of ATP to ADP and P_i, but this activity is latent in mitochondria. It is now generally accepted that in mitochondria the function of ATPase is reversed and that it is the terminal enzyme in ATP synthesis. The properties and structure of the complex have been reviewed extensively (26, 115, 144, 153, 185, 251, 276, 380, 437, 455) and only a brief account is given here.

The intact complex, which is called oligomycin-sensitive ATPase, can be separated and purified from mitochondria (155, 464, 472). The purified complex catalyzes ATP hydrolysis and an exchange reaction ATP $\leftrightharpoons$ P_i; both reactions require added phospholipid, suggesting that vesicles are required for activity (155). The ATPase activity of the purified intact complex is sensitive to the oxidative phosphorylation inhibitor oligomycin, and the exchange reaction is sensitive to uncouplers, oligomycin, and another oxidative phosphorylation inhibitor, DCCD. SDS gel electrophoresis separates the complex into 12–13 poly-peptide bands that range in molecular weight from 7,000 to 54,000 (155); the overall molecular weight is of the order of 400,000 to 500,000 (185). The complex consists of two parts, ATPase or F_1 and F_0. Binding these two portions of the complex together to form the energy-transducing complex are two proteins, OSCP and coupling factor F_6. In the electron microscope the complex is seen to be a tripartite structure, with a headpiece (F_1), a stalk (OSCP), and a base piece (F_0) (143, 300). The F_1 part projects into the mitochondrial matrix (115).[4]

The F_1 part of the complex can be isolated and purified (381 and references cited therein). It is composed of five subunits, α, β, γ, δ, and ϵ (276, 437) and is cold labile due to separation of the subunits at low temperature. There is some disagreement about the stoichiometry of the subunits. Some estimates suggest a structure $\alpha_2\beta_2\gamma_2\delta_2\epsilon_2$, whereas others suggest $\alpha_2\beta_2\gamma_2\delta_1\epsilon_1$ or $\alpha_3\beta_3\gamma_1\delta_1\epsilon_1$ (26, 276). The function of some of the subunits has not been clearly established. The α and β subunits are involved in the formation of the active site of the enzyme. The β subunit will bind nucleotides very tightly, and the complex can be inactivated by the covalent binding of the ATP analog, 8-nitreno-ATP, which suggests that the β subunit contains the active site (455). There are, however, numerous other nucleotide-binding sites on the α and β subunits, and their role is unknown; possibly these sites act as allosteric regulators (455). The γ and δ units appear to be involved in binding F_1 to the stalk (26), but the function of the ϵ subunit is unknown. In addition to the five subunits, a sixth, loosely bound subunit is found in some preparations of F_1. This subunit (molecular weight 10,000) is inhibitory and probably acts to regulate the complex (26, 437).

OSCP may be purified to homogeneity (438) as may F_6 (393). It has a molecular weight of 18,000 and is fairly hydrophobic. Although the binding site for oligomycin is on the membrane portion of the complex, OSCP is required for the inhibition of F_1 by oligomy-cin, suggesting that the inhibitory action of oligomycin is transferred via OSCP. F_6 is a low-molecular-weight (8,000) protein that appears to be involved in facilitating the binding of F_1–OSCP to the membrane (502).

The membrane component of the complex (F_0) can be isolated and purified (144, 275). F_0 appears to be composed of three subunits that range in molecular weight from approxi-mately 8,000 to 24,000. The smallest subunit is the proteolipid that constitutes the proton channel in the membrane (144). Proton permeability of mitochondrial vesicles is low but increases significantly when F_1 is removed. This suggests that F_0 acts as a channel for protons through the membrane, and is a proton transporter (330). This flow of protons through the membrane can be reduced by amounts of oligomycin in molecular equiva-lence with F_0 (330). Oligomycin, therefore, blocks the proton channel. The isolated trans-porter can be incorporated into phospholipid vesicles. Bacteriorhodopsin can also be incorporated into these vesicles, and a proton gradient can then be generated by light. This gradient is discharged by the proteolipid, indicating that it is indeed a proton transporter (107). Proton transport is blocked by oligomycin, indicating that the binding site for oligomycin is on the proteolipid. Similarly, DCCD will bind irreversibly to the proteolipid, possibly at the same site as oligomycin, and block the proton-conducting channel (276).

Despite this detailed knowledge of the ATPase, the mechanism by which the passage of a proton through the coupling factor (F_0) in the inner membrane allows ATP synthesis in the F_1 complex, which pro-trudes into the mitochondrial matrix, is not known. The number of

---

[4] See Chapter 1 for a discussion of this aspect of mitochondrial morphology. (Ed.)

protons required to pass through the coupling factor to form one molecule of ATP has been estimated. Earlier estimates put the ratio of H^+ to
ATP at two (75), but a more recent estimate suggests the ratio may be
three (3). The Hence four protons would be ejected from the mitochondrion
at each coupling site; three of these would flow back through the coupling factor to form one ATP molecule, and one would be used to
transport a P_i into the matrix (3). The ADP required for ATP synthesis is
transported into the matrix by the exchange reaction of ADP and ATP.
A ratio of H^+ to ATP of three has been measured in chloroplasts (Section VI,C,2). The Δ pH across the thylakoid membrane is 3.5 units, and
the passage of three protons across this gradient would give a free-
energy change of approximately 14 kcal/mol (294). This free-energy
change is required to maintain the phosphate potential in chloroplasts
and is probably also required in mitochondria. The reverse reaction, the
pumping of protons across membranes by isolated ATPases incorporated into liposomes, can, of course, take place using energy from the
hydrolysis of ATP (439). It is possible that a study of this reverse reaction
would increase our knowledge of energy transduction (see discussion in
Sections II,B and II,C).

IV. Electron Transport and ATP Synthesis in
Plant Mitochondria

A. THE ELECTRON TRANSPORT CHAIN

1. Introduction

Considering the enormous amount of information available on the
electron transport chain in animal mitochondria, our knowledge of
plant mitochondria is remarkably scant. This is unfortunate, because
although the basic electron transport chain of plant mitochondria is very
similar to that of animals, there are significant differences. For example,
unlike animal mitochondria, plant mitochondria can oxidize external
NADH directly by means of an NADH dehydrogenase, which is located
on the outside of the inner membrane. Plant mitochondria also have
alternate pathways for electron transport that bypass cytochrome oxidase.

The separation of the electron transport chain into distinct complexes equivalent to those in animal mitochondria has not been accomplished, although cytochrome oxidase has been purified from plants (301). This means that detailed observations of the individual complexes have not yet been made, and the chain has only been studied by measuring changes in light absorption or EPR in the whole chain. An additional difficulty in plant studies is that, because of the greater complexity of the chain due to alternate electron pathways, classical inhibitors are often less effective (481).

The main emphasis in plant research has been the determination of the nature and sequence of the electron transport chain components. The classical complexes described in the animal system may not, of course, be physically separable as they are in animal mitochondria, although they may function in the chain as parts of the complete unit. The components of the animal complexes, which are also represented in plant mitochondria, seem to operate in similar sequences. Therefore, for the purposes of this discussion, it is appropriate to proceed as though the same complexes exist in plants, even though they may not yet have been isolated. The following Sections IV,A,2–5 therefore treat complexes I–IV in parallel with and in similar detail to the treatment in Sections III,A,2–5. This will facilitate detailed comparison of the plant and animal systems. In addition, Sections III,A,6 and III,B deal with distinctive features of the plant system. Finally, plant mitochondrial AT-Pase is discussed in Section III,C.

A generalized interpretation of the whole system of electron transport chains in plants is shown in Fig. 9. It should be remembered that Fig. 9 is based on knowledge derived from animal systems, even though the organization or existence of individual complexes have not necessarily been demonstrated in plants. The fine detail is documented in distinctive type in Sections IV,A,2–6, with a final summary section on the role of UQ in Section IV,A,7.

2. Complex I (NADH : Ubiquinone Oxidoreductase)

Plants have three, possibly four, NADH dehydrogenases, of which only one is associated with complex I (370, 481). One is located on the outer mitochondrial membrane and is not part of the electron transport chain (481). A second is located in the inner mitochondrial membrane but is used to oxidize cytosolic NADH because the outer membrane, but not the inner, is permeable to NADH; it is not associated with ATP synthesis. A third dehydrogenase, which is insensitive to piericidin and which bypasses the ATP generation site, may also be associated with the inner membrane (370). The three enzymes, which are not associated with complex I, are discussed in Section III,B. The fourth dehydrogenase appears to be very similar to that in complex I of animal mitochondria; it is involved in the oxidation of NADH and the concurrent reduction of ubiquinone and synthesis of ATP.

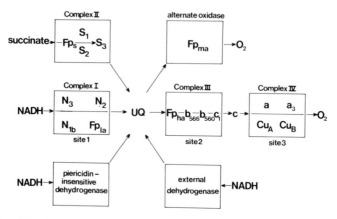

FIG. 9. Possible scheme for the electron transport chain in plant mitochondria, showing all postulated complexes and components; discussed in Section IVA, and B. Adapted from Storey (481). This figure should be compared with Fig. 7 (animal mitochondria).

The electron transport chain in plants contains a number of flavoproteins, although none of them have been isolated. They can be characterized by their absorbance and fluorescence and also by their midpoint potentials, although this latter distinction has led to a certain amount of confusion (481). Five flavoproteins have been identified in skunk cabbage (*Symplocarpus foetidus*), mainly on the basis of midpoint potential (475). Of these FP_{lf}, with a midpoint potential of -150 mV, is probably the flavoprotein of lipoate dehydrogenase. (The subscripts h and l refer to high and low midpoint potentials, and a and f to absorbing and fluorescing flavoproteins.) A second flavoprotein, FP_{la}, has a midpoint potential of -70 mV and has been placed by Storey in complex I (481). Because its redox potential is much more positive than that of the $NAD^+ : NADH$ couple, it has been placed on the product side of the energy-conserving step in complex I. This flavoprotein can be reduced by malate, a reduction that is sensitive to amytal, and by reversed electron flow from succinate (474), a process driven by ATP that transfers electrons between the components of complexes I and II (476). Little is known about flavoprotein FP_{la} or its correspondence to FMN of animal mitochondria, which has a much more negative potential and reacts directly with NADH.

Iron–sulfur centers in complex I have not been studied in detail in plants. Their presence is indicated by the inhibition of electron transport by rotenone and piericidin A, which inhibit complex I by combining with the iron–sulfur centers (370). EPR analysis of plant mitochondria has been done, but not in as much detail as in animal mitochondria and with more concern for the alternate pathways of electron transport such as that found in the *Arum maculatum* spadix. Centers N-1b, N-2, and N-3 have been identified (83, 410) and seem to be similar to those in animals (481). Center N-4 may also be present (83, 421). A more detailed analysis of these centers has been made, and it was suggested that N-2 in plants might also be involved in energy transduction (421). Center N-1a appears not to be present in plant mitochondria, which is somewhat surprising because it has been implicated in energy transduction in animals (481).

The mechanism of energy transduction by either proton pumping or energy-induced midpoint potential shift has not been studied in plants. Because complex I has some structural similarity to animal complex I, it may be assumed that their operations may also be similar.

3. Complex II (Succinate : Ubiquinone Oxidoreductase)

Plants rapidly oxidize succinate, and succinate dehydrogenase is often used as a mitochondrial marker. It can be assumed, therefore, that plants contain complex II. However, succinate dehydrogenase appears not to have been isolated and studied except for some preliminary data that are still quoted as the primary reference on this enzyme (210). There is no direct evidence that plant succcinate dehydrogenase is a flavoprotein, although it is usually assumed that succinate is linked to the respiratory chain through a flavin (481). Succinate oxidation is inhibited by 2-thioyltrifluoroacetone (413), which inhibits electron transfer from flavins to iron–sulfur centers, providing evidence for a flavin component (481).

The iron–sulfur centers of complex II have been studied more intensely by EPR analysis. As in animal mitochondria, there appear to be three centers, S-1, S-2, and S-3. The midpoint potentials of S-1 and S-2 are similar to those in animals (83, 410). Center S-3 has been studied in more detail because it was originally thought to be associated with the alternate oxidase. Center S-3 is paramagnetic in its oxidized state; it is a HiPIP-type of center (343, and see Section III,A,3). The signal is more complex, however, especially when measured in the presence of various inhibitors (343). Other HiPIP centers are absent from plant mitochondria, which simplifies study of this center (409). The plant center S-3 differs from that in animals. The midpoint potential is 65 mV in plants compared to 120 mV in animals. Center S-3 appears to transfer electrons to the main transport chain, obtaining electrons from succinate via centers S-1, S-2, and the flavoprotein (409). The alternate oxidase keeps S-3 partially oxidized, suggesting that it also accepts electrons from S-3. (The alternate oxidase is discussed in Section IV,B,3.)

Inhibition of both the normal and alternate electron transfer pathways fully reduces S-3, clearly showing that S-3 is not the autoxizable component of the alternate pathway (409), as had been suggested, nor is S-3 involved in the external NADH dehydrogenase (409). There appears to be an interaction between S-3 and ubisemiquinone (421), suggesting the possibility of a protonmotive Q cycle in plants (409). Center S-3 has been detected in a number of plants (414), but nothing is known of its molecular weight or subunit structure.

4. Complex III (Ubiquinone : Cytochrome c Oxidoreductase)

Much research in plants has been performed on what are thought to be components of this segment of the electron transport chain. There is, however, remarkably little consensus on the nature of these components or their sequence, although it appears that the chain differs significantly from that found in animal mitochondria. First, the prominent iron–sulfur center, the Rieske center, may not be present in plant mitochondria, or at least it cannot be detected by EPR spectroscopy (481). This center is an integral part of the animal chain, closely associated with cytochrome c_1. One assumes that its function is taken by another component in plants. Storey (481) has suggested that this segment of the chain contains a flavoprotein (FP_{ha}) with a midpoint potential of 110 mV, and Storey has placed it in a position to accept electrons from UQ and pass them to cytochrome b.

Information about the b cytochromes is remarkably confused with regard to their sequence in the chain and their function in energy conservation. It is generally accepted that there are three b cytochromes in plant mitochondria, b_{556}, b_{560}, and b_{565} (originally termed b_{553}, b_{557}, and b_{562}), with midpoint potentials of approximately 75, 42, and -77 mV, respectively. A fourth cytochrome, b_{558}, has also been reported in plant mitochondria (279). However, it is probable that this absorbance peak is due to cytochrome b_{565} and that there are in fact only three b cytochromes.

The midpoint potentials and absorption maxima of cytochrome b_{560} and b_{565} suggest that they are analogous to cytochromes b_k and b_T, respectively, in animal mitochondria. However, cytochrome b_{565} does not increase its midpoint potential on addition of ATP as does cytochrome b_T (129). As discussed in Section III,A,4, the change in potential of cytochrome b_T may be due either to a barrier between the substrates and the cytochrome or to reversed electron flow mediated by ATP, and a direct role for cytochrome b_T in energy transduction has been questioned (279, 370, 481). In this respect it now appears that the animal cytochrome may be anomalous (280) and, if so, the study of plant mitochondria will have shown how all mitochondria might function (370). However, the actual status of cytochrome b_{565} in plant mitochondria has not been determined. Measurements of its equilibration with other redox carriers indicate that it may not be part of the main electron transport chain, but that it may be in slow equilibrium with the chain (478) and associated with a flavoprotein FP_{vha} (very high potential) (479). The redox properties of cytochrome b_{565} are also strongly affected by antimycin A [reviewed by Storey (481)]. It has also been suggested that it may be involved in hydroxylation reactions (478).

Cytochrome b_{560} is oxidized rapidly in response to an oxygen pulse with a time course similar to that of the c cytochromes (481). This cytochrome, therefore, appears to be on the product side of complex III, donating electrons to cytochrome c_1. Cytochrome b_{556} is oxidized much more slowly, with a time course similar to that of UQ and the flavoprotein associated with this complex. Storey (481) has therefore suggested that UQ donates electrons first to the flavoprotein, then to b_{556}, and finally to b_{560} (see Fig. 9). Cytochrome b_{565} may equilibrate with b_{560} in a slow side reaction.

Cytochrome c_1 (c_{552}) has received practically no attention in plants. Unlike c_1 from animal mitochondria, it cannot be identified spectrophotometrically in intact mitochondria and can only be detected after all the cytochrome c has been removed by washing in phosphate buffer (281). It seems highly likely that cytochrome c_1 serves the same function in plants and animals. Storey has placed the energy conservation step between cytochrome b_{560} and cytochrome c_1 (481). A soluble cytochrome b–c_1 complex has been isolated from washed potato (*Solannum tuberosum*) mitochondria by treatment with bile salts (126).

5. Complex IV (Cytochrome c : Oxygen Oxidoreductase)

Cytochrome c oxidase has been isolated and purified from sweet potato (*Ipomoea batatas*) roots (301). Although the oxidase has been isolated before, this is the first description of purification to homogeneity. The cytochrome oxidase complex consists of at least five polypeptides with molecular weights ranging from 5,700 to 39,000; cytochrome oxidase from other sources typically consists of 6–8 polypeptides. Similar results have been obtained for pea cytochrome c oxidase, although here two other polypeptides may be weakly associated with it (317a). The complex contains heme a with an α band at 601 nm and a β band at 517 nm. It also contains phospholipids but requires additional phospholipid for full activity (301).

There appears to be no other current research on the components of complex IV in plants; the earlier work has been reviewed by Storey (481). Plant cytochrome oxidase contains cytochrome a and a_3 with midpoint potentials of 190 and 380 mV, respectively, (129), midpoint potentials very similar to those found in animal mitochondria. Cyanide complexes with the reduced form of cytochrome a_3, and the mechanism of inactivation of the complex has been determined (481). The copper content and copper properties in the plant complex do not appear to have been determined.

6. Cytochrome c

Cytochrome c is readily extracted from mitochondria and is regarded as a mobile carrier between complexes III and IV, acting in a fashion similar to that of UQ elsewhere in the chain. The structure and function of cytochrome c from many sources including plants have been extensively reviewed by Dickerson and Timkovich (119). The ready solubility of the cytochrome and its relatively simple structure have enabled its amino acid sequence to be determined from many sources, including a large number of plants. Because of this, cytochrome c has been used extensively to determine evolutionary relationships between species. There is considerable variation in the amino acid sequence from different plant species. These differences have been outlined by Boulter et al. (68) and have been used to construct a phylogenetic tree (68, 119). The variations in structure appear not to affect the function of the cytochrome because cytochromes can be interchanged between different plant and even animal mitochondria (481). Plant cytochrome c is composed of 111 residues in most plants (112 in three species), and 104 (481) in animals; the N-terminal sequence of the plant protein contains the extra 8 residues (481). The spectral and midpoint potentials of plant cytochrome c are identical with those described for the animal cytochrome (481).

7. Ubiquinone

Ubiquinone is the molecule that links complexes I and II with complex III (reviewed in 496a). UQ from plants is identical with UQ from animal mitochondria; each has a side chain of 10 isoprenyl residues (43). UQ can be reduced by both NADH and succinate, indicating its central role in the chain (43). In the succinate dehydrogenase complex, electrons are transferred to UQ via the small subunit of this complex, a process that involves iron–sulfur center S-3 (32, 360) and possibly a UQ-binding protein (548). Two molecules of UQ may act as an electron-binding pair in the membrane, and a model of a possible transfer system involving the UQ-binding protein has been presented (425).

Detailed kinetic study of UQ oxidation by a pulse of oxygen has been carried out by Storey and Bahr, and it clearly implicates UQ as a two-electron carrier between complexes II and III (482). However, the acceptor of electrons in complex III appears to be a flavoprotein (FP_{ha}) that has no counterpart in the animal system. UQ also accepts electron in a biphasic manner, indicating that it is also the electron acceptor for the external NADH dehydrogenase (481) and that it may link all complexes in that region of the chain that have the required binding site (481). The midpoint potential of UQ in plant mitochondria is 70 mV, which is very similar to that of animal UQ (477). There is an indication that a second small UQ pool with a lower potential may exist (477).

B. OTHER MITOCHONDRIAL COMPLEXES

1. The External NADH Dehydrogenase

Plant mitochondria will oxidize external NADH, and this is not due to damage of the inner mitochondrial membranes during extraction (342a, 370). The direct oxidation of NADH replaces a glycerol phosphate–dihydroxyacetone phosphate shuttle that accomplishes the oxidation of cytosolic NADH in animal systems. Fresh red beet (*Beta vulgaris*) roots have mitochondria that are exceptional in that they cannot oxidize external NADH: instead they use a malate–oxaloacetate shuttle (114). Mitochondria from red beet discs do, however, develop the capacity to oxidize external NADH after 48 hr (114). This capacity is a result of an external NADH dehydrogenase complex located on the outer surface of the inner mitochondrial membrane. How many plants have an external dehydrogenase and how many use shuttle systems are not known.

The most complete analysis of NADH dehydrogenases was carried out by Douce *et al.* (123). Using different methods of disrupting mitochondrial structure and using various electron acceptors with differing mitochondrial permeability, they were able to demonstrate three separate enzymes: (*a*) internal NADH dehydrogenase, located on the inner surface of the inner membrane and oxidizing NADH in the mitochondrial matrix; (*b*) external NADH dehydrogenase, located on the outer surface of the inner membrane and oxidizing exogenous NADH; and (*c*) a third enzyme located in the outer membrane. Both internal and external inner membrane dehydrogenases (*a* and *b* above) are sensitive to antimycin A, indicating that electrons are fed into the normal electron transport chain including coupling site 2. The external NADH feeds electrons into the electron transport chain at UQ and hence bypasses site 1 of energy coupling, but is is coupled to sites 2 and 3 (342a). External NADH dehydrogenase (*b*) is loosely attached to the membrane and is solublized on swelling the mitochondria. There is evidence that it is a flavoprotein, like internal NADH dehydrogense but unlike the outer membrane enzyme (123).

Little is known about the regulation of external NADH dehydrogenase. Divalent cations stimulate its activity, possible by attaching it to the external surface of the inner membrane (99). The calcium chelator EGTA is inhibitory, and the effect can be reversed by calcium. Citrate will also inhibit, an effect that can also be reversed by calcium. The site of inhibition of citrate and EGTA appears to be between the flavoprotein of the external dehydrogenase and UQ (99). An alternative possibility is that calcium may reduce the negative charge on the surface of the membrane, which may result in NADH being able to approach and interact with the external dehydrogenase (239).

2. The Piericidin-Insensitive Internal NADH Dehydrogenase

Plant mitochondria respond to some inhibitors differently than animal mitochondria. In particular, they appear to be less sensitive to compounds such as rotenone and piericidin, which bind to the iron–sulfur center in complex I, a site of ATP synthesis. Much higher concentrations of rotenone are required to inhibit electron flow through the internal NADH dehydrogenase, but this may be due to inaccessibility of the inhibitory site to the inhibitor (481). Brunton and Palmer, in a study of the effects of inhibitors on malate oxidation by wheat mitochondria, postulated the existence of two internal NADH dehydrogenases in plant mitochondria (77). One of the dehydrogenases is sensitive to piericidin, but the other is not. The insensitive dehydrogenase bypasses the complex-I site of ATP synthesis. The two respiratory chain dehydrogenases appear to receive electrons from different NADH-linked dehydrogenases in the matrix.

The possible role of the two dehydrogenases and the regulation of the piericidin-insensitive dehydrogenase have been discussed by Palmer (370). A salient difference between plant and animal mitochondria is that plant mitochondria are more heavily involved in reactions of intermediary metabolism, producing carbon skeletons for biosynthetic purposes (370) in addition to their role in the synthesis of ATP. Palmer speculates that what is called the nonphosphorylating electron transport chain confers an advantage to plant mitochondria by enabling them to conduct synthetic reactions in the presence of the high levels of ATP, which may be maintained by photosynthesis (370). Palmer illustrates his viewpoint by two informative diagrams [Figs. 2 and 3 of Palmer (370)].

3. The Alternate Oxidase of Plant Mitochondria[5]

As described earlier (Section III,A), the electron transport chain can be inhibited by compounds that interact with cytochrome oxidase, such as cyanide, and those that react with complex II, such as antimycin A. In many plants, a large number of microorganisms, and a few animals, the flow of electrons from NADH or succinate to oxygen is only partially inhibited by cyanide and, in some cases, it is not affected by cyanide at all (209). This pathway is often termed the cyanide-insensitive pathway or oxidase. Because the pathway can also be deomonstrated by the use of a number of other inhibitors, the term alternate oxidase is preferable and is used here. The normal pathway of electron transport is termed the

---

[5] Also called the alternative pathway. For a discussion of the physiological role of the alternate pathway, see Chapter 4, Section II,A and C. (Ed.)

cytochrome pathway to distinguish it from that of the alternate oxidase, which does not contain cytochromes.

The alternate pathway has received a large amount of attention and has been the subject of several reviews (113, 209, 278a, 282a, 342a, 463), but, except in arum lilies (*Arum* spp.) and skunk cabbage (*Symplocarpus foetidus*), in which it generates heat (318), the role of this respiratory pathway is unclear. In a given situation the ratio of alternate oxidase to cytochrome respiration may change rapidly. For example, respiration of potato tubers is unaffected by cyanide, whereas fresh-cut slices have oxygen uptake five times higher and that is sensitive to cyanide (113). This change is associated with a wound response. It has also been suggested that the alternate pathway is involved in fruit ripening and germination, where it might function in conjunction with the external NADH dehydrogenase to oxidize either cytosolic or mitochondrial NADH. In either case this would allow biosynthetic processes that generate NADH to continue even in the presence of high concentrations of ATP (318). The alternate pathway may, therefore, operate to balance the availability of reducing equivalents with high-energy intermediates in these non-thermogenic tissues (318) and may act as an overflow for excess reducing equivalents (278a).

The total respiration of a tissue or mitochondrial preparation can easily be measured by determining the total oxygen uptake. Because the alternate and cytochrome electron transport chains both transfer electrons to oxygen from the same substrates, the contribution of each pathway to the total is not as easy to measure. The maximum capacity of the alternate pathway can be measured by the addition of cyanide, which completely inhibits the cytochrome pathway. In 1971 Schonbaum *et al.* introduced aromatic hydroxamic acids as specific inhibitors of the alternate pathway (434). A range of these inhibitors (e.g. *m*CLAM and SHAM) are now used to study the alternate pathway. Bahr and Bonner, in an elegant series of experiments, established the relationship between the cytochrome and the alternate pathways (24, 25). They measured the effect of a series of concentrations of hydroxamic acid on the respiration rate of mitochondria from skunk cabbage and mung beans (*Vigna radiata*) in the presence of cyanide to determine the effect of the inhibitor on the alternate pathway alone. By then measuring the inhibition of total respiration (i.e., in the absence of cyanide), they could evaluate the contribution of the alternate pathway to total respiration because inhibitor does not affect the cytochrome pathway (434). They found that the alternate pathway is only fully active in the presence of cyanide, so total respiration is less than the sum of the maximum capacities of the two pathways. The cytochrome pathway operates at a rate determined by the energy state or phosphate potential (the ratio of ATP to ADP) of the mitochondria and is unaffected by the complete inhibition of the alternate pathway. When both pathways are active, the cytochrome pathway operates at the maximal rate allowed by the phosphate potential, whereas the alternate pathway operates at a variable fraction of its maximal rate.

The control of electron flow through the alternate pathway does not appear to be directly regulated by either the energy state of mitochondria or competition for a component common to both pathways. Bahr and Bonner (25) postulated that the first component of the alternate pathway is in electronic equilibrium with a component of the main cytochrome pathway. The redox potential of these two components might be different; the cytochrome-pathway component might have a potential compared to that of the alternate component more positive by at least 35 mV. At intermediate reduction states, therefore, the cytochrome-chain component would be much more reduced than the alternate component, and electron flow through the alternate pathway would be minimal. When the cytochrome chain is blocked by either cyanide or a high phosphate potential, the cytochrome-chain component would become fully reduced, and because it is in equilibrium with the alternate chain component, this would in time become reduced and allow the alternate pathway to operate. This explanation, showing how the activity of the cytochrome chain may regulate the flow of electrons through the alternate oxidase, seems to be generally (113) although not universally (117) accepted.

In order to clarify the control of the alternate pathway, it is necessary to determine where the branch point occurs between the two pathways and which components are involved (see Fig. 9). The alternate pathway is resistant to antimycin A, suggesting that the branch is before complex III. It does, however, involve the first energy-coupling site and is inhibited by rotenone (in some cases piericidin) when internal NADH is being oxidized (113). Therefore, the branch point appears to be between complexes I and III and the most likely candidate is UQ. UQ acts as a link between the other complexes in this part of the chain and could also interact with the alternate oxidase (113 and references cited therein). It has been suggested that there are two pools of UQ that interact with different dehydrogenases (421a) and that only one pool interacts with the alternative oxidase (421a). Some dehydrogenases may donate electrons only to the cytochrome pathway and others to the alternative pathway. However, more recent results indicate that a structural separation of UQ pools may not be correct (278a). The ability of dehydrogenases to donate electrons to the alternate oxidase appears also to vary in different tissues. Hence the external NADH dehydrogenase of the inner membrane in aroid spadices will donate electrons to the alternate oxidase, whereas this ability is limited in spinach, where the oxiation of external NADH is linked to the cytochrome chain (123a). This may again indicate two pools of UQ that are not in equilibrium.

A clear indication that the branch point is indeed ubiquinone came from the work of Storey on mitochondria from skunk cabbage (480). The mitochondria were subjected to anaerobic conditions and the chain components reduced by the addition of succinate. The cytochrome chain was blocked by carbon monoxide. After a pulse of oxygen, UQ and a flavoprotein were oxidized within 1 sec whereas the rate of oxidation of cytochromes a and a_3 was much slower and consistent with the dissociation time for carbon monoxide from this complex. The alternate pathway inhibitor mCLAM prevented the oxidation of UQ and the flavoprotein, indicating that the alternate pathway operated during the oxygen pulse. Under the same conditions white-potato (*Solanum tuberosum*) mitochondria, which do not have the alternate pathway, do not show the rapid UQ and flavoprotein response to oxygen. UQ was thus identified as the component common to both pathways. It appears to be in rapid equilibrium with the flavoprotein. This flavoprotein has a potential that is 50 mV more negative than that of UQ and thus has the characteristics of the first component of the alternate pathway as predicted by Bahr and Bonner (25). It could act as a switch between the main pathway and that of the alternate oxidase.

The possibility that an iron–sulfur center may be involved with the alternate oxidase has been considered. Detailed studies of the EPR signals from ubisemiquinone and center S–3 by Rich and Bonner (409, 414) showed that neither ubisemiquinone nor center S–3 is the autoxidizable component of the alternate oxidase. However, they did confirm that UQ is the branching point of the pathway (113).

Because the branch point appears to occur at UQ, the operation of the alternate oxidase bypasses, ATP synthesis at complexes III and IV. Oxidation of succinate by the alternate pathway in *Arum maculatum* mitochondria does not result in ATP synthesis, whereas oxidation of matrix NADH allows ATP synthesis, but less than when the cytochrome chain is not inhibited by antimycin A (378). Energy coupling at site 1 still occurs, therefore, in the presence of antimycin A or cyanide. External NADH can be oxidized by *A. maculatum* mitochondria in the presence of these inhibitors, but then all the energy-transducing sites are bypassed although this appears not to be the case in spinach leaves (123a).

There is evidence that the UQ protonmotive Q cycle occurs in plant mitochondria (412), generating a protonmotive force across the inner mitochondrial membrane (Section II,B,I,a; see also Fig. 5). The alternate oxidase appears to interact with this cycle at some point (412). However, a detailed study of the protonmotive force that develops during succinate oxidation by the alternate pathway in mitochondria from many plants showed that protonmotive force does not increase on succinate oxidation and is not decreased by ADP. This indicates that proton pumping is not associated with the alternate pathway. Furthermore, the protonmotive force is not affected by hydroxamic acids (342). In contrast, the oxidation of malate by the alternate pathway does increase the protonmotive force, and this increase is reduced slightly by ADP and abolished by hydroxamic acids. This clearly indicates that energy conservation under these conditions does occur at Site 1.

Although these data indicate that energy conservation is not associated with the alternate oxidase, this conclusion is not universally accepted. Wilson, in a careful study of ADP phosphorylation and other energy-linked functions of mung bean mitochondria, claimed to have demonstrated energy transduction in association with the alternate oxidase under conditions in which sites 1, 2, and 3 of the cytochrome chain are not operating (536). He suggested that the high concentrations of cyanide used to block cytochrome oxidase may uncouple mitochondria and so prevent energy conservation. A similar demonstration of energy conservation by an alternate oxidase has been made using *Paramecium* (124). However, if energy conservation is associated with the alternate oxidase, it is difficult to envisage a function for it because it appears to be duplicating the cytochrome pathway. Such conservation would also limit the thermogenic capacity of the alternate oxidase and its ability to balance reducing equivalents and phosphate potential during cellular biosyntheses.

The composition of the alternate oxidase is still unknown. That hydroxamic acids chelate transition metals suggests that these metals may be involved. But the discovery of nonchelating inhibitors that are effective at much lower concentrations (113) and the discovery that known iron and copper chelators have no effect indicate that this may not be so (113). Furthermore, a detailed survey of the EPR spectrum of mitochondria known to possess the alternate pathway revealed no signals that could be associated with the alternate oxidase (409, 414).

Attempts to purify the alternate oxidase have been hampered by its lability (408). The use of quinols in the assay for the alternate oxidase aided in its purification and led to the conclusion that the oxidase is a protein and not merely an autoxidizable quinone pool (408). Huq and

Palmer solubilized the alternate oxidase from *Arum* mitochondria by use of the detergent Lubrol and partially purified it (229). Their preparation was free of cytochromes and had no EPR signal, that is, it had no iron–sulfur center. It did, however, contain a flavoprotein and possibly copper, although the latter has been disputed (113). The presence of a flavoprotein supports the results of Storey, who worked with intact mitochondria (480).

The end product of electron transport through the alternate oxidase pathway appears to be water, because water is the only product found when *Arum* submitochondrial particles oxidize succinate (230). Small amounts of H_2O_2 and superoxide (i.e., O_2^-) are produced when these particles oxidize NADH, but at too low a rate to be primary products of the alternate oxidase pathway (230), as had been suggested (411).

Neither the function nor the composition of the alternate oxidase has been elucidated. The development of techniques to purify this oxidase are urgently needed. Because the contribution of the oxidase varies at different stages of development, it may be presumed to play some important role in plant metabolism other than that of a thermogenic pathway.

C. The Coupling Factor: ATP Synthesis

The coupling factor complex in animal mitochondria has been extensively studied, and many of its properties are known. The plant complex has not been studied so intensively, and its structure and function are less well known. Because the function of the complex is the same in plant and animal mitochondria, namely, ATP synthesis, it seems likely that the structures of the complexes are similar, even though other functions of the mitochondria may differ.

ATPase activity has been found in intact and sonicated mitochondria of cauliflower (*Brassica oleracea* var. *botrytis*) (245), Jerusalem artichokes (*Helianthus tuberosus*) (379), and castor bean (*Ricinus communis*) endosperm (486), and it is stimulated by DNP in most cases. For maximal activity, the membrane may have to be disrupted to allow ATP to enter the mitochondria and reach the active site of the ATPase (245, 379, 486). The ATPase is inhibited by oligomycin and can be stimulated by trypsin or high pH (486).

Soluble plant ATPase was first isolated from cabbage (*Brassica oleracea* var. *capitata*) and reported to be cold stable, unlike the animal complex (382). However, a later report on ATPase from castor bean endosperm showed that the complex is cold labile (545). It is stable at higher temperatures in the presence of ATP and will hydrolyze ITP in addition to ATP. The soluble enzyme is not inhibited by oligomycin, but its activity is increased by DNP and magnesium (545).

The ATPase from germinating pea (*Pisum sativum*) seeds has been solubilized and purified (303). It has a pH optimum between 8.0 and 8.5 and requires the presence of ATP and magnesium in a ratio of 1 : 1 for optimal activity. The complex is also stimulated by sodium and potassium ions, and both ions are required for maximal activity (303). The complex is partially inhibited by oligomycin, indicating the presence of the membrane-located F_o portion (see Section III,B). The soluble ATPase is stimulated by a mixture of ethylene and carbon dioxide when sodium and potassium are present (304). This may be important in regard to the physiological effect of ethylene on plant growth and development (304).

The activity of ATPase is probably due to the F_1 portion of the mitochondrial coupling factor. This has been demonstrated by comparing the properties of partially purified ATPase from pea cotyledons with animal F_1 (173). The plant enzyme is cold labile and is inhibited by sodium azide but not by oligomycin. The enzyme will hydrolyze GTP and ITP in addition to ATP, the rate of GTP hydrolysis being even greater than that of ATP hydrolysis.

The complex is also greatly stimulated by anions, especially bicarbonate ions, which appear to act directly on the ATPase rather than by removing inhibitor protein (175). Plant ATPase differs from the animal complex in that it is greatly stimulated by calcium ions (173), which may have a regulatory function, especially as plant mitochondria accumulate these ions (120). The uptake and output of calcium by plant mitochondria appear to be controlled by phytochrome bound to the mitochondrial membrane (419). Red-light activation of phytochrome increases calcium efflux, and far-red light reverses it. The concentration of calcium in the cytosol modulates the activity of NAD kinase and possibly other enzymes via the calcium-binding protein calmodulin (419). The calcium-activated ATPase described by Spencer and co-workers (173, 303, 304) may be involved in this effect.

More recently the plant mitochondrial coupling factor has been studied in submitochondrial particles. These particles will catalyze oxidative phosphorylation (174). Synthesis of ATP is inhibited by potassium cyanide, uncouplers, and oligomycin. These submitochondrial particles, like those from animal mitochondria, turn inside out during their preparation (174). The ATPase activity is low in these particles but can be increased by trypsin treatment, which indicates the presence of an inhibitor protein (176, 246). The ATPase can also be activated by aging (176). Sodium chloride or bicarbonate activate freshly prepared ATPase of submitochondrial particles (176). This is in contrast to the ATPase in submitochondrial particles from rat liver, which is stimulated by bicarbonate but inhibited by chloride.

Nevertheless, the plant mitochondrial coupling factor complex is very similar to the animal complex. No details are available yet on the subunit composition of the complex or its regulation. There are, however, some differences between plant and animal ATPase, as in their activation by anions such as chloride and bicarbonate. Calcium activation is also different, and this may be important in the general regulation of metobolic activity in plants.

D. SUMMARY

Despite all the recent work, the knowledge of plant mitochondria as organelles for electron transport and ATP synthesis is still much more limited than that of animal mitochondria. Although the essential components of the plant mitochondrial system are regarded as similar to those that operate in animals, their isolation and structural characterization have scarcely begun. The many similarities between the two systems should not, however, obscure the important differences to which attention has been drawn. This may relate to differences in functions of plants and animals. Whereas animal mitochondria are specialized primarily to synthesize ATP, plant mitochondria also take part in many biosynthetic pathways by furnishing necessary intermediates for cellular metabolism, photosynthetic pathways, protein synthesis, etc. In performing these functions plant mitochondria must respond to a variety of environmental and other stimuli that link mitochondria to developmental events. This is illustrated by the response of plant mitochondria to ethylene, to red–far-red stimuli through their association with phytochrome, and to factors such as the supply of calcium and other nutrients. Thus the apparent similarities between the electron transport chains of plant and animal mitochondria should not be allowed to discourage vigorous further research on plant mitochondria.

Figure 9 shows an overview of the presumed operation of plant mitochondrial electron transport systems. Whereas Fig. 9 shows how the components of the systems may be interconnected, it does not show the complexity of the controls necessary for their proper integration. Until progress is made in this direction it will be difficult to see how mitochondria are involved in the control of plant development.

V. Photosynthetic Electron Transport

A. INTRODUCTION

The essential energetic process of photosynthesis is the reduction of $NADP^+$ by electrons from water, the energy for which is provided by light. Theoretically, the energy available in a quantum of red light is sufficient to transfer one electron from water to $NADP^+$. Such calculations are based on the standard redox potentials for the oxidation of water (0.81 V) and the reduction of $NADP^+$ (− 0.32 V). But, as indicated earlier, calculations based on standard redox potentials do not always provide a complete description of the process. In order for the equilibrium of the $NADP^+$-reduction reaction to favor NADPH, a negative potential much more than −0.32 V must be generated and, similarly, positive potential more than 0.81 V must also be available for the effective oxidation of water. In addition the generation of ATP, which also requires energy input, must be integrated into the process of $NADP^+$ reduction.

A further factor is that the initial process of photosynthesis is production of a charge separation within the thylakoid membrane. To stabilize the charge separation, the charges must be immediately transferred to acceptor molecules. This requires that the potentials developed by the photosystems must be much more negative than the standard redox potential for the reduction of $NADP^+$ and much more positive than the standard redox potential for the oxidation of water. It is now generally believed that two quanta of red light are required in the process.

The light reactions of photosynthesis involve two photosystems connected by an electron transport chain, an arrangement commonly referred to as the Z scheme. The historical development of this scheme has been reviewed on many occasions (e.g., 10, 44) and is not dealt with here. The scheme most commonly accepted is that shown in Fig. 10, although other schemes have been proposed (13a, 270). In this scheme energy is trapped by light-harvesting pigments in the two photosystems (PSI and PSII) and passed to the reaction centers where charge separation occurs. Associated with each photosystem are electron accepting and donating molecules, which can rapidly stabilize the charge separation and return the reaction center to the ground state, after which they can accept a further input of energy. The reaction centers consist of chlorophyll a molecules in specialized environments, the chlorophyll a molecules possibly existing as dimers.

When excited, PSI transfers an electron to an acceptor molecule

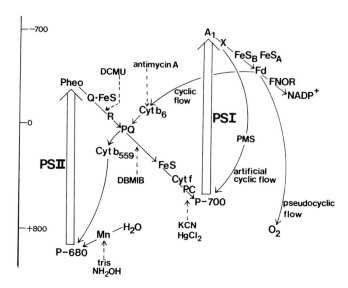

FIG. 10. Pathways of electron flow in photosynthesis. Alignment of components with the redox scale (left) is approximate. The sites of action of a number of inhibitors of electron flow are also shown. Oxidation of H_2O, shown at left, is associated with PSII. The production of O_2 (not shown) would take place at the bottom of PSII. PSI, which is associated with the reduction of $NADP^+$, is shown on the right. Noncyclic electron transport occurs between photosystems II and I. Cyclic flow around PSI occurs via Fd, Cyt(ochrome) b_6, PQ, and P-700. Cyclic electron flow may also occur around PSII via Cyt(ochrome) b_{559}. Artificial cyclic flow (through added electron-transfer agents) and pseudocyclic flow to O_2 (converting O_2 to H_2O) are also shown associated with PSI. FeS, iron–sulfur center; FNOR, ferredoxin : $NADP^+$ oxidoreductase; PC, plastocyanin; pheo, pheophytin; O, R, A, X, postulated intermediates of undetermined structure.

and eventually to $NADP^+$. The resulting positive charge at the reaction center of PSI is neutralized by an electron from the excited state of PSII by way of an electron transport chain. The positive charge at the reaction center of PSII is ultimately neutralized by an electron from water.

The electron transport chain transfers electrons energetically downhill from PSII to PSI through carriers such as PQ, an iron–sulfur center, cytochrome f, and plastocyanin (161). The transfer of electrons down this chain results in the movement of protons across the thylakoid membrane from the stroma to the lumen. The protons from the gradient developed by the electron transport chain, in addition to the protons released by the oxidation of water, are ultimately used for the synthesis of ATP by a chemiosmotic mechanism involving the chloroplast coupling factor complex. In addition to the flow of electrons from photosystems II to I, under some circumstances a cyclic flow of electrons can occur around PSI, which is known as cyclic photophosporylation. A further possibility is a flow of electrons from PSI to molecular oxygen, which is often termed pseudocyclic flow or the Mehler reaction.

The remainder of this chapter is concerned with the components involved in and the mechanism of energy transduction in chloroplasts. Section V provides details of the photosynthetic electron transport chains and their individual components. Section VI deals primarily with photophosphorylation and the way in which it is linked to the various electron transport pathways. Throughout these discussions reference should be made as necessary to the summary Fig. 10 and its explanatory legend.

B. PRIMARY ACCEPTORS

1. The Primary Acceptor for Photosystem I

The data on primary electron acceptors has been reviewed (30, 50, 257). As already described, EPR spectroscopy has been used extensively to determine iron–sulfur centers in mitochondrial electron transport chains and to demonstrate their importance. This technique has also been used in photosynthetic systems to show the importance of iron–sulfur centers in energy transduction in chloroplasts. Plant Fd is an iron–sulfur protein containing two labile sulfur and two iron atoms per molecule and is the soluble component connecting PSI with $NADP^+$ via $NADP^+$: ferredoxin oxidoreductase (see Section V,F). It was originally thought that Fd, because of its low standard redox potential (-432 mV), might actually be the primary acceptor of an electron from PSI (9). However, PSI develops a very negative redox potential, suggesting that a component with a redox potential more negative than Fd is the primary acceptor. In fact at least four components have been postulated to occur between P-700 and Fd.

The photosystems, with their associated electron donors and acceptors, must be organized in such a manner that all the components of each system are closely associated, and the primary transfer of an electron should therefore be possible at low temperatures (25–75°K), at which diffusion is virtually eliminated and normal chemical reactions are very slow. It can be assumed that the primary donors and acceptors in the two photosystems will function at these temperatures, whereas components less closely associated would not. Measurement of absorption changes at low temperature can, therefore, be used to determine which components are the primary acceptors.

The EPR spectrum of Fd can be used to determine the redox state of the molecule because EPR peaks are only given by Fd in the reduced state. In order to determine whether Fd is the primary acceptor in PSI, the EPR spectrum has been measured in chloroplasts illuminated at 77°K and measured at 25°K (306). A light-induced change in the EPR spectrum typical of reduced Fd is found, implicating Fd as the primary acceptor. However, an identical spectrum can be obtained from washed, broken chloroplasts from which soluble Fd has been removed and which are incapable of reducing $NADP^+$ without added Fd. This suggests that some Fd, tightly bound in the thylakoid membranes, may act as the primary electron acceptor and, in turn, pass the electron to the soluble pool of Fd.

Further evidence for the role of a ferredoxin-like iron–sulfur protein as the primary electron acceptor in PSI is provided by the observation that the EPR signal can be obtained at 25°K when the chloroplasts are illuminated with far-red light (750 nm), which activates only PSI. Digitonin is used to fragment the chloroplasts and produce fragments enriched in PSI (28). These fragments are also enriched in iron–sulfur protein and show the EPR signal of Fd when illuminated at 25°K (28). The primary event, therefore, in PSI is the transfer of an electron from P-700 to an iron–sulfur center leaving P-700 oxidized. Hence, P-700 becomes a free radical with one unpaired electron that gives a strong EPR signal. It is possible to measure simultaneously the appearance of the EPR signal from both the iron–sulfur center and P-700 in the same chloroplast preparation. The iron–sulfur centers and P-700 were thought to be present in equal amounts (29), but a more detailed examination suggests that two iron–sulfur centers (A and B) are involved, possibly representing two active sites on a single protein (141).

Redox titrations have also shown that there are at least two iron–sulfur centers and that they have the very negative potentials required by an electron acceptor of PSI (138, 259). The midpoint potential of center A is −553 mV, and that of center B is −594 mV (138). A difference between the bound iron–sulfur centers and Fd is that Fd has two sulfur and two iron atoms per mole, whereas the bound molecules may have four of each, like bacterial Fd (82, 136).

It was first thought that either center A or B was the primary acceptor for PSI. However, after photooxidation of P-700 at low temperatures, there is a partial return to the reduced state in the dark, but the EPR spectrum of centers A and B remains unchanged. This means that the reduction of P-700 is not a result of reversed electron flow from these centers (297). A search of other regions of the EPR spectrum revealed a new component (X) that shows the same kinetics as P-700, becoming oxidized when P-700 is reduced. This indicates that X is the primary acceptor (297). Under anaerobic conditions, in the presence of an oxygen-trapping system, the photoreduction of X with water as the sole electron donor can be demonstrated (221). In the presence of NADP$^+$ or soluble Fd, the EPR signal is very much reduced, indicating an electron transfer from X to those components and placing it in the main electron transport chain in the chloroplast (221), as shown in Fig. 10. The EPR spectrum is different from Fd, and X may be an iron–sulfur center in a specialized environment.

If centers A and B are fully reduced by dithionite, P-700 can still be photooxidized, but now the process is almost completely reversible in the dark. This indicates that X remains reduced when A and B are reduced; it cannot transfer the electrons under these conditions, a result that is consistent with the intermediate position of X (137, 139, 140). It has been suggested that transfer of electrons to centers A and B is irreversible, that they act as a trap. Centers A and B, P-700, and X are all present in chloroplasts in equal amounts (203, 204, 533), which suggests that they are part of a sequence [P-700 → X → center B → center A (203)].

Iron–sulfur centers abosrb at the blue end of the spectrum, and the intense absorbance of chlorophyll at these wavelengths makes it difficult to detect them by spectroscopy. However, an absorption change at 430 nm has been detected, the formation of decay kinetics of which closely resemble those of P-700 (219, 220), suggesting that this absorption change is associated with a primary acceptor (P-430) in PSI. P-430 may be an iron–sulfur center (258), possibly center A (162). A further absorption change (A$_2$) has been identified that may also be an iron–sulfur center, probably X (162, 444).

Another component has been detected between P-700 and X as an absorbance change in chloroplast particles enriched in PSI; it has been termed A$_1$ (317). An EPR signal possibly associated with A$_1$ has been detected (202), and there are indications that A$_1$ may be closely associated with X (201). It has been suggested that A$_1$ is a dimer of chlorophyll that is very closely associated with P-700 (444, 445, 446). However, other data indicate that it may be a chlorophyll a monomer (27, 202) or even pheophytin (202).

The primary acceptor complex in PSI is very complex. Component A$_1$ appears to be the primary acceptor and may be a chlorophyll a dimer or monomer. Component X, an iron–sulfur center, accepts electrons from A$_1$ and passes them to iron–sulfur centers A and B, which may be sequentially arranged or, possibly, because their redox potentials are similar, arranged in parallel. There are therefore at least four components between P-700 and Fd, and their exact natures are as yet unknown.

2. *The Primary Acceptor for Photosystem II*

A number of approaches have been used in attempts to identify the primary electron acceptor in PSII (reviewed in 30, 50, 271). This has resulted in a confused situation that has only recently been resolved to any extent. It now appears that PQ acts as a primary acceptor in addition to acting as a pool of oxidizing power. It is also very probably involved in proton transport (494).

On illumination the intensity of chlorophyll fluorescence in the reaction center of PSII is initially low, but then it rapidly increases. This increase is closely paralleled by an increase in oxygen consumption (240). It has been suggested that the rise in fluorescence may be a result of the photoreduction of a primary oxidant (Q) in PSII (313). The oxidized form of Q would, therefore, quench the initial fluorescence of the chlorophyll by removing an electron from P-680 in PSII. On the other hand, when Q is reduced, an electron is not removed from P-680, and the chlorophyll can fluoresce. Component Q was thought to be directly associated with the photosystem. It could then pass electrons on to a second component in the electron transport chain between the two photosystems (312).

The increase in fluorescence is in two steps, an initial rapid rise in fluorescence followed by a slower increase, eventually reaching a maximum (145). The initial phase is a result of partial reduction of Q. The second phase is due to the maintenance of Q in partially oxidized form by a secondary electron acceptor. When this secondary acceptor is fully reduced, Q also becomes fully reduced and fluorescence reaches a maximum. In the presence of DCMU fluorescence reaches the maximum very rapidly (470). This suggests that DCMU inhibits the transfer of electrons from Q to the secondary acceptor (145). It has been estimated that the ratio of Q to the secondary acceptor is 1:18 and that Q is present in amounts equivalent to P-680 (145).

A pool of PQ, comprising 6–10 molecules of PQ for every one of P-680, acts as an electron carrier between photosystems II and I. Plastoquinone can be detected by its characteristic UV spectrum between 220 and 340 nm. When PSII is excited by a long flash of light, the difference spectrum of PQ can be found, indicating a partial reduction of PQ due to transfer of electrons from PSII to PQ. When much shorter flashes are used (10^{-5} sec), a new spectrum is found that has a half-life of 5×10^{-4}

sec, which does not correspond to that of PQ (223). It has been suggested that the initial reaction in PSII is the transfer of an electron to a specialized plastoquinone to form the semiquinone (PQ⁻), and the new absorption spectrum might be due to PQ^-. Because this absorption peak is at 320 nm, it has been termed X-320. The half-life of X-320 is approximately 0.6 msec (471). The initial formation of X-320 is very fast (<20 nsec).

The identification of X-320 as a plastosemiquinone radical was supported by the study of pure PQ in methanol. The semiquinone can be generated by radiolysis and the difference spectrum ($PQ - PQ^-$) determined. It was found that the absorption spectrum of the plastosemiquinone is almost identical with that of X-320 (39).

Further evidence that X-320 is the primary acceptor was obtained when it was found that this absorbance peak could still be detected at 140°K (540). If DCMU is added to the preparation the half-life of X-320 is greatly increased. DCMU, therefore, appears to inhibit the transfer of electrons from X-320 to the rest of the electron transport chain, and X-320 remains in the reduced form (540). It had been shown previously that oxidation of Q (the fluorescence quencher of PSII) by PQ is inhibited by DCMU (470). The kinetics of formation of X-320 and Q are also very similar, and it has been proposed that Q is X-320 (540), the semiquinone of PQ. PQ, therefore, has two functions; it is a molecule specialized to accept electrons from PSII, and it acts as a pool of oxidizing power to accept electrons from Q.

Subchloroplast fragments prepared with deoxycholate also show the absorption difference spectrum associated with the primary acceptor X-320. The shape and amplitude of the spectrum indicate that the primary reaction is the reduction of one membrane-bound PQ molecule to its semiquinone, and one bound PQ per PSII has been found (504). Kinetic studies of the primary acceptor have used Tris-washed chloroplasts to inactivate the oxygen-producing side of PSII (407; Section V,D,2). These studies reveal that the electron transfer to X-320 occurs within less than 1 μsec. The reoxidation of X-320 in the dark has a half-life of 0.1–0.2 msec.

An intermediate between P- and Q-680 has been suggested. This acceptor appears to be pheophytin, which has the very negative redox potential of −450 mV (263–265). Q, which may be complexed with an iron–sulfur center, is closely associated with the pheophytin. The loss of energy in transferring an electron from pheophytin to Q·FeS would make Q·FeS a stable primary acceptor (264) (see Fig. 10).

One problem in relating Q to the rest of the electron transport chain is that Q transfers one electron, whereas PQ needs two electrons for reduction. Stiehl and Witt proposed that two reaction centers interact with one molecule of PQ (471, 538). Other models have suggested an intermediate component receives an electron from Q. This component, labeled either B (63) or R (507), retains its charge until Q is photoreduced a second time, whereupon two electrons are simultaneously transferred from Q and R to PQ (507). DCMU would inhibit electron transfer between Q and R (or B) (507). It has also been suggested that R is a specialized plastoquinone molecule that acts as a negative charge accumulator (390). After one flash of light, an electron is transferred from Q^- to R to give Q and R^-, which is fairly stable. After a second light flash a second electron is transferred from Q to R^- to give R^{2-}. Thereafter two electrons pass from R^{2-} to PQ (507).

The acceptor site in PSII may be activated bicarbonate ions, which may affect the oxidation and reduction of R (448), and a complex of R and HCO_3^- may be the electron-transferring component between Q and R (467). Alternatively, there may be two Q molecules (Q_1 and Q_2), which may react at the same or at different reaction centers with different efficiencies (243).

Obviously, the prime electron-acceptor site in PSII is a complex like PSI, with both primary and secondary acceptors. In PSII the acceptor appears to be PQ complexed with other components (R and an iron–sulfur center) so that it may act as either a one- or two-electron-transferring agent.

C. INDICATORS OF THE PRIMARY EVENTS

The behavior of some of the components of the electron transport pathways is difficult to monitor. Nevertheless, it is sometimes possible to determine their state (i.e., oxidized or reduced) from the behavior of other components that are associated with but not necessarily part of the electron transport pathway. Absorbance changes at 550 and 515 nm are examples. Absorbance changes at 550 nm (component C-550) indicate the redox state of Q, which is the primary electron acceptor from PSII. Absorbance changes at 515 nm (C-515) measure charge separation across the thylakoid membranes resulting from electron transport.

1. C-550: An Indicator for the Redox State of Q

In 1969 Knaff and Arnon found an absorbance decrease at 550 nm that occurred when chloroplasts from various sources were exposed to actinic light at 647 nm (269). The component assumed to be responsible for this absorbance change was termed C-550. The absorbance change occurred at $-196°C$, indicating a close association with the primary photoact. The absorbance change was elicited by light of 647 but not 715 nm, indicating that the component was associated only with PSII. Because the absorbance peak was at 550 nm, it was suggested that it might be associated with a c-type cytochrome.

The discovery of C-550 and the associated demonstration of the oxidation of cytochrome b_{559} at $-189°C$ by short-wavelength light (268) led to the suggestion by Knaff and Arnon that there are two light reactions associated with PSII (270). In this scheme PSII was responsible for the oxidation of water and the reduction of $NADP^+$ using two quanta of light. Photosystem I was considered to be an independent system involved solely with cyclic photophosphorylation (270).

This concept of three light reactions was challenged by Erixon and Butler (133, 134), who confirmed the photoreduction of C-550 and the photooxidation of cytochrome b_{559} at $-196°$ by PSII but suggested that these processes are associated with the accepted PSII and not the double PSII proposed by Knaff and Arnon. There is a close association of C-550 with the primary acceptor Q; a redox titration of chloroplasts has shown that the appearance of C-550 and the enhancement of fluorescence as Q becomes reduced have the same midpoint potential (133). These data initially suggested that C-550 could also be the primary acceptor in PSII. However, more detailed observations indicate that the changes in C-550 probably result from a change in the environment of the molecule rather than from a change in its redox state (81).

Hexane will extract PQ and β-carotene from lyophylized chloroplasts, and Hill activity and the light-induced changes of C-550 and cytochrome b_{559} at $-196°C$ (362) are then lost. The absorbance changes due to C-550 can be restored by the addition of β-carotene, but both β-carotene and PQ are required to restore full activity. This indicates that removal of PQ interrupts electron transport and that removal of β-carotene interferes with the primary photochemical activity of PSII. It was therefore suggested that C-550 is a carotene that indicates the redox state of the true primary electron acceptor (362). Cox and Bendall (103) confirmed that β-carotene is closely associated with C-550 absorbance by extracting the chloroplasts with heptane, and they concluded that C-550 indicates the redox state of the primary acceptor but that it is not an obligatory component of electron transport from PSII. This was confirmed by extraction of lyophilized PSII chloroplast fragments with 0.2% methanol in hexane to remove almost all PQ (272). The extracted fragments showed neither light-induced C-550 absorbance change nor oxidation of P-680. The P-680 oxidation can be restored by PQ but not by β-carotene, and carotene can restore the C-550 absorbance change only when added in conjunction with PQ. This shows that PQ is the primary acceptor (Q) and that carotene plays no direct role in electron transfer although it is required for the C-550 absorbance change.

The conclusion is that the absorbance change at 550 nm is due to a component, probably a carotenoid that is closely associated with the primary acceptor in PSII, that shows an absorbance-band shift on reduction of the primary acceptor. C-550 can, therefore, be used to monitor the redox state of the primary acceptor Q even though it is not itself reduced (243).

2. C-515: The Electrochromic Shift

An absorbance increase, different from that of C-550, occurs at 515 nm; it is called the electrochromic shift. It is due to an absorbance change in a pigment molecule, possibly chlorophyll b, but in this case the absorbance change is a result of a charge separation across the membrane. It is considered at this point because of its similarities to C-550, which can be confusing even though C-515 is not directly involved with the primary acceptors.

The nature of C-515 was first explored by Junge and Witt, who demonstrated that the increase in absorbance at 515 nm was extremely rapid after a light flash (250). The decay of C-515 is biphasic, and the slow phase is correlated with phosphorylation of ADP. The rate of decay of C-515 can be greatly increased by the inclusion of K^+ and the potassium transporter, gramicidin, in the medium. This indicates that the absorbance peak is due to charge separation across the membranes. The fast phase of decay is the result of damage to the membrane; the slow phase is due to proton transport across the membrane to neutralize the charge separation (250).

Reich et al. (403) applied an electric charge to artificial membranes containing layers of pigments. They found a C-515 absorbance shift that most closely resembled the natural one when the pigments had a chlorophyll to carotenoid ratio of 5 : 1, similar to the ratio found in vivo. The absorbance change was proportional to the electric field strength. They concluded that the absorbance change at 515 nm is a measure of the number of charges that cross the thylakoid membrane. It is called the electrochromic shift because the change in absorbance is caused by an electrical potential across a membrane.

There is a linear relationship between the amount of absorbance change at 550 and that at 515 nm, indicating that the charge separation is related to the formation of Q^- from Q, that is, to the transport of electrons between the two photosystems (see Fig. 10). Recently the C-515 absorbance change has been used to measure the contribution of cyclic electron transport to the generation of a charge separation and proton gradient across the thylakoid membrane (109). However, the interpretation of flash-induced changes at these wavelengths can be complicated by other absorbance changes (156).

D. ELECTRON DONORS TO THE PHOTOSYSTEMS

1. Plastocyanin: Electron Donor to P-700

The structure and function of plastocyanin have been reviewed extensively (67, 254). The role of plastocyanin as an intermediate electron-transfer component between the two photosystems has long been recognized (492). It is the only component in this electron transfer chain that contains copper. Plastocyanin is a small protein with a molecular weight of about 10,500 and contains one copper atom per molecule (401). The protein has been purified to homogeneity from a number of plant sources and has been shown in all cases to be very similar (254). It absorbs red light of 597 nm, so the purified protein is blue (254) and crystallizes as deep-blue crystals (91, 96).

The amino acid composition of plastocyanin from several sources has been published and the protein sequenced (67, 254). It is a single polypeptide chain of 99 residues in higher plants, and the sequences of plastocyanins from a great variety of sources are similar, with two regions highly conserved (67). The midpoint potential is approximately 350 mV, which is considerably higher than the redox potential of the Cu(II)–Cu(I) couple in aqueous solution (97). This indicates that the copper atom is coordinated in the molecule in such a way that Cu(I) is stabilized with respect to Cu(II) to give the higher midpoint potential (97).

The structure of plastocyanin has been determined by X-ray crystallography (97). The molecule is a flattened cylinder in which the wall of the cylinder is formed by eight strands of polypeptide parallel to the long axis. The core of the molecule is hydrophobic. The copper atom is embedded in the molecule and coordinated to the protein backbone by the sulfur atoms of a residue each of cysteine and methionine and the nitrogen atoms of two histidines. The environment around the copper atom is probably optimal for its redox function in the electron transport chain (97). The copper atom is separated on one side from the medium by a histidine ring, whereas on the other side a hydrophobic channel is found in all plant plastocyanins. This channel is therefore probably essential to the function of electron transfer.

The solvent-accessible site and the hydrophobic site may give direction to electron transfer in the molecule. Cytochrome f may thus bind to one site and transfer an electron to the copper. This electron is then transferred through the molecule to P-700 (97). Hydrophilic and hydrophobic patches on the outside of the molecule may provide specific sites for the interaction of plastocyanin and its redox partners. The binding sites for these partners have been identified by NMR spectroscopy and have been shown to be different (98). One is close to the copper atom, and one is removed by at least 10–15 Å, confirming that two electron pathways (possibly two mechanisms) are involved (98). There is evidence that complexes between the redox partners must form before electron transfer can take place (282). Again, two sites appear to be involved.

Five different models for the sequence of electron carriers in this part of the chain have been proposed (254). A sequence of cytochrome f to plastocyanin to P-700 now appears to be usually accepted (see Fig. 10). Most of the data on the sequence of components have been derived by measuring the absorption changes of plastocyanin and cytochrome f associated with the reduction of P-700 after its oxidation by a short flash of light. This measurement is complicated by the reduction of P-700 in three phases: a fast phase taking 10–20 μsec, an intermediate phase of 200 μsec, and a slow phase of 10 msec (182). This last phase is due to the flow of electrons from PSII. The kinetics of the intermediate phase coincide exactly with the oxidation kinetics of plastocyanin, that is, it represents a transfer of electrons from plastocyanin to P-700. Measurement of oxidation of both plastocyanin and cytochrome f in this time span have led to the suggestion that cytochrome f and plastocyanin act in parallel to supply electrons to P-700 (180). In contrast, Bouges-Bocquet has proposed from similar measurements that there is a linear flow of electrons from cytochrome f to plastocyanin and then to P-700, because cytochrome f is oxidized more slowly than plastocyanin and there is a time lag for the start of oxidation (64). However, the relationship between cytochrome f and plastocyanin, as measured by the reduction kinetics of these carriers (65), may be complicated by the absorbance of a number of components (including C-515) at the wavelengths of the absorbance of cytochrome f and plastocyanin (64).

The very rapid phase has been more difficult to interpret, and it has been suggested that there is an unknown component between plastocyanin and P-700 (66). However, other studies (182) with inhibitors of plastocyanin (232) indicate that plastocyanin is involved in the fast transfer and that it is the direct electron donor to P-700.

From the effects of salt, sorbitol, and KCN on the electron transfer from plastocyanin to P-700, Haehnel *et al.* (182) concluded that there is a tight association between plastocyanin and PSI. The reaction center of PSI consists of six subunits. One of these, subunit III, is required for electron transfer from plastocyanin to PSI (38), although this subunit is not itself an electron carrier (181). Subunit III may act as a recognition site for the binding of plastocyanin (182). Kinetic analysis at low temperature also provides evidence of a direct transfer of an electron from plastocyanin to P-700 without intermediates (365). However, the oxidation of cytochrome f by P-700 has been shown to be inhibited by low ionic strength, which appears to disrupt the interaction between plastocyanin and P-700. This suggests that an intermediate may be mobile or easily solubilized (291).

The interaction of electron donors and P-700 is complex and seems to involve both tightly bound and mobile carriers (291). It may be concluded that electrons from cytochrome f are transferred to P-700 via plastocyanin and that bound plastocyanin is the immediate donor to P-700. Whether soluble plastocyanin is an intermediate between cytochrome f and bound plastocyanin or is merely in electronic equilibrium with it is not yet clear (182).

2. The Electron Donor to Photosystem II: Photolysis of Water

The production of oxygen by the photooxidation of water by PSII is one of the most important reactions in nature (reviewed in 121, 430). However, despite the attention this topic has received, the mechanism by which oxygen is released from water is still obscure. To produce a molecule of oxygen, four electrons must be removed from two molecules of water

$$2 \ H_2O \rightleftharpoons O_2 + 4H^+ + 4e^-$$

The protons are liberated into the lumen of thylakoids as part of the mechanism by which a protonmotive force is developed, and the electrons are passed by way of the two photosystems to $NADP^+$ and finally to CO_2. Because oxygen must be liberated as the dioxygen molecule, there must be a mechanism by which two molecules of water are bound to the reaction site while four electrons are removed and a molecule oxygen is produced.

The midpoint potential of this reaction is 0.81 V at pH 7.0 (121, 430). Therefore, a powerful oxidizing agent must be produced by PSII to remove the electrons from water. Because four oxidizing equivalents are required before the two molecules of water are oxidized, there must be a mechanism for stabilizing the oxidizing intermediates, which are very reactive. Research is just beginning to determine how this is achieved.

It has long been known that manganese is essential for the water-splitting reaction of PSII (121, 430). Manganese can assume multiple oxidation states, some of which have fairly high reduction potentials. This makes manganese a suitable atom for the oxidation of water (430). There are two pools of manganese in the chloroplast. One is not involved in oxygen evolution and can readily be removed by EDTA; the other is firmly bound in the membrane. The latter pool consists of two portions. Two-thirds of it can be removed by various means such as heat, aging at 35°C, Tris buffer at pH 8.0, and NH_2OH (92, 93), with concomitant loss of oxygen-evolving activity. The remainder of the pool is firmly bound and can still accept electrons from artificial donors (93). The restoration of the oxygen-evolving site by incorporation of manganese requires light (94, 544) and can reach 100% of the control level by suitable manganese treatment (47). Estimates of the number of manganese atoms associated with each site range from 2.5 (93) to 6 (95). EPR spectroscopy has been used to study the membrane-bound manganese. The manganese has been shown to be tightly bound; it does not exhibit an EPR signal until after its release into the lumen of the thylakoids by the action of Tris buffer (51).

Details of the structure of the oxygen-evolving complex will not be known until it has been isolated and characterized. A manganese-containing protein has been isolated from thylakoid membranes (465). Membranes without the protein do not evolve oxygen, but they have active PSII centers as determined by electron transport between artificial donors and acceptors. When the depleted membranes are reconstituted with the manganese protein, oxygen evolution is resumed. The isolated protein shows some of the properties of the intact system, being inactivated by Tris buffer at pH 8.0, a procedure which results in the removal of the bound Mn^{2+}. The molecular weight of the protein is 65,000, and it contains two manganese atoms per molecule. Tris treatment of the depleted membranes before they are reconstituted does not affect subsequent activity, indicating that the manganese–protein is the site where oxygen evolution is inhibited by Tris in intact chloroplasts (465).

Attempts have been made to study the donor site to PSII by means of EPR spectroscopy. An EPR signal (signal IIf) can be observed in chloroplasts in which the water oxidation mechanism has been inhibited by treatments such as Tris washing (22, 23). This signal has been interpreted as originating from the primary donor to PSII, which has been termed Z. It has been suggested that Z accepts electrons from water or artificial reductants. Reduction by artificial donors is slower than that by water, and hence the signal can be seen more readily in Tris-washed chloroplasts. An EPR signal with a spectrum similar to signal IIf is observed in uninhibited chloroplasts, which can oxidize water. This signal (signal IIvf) decays much more rapidly than signal IIf, and it has been suggested that it arises from the action of Z in its normal physiological role as an electron transfer agent between the water-splitting complex Y (see following discussion) and P-680 (48). Detailed kinetic studies of the rise time of signal IIvf show that its formation is slower than that required for an immediate donor to P-680, and an intermediate component between Z and P-680 has been proposed (49). The nature of these components, detected by EPR spectroscopy, and their relationship to the manganese–protein complex, are at present unknown.

In addition to manganese, the reactive center of the electron donor to PSII appears to require chloride ions (218). The chloride effect is found only in chloroplasts that have been sensitized to chloride by uncouplers such as EDTA (231). The chloride effect is not found when artificial electron donors are used to supply electrons to PSII, indicating that chloride ions directly affect the water-splitting reaction (231). Chloride can be replaced to some extent by anions of other strong acids (218), and anions have been incorporated into models of oxygen evolution by PSII (430).

In early reports the water-splitting enzyme complex was termed Y. The mechanism of oxygen evolution by this complex has been studied by measuring the release of oxygen after very short flashes of light designed to activate each reaction center with one photon (single-turnover flashes). An essential component for the study of oxygen evolution under these conditions was the development of an extremely sensitive oxygen monitor with a very rapid response time. A polarographic method for detecting oxygen production was developed by Joliot and Joliot (242) and was used to determine oxygen output by dark-adapted chloroplasts and algae in response to single-turnover light flashes (241). It was found that little oxygen is released after a single light flash, but after a series of flashes oxygen is evolved with a periodicity corresponding to four light flashes. This evolution becomes dampened and fairly rapidly reaches a constant evolution after each flash. It has been concluded that there are two electron donors to the oxygen-evolving center that donate two electrons each. A switching mechanism between the two donors connects them to the complex to allow the evolution of single oxygen molecules (241).

A detailed study of the oxygen evolution using isolated chloroplasts by Kok *et al.* (274) led to the suggestion that each water-splitting complex Y accumulated four oxidizing equivalents

$$S_0 \xrightarrow{h\nu} S_1{}^+ \xrightarrow{h\nu} S_2{}^{2+} \xrightarrow{h\nu} S_3{}^{3+} \xrightarrow{h\nu} S_4{}^{4+} \xrightarrow{2\,H_2O} S_0 + O_2 + 4\,H^+$$

The sequence S_0 to S_4 represents the four S states or oxidation states of the complex. The mechanism is, therefore, a linear, four-step sequence. The S_4 state reacts with water to liberate oxygen. The S_2 and S_3 states are thought to be unstable in the dark and revert to the S_1 state so that only S_0 and S_1 are found in the dark. This assumption is required to explain the occurrence of the maximum yield of oxygen on the third flash (274). This mechanism has provided the basis for further models of oxygen evolution by PSII (e.g., 405, 430). The rate of transformation of the various S states and the potentials developed by each have been studied on a number of occasions (e.g., 62) and have been reviewed extensively (121, 430). Until the structure and component parts of the oxygen evolving complex are determined, the identity of the various S states cannot be clarified.

It appears that some compounds uncouple photophosphorylation by deactivating the water-splitting enzyme complex Y (404). These compounds have been termed ADRY (Acceleration of the Deactivation Reactions of the water splitting-enzyme system Y) reagents. ADRY compounds appear to act specifically on the S_2 and S_3 states of the complex by short-circuiting electrons either from the PQ pool (Section V,E,3) or from some other electron-transfer component back to the electron holes[6] in the water-splitting complex (406).

In the system of S states, it would appear that four oxidizing equivalents are accumulated and that two molecules of water are then oxidized. This would result in the simultaneous liberation of a molecule of oxygen and four protons. The use of fast and sensitive glass electrodes allows the measurement of proton release in response to single-turnover flashes, and the results indicate that proton release oscillates with a periodicity similar to that of oxygen yield. This at first suggested that water is indeed decomposed in a final, concerted oxidation with the liberation of four protons (150). More-detailed analyses of proton release, however, suggested that the situation is more complex, that only two protons are

---

[6] That is, the sites from which electrons have been removed. (Ed.)

released on the conversion of S_4 to S_0 (149, 248, 427). The other two protons are released singly at prior oxidation steps; but there is no agreement as yet on which oxidation steps release protons (121). The uncertainty may be due to the fact that the protons are released inside the thylakoid lumen (18, 166, 248, 427). This means that the release of protons can only be detected after their passage across the thylakoid membrane. This passage may be accelerated by the use of uncouplers (150) or by the use of thylakoid-permeating pH indicators (248, 427), but measurements are still difficult.

Substantial evidence shows that the oxygen-evolving site of PSII is located on the inner surface of thylakoid membranes. In this way proton release from PSII occurs inside the thylakoid lumen and contributes to the protonmotive force for ATP synthesis. Oxygen is also released into the inner thylakoid space. Evidence bearing on the location of the oxygen-evolving complex has been summarized by Diner and Joliot (121).

Early work on oxygen evolution in photosynthesis indicated that carbon dioxide or bicarbonate was directly involved in the photoreaction (514). Although the original interpretation of these data is obviously incorrect, reports by Stemler have suggested that CO_2 or HCO_3^- is bound to the reaction center of PSII. Chloroplast fragments will tightly bind one HCO_3^- molecule per 380–400 chlorophyll molecules, and the removal of the bound HCO_3^- inhibits oxygen evolution (466). The inhibition is reversed by the addition of HCO_3^-. The HCO_3^- appears to bind in a dark reaction and to be released in a light-requiring reaction (467). More recent refinements in studies of this cyclic activity of HCO_3^- have indicated that CO_2, not HCO_3^-, is bound to the reaction center and is subsequently hydrated to $HCO_3^- + H^+$ on the surface of the complex (469). Formate inhibits the binding of CO_2 and also inhibits oxygen evolution. It also increases the duration of two of the four S-state transitions (468). From these observations a model has been proposed in which CO_2 acts catalytically in the oxidation of water. On two occasions during the development of four oxidizing equivalents at the catalytic center, CO_2 is bound to the complex. On each occasion CO_2 is then hydrated, a process that acts to bring two molecules of water to the reaction center. On each occasion, after activation by two quanta of light, CO_2 is released together with a proton, and an oxygen atom remains bound to one of the two manganese atoms in the complex (468).

Whether the model involving CO_2 (468) or other models such as those proposed by Sauer (430) or Renger (405) eventually prove to be correct will only be determined when pure preparations of the complex are made. It is clear that manganese is the agent involved in the oxidation–reduction reaction, but how it interacts with PSII or water has not been determined. The development of techniques to produce a pure complex (465) should lead to clarification of these interactions.

E. Noncyclic Electron Transport

Electrons that have been excited by PSII are transported to the acceptor site of PSI via an electron transport chain that is similar in many ways to the mitochondrial system. This noncyclic path, which is shown in Fig. 10, consists of electron acceptors Q and R, PQ, an iron–sulfur center, cytochrome f, and plastocyanin, which passes electrons to the PSI acceptor. The major transport agents are described in the following subsections. The systems that donate electrons to or accept them from the photosystems have already been described (Section V,B–D).

1. Cytochrome f

The importance of cytochrome f in the electron transport chain was recognized as early as 1960 by Hill and Bendall, who associated it directly with what would now be called PSI (214). The central role of cytochrome f as an electron transport component between the two photosystems was also recognized in 1961 by Duysens et al. (131). It is surprising that after these early observations and numerous other reports on the function of cytochrome f its exact role in chloroplast electron transport is still debated (106) and reviewed (36, 104, 106).

Cytochrome f is a c-type cytochrome bound in the thylakoid membrane (542). It is characterized by an α band at 554–555 nm and a β band for the reduced cytochrome at 421–422 nm. It has a midpoint potential of about 360 mV (542). Cytochrome f has been purified to homogeneity from spinach (Spinacia oleracea) chloroplasts (222). The purified cytochrome has a molecular weight of 36,000 and aggregates to form octamers (222). Cytochrome f from spinach appears to be quite similar to that cytochrome from a cyanobacterium, suggesting that the structure has been conserved during evolution. It has been suggested that cytochrome f is maternally inherited through chloroplast DNA (167a).

It is now well recognized that cytochrome f is an essential component in the linear flow of electrons from PQ to P-700 in noncyclic electron flow (e.g., 520) and also of cyclic electron flow (456, 457). Because the rate of reduction of cytochrome f is increased when PSII is activated, the electrons for its reduction may come from both cyclic and noncyclic electron flow (45). Cytochrome f, therefore, may act as an electron acceptor from PQ for both cyclic and noncyclic electron flow and may be involved in the formation of protonmotive force by both these pathways (Section VI,A). Cytochrome f remains oxidized in the light, indicating that it follows a rate-limiting step, which is the generation of a proton gradient between PQ and cytochrome f (316). Hence the rate of reduction of cytochrome f is enhanced by the uncoupler NH_4Cl (100). The PQ antagonist DBMIB blocks the photoreduction of cytochrome f, but not its photooxidation (58, 198, 227, 543). The reduction of cytochrome f by electrons from PSII is also blocked by DCMU, which prevents electron transfer from Q to PQ (100). All of this evidence clearly establishes the location of cytochrome f between the two photosystems; it receives electrons from PQ and passes them to P-700. The precise location of cytochrome f between PQ and P-700 must now be considered.

The close association of cytochrome f with PSI is indicated by its rapid oxidation when chloroplasts are illuminated with red light that energizes only PSI. It has been frequently suggested that cytochrome F may donate electrons directly to P-700 because it is oxidized rapidly (60–150 μsec), as shown by Hildreth (211) using laser light and by Dolan and Hind (122).

The stoichiometry between the oxidation of cytochrome f and reduction of P-700 is 1 : 1, and the amplitude of the oxidation–reduction changes is compatible with the participation of one cytochrome f per electron transport chain (520). This conflicts with an earlier suggestion that cytochrome f does not transfer electrons to P-700 in the dark (178, 180). Although this transfer can take place in the light, the more direct route from PQ was thought to be via plastocyanin. Cytochrome f and plastocyanin would, in this view, act in parallel, with plastocyanin being the more important component (178, 180). Although there is still some controversy about this, the most probable scheme is a linear sequence of PQ to cytochrome f to plastocyanin to P-700. The course of cytochrome f oxidation is sigmoidal with a half-time of 100 μsec, which is slower than the oxidation time of plastocyanin (64). This places cytochrome f between plastocyanin and P-700. In addition, cytochrome f is reduced more rapidly than plastocyanin, suggesting that it is closer to the source of electrons from PSII than plastocyanin (65).

The isolation of mutants of *Chlamydomonas reinhardtii* lacking either cytochrome f or plastocyanin provided a most useful tool for the study of electron carriers (163, 288). In both mutants noncyclic electron transfer from water to $NADP^+$ is blocked. This shows that both carriers are essential for the transfer of electrons between the photosystems and that they do not operate in parallel. P-700 can be reduced by artificial donors in the mutant that lacks cytochrome f but not in the one that lacks plastocyanin. This places plastocyanin close to P-700. Cyclic photophosphorylation mediated by phenozine methosulfate is blocked in the mutant that lacks plastocyanin but not in the one that lacks cytochrome f, which confirms the transfer of electrons from plastocyanin to P-700. Also, Wood showed that purified reduced cytochrome f, but not other cytochromes or hydroquinone, will transfer electrons to purified plastocyanin in solution at a very rapid rate (541). All this justifies placing plastocyanin between cytochrome f and P-700, as shown in Fig. 10.

When plastocyanin is removed from chloroplasts by treatment with the detergent Triton X-100, the rate of cytochrome-f oxidation by flashes of light that activate PSI is reduced (215). The rate is restored by the addition of plastocyanin. Also, plastocyanin, loosely attached to the inner surface of the thylakoid membrane, can easily be removed by sonication (384). Sonicated chloroplasts that contain only 10% of their original plastocyanin will not oxidize cytochrome f when subjected to far-red light, and addition of plastocyanin restores this oxidation. Finally, high concentrations of KCN inhibit electron transport in chloroplasts by inactivating plastocyanin, but KCN does not react with cytochrome f. In a detailed investigation of the effect of KCN, Izawa et al. showed that KCN blocks both the reduction of photooxidized P-700 and the photooxidation of cytochrome f (232).

All these effects are consistent with a close connection between cytochrome f and PSI mediated by plastocyanin, and measured in situ redox potentials of cytochrome f and plastocyanin are compatible with this view (310). Although the sequence of these components is now clear, the precise mechanisms of their mutual operation remain to be determined.

2. The High-Potential Rieske Iron–Sulfur Protein

Iron–sulfur proteins have been identified in a number of electron transfer systems, and they may have a universal function in energy transduction (308). The particular iron–sulfur protein involved in noncyclic electron transport is very similar to the Rieske protein first identified in complex III of mitochondria (415; see Section III,A,4). This center can be detected in broken, washed chloroplasts that have been oxidized with ferricyanide and reduced with dihydroquinone (305). It is tightly bound to the membrane. It has a midpoint potential of 290 mV, which suggests that it is between PQ and cytochrome f in the photosynthetic electron transport chain.

There are several lines of evidence that bear on the location of the iron–sulfur center in the electron transport chain. First, lipophilic iron chelators like bathophenanthroline inhibit iron–sulfur centers. These compounds were found to stop electron transfer between PQ and cytochrome f (40). Second, a mutant of *Lemna perpusilla* that has a block in the electron transfer chain between PQ and cytochrome f was found to lack Rieske protein (440). Third, duroquinol (tetramethyl-*p*-hydroquinone), which feeds electrons into the PQ pool, causes the reduction of the iron–sulfur center, cytochrome f, plastocyanin, and P-700 (311). Further work with the electron transport inhibitors DCMU and DBMIB is compatible with the linear sequence of these components (517). Finally, the redox state of the Rieske protein can be detected by EPR measurements (521). The center is reduced in chloroplasts that are illuminated with red light. In the presence of DCMU, which blocks the transfer of electrons from PSII, the center is oxidized.

All of these data support the placement of the iron–sulfur center in the linear sequence of carriers in the noncyclic electron transport chain between PQ and cytochrome f, as shown in Fig. 10. Because iron–sulfur centers are widespread, further elucidation of their structure and function would seem to be of utmost importance.

3. Plastoquinone

The history and function of PQ as an electron carrier between the two photosystems has been reviewed (6, 36, 495). Several types of PQ are found in plants; the most abundant and probably most important is PQ_A which is located in the thylakoid membranes. Some PQ is also found in the osmiophilic granules of chloroplasts, but this is probably not directly involved in photosynthesis. PQ requires two electrons for full reduction, but it can accept one electron and form the semiquinone anion, which has a characteristic absorption spectrum (39). The fully oxidized molecule has an absorption band at 260 nm, and the fully reduced molecule (PQH_2) a weaker band at 290 nm. The midpoint potential of the $PQ-PQH_2$ couple is 113 mV (6).

It was shown early that a pool of PQ molecules can be reduced by PSII and oxidized by PSI (420). The primary acceptor of PSII, which is probably a semiquinone, passes electrons to a pool of seven or eight molecules of PQ (5, 470, 471). If the PQ pool is removed by organic solvents, PSII activity is lost, but it is restored by the addition of pure PQ (5). The high efficiency of PQ oxidation and reduction and the kinetics of all the chain components suggest that the only component in the chain that is present as a pool is PQ (7). The relative amounts of PQ to P-700 found in this study were 15 : 1 in terms of electron equivalents, that is, 7–8 molecules of PQ to every one of P-700 (7). Understanding of the operation of the photosynthetic electron transport chain and, particularly, the role of PQ requires recognition that all other components of the chain are fixed and oriented in a membrane in such a way that interactions can take place only between individual molecules. In other words, parallel chains act independently; in each one electrons are transported from molecule to molecule. PQ is different in that it exists as pools of seven to eight molecules each (7); each pool has access to at least ten individual and parallel chains (449). PQ therefore acts to bridge these individual chains, coordinating their operation and acting as an electron "buffer" in the electron transport system (495).

In Mitchell's formulation of chemiosmotic coupling (see Section II,B), PQ acts as a redox loop receiving electrons from Q and protons from the chloroplast stroma (328). In turn, PQ is oxidized on the inside of the thylakoid by cytochrome f, and it transfers two protons into the vesicle of the thylakoid (328). The effect of ADP and ATP and the uncoupler NH_4Cl on the redox state of cytochrome f and PQ shows that there is a coupling site for energy transduction between two components (59). There appears to be a 1 : 1 ratio between external proton uptake and electron uptake by the PQ pool (491), and the rate of internal H^+ release is identical to the rate of PQ oxidation. All these data show that PQ acts as a proton pump in noncyclic electron flow by transporting protons vectorially across thylakoid membranes (491).

Plastoquinone also takes part in cyclic electron flow in intact chloroplasts where it also acts as a proton pump across the membranes. There a more complex arrangement of cytochrome b_6 and bound PQ may form a type of Q cycle, as had been proposed earlier by Mitchell (334). A Q loop has also been proposed for noncyclic electron flow to account for measurements showing that possibly more than one proton is transported for every electron passing between PQ and PSI (505). Proton transport associated with electron transport is discussed in more detail in Sections VI,A and B.

The PQ antagonist DBMIB (125) inhibits electron flow from PQ to the rest of the chain, but it does not affect the transfer of electrons from Q to PQ (495). It has been used to locate a second site of energy transduction, where protons are generated within thylakoid vesicles by the water-splitting activity of PSII (495). Inhibition by DBMIB of electron flow between the two photosystems can be relieved by TMPD, which acts as a bypass of the DBMIB block and feeds electrons directly to plastocyanin, though without proton transport (495).

There is some evidence from studies with artificial electron acceptors and donors that PQ may also be involved in the water-splitting reactions of PSII (424). The midpoint potential of PQ is not compatible with this role, but it is possible that molecules of PQ involved in this process are in a specialized environment (424).

Plastoquinone, therefore, plays an important role in energy transduction in photosynthesis. It transports electrons between the two photosystems and in the process transports protons across the thylakoid membrane either by a simple redox loop or by a more complex Q cycle. PQ may perform this function in both cyclic and noncyclic electron transport. In addition, it acts as a primary acceptor (Q) in PSII and, possibly, as secondary acceptor between the primary acceptor and the PQ pool. Finally, PQ may be involved in the water splitting-reactions of PSII.

F. THE REDUCTION OF $NADP^+$

1. Ferredoxin

Ferredoxin is the redox component that links the reducing power produced by PSI to other biochemical reactions in the chloroplast (reviewed in 183). Arnon, a pioneer in Fd research, has reviewed its history and has evaluated its role in chloroplast metabolism (10). Not only is Fd the link between PSII and $NADP^+$ reduction, it also serves as the intermediate in ATP synthesis by cyclic electron transport (10, 11, 55). It also provides the reducing power for converting nitrite to ammonia and sulfite to sulfide (10).[7] An entirely different role for Fd in the control and integration of cellular metabolism is its ability to activate several chloroplast enzymes by way of thioredoxin (78, 79).

The ability of Fd to perform these various tasks is dependent upon its location in thylakoid membranes. Most of the Fd is loosely bound to the outer surface of the thylakoid membrane and is readily lost when chloroplasts are isolated, as are functions such as cyclic photophosphorylation. There appears, however, to be a small amount of tightly bound Fd that will support a minimal rate of cyclic photophosphorylation (55). Ferredoxin can, therefore, interact with the tightly bound iron–sulfur centers that act as electron acceptors from PSI (Section V,B,1), but it also interacts with other membrane electron carriers in addition to soluble components. The rest of the electron carriers in thylakoid membranes are tightly bound (including the $NADP^+$ oxidoreductase; see Section V,B,2), and they could not serve this function.

A ferredoxin was first isolated by Mortenson et al. from Clostridium pasteurianum during studies on nitrogen fixation by this organism (345). Previously, several soluble factors had been isolated from chloroplasts that stimulated $NADP^+$ reduction (10). However, it was not until 1962, when Tagawa and Arnon (485) showed that Fd from C. pasteurianum had

[7] The role of ferredoxin in nitrogen fixation will be treated in Volume VIII, Chapter 1. (Ed.)

the same function as the soluble factors from chloroplasts, that it was recognized that all these factors were in fact plant Fd, which is very similar in structure to the bacterial protein (345).

Ferredoxin belongs to the general class of iron–sulfur proteins (reviewed in 369). It is a soluble iron–sulfur protein that has two nonheme iron atoms and two labile sulfur atoms at the active center, that is, 2Fe–2S (183). This is in contrast to the bacterial protein, which is 4Fe–4S at the active center. The protein has been purified to homogenity from a number of sources and consists of a single polypeptide chain of 96–98 amino acids with a molecular weight of approximately 11,000. The Fd from wheat (*Triticum aestivum*) has been sequenced by Takruri *et al.* (487) and has been shown to contain only four cysteine residues, which are present in all ferredoxins and are required for the binding of the two iron atoms. Takhuri *et al.* (487) present a table comparing the various plant ferredoxins. There is a great deal of homology between the proteins in addition to the four cysteines (183, 369).

The structure of the active center of ferredoxin has been closely studied (369). It contains two iron atoms linked by sulfide bridges. Each iron atom is externally coordinated by two mercaptides (from two of the invariant cysteines in the polypeptide). Dilute acid liberates H_2S from the noncysteine sulfur of Fd (183). An apoprotein can be prepared and native Fd reconstituted from it, using iron, sodium sulfide, and a mercaptan (369). The oxidized molecule does not give an EPR signal. On reduction, Fd gains one electron, and one of the iron atoms is converted from Fe^{+3} to Fe^{+2}. In this state the protein gives an EPR signal, which is due to a net unpaired electron (369).

Ferredoxin has the very negative midpoint potential of -420 mV, close to that of the hydrogen electrode. It is red-brown in color and has absorbance peaks at 330, 420, and 460 nm. The absorbance at 420 nm decreases by 50% when Fd is reduced.

Ferredoxin, therefore, acts as a link between the reducing power generated by PSII and various other pathways in the chloroplast. Its primary role in photosynthetic electron transport is to transfer electrons from PSI to $NADP^+$ (see Fig. 10). Its involvement in cyclic and pseudocyclic electron transport is discussed in Sections V,G and VI. It is a well-characterized redox component because it can easily be solubilized and purified, unlike other components of the thylakoid that are firmly bound in the membrane.

2. Ferredoxin: NADP$^+$ Oxidoreductase

The transfer of electrons from Fd to $NADP^+$ is catalyzed by the enzyme ferredoxin:NADP oxidoreductase (reviewed in 146). The enzyme was first purified by Avron and Jagendorf, but they did not establish its involvement in photosynthetic electron transport (19). Since then the enzyme has been purified several times and its function clearly recognized (550). It contains 1 mol of FAD, which is not covalently bound because it can be removed by urea when it is reduced. There are one disulfide bridge and four free SH groups, one of which is not readily accessible to SH reagents and is essential for catalytic activity (146). Incubation of the enzyme with NADPH inactivates it; the activity can be restored by NADP$^+$. The full reduction of the enzyme makes the binding of FAD to the protein less stable and more sensitive to SH reagents (147). It also forms a stable 1 : 1 complex with Fd (148), which affects the kinetics of the enzyme (146).

When purified, the enzyme is a monomer with a molecular weight of approximately 40,000 (146). Multiple forms of the enzyme have been reported (152, 164), and heterodimers have also been found (152, 432) that may be quite stable (432). The tight binding between Fd and the enzyme has been used to purify the enzyme. An affinity column of Fd bound to Sepharose 4B will bind the enzyme, and it can then be eluted by high salt concentrations (442). After purification on the affinity column the reductase can be separated into two fractions by DEAE-cellulose (442). However, during SDS-gel electrophoresis only one band is found, and it has been suggested that one fraction is a homodimer of the other (442). The presence of a homodimer has been confirmed for the spinach leaf enzyme (549). This dimer tends to dissociate and is lost on purification (549). The enzyme is thought to be bound in a crevice below the surface of the membrane (41) and held to the surface by ionic interactions (432).

G. CYCLIC ELECTRON TRANSPORT

Cyclic electron transport coupled with ATP synthesis is a device for producing ATP without the concomitant reduction of NADP$^+$. This is achieved by passing an electron from a photosystem, energized by the absorption of light, in a cyclic pathway via an electron transport chain back to the same photosystem. ATP is synthesized *en route*.

There has been much controversy about the need for ATP production by cyclic photophosphorylation and about the very existence of the pathway. However, its existence now seems to be well established. The photosynthetic carbon reduction cycle requires three ATPs per two NADPH, and the operation of noncyclic electron transport (shown in Fig. 10) probably produces only two. Furthermore, other photosynthetic cycles and reactions (e.g., C_4 photosynthesis and photorespiration; see Chapter 3, Sections III and IV) require additional ATP. Finally, other cellular

235

236   D. T. DENNIS

syntheses (including nitrogen metabolism, which in darkness would be driven by respiratory ATP) are thought to be driven, at least in part, by photosynthetically produced ATP in the light (see Chapter 3, Section VI; Chapter 4, Section VII). It is probable that the proportion of cyclic and noncyclic photophosphorylation is governed by the relative demand for ATP and NADPH, the products of the two cycles.

The following discussion will present evidence for the formulation of cyclic electron transport. The main pathway in Fig. 10 shows electrons leaving the activated (top in Fig. 10) site of PSI and passing via Fd, cytochrome b_6, PQ, cytochrome f, and plastocyanin to the electron acceptor site (P-700) of the same photosystem. The cycle is repeated when the photosystem is reactivated by light. A second and more controversial system is shown associated with PSII, in which electrons pass from the top of PSII via PQ and cytochrome b_{559} to the acceptor site of PSII. These two systems will now be described in terms of their characteristic cytochromes, b_6 and b_{559}.

1. Cytochrome b_6

A 563-nm absorbance peak first observed in chloroplasts by Davenport in 1952 was designated cytochrome b_6 (111). It was further characterized in 1954 by Hill (213). This cytochrome is, however, very tightly bound into the thylakoid membrane and was not isolated and purified until 1973 (483). Cytochrome b_6 is a strongly hydrophobic protein that tends to aggregate even when purified. It has a typical b-type cytochrome absorption spectrum with peaks at 434, 536, and 563 nm, except that at $-199°C$ the α peak splits into two peaks of equal intensity at 561 and 557 nm (483). Purified cytochrome b_6 is a lipoprotein, with one-third of the molecule being lipid (484). The protein portion consists of three different polypeptide chains (484).

Cytochrome b_6 is also similar to the mitochondrial b cytochrome that is closely associated with an iron–sulfur center and a c-type cytochrome, cytochrome f (353). Isolated particles with PSI activity contain cytochromes b_6 and f in the ratio 2 : 1 (52). Cytochromes b_6 and f do not separate after a series of purification steps of the PSI particle, indicating that they are closely associated (353).

The midpoint potentials of cytochrome b_6 and other associated electron transport enzymes have been measured in fresh chloroplasts and in various chloroplast particle preparations (61, 106, 260, 353). A value of 0 mV is normally accepted for cytochrome b_6 in intact, fresh chloroplasts (61), but much higher or lower values have been found in other preparations. It has been suggested that the hemes of two cytochrome b_6 molecules interact strongly and function as a two-electron transport system in intact chloroplasts (61). If chloroplasts are damaged, the two molecules separate and their individual redox potentials differ, which may give rise to the various values found in particles or damaged chloroplasts.

Cytochrome b_6 probably functions in cyclic photophosphorylation, as was first proposed by Arnon et al. to account for the inhibition of ferredoxin-mediated cyclic photophosphorylation by antimycin A, which specifically inhibits b-type cytochromes (13). Various cycles involving cytochrome b_6 and PSI have been proposed, each involving at least one coupling site for ATP synthesis.

Isolated PSI particles contain cytochrome b_6. Cytochrome b_6 is also reduced by red light that activates only PSI (105, 122, 267). Cytochrome b_6 is therefore probably associated with

PSI. The half-rise time for reduction of cytochrome b_6 has been reported as 1.3 msec (122), although earlier reports had claimed that it was a faster reaction. This is much longer than the half-rise time for oxidation of cytochrome f (122), indicating that cytochrome b_6 is not a primary acceptor of electrons from PSI. The half-rise time for oxidation of cytochrome b_6 is 35 msec, whereas that for the reduction of cytochrome f is biphasic with times of 7.3 and 83 msec (122). Cytochrome b_6 may, therefore, be the donor of electrons for the slower phase of cytochrome-f reduction.

These data and those described by Hildreth for *Chlamydomonas reinhardtii* (212) show that there is a slow step in the electron transport chain between cytochromes b_6 and f. The dark oxidation of cytochrome b_6 is inhibited by the quinone analog DBMIB and is restored by PQ, showing that PQ is involved in the pathway between cytochromes b_6 and f (60). Cytochrome b_6 can be reduced by diaminodurol in the dark. Its subsequent oxidation by PSI in the light is enhanced by the uncoupler (NH_4Cl) and ADP. In contrast, the rate of reduction of cytochrome f is enhanced under these conditions (60). These data indicate that there is a coupling site for ATP synthesis between cytochromes b_6 and f and that PQ is involved in the pathway.

These data and other work with DBMIB (184) indicate that cyclic and noncyclic photophosphorylation share the same pathway of electron carriers between PQ and P-700 (see Fig. 10). The kinetics of the oxidation of cytochrome f and reduction of cytochrome b_6 in response to a flash of light indicate the transfer of an electron from cytochrome b_6 to f; the relaxation times of both are also similar (458). The uncoupling agent NH_4CL accelerates b_6 oxidation and f reduction, supporting the presence of a coupling site between them (458). As mentioned previously, the rise of fluorescence of chlorophyll in PSII is biphasic; the initial phase is due to partial reduction of the primary acceptor Q, and the second phase is due to an increase in the reduction of the PQ pool. Antimycin A, which inhibits cyclic but not noncyclic electron flow, inhibits the second phase of fluorescence increase without affecting the initial phase (458). From this result it has been concluded that, because antimycin A acts on b cytochromes, the inhibition of the second phase of fluorescence is due to an inhibition of flow of electrons from cytochrome b_6 to PQ (458). The coupling site of cyclic photophosphorylation is, therefore, postulated to lie between PQ and cytochrome f as it does in noncyclic photophosphorylation, and the electron flow in cyclic photophosphorylation is coupled to ATP synthesis at this point (458). Cytochrome b_6 in the cyclic pathway, therefore, interacts with the PQ pool, and this transfer is unaffected by DCMU (60). Potassium cyanide inhibits the transfer of electrons from cytochrome f to P-700, probably by acting on plastocyanin (Section V,E,1). The photoreduction of cytochrome b_6 by far-red light is not inhibited by KCN, but its photooxidation is severely inhibited, confirming the location of cytochrome b_6 in a cyclic electron flow around PSI (53).

Experiments with the inhibitor monensin show that proton pumping is brought about by the cyclic flow of electrons in intact isolated chloroplasts. Monensin uncouples photophosphorylation by facilitating H^+–Na^+ exchange across the thylakoid membrane, thus destroying the proton gradient (326). Cytochromes b_6 and f are on opposite sides of the coupling site in photophosphorylation, and monensin increases the rate of electron transfer from b_6 to f sevenfold due to collapse of the pH gradient (456), showing that this site is coupled to the generation of a pH gradient. The inhibition of electron transport by DBMIB shows that PQ is involved in proton transport in this site (60, 184, 456).

For maximal cyclic flow of electrons, the carriers in the pathway must be poised at a certain potential. This can be achieved by adding DCMU, an inhibitor of electron transfer from Q to PQ, to the chloroplasts; this enhances cyclic photophosphorylation (459). The transfer of electrons from cytochrome b_6 to PQ is also enhanced in the presence of DCMU (459). DCMU more than doubles the antimycin-A-sensitive (i.e., cyclic) contribution to the overall phosphorylation process (459).

Although there is good evidence that cytochrome b_6 acts as a simple electron transfer agent between PSI and PQ in cyclic photophosphorylation, an alternative function for this cytochrome has been proposed. In 1976 Mitchell modified his earlier scheme for a protonmotive Q loop to involve cytochromes b_{559} and b_{563} (334). In this scheme, (similar to Fig. 5) PQ receives one electron from PSII and one from the loop involving cytochrome b_{559} and in the process removes two protons from the outside of the thylakoid membrane. On the inside of the thylakoid membrane one electron is passed from PQ to cytochrome f and the other to cytochrome b_{563}, two protons being released to the inside of the thylakoid. Finally, cytochrome b_{563} returns one electron to cytochrome b_{559}. This results in the transport of two protons per electron, and a value of two has been reported for the electron transport chain between the two photosystems activated by flashing light (151). The two b cytochromes would, therefore, act in a manner similar to that of the two b cytochromes in the proposed Q cycle in mitochondria (334).

There is increasing evidence that cytochrome b_6 plays a more complex role than just that of an electron transfer component between Fd and PQ. Heber *et al.* (205) reported that cytochrome b_6 can be reduced by PSII when it is preferentially activated by red light and that this electron transfer is inhibited by DCMU. A similar conclusion has been reached by other workers, who have suggested that the redox state of PQ influences the reduction of cytochrome b_6 and hence the rate of cyclic electron flow (55, 56). The relative rates of electron flow through the linear and cyclic pathways appear to be regulated by the common PQ pool (505).

Velthuys measured cytochrome-b_6 reduction in dark-adapted chloroplasts after flash illumination and obtained data showing that under these conditions electrons are transferred to cytochrome b_6 from PSII via PQ (505). The oxidation of cytochrome b_6 is again dependent on the redox state of the PQ pool. From this observation a new scheme involving cytochrome b_6 as an intermediate in a cycle similar to the Q cycle was devised (505).

A similar scheme was devised by Crowther and Hind (110) from flashing-light experiments in which PSII was inhibited by DCMU and cyclic flow induced by treatment with sodium dithionite. They concluded that although a simple PQ shuttle could explain proton transport in linear electron transport, some type of Q cycle is required for the cyclic pathway (110). They suggested that an unknown component V is involved in the shuttle and that this component is in electronic equilibrium with cytochrome b_6. The function of cytochrome b_6 is to store an electron so that PQ can be reduced by two electrons, one from V and one from cytochrome b_6 via V. In this process two protons are taken from the stroma (110).

Cytochrome b_6 is linked to PSI via Fd, as has been demonstrated by Arnon and co-workers (10). Ferredoxin is required for the photoreduction of cytochrome b_6 in chloroplast particles from which Fd has been removed (54). Antibodies against Fd inhibit electron flow from PSI to cytochrome b_6, whereas antibodies against ferredoxin:NADP$^+$ reductase do not (55). Hence Fd, but not the reductase, is involved in cyclic flow. The reduction of cytochrome b_6 by PSII does not, however, involve Fd (57).

It has been suggested that electrons pass from Fd to cytochrome b_6 via an unknown membrane-bound, hydrophobic component that may operate in some type of Q cycle (101). The evidence is that cytochrome b_6 cannot be reduced by dithionite, which normally reduces other chloroplast cytochromes, except in the presence of a hydrophobic electron carrier. Ferredoxin, on the other hand, does not mediate electron transfer between dithionite and cytochrome b_6. However, this suggestion does not seem to have been generally accepted.

In conclusion, cytochrome b_6 is probably a component of the cyclic electron transfer pathway. It receives electrons from the reducing side of PSI and returns them to the oxidizing side of PSI via PQ and cytochrome f. This pathway involves the transport of protons across the thylakoid membrane, at a coupling site where ATP is synthesized. However, cytochrome b_6 can also receive electrons from PSII, and detailed kinetic study of its oxidation and reduction suggests that it is not just a simple link between Fd and PQ, for it may be involved in some type of Q cycle.

2. Cytochrome b_{559}

Although there is still some controversy about the exact role of cytochromes b_6 and f, there is general agreement about the approximate location at which they function in photosynthetic electron transport. However, the same cannot be said for cytochrome b_{559} despite a voluminous literature (reviewed in 104, 106). Cytochrome b_{559} has an α absorption peak at 559 nm and was first observed by Lundegårdh in 1962 (292). This cytochrome was at first called cytochrome b_3 by Lundegårdh, but because there is also a nonchloroplast cytochrome b_3 that absorbs at 559 nm, Boardman and Anderson called the chloroplast cytochrome b_{559} (52).

Cytochrome b_{559}, which is very unstable, has been purified to homogeneity (157). The molecular weight of the complex is 111,000 per heme, but because only 41% of this complex is protein, the molecular weight of the protein component is 45,900 per heme (158). The rest of the complex consists of four lipid components, chlorophyll a (4 molecules per heme), carotene (3 molecules per heme), and two unknown polar lipids. The protein component is composed of eight polypeptide chains of at least three different types (158).

Cytochrome b_{559} was found to be associated with PSII particles (52), and it can be photoreduced by PSII (261). Determining the location of cytochrome b_{559} is complicated by the presence of high (HP) and low (LP) potential forms of the cytochrome (35). It has been suggested that these two forms are associated with photosystems II and I, respectively (8). No light-induced redox changes associated with the LP form have been found, and its importance in PSI has not been established (205). The identification of cytochrome b_{559} is complicated because the α absorption peak of cytochrome b_6 is split and produces a peak that could be mistaken for b_{559} (106, 483). In addition, fully active PSI particles have been prepared that reduce cytochrome b_6 and oxidize cytochrome f, but that do not contain cytochrome b_{559} (54). The evidence suggests, therefore, that cytochrome b_{559} is not associated with PSI, but this is not universally accepted (207). In contrast, the association of cytochrome b_{559HP} with PSII has been well established (104, 106).

The function of cytochrome b_{559} has not been determined with any degree of certainty. There are three basic reasons for this. First, cytochrome b_{559} appears to exist in two forms with different potentials, b_{559HP} and b_{559LP}, as described previously. Second, the midpoint potential of the cytochrome can be changed from the high value (0.35 V) to the low value (0.08 V) by treatments that disturb the structure of the membrane (102, 512). Removing PQ from thylakoid membranes with heptane also converts the HP form to the LP form (103). This change can be reversed by adding PQ to the extracted chloroplasts (103). After purification only cytochrome b_{559LP} is found (157). These data suggest that the midpoint potential of the cytochrome depends upon its environment. It is also possible that the LP form is an artifact of the assay because redox compounds such as ascorbate are unable to reach the cytochrome, especially after extraction by lipid solvents (157). Finally, redox changes associated with cytochrome b_{559} are usually small and slow in intact, CO_2-fixing chloroplasts. The extent of oxidation and reduction can be increased only by reagents that disrupt the structure or function of the thylakoid membrane (104).

Nevertheless, a number of positions or functions have been proposed for cytochrome b_{559}: (a) in the main electron transport chain between photosystems II and I; (b) in the oxidation of water, (c) in a cycle around PSII; and (d) in proton transport.

Cytochrome b_{559} is oxidized by far-red light (PSI) and reduced by red light (PSII), suggesting that it is a component of the main electron transport chain between the two photosystems (518 and references cited therein). Its oxidation is also inhibited by DBMIB, which inhibits the transfer of electrons from PQ to cytochrome f (Section V,E,3), thereby locating cytochrome b_{559} between PSII and PQ (58). However, the rate of reduction of cytochrome b_{559} by red light after its oxidation by far-red light is one-tenth that of PQ, indicating that it is not likely to be involved in electron transfer from PSII to PQ (518). The inhibitor DCMU, which inhibits electron transport between Q and PQ (Section V,B,2), also inhibits the photoreduction of cytochrome b_{559} (34). Detailed kinetic analysis of this inhibition indicates that cytochrome b_{559} obtains electrons from the PQ pool (519), but it is a on branch pathway separate from the main pathway of electron flow between photosystems II and I (518, 519). Current evidence suggests that cytochrome b_{559} is not on the main pathway.

A function for cytochrome b_{559} in water oxidation was first proposed by Lundegårdh in 1965 (293). It was shown that cytochrome b_{559} is closely associated with the oxidizing side of PSII by Knaff and Arnon, who demonstrated a short-wavelength oxidation of cytochrome b_{559}, even at $-189°C$ (269). This indicates a very close association between the photosystem and the cytochrome, and it was suggested that cytochrome b_{559} may be the carrier of the oxidation equivalents called S states in PSII (37). However, later work indicated that treatments with Tris buffer and hydroxylamine, which inhibit oxygen evolution, have no effect on cytochrome b_{559} (102). Neither is oxygen evolution correlated with the conversion of cytochrome b_{559HP} to b_{559LP}, which is caused by a variety of treatments (102). Also, pea (Pisum sativum) plants grown under intermittent light contain only one-fifth of the normal content of cytochrome b_{559HP} and yet they are fully capable of oxidizing water (226). Finally, the oxygen-evolving activity of etiolated barley (Hordeum vulgare) plants develops much more rapidly than does cytochrome b_{559HP} during greening (208). From all this it was concluded that cytochrome b_{559} has no direct function in oxygen evolution. However, its oxidation at low temperature and its more completely oxidation in the S_2 and S_3 states than in S_0 or S_1 indicate its close association with the reaction center (508). Cytochrome b_{559} may, therefore, act as an indicator of the oxidation state of the PSII reaction center (226).

The association of cytochrome b_{559} with the reducing side of PSII via PQ and its close association with the oxidizing side of this photosystem suggest that cytochrome b_{559} operates in a loop around PSII (33, 407). Cytochrome b_{559HP} can be photooxidized by red light with a high quantum efficiency if the electron flow from water to PSII is inhibited by the ADRY reagents FCCP or CCCP (207). These uncouplers have a pronounced effect on the PSII reaction center and accelerate decay of the S_2 and S_3 states (see Section V,D,2). The photooxidation of cytochrome b_{559} decreases when the inhibition of water oxidation by CCCP and FCCP is prevented by dithiothreitol, which reacts with them. Similarly, high light intensities overcome the effect of these reagents and will in turn prevent the oxidation of cytochrome b_{559}. These effects are observed even when the main electron transport pathway is blocked by DBMIB, showing that the photooxidation is a property of PSII, that PSI is not involved. From these data it was concluded that there is competition between cytochrome b_{559} and water to donate electrons to the oxidizing side of PSII (207; see Fig.

10). Under normal conditions photooxidation of cytochrome b_{559} is not observed because water is the preferred donor and competes effectively with cytochrome b_{559}. The photooxidation of cytochrome b_{559} is only seen when electron donating from water is reduced (207). In this scheme the function of cytochrome b_{559} is to donate electrons to PSII when the rate of generation of oxidizing capacity exceeds the rate of the water-splitting reaction. This might occur under high light intensities. This would initiate a cyclic flow of electrons around PSII and prevent the buildup of the S_2 and S_3 states, which are powerful oxidants that could damage the photochemical apparatus.

A scheme has been proposed in which cytochrome b_{559} acts as a proton pump across membranes (80). Its midpoint potential has been reported to decrease as pH is lowered to 5.0 (224, 225). Lowered pH also increases the ability of PSI to oxidize cytochrome b_{559}. This indicates that its midpoint potential depends on its state of protonation. But questions remain whether a proton-pumping scheme of this kind operates or whether it is important in energy conservation in chloroplasts.

The position of cytochrome b_{559} in photosynthetic electron transport is still uncertain. The probability is that it does not function directly in the main electron transport chain between the photosystems or in the oxidation of water. Its slow rate of reaction precludes a direct role in the photochemical reactions of thylakoids. Nevertheless, cytochrome b_{559} must be closely associated with PSII because it can be reduced and oxidized by that system.

In conclusion, cytochrome b_{559} is most likely to be in a cyclic pathway around PSII. In this role it would prevent the buildup of damaging oxidizing potential in PSII by short-circuiting electrons from PQ and feeding them back into the oxidizing side of PSII. This would account for its slow reaction rate and its preferential oxidation by the S_2 and S_3 states in the PSII electron-accepting center.

VI. Photophosphorylation

The primary function of electron-transport chains, which has been considered in depth, is to make ATP. When these chains are activated by light in the chloroplast, the process is termed photophosphorylation. Photophosphorylation will be discussed under the broad headings of proton transport (an important part of the mechanism transferring energy from electron-transport systems to ATPase where ATP is made), pseudocyclic electron flow (questions have been raised about the possibility that photophosphorylation is associated with a transfer of electrons from PSI to oxygen), chloroplast coupling factor (or ATPase), and the extent to which the system can be studied in preparations reconstituted from isolated and purified fractions.

It should be noted that although there are many similarities in the structure and function of mitochondrial oxidative phosphorylation and photophosphorylation, there are nevertheless important differences. It is therefore instructive to compare this section to Section III on mito-chondrial electron transport and ATP synthesis.

A. PROTON TRANSPORT

In the light there is an accumulation of protons within the thylakoids of chloroplasts and a corresponding rise in external pH. The evidence for this proton accumulation and its relationship to ATP synthesis is discussed in Section II,B,2 and it has been reviewed (108, 200, 233, 402, 418, 506 and references cited therein). The factors that relate proton gradients to energy conservation in chloroplasts are (a) the orientation of the photosystems and accompanying electron acceptors and donors, which determine the initial charge separation that can be translated into a proton gradient; (b) the number of protons transferred for every two electrons traversing the electron-transport chain, which in turn indicates the number of sites of proton transport; and (c) the contribution made to the proton gradient by cyclic flow of electrons. The function of pseudocyclic electron flow is described in subsection B.

1. Arrangement of the Photosystems and Electron Transfer Components

The asymmetry of thylakoid membranes is well known (200, 233, 492). The structure of the membrane is also known in some detail (15, 347, 426). Both photosystems are orientated so that when they are ener-gized, at each reaction center there is a charge separation across the membrane. These arrangements are shown in Fig. 11.

The charge separations can be detected by absorption changes due to the electrochromic effect (Section V,C,2). The negative side faces the stromal surface and the positive side faces the thylakoid lumen. The positive charge generated by PSII is neutralized by electrons from water with the release of protons into the lumen of the thylakoid. The negative charge on PSII is ultimately neutralized by proton uptake by PQ. In PSI the positive charge is neutralized by electrons from PQ with the positive charge appearing as protons in the lumen of the thylakoid. The negative charge from PSI on the stromal side of the membrane is neutralized by the reduction of $NADP^+$. All these reactions are illustrated in Fig. 11.

The orientation of the thylakoid components has been evaluated by flashing-light experiments to determine proton release, by the accessibil-

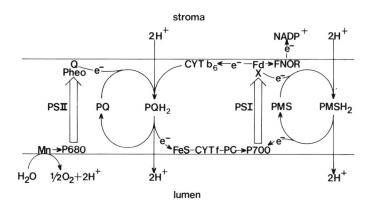

FIG. 11. Probable arrangement of the photosystems and electron transfer components in the thylakoid membrane. The sites of proton transfer are also shown, as is artificial proton transfer by PMS. The stoichiometry between electron flow and proton translocation is not shown. Adapted from Witt (539). Abbreviations, see Fig. 10.

ity of lipophilic and lipophobic electron donors and acceptors to the various components of the thylakoid, and by the ability of antibodies to block specific sites in intact thylakoids (42).

The electric field across the membrane is generated in less than 20 nsec and is simultaneous with photooxidation of the chlorophylls at the reaction centers (539). That the positive charge is on the inside of the thylakoid means the porphyrin rings of those chlorophylls are located at the inside surface of the thylakoid membrane (539). The electron donor to PSII is also located on the inside of the thylakoid membrane, as indicated by the oxidation of water at this location (demonstrated by the liberation of protons into the thylakoid lumens) (Section V,D,2; 200, 492, 539). Artificial electron-acceptors for PSII have to be lipophilic, which shows that the acceptor site is somewhat buried in the membrane (492). Other evidence from the use of antibodies suggests a stromal location for this site (492), but the significance of this finding is not clear.

Experiments with antibodies to cytochrome f and plastocyanin show that the electron donors to PSI are located on the lumen side of the thylakoid membrane (347, 539) and that hydrophilic electron donors will not interact with this site. In contrast, antibodies to Fd and ferredoxin:$NADP^+$ oxidoreductase show that these components are on the stromal side of the membrane (492, 539). Both lipophilic and polar electron acceptors are easily reduced by PSI, indicating that the electron-accepting site for PSI is on the outside of the membrane.

These results lead to the currently accepted scheme for the arrangement of the photosystems and vectorial electron transport components (shown in Fig. 11). In this scheme light energy is collected by the "antennae" or light-gathering pigments of the two photosystems, and excitation of P-680 and P-700 at the reaction centers takes place by energy migration.[8] At the reaction centers there is a vectorial removal of electrons so that the acceptors on the stromal side of the membrane become negative whereas the inside of the membrane becomes positive. The resulting electric potential difference across the membrane can be detected by the electrochromic shift (see Section V,C,2). This localized field is dispersed by ion movements on the surface of the membrane. Protons are then absorbed by the acceptors and released by the donors to the photosystems so that the electric potential across the membrane is transformed into a proton gradient in the bulk phase. This energized state is discharged through the proton-transporting ATPase to form ATP.

2. The Number of Sites for Proton Transport

Measurements of the ratios of H^+ to e^-, H^+ to ATP, and P to $2e^-$ (especially the first) aid the determination of the number of sites involved in proton translocation. The measurement of the ratio of H^+ to e^- should be simple because the pH gradient generated by the chloroplast can be measured by sensitive glass electrodes, and the electron flux through the chain can be measured by suitable electron acceptors. In practice the measurements are not so easy due to proton leakage from the thylakoids and buffering effects of various additives such as electron acceptors. The first problem can be overcome by the use of very short flashes of light during which proton leakage is negligible, and the second by taking account of the buffering capacity.

Early determinations of the ratio of H^+ to e^- gave values of two, which is consistent with two coupling sites in the electron transport chain (200, 233, 539). This value is also consistent with the transfer of two negative charges across the thylakoid membrane by the two photosystems and the replacement of the two internal-membrane positive charges by two protons in the bulk phase of the lumen of the thylakoid. Later measurements indicated higher values for the ratio of H^+ to e^-, and a value of four was found by Fowler and Kok (151). However, later data [Saphon

---

[8] In the present view, each photosystem assembly comprises one photoreactive center and a number of additional chlorophyll molecules. These are so arranged that when one of them is excited by a photon, the excitation energy migrates from molecule to molecule without loss until it finds its target in the reaction center, either P-700 or P-680, depending on whether it is in PSI or PSII. (Ed.)

and Crofts (428)] suggest that the overall value of the ratio is indeed two, and the value of four found by Fowler and Kok might have resulted from a failure to take account of the buffering of the chloroplast suspension when uncoupling concentrations of methylamine were added. But this latter interpretation has been questioned (506).

The first site of proton ejection into the lumen of the thylakoid is clearly on the oxidizing side of PSII, and the stoichiometry of one proton ejected per electron moved across the membrane is accepted (47, 149; Section V,D,2). A second site of ejection is due to oxidation of PQ on the lumen side of the membrane (18). PQ may act as a simple vectorial translocator of protons (492), that is, it may be reduced by two electrons from Q on the outside of the membrane and in the process take up two protons. This would result in a ratio of H^+ to e^- of one. On the inside of the membrane, PQ would be oxidized and lose protons to the lumen with ratio of H^+ to e^- of one. A ratio of one between proton and electron uptake by the PQ pool and for proton and electron release by this pool on the inside of the membrane has been measured (109). As a result, a simple oscillatory movement of PQ, acting as a proton pump across the membrane, has been proposed (109).

Since the earlier work of Fowler and Kok (151), the suggestion that more than one transporting site between the two photosystems, involving some type of Q cycle, has frequently been made (reviewed in 506). Crowther and Hind have suggested that a simple PQ shuttle may operate in linear electron transport but that a Q cycle may be functional in cyclic electron transport (109). These authors discuss in detail the merits of various Q cycles; the exact arrangement of Q-cycle components has not yet been determined (if in fact such a cycle does operate).

The number of protons that must be transported through ATPase to produce one ATP is the ratio of H^+ to ATP. This ratio is of necessity a whole number. The number of coupling sites can be measured by comparing the ratio of H^+ to ATP with that of ATP to $2e^-$ (i.e., the number of ATP molecules formed for every two electrons transported from water to $NADP^+$). However, because the development of a proton gradient is independent of ATP synthesis, the second ratio of ATP to $2e^-$ need not be a whole number.

Unfortunately, measuring these ratios is as difficult as measuring that of H^+ to e^-. Furthermore, there is a base level of electron flow that does not generate ATP for which corrections must be made. The most widely accepted value for the ratio of H^+ to ATP is 3.0 (385, 386; reviewed in 233, 402), although a value of 2.4 has been reported that, when corrected, might indicate a ratio of 2 (167). Ratios of ATP to $2e^-$ between 1.0 and 2.0 have been measured (402). If a proton is transported for every electron passing a coupling site, 4 protons would be transferred as 2 electrons pass 2 coupling sites. If 3 protons are required for each molecule of ATP synthesized, one might expect a ratio of ATP to $2e^-$ of 1.3; this value has been obtained by Portis and McCarthy (386). However, a ratio of ATP to $2e^-$ of 2 has been measured more recently in spinach chloroplasts (416), so the ratio is still uncertain.

Hence it cannot be stated precisely how many sites are involved in proton transport. The site associated with photooxidation of water clearly generates one proton in the lumen for each electron transferred across the membrane by PSII. There is also clearly a vectorial transport of protons by PQ across the membrane from Q to the iron–sulfur center (see Fig. 11). But it is not clear whether there is a second site resulting from some type of Q cycle between the photosystems that plays a part either in noncyclic or cyclic electron flow.

3. Proton Transport during Cyclic Electron Flow

A cyclic flow of electrons coupled to phosphorylation around PSI was first shown by Arnon and co-workers in 1955 (14, 516). It has now become clear, mainly through the work of Trebst, that there are two forms of energy conservation associated with cyclic electron transport. One of these is the naturally occurring cycle involving Fd and having PQ as the proton carrier, and the other is an artificial loop in which an artificial electron transporter is added to the system (reviewed in 200, 492, 493). Compounds that potentiate electron flow around PSI must be lipophilic, and to catalyze photophosphorylation they must also be protolytic on oxidation (200). The most effective catalyst of cyclic photophosphorylation is PMS, which is both lipophilic and a proton transporter. PMS is not only reduced by the electron donor in PSI, it can also interact directly with oxidized P-700 and bypass plastocyanin. During these reactions protons are transported across the thylakoid membrane and a proton gradient is formed that can be used for ATP synthesis (200). Other proton and electron carriers such as DAD interact with the oxidizing side of PSI via plastocyanin (492). In contrast, the electron carrier TMPD will catalyze cyclic electron flow around PSI but will not support photophosphorylation because it is not a proton carrier (200).

Cyclic phosphorylation is probably important in photosynthesis because it supplies the extra ATP over that produced by the noncyclic electron transport system required by the carbon-reduction cycle. Earlier doubts about *in vivo* operation of the cycle resulted from the use of broke chloroplasts that required the supply of electron carriers such as PMS or Fd. However, more recent work on intact chloroplasts to which no redox carriers were added has shown that cyclic electron flow contributes to the proton gradient across the membrane. For example, at concentrations of DCMU that inhibit noncyclic electron flow, a membrane potential is still developed by cyclic flow (459). Similarly, under optimal CO_2-fixing conditions antimycin A, which binds to cytochrome b_6 and inhibits cyclic electron flow, will inhibit CO_2 fixation (326). A proton gradient has been shown to be developed by cyclic flow around PSI in intact chloroplasts (110). The evolution of oxygen and turnover of cytochromes b_6 and f after flash illumination have indicated that cyclic and noncyclic flow can occur at the same time and that cytochrome f receives electrons from both pathways (457). The involvement of cytochrome b_6 in some type of Q cycle (to explain the kinetics of proton translocation in chloroplasts) has already been discussed (Section V,G,1).

If the function of cyclic photophosphorylation is to complement noncyclic flow in CO_2 fixation (to maintain the ratio of ATP to NADPH at an optimal value), the flux of electrons through the cyclic pathway must be controlled. In other words, once NADP has become maximally reduced there will be an inhibition of electron flux through the energy-transducing sites with a concomitant reduction in the protonmotive force unless the force can be maintained by cyclic electron flow.

The cyclic pathway appears to be regulated by the redox state of the electron carriers that are common to both pathways. The maximal rate of flux through the cyclic pathway is dependent upon the poise of these carriers at a particular potential (11). This can be achieved by partial, but not total, inhibition of electron flow from PSII into the common pool of PQ by inhibitors such as DCMU (11, 326). If, however, DCMU inhibition is too great, cyclic flow will also be inhibited, so the cyclic carriers must be partially reduced for optimal flux (11, 326).

The *in vivo* control of this poising is not clear. The reduction of the primary acceptor (Q) of PSII by NADPH has been described in a pathway involving Fd, ferredoxin:$NADP^+$ oxidoreductase, and cytochrome b_6 (11, 12, 323). The redox state of Fd is crucial in determining the flow of electrons round the cycle, which does not operate until most Fd is reduced; possibly because of the effect of Fd on the electron acceptors in PSII (498). Thus, when the ratio of NADPH to ATP becomes too high for optimal CO_2 fixation, NADPH could reduce the primary acceptor of PSII and thereby inhibit the flow of electrons from PSII to the intermediates of the common pathway. In this way the intermediates could become poised at a potential that enhances cyclic flow (85, 457).

It is now well established that cyclic electron transport is involved in the production of ATP for CO_2 fixation and serves to maintain the ratio of ATP to NADPH at the optimum value. Furthermore, it is clear that the cyclic pathway is controlled and operates only when the concentration of NADPH becomes excessive. Control is exercised through the state of reduction of common electron carriers, especially PQ. The details of the control mechanism have not yet been determined.

B. Pseudocyclic Electron Flow

Under some conditions Fd can donate electrons to oxygen and generate hydrogen peroxide, which is converted to water, in a reaction first described by Mehler in 1951 (319; reviewed in 159). Electrons are derived from water in PSII and transported via the electron transport chain to PSI and on to Fd (see Fig. 10). Electrons are then returned from Fd directly to oxygen, making water. Because the electrons have passed from water to water over an extended pathway, they appear to have traveled in a cycle. To distinguish this pathway from true cyclic flow such as that mediated by cytochrome b_6 and PQ (see Fig. 10), it is called pseudocyclic electron flow.

It has been proposed that this process can perform the same function as true cyclic phosphorylation—synthesis of ATP when NADPH reaches a high concentration. ATP synthesized by this pathway has been observed under some conditions (132). In this scheme oxygen competes with $NADP^+$ for reduced Fd, and the pseudocyclic flow of electrons to oxygen becomes significant when NADPH is high (4, 396). However, pseudocyclic photophosphorylation in isolated chloroplasts is only demonstrable when high concentrations of Fd are added, and the concentrations of Fd found *in vivo* would probably not support this pathway (11). Pseudocyclic flow is also insignificant as determined by flash experiments during steady-state photosynthesis, and it can be detected only when terminal acceptors such as bicarbonate or 3-phosphoglycerate are omitted (457).

Therefore $NADP^+$ is an effective trap for electrons, and while it is available it inhibits both cyclic flow and the flow of electrons to oxygen (206). When $NADP^+$ becomes reduced, electrons are diverted to these pathways (206). It has been suggested that the function of pseudocyclic flow is to remove electrons from the cyclic pathway to prevent it from becoming over-reduced, that it functions to poise the cyclic pathway at optimum levels (206). The cyclic pathway, therefore, is the principal means by which additional ATP required for CO_2 fixation is formed. When electron carriers are so strongly reduced that cyclic flow becomes inhibited, electrons are passed to oxygen to maintain the optimal redox poise of the cyclic system. Chloroplasts, therefore, appear to possess an elegant control system to ensure that the correct ratio of ATP to NADPH is maintained.

C. Chloroplast Coupling Factor

1. Structure of the Complex

The structure and properties of the chloroplast coupling factor (ATPase) have been extensively reviewed (26, 294, 348, 349, 441). Its structure is very similar to that of mitochondrial ATPase (Section III, B). Chloroplast coupling factor consists of two parts, the ATPase itself, referred to as CF_1, and the thylakoid-membrane-located hydrophobic part, CF_0. The coupling factor is seen in the electron microscope after negative staining as a series of spherical structures, 100 Å in diameter, on the stroma side of the membrane (26). The spheres can be removed from the membrane by chelating agents and restored to the membrane when the depleted membranes are combined with purified CF_1, indicating that the spheres seen in the electron microscope are indeed CF_1.[9]

Removal of CF_1 from the thylakoid membrane uncouples photophosphorylation from electron flow; photophosphorylation can be restored by adding back CF_1 to the preparation. ATPase activity of the isolated CF_1 is latent but can be greatly enhanced by treatment with trypsin (294). The molecular weight of solubilized CF_1 is approximately 325,000 (26). It is made up of five different subunits, α, β, γ, δ, and ϵ, ranging in molecular weight from 59,000 (α) to 13,000 (ϵ). Details of the properties and spatial arrangement of these subunits have been given by Baird and Hammes (26). The probable stoichiometry of the subunits is $\alpha_2\beta_2\gamma\delta\epsilon_2$ (46). The two α subunits appear to be adjacent and in contact with the two β subunits. There is evidence for a $\gamma\epsilon_2$ unit that will also bind to the α and β subunits. The δ subunit will bind only to the α and β subunits (26). This subunit is required for the binding of CF_1 to the membrane and may represent the stalk seen in electron micrographs (546).[10] It probably has the structure of a flexible rod (431). Antisera to all these subunits have been made, and each subunit will inhibit photophosphorylation and electron transport (273).

As mentioned previously, the ATPase activity of solubilized CF_1 is very low, but it can be greatly increased by partial trypsin digestion, addition of sulfhydryl compounds, or heat treatments (26). Activation

[9] For a discussion of the morphological interpretation of these structures, see Chapter 1. (Ed.)
[10] See Chapter 1, Fig. 30. (Ed.)

can occur with minor alterations to the structure that cause two additional sulfhydryl groups on the γ subunit to be exposed. Inhibition of the complex is caused by the ϵ subunit, which is hydrophobic and binds tightly to CF_1 (351). This subunit may regulate activity of CF_1. Trypsin activates the complex by digesting the ϵ subunit. Prolonged digestion with trypsin leaves only the α and β subunits, which are still active but no longer inhibited by the ϵ subunit (116).

The α and β subunits that remain after prolonged trypsin digestion retain ATPase activity (116). This $\alpha_2\beta_2$ tetramer is cold-labile, a property of complete CF_1, indicating that cold lability of CF_1 is a result of instability of hydrophobic interactions between the subunits when cold. The active site of CF_1 appears to be formed by an interaction between the two β subunits (see following discussion).

In the chloroplast, CF_1 is attached to CF_0 in the thylakoid membrane to form the native proton-transport coupling factor complex. These complexes are located on all thylakoid membranes that are adjacent to the stroma, but they are absent from thylakoid membranes that form the interfaces between the grana stacks[11] (347). The complexes are highly mobile in the membrane and can be aggregated *in situ* when incubated with CF_1 antisera (347).

The entire ATPase (i.e., CF_1 plus CF_0) can be extracted from thylakoid membranes with detergents (383). The approximate molecular weight of the entire complex is 435,000, and it consists of the five subunits of CF_1, a proteolipid, and three additional components of unknown function, one of which is very hydrophobic. The complex is inhibited by DCCD, which binds to the proteolipid. The entire complex can be incorporated into phospholipid vesicles where it will then catalyze ATP synthesis on acid–base transition (383). The DCCD-binding proteolipid has been isolated and purified and has been shown to have a molecular weight of $\sim$8,300 (350). There are four to six of these proteolipid subunits in CF_0, and when they are incorporated into phospholipid vesicles they mediate proton transport, indicating that they form a proton channel in the membrane (350). When DCCD is bound to the proteolipid this channel is blocked. The function of the other subunits of CF_0 is still unresolved.

An interesting difference between the chloroplast and mitochondrial coupling factors is that all the subunits of mitochondrial F_1 are synthesized on 80 S ribosomes in the cytosol and transported into the mitochondrion. In contrast, all subunits of CF_1 except δ are synthesized in the chloroplast (352). Subunits I and III (the proteolipid) of CF_0 are also

[11] See Chapter 1, Section VI,B and Fig. 30. (Ed.)

synthesized in the chloroplast. Subunits δ of CF_1 and II of CF_0 are synthesized in the cytosol as precursor molecules and then transported into the chloroplast by a vectorial processing mechanism (352). It has been suggested that cytosolic synthesis of these two subunits plays a regulatory role by preventing the ATPase from becoming active, and the proteolipid from forming proton channels, until the complex is completely assembled (352).

The coupling factor of the thylakoid membrane consists, therefore, of two distinct parts: CF_1, which is a latent ATPase and presumably acts as an ATP synthase *in vivo*, and CF_0, which is a proton channel through the membrane. The proton channel is normally blocked by CF_1, probably by the δ subunit. The channel through CF_0 is specific for protons. Removal of CF_1 increases the permeability of the membrane to protons but does not affect chloride or potassium permeability (363). This suggests that CF_0 does not merely form an aqueous channel through the membrane, because the hydrated diameter of a proton is three times that of a hydrated potassium ion (363). The proton may travel through the membrane in an unhydrated state, losing its water of hydration as it enters a channel. It may also move through a region of structured water on the surface of the membrane into the channel, a very rapid process (363). The movement of protons through this channel must affect CF_1 so that ATP can be synthesized. The close connection between CF_0 and CF_1 is demonstrated by the inhibition of the CF_1 ATPase activity by the binding of DCCD to the proteolipid of CF_0. Details of the shape of the complex are not available because crystals large enough for X-ray analysis have not yet been prepared. The conformation of CF_1 in solution has been determined by light scattering and small-angle X-ray scattering, by which the molecular weight has been confirmed as 325,000 and the maximum dimension as 115.0 Å (377).

2. The Active Site of CF_1

When CF_1 is isolated, nucleotides are found to be firmly bound to it (reviewed in 26, 294, 348). There appear to be three binding sites on the complex (26), two of which bind ADP and its analogs. Binding of nucleotides to these sites causes conformational changes in the complex, and they are probably involved in the regulation of the complex rather than being at the active site (26).

The third site binds ADP very tightly, and it cannot be removed by dialysis or gel filtration. However MgATP enters this site rapidly to be hydrolyzed to ADP and P_i. The P_i dissociates from the activated complex but not from the latent complex (26). Photoaffinity labeling of CF_1 indicates that this site is on the β subunit, although how one site is formed from the two β subunits is not clear (26). It is most likely that this site represents the active site of the complex.

Specific amino-acid-modifying reagents have been used to probe the active center of the complex. The modification of a single tyrosine on each of the β subunits inhibits ATPase activity of the complex (116) and increases the reactivity of sulfhydryl groups, showing that the modification also causes a conformational change in the complex (26). ATPase activity of CF_1 is also inhibited by the arginine reagents phenylglyoxal and 2,3-butanedione (503). These reagents also inhibit cyclic photophosphorylation, showing that arginine is involved in the catalytic reactions *in vivo* (503). A sulfhydryl reagent, *N*-ethylmaleimide, also inhibits the enzyme in the light but not in the dark (296). This inhibitor is not incorporated into the active site but is associated with the γ subunit (295), indicating the importance of conformational changes in the activity of the complex. Similarly, modification of lysine residues in the light inactivates the complex (364).

There are, therefore, two types of inhibitor of the chloroplast coupling factor. There are those that directly affect the active site and inhibit ATPase and, usually, the photophosphorylation activity of the complex by binding at the active site. These inhibitors indicate that the active site is on the β subunit. Whether there are one or two active sites (i.e., one associated with each β subunit) is not clear. Two sites would be required for Boyer's alternating site theory of photophosphorylation.[12] The second group of compounds inhibit only the chloroplast coupling factor when the complex is activated by light or an acid–base transition. These inhibitors may not always bind at the active site, but they affect the conformational changes associated with activity. Two other sites of inhibitor binding are associated with the complex, one on the α and one on the β subunit (26). These are probably involved in the regulation of the complex.

3. Regulation of the Chloroplast Coupling Factor

In the dark the chloroplast coupling factor (CF_1–CF_0) is inactive and will not catalyze ATP hydrolysis, ATP-mediated proton transport, reverse electron transport, or photophosphorylation. The ATPase can be activated to perform photophosphorylation by imposing a proton gradient across the membrane (461) or by intense illumination, which causes

---

[12] See Section II,D. (Ed.)

an electrical potential across the membrane (324). Activation by flashing light may be due to removal of the inhibitory subunit from the complex (186). ATPase activity gradually decays again in the dark (324).

Isolated coupling factor contains tightly bound nucleotides that cannot be readily exchanged. Exchange is enhanced in the light (188). This has been cited as evidence for conformational changes in the complex. More adenine nucleotide is bound in the dark than in the light, but the kinetics of activation and loss of nucleotide are very similar (443). Nucleotide binding and loss of activity in the dark are also correlated (443). The addition of ADP to the light-activated coupling factor accelerates its decay, and, again, ADP binding and loss of activity follow the same kinetics (128).

When activated, the coupling factor can be shown to hydrolyze ATP and drive reversed electron transport in the chloroplast membrane by means of an increase in ΔpH across the membrane (20). In these experiments the thiol compound dithiothreitol promotes activation (20). However, more recently it has been shown that dithiothreitol is not absolutely required for activation (435). It has been suggested that the thioredoxin system (78, 79)[13] may replace dithiothreitol *in vivo* (435). As mentioned in Section VI,C,1, the latent coupling factor can be activated by thiol compounds (26). This activation process in broken chloroplasts can also occur following the addition of purified ferredoxin: thioredoxin reductase and thioredoxin (325). It is concluded that light activates the coupling factor through a thiol-mediated photoreduction.

It may be concluded that light has a dual function in the activation of the coupling factor (325). First, it generates a proton gradient across the membrane, and second, it reduces thioredoxin. It has been reported that the δ subunit of CF_1 does in fact have the required thioredoxin activity (299).

It is clear, therefore, that the chloroplast coupling factor is highly regulated *in vivo*, probably by means of the regulatory nucleotide-binding sites and the inhibitory protein. Inhibition can be relieved by energizing the thylakoid membrane with a proton gradient that causes a conformational change and that displaces bound nucleotides. In addition, there appears to be a reduction of the inhibitory protein, probably by a thioredoxin-mediated photoprocess. Regulation of this complex is obviously of paramount importance because it contains ATPase and a proton channel. These activities only operate safely when the membrane is energized in such a way that ATP can be synthesized by reversal of the ATPase activity.

---

[13] See Chapter 3, Section II,D. (Ed.)

D. RECONSTITUTION EXPERIMENTS

Reconstitution of oxidative phosphorylation has been useful both in showing the importance of proton gradients in energy transduction and in characterizing the components involved. The ability to perform similar experiments in photophosphorylation would be very valuable. A model system would require only three components: a liposome into which is embedded one of the photosystems (to allow light-induced charge separation to occur across the membrane), a redox component such as PMS (to convert this charge separation into a proton gradient), and the chloroplast coupling factor complex (to use the energy of the proton gradient for ATP synthesis).

In an early attempt to reconstitute photophosphorylation, Jaynes *et al.* (237) prepared PSI particles from spinach (*Spinacia oleracea*) chloroplasts and incorporated them into vesicles, using phospholipids from soybeans (*Glycine max*). These vesicles are capable of PMS-mediated photophosphorylation. These particles also contain the chloroplast coupling factor, because removal of CF_1 by EDTA inhibits phosphorylation, which can be restored by adding purified CF_1 (237). This system, although it indicates that reconstitution is possible, contains impure PSI particles and coupling factor. Later, purified chloroplast coupling factor was incorporated into phospholipid vesicles that contained bacteriorhodopsin to generate a proton gradient (537). When these vesicles were illuminated, ATP synthesis was observed that was sensitive to uncouplers and to energy-transfer inhibitors such as DCCD (537).

Purified PSI reaction centers can be prepared that consist of six polypeptides. These purified centers have been incorporated into phospholipid vesicles (366). When these vesicles are energized by light in the presence of PMS (a proton transporter) protons are pumped across the membranes. Unfortunately, measurements are difficult to make because two populations of vesicles appear to be formed. Some vesicles have photosystems oriented in the native configuration so that protons are pumped into them, but others are inside out so that protons are extruded. More effective proton-pumping activity can be obtained with an even more purified system. PSI reaction centers can be fractionated into a large subunit and five smaller subunits by SDS treatment (237). The large subunit contains P-700 and the primary electron acceptor A_1, which may be a chlorophyll a molecule (Section V,B,1 and Fig. 10). When this large subunit is incorporated into phospholipid vesicles it will catalyze a light-induced proton gradient across the membrane if PMS is present, indicating that the large subunit alone induces a charge separation across the membrane (197). When purified chloroplast coupling factor is added to phospholipid vesicles containing complete PSI reaction centers (the six-subunit complex) it is incorporated into the vesicles, and light-induced ATP synthesis can be detected that is sensitive to DCCD and uncouplers (199). As yet this system operates at only 10% of the rate in a comparable chloroplast system, possibly due to inactivation of the coupling factor (199).

A useful start has been made, therefore, in reconstituting photophosphorylation from purified components. A problem that remains is to ensure that orientation of the reaction centers can be controlled. However, rapid progress in this type of experiment can be expected.

VII. Conclusions

A very large amount of information has been accumulated on energy transduction in both plants and animals. This chapter has presented it only in outline and in summary form. Nevertheless, it is hoped that the information has been coordinated to provide an overview of the molecular processes as they occur in various organelles and in different kinds of organisms.

Energy transduction in both mitochondria and chloroplasts apparently requires a great amount of complexity. A conspicuous feature is that nature seems to require such complicated machinery to achieve what seem to be relatively simple end results. But the complexity seems to be necessary to provide increased efficiency and the selective controls that make possible the effective functioning of organisms.

Facing these evident complexities, the progress that has been made owes much to the development of modern experimental technology. Throughout, much use has been made of spectrophotometric and related techniques, which have led to identification of substances that occur in very small amounts (some of which have not been isolated) and which may function over very brief intervals of time. A few years ago this would have seemed inconceivable. Other techniques that have contributed to the rapid development include electronic devices for measuring rapid and minute changes in physical parameters and many refined biochemical techniques for separating and identifying complexes and their individual components. But perhaps the most important development has been the linkage between biophysical or biochemical phenomena and microscopic and submicroscopic morphology of the structure in which they occur *in vivo*. This has led to a degree of cohesion and integration in interpretation that was hitherto lacking. This integration should be further encouraged.

Many specific developments are worthy of special mention. Notably, the concept of proton gradients as the energetic driving force in ATP synthesis has had a unifying influence over the entire field. Similarly, new information on the structure and function of the coupling factors in chloroplasts and mitochondria is crucial to the understanding of mechanisms of energy transduction and ATP synthesis. In chloroplasts, the structures of the donor and acceptor complexes associated with the photosystems are rapidly being clarified. It is of interest to note their similarity to comparable systems in photosynthetic bacteria. The mechanism for establishing proton gradients in thylakoids during cyclic and noncyclic electron flow is also rapidly being elucidated, although the reactions of water splitting remain unknown. Progress is being made in clarifying the differences and similarities in structure and function of plant and animal mitochondria. When one considers the number of people working with plants compared to the number working on animal systems and the relative difficulties inherent in the materials, the achievements with plants have been quite remarkable.

Of necessity this chapter has dealt with the minutiae of many reactions in cellular and subcellular metabolism. A great task for the future is to achieve a synthesis of all these reactions and to integrate them with metabolism, growth, development, and the controls that enable whole organisms to function coherently in their environments.

References

1. Albracht, S. P. J., Leeuwerik, F. J., and van Swol, B. (1979). The stoichiometry of the iron-sulphur clusters 1a, 1b and 2 of NADH:Q oxidoreductase as present in beef-heart submitochondrial particles. *FEBS Lett.* **104**, 197–200.
2. Alexandre, A., Galiazzo, F., and Lehninger, A. L. (1980). On the location of the H^+-extruding steps in Site 2 of the mitochondrial electron transport chain. *J. Biol. Chem.* **255**, 10721–10730.
3. Alexandre, A., Reynatarje, B., and Lehninger, A. L. (1978). Stoichiometry of vectorial H^+ movements coupled to electron transport and to ATP synthesis in mitochondria. *Proc. Nat. Acad. Sci. USA* **75**, 5296–5300.
4. Allen, J. F. (1975). Oxygen reduction and optimum production of ATP in photosynthesis. *Nature (London)* **256**, 599–600.
5. Amesz, J. (1973). The function of plastoquinone in photosynthetic electron transport. *Biochim. Biophys. Acta* **301**, 35–51.
6. Amesz, J. (1977). Plastoquinone. *In* "Encyclopedia of Plant Physiology, New Series" (A. Pirson and M. H. Zimmermann, eds.), Vol. 5, Photosynthesis I. Photosynthetic Electron Transport and Photophosphorylation (A. Trebst and M. Avron, eds.), pp. 238–246. Springer-Verlag, Berlin.
7. Amesz, J., Visser, J. W. M. van den Engh, G. J., and Dirks, M. P. (1972). Reaction kinetics of intermediates of the photosynthetic chain between the two photosystems. *Biochim. Biophys. Acta* **256**, 370–380.

8. Anderson, J. M., and Boardman, N. K. (1973). Localization of low potential cytochrome b_{559} in photosystem I. *FEBS Lett.* **32**, 157–160.
9. Arnon, D. I. (1965). Ferredoxin and photosynthesis. *Science* **149**, 1460–1470.
10. Arnon, D. I. (1977). Photosynthesis 1950–75: changing concepts and perspectives. *In* "Encyclopedia of Plant Physiology, New Series" (A. Pirson and M. H. Zimmermann, eds.) Vol. 5, "Photosynthesis I. Photosynthetic Electron Transport and Photophosphorylation (A. Trebst and M. Avron, eds.), pp. 7–56. Springer-Verlag, Berlin.
11. Arnon, D. I., and Chain, R. K. (1975). Regulation of ferredoxin-catalyzed photosynthetic phosphorylations. *Proc. Natl. Acad. Sci. USA* **72**, 4961–4965.
12. Arnon, D. I., and Chain, R. K. (1979). Regulatory electron transport pathways in cyclic photophosphorylation: reduction of C-550 and cytochrome b_6 by ferredoxin in the dark. *FEBS Lett.* **102**, 133–138.
13. Arnon, D. I., Tsujimoto, H. Y., and McSwain, B. D. (1967). Ferredoxin and photosynthetic phosphorylation. *Nature (London)* **214**, 562–566.
13a. Arnon, D. I., Tsujimoto, H. Y., and Tang, G. M. S. (1981). Proton transport in photooxidation of water: a new perspective on photosynthesis. Proc. Natl. Acad. Sci. USA **78**, 2942–2946.
14. Arnon, D. I., Whatley, F. R., and Allen, M. B. (1955). Vitamin K as a cofactor of photosynthetic phosphorylation. *Biochim. Biophys. Acta* **16**, 607–608.
15. Arntzen, C. J. (1978). Dynamic structural features of chloroplast lamellae. *Curr. Top. Bioenerg.* **8**, 111–160.
16. Atkinson, D. E. (1977), "Cellular Energy Metabolism and Its Regulation." Academic Press, New York.
17. Äuslander, W., Heathcote, P., and Junge, W. (1974). On the reduction of chlorophyll-A1 in the presence of the plastoquinone antagonist dibromothymoquinone. *FEBS Lett.* **47**, 229–235.
18. Äuslander, W., and Junge, W. (1975). Neutral red, a rapid indicator for pH-changes in the inner phase of thylakoids. *FEBS Lett.* **59**, 310–315.
19. Avron, M. and Jagendorf, A. T. (1957). Some further investigations on chloroplast TPNH diaphorase. *Arch. Biochem. Biophys.* **72**, 17–24.
20. Avron, M., and Schreiber, U. (1979). Properties of ATP-induced chlorophyll luminescence in chloroplasts. *Biochim. Biophys. Acta* **546**, 448–454.
21. Azzone, G. F., Pozzan, T., and Di Virgilio, F. (1979). H^+/Site, charge/site, and ATP/site ratios at coupling site III in mitochondrial electron transport. *J. Biol. Chem.* **254**, 10206–10212.
22. Babcock, G. T., and Sauer, K. (1975). A rapid light-induced transient in electron paramagnetic resonance signal II activated upon inhibition of photosynthetic oxygen evolution. *Biochim. Biophys. Acta* **376**, 315–328.
23. Babcock, G. T., and Sauer, K. (1975). Two electron donation sites for exogenous reductants in chloroplast photosystem II. *Biochim. Biophys. Acta* **396**, 48–62.
24. Bahr, J. T., and Bonner, W. D., Jr. (1973). Cyanide-insensitive respiration. I. The steady states of skunk cabbage spadix and bean hypocotyl mitochondria. *J. Biol. Chem.* **248**, 3441–3445.
25. Bahr, J. T., and Bonner, W. D., Jr. (1973). Cyanide-insensitive respiration. II. Control of the alternate pathway. *J. Biol. Chem.* **248**, 3446–3450.
26. Baird, B. A. and Hammes, G. G. (1979). Structure of oxidative- and photo-phosphorylation coupling factor complexes. *Biochim. Biophys. Acta* **549**, 31–53.
27. Baltimore, B. G., and Malkin, R. (1980). Spectral characterization of the intermediate electron acceptor (A_1) of photosystem I. *FEBS Lett.* **110**, 50–52.

28. Bearden, A. J., and Malkin, R. (1972). The bound ferredoxin of chloroplasts: a role as the primary electron acceptor of photosystem I. *Biochem. Biophys. Res. Commun.* **46,** 1299–1305.
29. Bearden, A. J., and Malkin, R. (1972). Quantitative EPR studies of the primary reaction of photosystem I in chloroplasts. *Biochim. Biophys. Acta* **283,** 456–468.
30. Bearden, A. J., and Malkin, R. (1974). Primary photochemical reactions in chloroplast photosynthesis. *Q. Rev. Biophys.* **7,** 131–177.
31. Bearden, A. J., and Malkin, R. (1976). Correlation of reaction-center chlorophyll (P-700) oxidation and bound iron–sulfur protein photoreduction in chloroplast photosystem I at low temperatures. *Biochim. Biophys. Acta* **430,** 538–547.
32. Beinert, H., Ackrell, B. A. C., Vinogradov, A. D., Kearney, E. B., and Singer, T. P. (1977). Interrelations of reconstitution activity, reactions with electron acceptors and iron–sulfur centers in succinate dehydrogenase. *Arch. Biochem. Biophys.* **182,** 95–106.
33. Ben-Hayyim, G. B. (1974). Light-induced absorbance changes of the high-potential cytochrome b$_{559}$ in chloroplasts. *Eur. J. Biochem.* **41,** 191–196.
34. Ben-Hayyim, G. B., and Avron, M. (1970). Cytochrome b of isolated chloroplasts. *Eur. J. Biochem.* **14,** 205–213.
35. Bendall, D. S. (1968). Oxidation–reduction potentials of cytochromes in chloroplasts from higher plants. *Biochem. J.* **109,** 46P–47P.
36. Bendall, D. S. (1977). Electron and proton transfer in chloroplasts. *Int. Rev. Biochem.* **13,** 41–78.
37. Bendall, D. S., and Sofrová, D. (1971). Reactions at 77°K in photosystem 2 of green plants. *Biochim. Biophys. Acta* **234,** 371–380.
38. Bengis, C., and Nelson, N. (1977). Subunit structure of chloroplast photosystem I reaction center. *J. Biol. Chem.* **252,** 4564–4569.
39. Bensasson, R., and Land, E. J. (1973). Optical and kinetic properties of semireduced plastoquinone and ubiquinone: electron acceptors in photosynthesis. *Biochim. Biophys. Acta* **325,** 175–181.
40. Bering, C. L., Dilley, R. A., and Crane, F. L. (1976). Inhibition of energy-transducing functions of chloroplast membranes by lipophilic iron chelators. *Biochim. Biophys. Acta* **430,** 327–335.
41. Berzborn, R. (1968). Uberlösliche und unlösliche Chloroplasten-Antigene Nachweis der Ferredoxin-NADP-Reduktase in der Oberfläche des Chloroplasten-Lamellarsystes mit Hilfe Spezifischer Antikörper. *Z. Naturforsch.* **23B,** 1096–1104.
42. Berzborn, R. J., and Lockau, W. (1977). Antibodies. *In* "Encyclopedia of Plant Physiology, New Series" (A. Pirson and M. H. Zimmermann, eds.), Vol. 5, Photosynthesis I. Photosynthetic Electron Transport and Photophosphorylation (A. Trebst and M. Avron, eds.), pp. 283–296. Springer-Verlag, Berlin.
43. Beyer, R. E., Peters, G. A., and Ikuma, H. (1968). Oxidoreduction states and natural homologue of ubiquinone (Coenzyme Q) in submitochondrial particles from etiolated Mung bean (*Phaseolus aureus*) seedlings. *Plant Physiol.* **43,** 1395–1400.
44. Bidwell, R. G. S. (1979). "Plant Physiology" (2nd ed.). Macmillan, New York.
45. Biggins, J. (1973). Kinetic behaviour of cytochrome f in cyclic and noncyclic electron transport in *Porphyridium cruentum. Biochemistry* **12,** 1165–1170.
46. Binder, A., Jagendorf, A., and Ngo, E. (1978). Isolation and composition of the subunits of spinach chloroplast coupling factor protein. *J. Biol. Chem.* **253,** 3094–3100.
47. Blankenship, R. E., Babcock, G. T., and Sauer, K. (1975). Kinetic study of oxygen

evolution parameters in tris washed, reactivated chloroplasts. *Biochim. Biophys. Acta* **387**, 165–175.

48. Blankenship, R. E., Babcock, G. T., Warden, J. T., and Sauer, K. (1975). Observation of a new EPR transient in chloroplasts that may reflect the electron donor to photosystem II at room temperature. *FEBS Lett.* **51**, 287–293.

49. Blankenship, R. E., McGuire, A., and Sauer, K. (1977). Rise time of EPR signal II$_{vf}$ in chloroplast photosystem II. *Biochim. Biophys. Acta* **459**, 617–619.

50. Blankenship, R. E., and Parson, W. W. (1978). The photochemical electron transfer reactions of photosynthetic bacteria and plants. *Annu. Rev. Biochem.* **47**, 635–653.

51. Blankenship, R. E., and Sauer, K. (1974). Manganese in photosynthetic oxygen evolution. I. Electron paramagnetic resonance study of the environment of manganese in Tris-washed chloroplasts. *Biochim. Biophys. Acta* **357**, 252–266.

52. Boardman, N. K., and Anderson, J. M. (1967). Fractionation of the photochemical systems of photosynthesis. II. Cytochrome and carotenoid contents of particles isolated from spinach chloroplasts. *Biochim. Biophys. Acta* **143**, 187–203.

53. Böhme, H. (1975). Inhibition of cytochrome b$_6$ oxidation by KCN. *FEBS Lett.* **60**, 51–53.

54. Böhme, H. (1976). Photoreactions of cytochrome b$_6$ and cytochrome f in chloroplast photosystem I fragments. *Z. Naturforsch.* **31C**, 68–77.

55. Böhme, H. (1977). On the role of ferredoxin and ferredoxin-NADP$^+$ reductase in cyclic electron transport of spinach chloroplasts. *Eur. J. Biochem.* **72**, 283–289.

56. Böhme, H. (1979). Photoreactions of cytochrome b$_{563}$ and f$_{554}$ in intact spinach chloroplasts: regulation of cyclic electron flow. *Eur. J. Biochem.* **93**, 287–293.

57. Böhme, H. (1980). On the pathways of cytochrome b-563 photoreduction in spinach chloroplasts. *FEBS Lett.* **112**, 13–16.

58. Böhme, H., and Cramer, W. A. (1971). Plastoquinone mediates electron transport between cytochrome b$_{559}$ and cytochrome f in spinach chloroplasts. *FEBS Lett.* **15**, 349–351.

59. Böhme, H., and Cramer, W. A. (1972). Localization of a site of energy coupling between plastoquinone and cytochrome f in the electron-transport chain of spinach chloroplasts. *Biochemistry* **11**, 1155–1160.

60. Böhme, H., and Cramer, W. A. (1973). The role of cytochrome b$_6$ in cyclic electron transport: evidence for an energy-coupling site in the pathway of cytochrome b$_6$ oxidation in spinach chloroplasts. *Biochim. Biophys. Acta* **283**, 302–315.

61. Böhme, H., and Cramer, W. A. (1973). Uncoupler dependent decrease in midpoint potential of the chloroplast cytochrome b$_6$. *Biochim. Biophys. Acta* **325**, 275–283.

62. Bouges-Bocquet, B. (1973). Limiting steps in photosystem II and water decomposition in *Chlorella* and spinach chloroplasts. *Biochim. Biophys. Acta* **292**, 772–785.

63. Bouges-Bocquet, B. (1973). Electron transfer between the two photosystems in spinach chloroplasts. *Biochim. Biophys. Acta* **314**, 250–256.

64. Bouges-Bocquet, B. (1977). Cytochrome f and plastocyanin kinetics in *Chlorella pyrenoidosa*. I. Oxidation kinetics after a flash. *Biochim. Biophys. Acta* **462**, 362–370.

65. Bouges-Bocquet, B. (1977). Cytochrome f and plastocyanin kinetics in *Chlorella pyrenoidosa*. II. Reduction kinetics and electric field increase in the 10 ms range. *Biochim. Biophys. Acta* **462**, 371–379.

66. Bouges-Bocquet, B., and Delosme, R. (1978). Evidence for a new electron donor to P-700 in *Chlorella pyrenoidosa*. *FEBS Lett.* **94**, 100–104.

67. Boulter, D., Haslett, B. G., Peacock, D., Ramshaw, J. A. M., and Scawen, M. D. (1977). Chemistry, function, and evolution of plastocyanin. *Int. Rev. Biochem.* **13**, 1–40.

68. Boulter, D., Ramshaw, J. A. M., Thompson, E. W., Richardson, M., and Brown, R. H. (1972). A phylogeny of higher plants based on the amino acid sequences of cytochrome c and its biological implications. *Proc. R. Soc. London Ser. B.* **181**, 441–455.

69. Boyer, P. D. (1975). A model for conformational coupling of membrane potential and proton translocation to ATP synthesis and to active transport. *FEBS Lett.* **58**, 1–6.

70. Boyer, P. D. (1977). Coupling mechanisms in capture, transmission and use of energy. *Annu. Rev. Biochem.* **46**, 957–966.

71. Boyer, P. D. (1977). Conformational coupling in oxidative phosphorylation and photophosphorylation. *Trends Biochem. Sci.* **2**, 38–41.

72. Boyer, P. D. (1979). The binding-change mechanism of ATP synthesis. In "Membrane Bioenergetics" (C. P. Lee, G. Schatz, and L. Ernster, eds.), pp. 461–479. Addison-Wesley, Reading, Massachusetts.

73. Boyer, P. D., Cross, R. L., and Momsen, W. (1973). A new concept for energy coupling in oxidative phosphorylation based on a molecular explanation of the oxygen exchange reactions. *Proc. Nat. Acad. Sci. USA* **70**, 2837–2839.

74. Boyer, P. D., Gresser, M., Vinkler, C., Hackney, D., and Choate, G. (1977). Nucleotide binding and ATPase subunit cooperativity in energy transduction by mitochondria and chloroplasts. In "Structure and Function of Energy Transducing Membranes" (K. van Dam and B. F. van Gelder, eds.), pp. 261–274. Elsevier/North-Holland Biomedical Press, Amsterdam.

75. Brand, M. D., and Lehninger, A. L. (1977). H^+/ATP ratio during ATP hydrolysis by mitochondria: modification of the chemiosmotic theory. *Proc. Natl. Acad. Sci. USA* **74**, 1955–1959.

76. Brand, M. D., Regnafarje, B., and Lehninger, A. L. (1976). Stoichiometric relationship between energy-dependent proton ejection and electron transport in mitochondria. *Proc. Natl. Acad. Sci. USA* **73**, 437–441.

77. Brunton, C. J., and Palmer, J. M. (1973). Pathways for the oxidation of malate and reduced pyridine nucleotide by wheat mitochondria. *Eur. J. Biochem.* **39**, 283–291.

78. Buchanan, B. B. (1980). Role of light in the regulation of chloroplast enzymes. *Annu. Rev. Plant Physiol.* **31**, 341–374.

79. Buchanan, B. B., Wolosiuk, R. A., and Schürmann, P. (1979). Thioredoxin and enzyme regulation. *Trends Biochem. Sci.* **4**, 93–96.

80. Butler, W. L. (1978). On the role of cytochrome b_{559} in oxygen evolution in photosynthesis. *FEBS Lett.* **95**, 19–25.

81. Butler, W. L., and Okayama, S. (1971). The photoreduction of C_{550} in chloroplasts and its inhibition by lipase. *Biochim. Biophys. Acta* **245**, 237–239.

82. Cammack, R., and Evans, M. C. W. (1975). E.P.R. spectra of iron–sulphur proteins in dimethylsulphoxide solutions: evidence that chloroplast photosystem I particles contain 4Fe–4S centers. *Biochem. Biophys. Res. Commun.* **67**, 544–549.

83. Cammack, R., and Palmer, J. M. (1977). Iron–sulphur centres in mitochondria from *Arum maculatum* spadix with very high rates of cyanide-resistant respiration. *Biochem. J.* **166**, 347–355.

84. Capaldi, R. A., Sweetland, J., and Merli, A. (1977). Polypeptides in the succinate-coenzyme Q reductase segment of the respiratory chain. *Biochemistry* **16**, 5707–5710.

85. Chain, R. K., and Arnon, D. I. (1977). Quantum efficiency of photosynthetic energy conversion. *Proc. Nat. Acad. Sci. USA* **74**, 3377–3381.

86. Chance, B. (1972). The nature of electron transfer and energy coupling reactions. *FEBS Lett.* **23**, 3–20.

87. Chance, B. (1977). Electron transfer: pathways, mechanisms and controls. *Annu. Rev. Biochem.* **46**, 967–980.

88. Chance, B. (1979). Possible structures of cytochrome oxidase oxygen intermediates and their reactivity toward cytochrome *c*. *In* "Membrane Bioenergetics" (C. P. Lee, G. Schatz, and L. Ernster, eds.), pp. 1–12. Addison-Wesley, Reading, Massachusetts.

89. Chance, B., Saronio, C., and Leigh, J. S., Jr. (1975). Functional intermediates in the reaction of membrane-bound cytochrome oxidase with oxygen. *J. Biol. Chem.* **250**, 9226–9237.

90. Chance, B., Wilson, D. F., Dutton, P. L., and Erecinska, M. (1970). Energy-coupling mechanisms in mitochondria: kinetic, spectroscopic and thermodynamic properties of an energy-transducing form of cytochrome b. *Proc. Natl. Acad. Sci. USA* **66**, 1175–1182.

91. Chapman, G. V., Colman, P. M., Freeman, H. C., Guss, J. M., Murata, M., Norris, V. A., Ramshaw, J. A. M., and Venkatappa, M. P. (1977). Preliminary crystallographic data for a copper-containing protein, plastocyanin. *J. Mol. Biol.* **110**, 187–189.

92. Cheniae, G. M., and Martin, I. F. (1967). Studies on the function of manganese in photosynthesis. *Brookhaven Symp. Biol.* **19**, 406–417.

93. Cheniae, G. M., and Martin, I. F. (1970). Sites of function of manganese within photosystem II. Roles in O_2 evolution and system II. *Biochim. Biophys. Acta* **197**, 219–239.

94. Cheniae, G. M., and Martin, I. F. (1971). Photoactivation of the manganese catalyst of O_2 evolution. I. Biochemical and kinetic aspects. *Biochim. Biophys. Acta* **253**, 167–181.

95. Cheniae, G. M., and Martin, I. F. (1971). Effects of hydroxylamine on photosystem II. I. Factors affecting the decay of O_2 evolution. *Plant Physiol.* **47**, 568–575.

96. Chirgadze, Y. N., Garger, M. B., and Nikonov, S. V. (1977). Crystallographic study of plastocyanins. *J. Mol. Biol.* **113**, 443–447.

97. Colman, P. M., Freeman, H. C., Guss, J. M., Murata, M., Norris, V. A., Ramshaw, J. A. M., and Venkatappa, M. P. (1978). X-ray crystal structure analysis of plastocyanin at 2.7 Å resolution. *Nature (London)* **272**, 319–324.

98. Cookson, D. J., Hayes, M. T., and Wright, P. E. (1980). Electron transfer reagent binding sites on plastocyanin. *Nature (London)* **283**, 682–683.

99. Cowley, R. C., and Palmer, J. M. (1978). The interaction of citrate and calcium in regulating the oxidation of exogenous NADH in plant mitochondria. *Plant Sci. Lett.* **11**, 345–350.

100. Cox, R. P. (1975). The properties of cytochrome f and P_{700} in chloroplasts suspended in fluid media at sub-zero temperatures. *Eur. J. Biochem.* **55**, 625–631.

101. Cox, R. P. (1979). Chloroplast cytochrome b_{563}. Hydrophobic environment and lack of direct reaction with ferredoxin. *Biochem. J.* **184**, 39–44.

102. Cox, R. P., and Bendall, D. S. (1972). The effects on cytochrome b_{559HP} and P546 of treatments that inhibit oxygen evolution by chloroplasts. *Biochim. Biophys. Acta* **283**, 124–135.

103. Cox, R. P., and Bendall, D. S. (1974). The functions of plastoquinone and β-carotene in photosystem II of chloroplasts. *Biochim. Biophys. Acta* **347**, 49–59.

104. Cramer, W. A. (1977). Cytochromes. *In* "Encyclopedia of Plant Physiology, New Series" (A. Pirson and M. H. Zimmermann, eds.), Vol. 5, Photosynthesis I. Photo-

synthetic Electron Transport and Photophosphorylation" (A. Trebst and M. Avron, eds.), pp. 227–237. Springer-Verlag, Berlin.

105. Cramer, W. A., and Butler, W. L. (1967). Light induced absorbance changes of two cytochrome b components in the electron transport system of spinach chloroplasts. *Biochim. Biophys. Acta* **143,** 332–339.

106. Cramer, W. A., and Whitmarsh, J. (1977). Photosynthetic cytochromes. *Annu. Rev. Plant Physiol.* **28,** 133–172.

107. Criddle, R. S., Packer, L., and Shieh, P. (1977). Oligomycin-dependent ionophoric protein subunit of mitochondrial adenosinetriphosphatase. *Proc. Natl. Acad. Sci. USA* **74,** 4306–4310.

108. Crofts, A. R., and Wood, P. M. (1978). Photosynthetic electron-transport chains of plants and bacteria and their role as proton pumps. *Current Topics Bioenerg.* **7,** 175–244.

109. Crowther, D., and Hind, G. (1980). Partial characterization of cyclic electron transport in intact chloroplasts. *Arch. Biochem. Biophys.* **204,** 568–577.

110. Crowther, D., Mills, J. D., and Hind, G. (1979). Protonmotive cyclic electron flow around photosystem I in intact chloroplasts. *FEBS Lett.* **98,** 386–390.

111. Davenport, H. E. (1952). Cytochrome components in chloroplasts. *Nature (London)* **170,** 1112–1114.

112. Davis, K. A., and Hatefi, Y. (1971). Succinate dehydrogenase. I. Purification, molecular properties and substructure. *Biochemistry* **10,** 2509–2516.

113. Day, D. A., Arron, G. P., and Laties, G. C. (1980). Nature and control of respiratory pathways in plants: the interaction of cyanide-resistant respiration with the cyanide-sensitive pathway. *In* "The Biochemistry of Plants" (E. E. Conn and P. K. Stumpf, eds.), Vol. 2, pp. 197–241. Academic Press, New York.

114. Day, D. A., Rayner, J. R., and Wiskich, J. T. (1976). Characteristics of external NADH oxidation by beetroot mitochondria. *Plant Physiol.* **58,** 38–42.

115. DePierre, J. W., and Ernster, L. (1977). Enzyme topology of intracellular membranes. *Annu. Rev. Biochem.* **46,** 201–262.

116. Deters, D. W., Racker, E., Nelson, N. and Nelson, H. (1975). Partial resolution of the enzymes catalyzing photophosphorylation. XV. Approaches to the active site of coupling factor 1. *J. Biol. Chem.* **250,** 1041–1047.

117. de Troostembergh, J. C., and Nyns, E. S. (1978). Kinetics of the respiration of cyanide-insensitive mitochondria from the yeast *Saccharomycopsis lipolytica. Eur. J. Biochem.* **85,** 423–432.

118. DeVault, D. (1976). Theory of iron–sulfur center N-2 oxidation and reduction by ATP. *J. Theor. Biol.* **62,** 115–139.

119. Dickerson, R. E. and Timkovich, R. (1975). Cytochromes c in "The Enzymes" (P. D. Boyer, ed.), Vol. XI, pp. 397–547. Academic Press, New York.

120. Dieter, P., and Marmé, D. (1980). Ca^{2+} transport in mitochondrial and microsomal fractions from higher plants. *Planta* **150,** 1–8.

121. Diner, B. A., and Joliot, P. (1977). Oxygen evolution and manganese. *In* "Encyclopedia of Plant Physiology, New Series" (A. Pirson and M. H. Zimmermann, eds.), Vol. 5, "Photosynthesis I. Photosynthetic Electron Transport and Photophosphorylation" (A. Trebst and M. Avron, eds.), pp. 187–205. Springer-Verlag, Berlin.

122. Dolan, E., and Hind, G. (1974). Kinetics of the reduction and oxidation of cytochromes b_6 and f in isolated chloroplasts. *Biochim. Biophys. Acta* **357,** 380–385.

123. Douce, R., Mannella, C. A., and Bonner, W. D. Jr. (1973). The external NADH dehydrogenases of intact plant mitochondria. *Biochim. Biophys. Acta* **292,** 105–116.

123a. Douce, R., Moore, A. L. and Neuberger, M. (1977). Isolation and oxidative properties of intact mitochondria isolated from spinach leaves. *Plant Physiol.* **60**, 625–628.

124. Doussière, J., Sainsard-Chanet, A., and Vignais, P. V. (1979). The respiratory chain of *Paramecium tetraurelia* in wild type and the mutant Cl_1. II. Cyanide-insensitive respiration. Function and regulation. *Biochim. Biophys. Acta* **548**, 236–252.

125. Draber, W., Trebst, A., and Harth, E. (1970). On a new inhibitor of photosynthetic electron-transport in isolated chloroplasts. *Z. Naturforsch.* **25B**, 1157–1159.

126. Ducet, G. and Diano, M. (1978). On the dissociation of the cytochrome b-c_1 of potato mitochondria. *Plant Sci. Lett.* **11**, 217–226.

127. Duggleby, R. G., and Dennis, D. T. (1974). Nicotinamide adenine dinucleotide-specific glyceraldehyde 3-phosphate dehydrogenase from *Pisum sativum:* assay and steady state kinetics. *J. Biol. Chem.* **249**, 167–174.

128. Dunham, K. R. and Selman, B. R. (1981). Regulation of spinach chloroplast coupling factor 1 ATPase activity. *J. Biol. Chem.* **256**, 212–218.

129. Dutton, P. L. and Storey, B. T. (1971). The respiratory chain of plant mitochondrion. IX. Oxidation–reduction potentials of the cytochromes of mung bean mitochondria. *Plant Physiol.* **47**, 282–288.

130. Dutton, P. L., and Wilson, D. F. (1974). Redox potentiometry in mitochondrial and photosynthetic bioenergetics. *Biochim. Biophys. Acta* **346**, 165–212.

131. Duysens, L. N. M., Amesz, J., and Kamp, B. M. (1961). Two photochemical systems in photosynthesis. *Nature (London)* **190**, 510–511.

132. Egneus, H., Heber, U., Matthiesen, U., and Kirk, M. (1975). Reduction of oxygen by the electron transport chain of chloroplasts during assimilation of carbon dioxide. *Biochim. Biophys. Acta* **408**, 252–268.

133. Erixon, K., and Butler, W. L. (1971). The relationship between Q, C-550 and cytochrome b_{559} in photoreactions at $-196°$ in chloroplasts. *Biochim. Biophys. Acta* **234**, 381–389.

134. Erixon, K., and Butler, W. L. (1971). Light-induced absorbance changes in chloroplasts at $-196°C$. *Photochem. Photobiol.* **14**, 427–433.

135. Ernster, L. (1977). Chemical and chemiosmotic aspects of electron transport-linked phosphorylation. *Annu. Rev. Biochem.* **46**, 981–995.

136. Evans, E. H., Rush, J. D., Johnson, C. E., and Evans, M. C. W. (1979). Mössbauer spectra of photosystem-I reaction centres from the blue-green alga *Chlorogloea fritschii.* *Biochem. J.* **182**, 861–865.

137. Evans, M. C. W., and Cammack, R. (1975). The effect of the redox state of the bound iron–sulphur centres in spinach chloroplasts on the reversibility of P700 photooxidation at low temperatures. *Biochem. Biophys. Res. Commun.* **63**, 187–193.

138. Evans, M. C. W., Reeves, S. G., and Cammack, R. (1974). Determination of the oxidation–reduction potential of the bound iron–sulphur proteins of the primary electron acceptor complex of photosystem I in spinach chloroplasts. *FEBS Lett.* **49**, 111–114.

139. Evans, M. C. W., Sihra, C. K., Bolton, J. R. and Cammack, R. (1975). Primary electron acceptor complex of photosystem I in spinach chloroplasts. *Nature (London)* **256**, 668–670.

140. Evans, M. C. W., Sihra, C. K., and Cammack, R. (1976). The properties of the primary electron acceptor in the photosystem I reaction centre of spinach chloroplasts and its interaction with P700 and the bound ferredoxin in various oxidation–reduction states. *Biochem. J.* **158**, 71–77.

141. Evans, M. C. W., Telfer, A., and Lord, A. V. (1972). Evidence for the role of a

bound ferredoxin as the primary electron acceptor of photosystem I in spinach chloroplasts. *Biochim. Biophys. Acta* **267**, 530–537.

142. Eytan, G. D., and Schatz, G. (1975). Cytochrome *c* oxidase from bakers' yeast. V. Arrangement of the subunits in the isolated and membrane-bound enzyme. *J. Biol. Chem.* **250**, 767–774.

143. Fernández-Morán, H., Oda, T., Blair, P. V., and Green, D. E. (1964). A macromolecular repeating unit of mitochondrial structure and function. *J. Cell. Biol.* **22**, 63–100.

144. Fillingame, R. H. (1980). The proton-translocating pumps of oxidative phosphorylation. *Annu. Rev. Biochem.* **49**, 1079–1113.

145. Forbush, B. and Kok, B. (1968). Reaction between primary and secondary electron acceptors of photosystem II of photosynthesis. *Biochim. Biophys. Acta* **162**, 243–253.

146. Forti, G. (1977). Flavoproteins. *In* "Encyclopedia of Plant Physiology, New Series" (A. Pirson and M. H. Zimmermann, eds.), Vol. 5, Photosynthesis I. Photosynthetic Electron Transport and Photophorylation (A. Trebst and M. Avron, eds.), pp. 222–226. Springer-Verlag, Berlin.

147. Forti, G., and Sturani, E. (1968). On the structure and function of reduced nicotinamide adenine dinucleotide phosphate-cytochrome f reductase of spinach chloroplasts. *Eur. J. Biochem.* **3**, 461–472.

148. Foust, G. P., Mayhew, S. G., and Massey, V. (1969). Complex formation between ferredoxin triphosphopyridine nucleotide reductase and electron transfer proteins. *J. Biol. Chem.* **244**, 964–970.

149. Fowler, C. F. (1977). Proton evolution from photosystem II. Stoichiometry and mechanistic considerations. *Biochim. Biophys. Acta* **462**, 414–421.

150. Fowler, C. F., and Kok, B. (1974). Proton evolution associated with the photooxidation of water in photosynthesis. *Biochim. Biophys. Acta* **357**, 299–307.

151. Fowler, C. F., and Kok, B. (1976). Determination of H^+/e^- ratios in chloroplasts with flashing light. *Biochim. Biophys. Acta* **423**, 510–523.

152. Fredricks, W. W., and Gehl, J. M. (1976). Multiple forms of ferredoxin-nicotinamide adenine dinucleotide phosphate reductase from spinach. *Arch. Biochem. Biophys.* **174**, 666–674.

153. Futai, M., and Kanazawa, H. (1980). Role of subunits in proton-translocating ATPase (F_0-F_1). *Current Topics Bioenerg.* **10**, 181–215.

154. Galante, Y. M., and Hatefi, Y. (1979). Purification and molecular and enzymic properties of mitochondrial NADH dehydrogenase. *Arch. Biochem. Biophys.* **192**, 559–568.

155. Galante, Y. M., Wong, S. Y., and Hatefi, Y. (1979). Composition of complex V of the mitochondrial oxidative phosphorylation system. *J. Biol. Chem.* **254**, 12372–12378.

156. Garab, G. I., Paillotin, G., and Joliot, P. (1979). Flash-induced scattering transients in the 10 μs–5s time range between 450 and 540 nm with *Chlorella* cells. *Biochem. Biophys. Acta* **545**, 445–453.

157. Garewal, H. S., and Wasserman, A. R. (1974). Triton X-100-4*M* urea as an extraction medium for membrane proteins. I. Purification of chloroplast cytochrome b_{559}. *Biochemistry* **13**, 4063–4071.

158. Garewal, H. S., and Wasserman, A. R. (1974). Triton X-100-4*M* urea as an extraction medium for membrane proteins. II. Molecular properties of pure cytochrome b_{559}: a lipoprotein containing small polypeptide chains and a limited lipid composition. *Biochemistry* **13**, 4072–4079.

159. Gimmler, H. (1977). Photophosphorylation *in vivo. In* "Encyclopedia of Plant Physi-

ology, New Series" (A. Pirson and M. H. Zimmermann, eds.), Vol. 5, "Photosynthesis I. Photosynthetic Electron Transport and Photophosphorylation" (A. Trebst and M. Avron, eds.), pp. 448–472. Springer-Verlag, Berlin and New York.

160. Goldbeck, J. H., and Kok, B. (1978). Further studies of the membrane-bound iron–sulfur proteins and P700 in a photosystem I subchloroplast particle. *Arch. Biochem. Biophys.* **188**, 233–242.

161. Goldbeck, J. H., Lien, S., and San Pietro, A. (1977). Electron transport in chloroplasts. In "Encyclopedia of Plant Physiology" (A. Pirson and M. H. Zimmermann, eds.), Vol. 5, Photosynthetic Electron Transport and Photophosphorylation (A. Trebst and M. Avron, eds.), pp. 94–116. Springer-Verlag, Berlin.

162. Goldberg, J. H., Velthuys, B. R., and Kok, B. (1978). Evidence that the intermediate electron acceptor, A_2, in photosystem I is a bound iron–sulfur protein. *Biochim. Biophys. Acta* **504**, 226–230.

163. Gorman, D. S., and Levine, R. P. (1965). Cytochrome f and plastocyanin: their sequence in the photosynthetic electron transport chain of *Chlamydomonas reinhardi*. *Proc. Natl. Acad. Sci. USA* **54**, 1665–1669.

164. Gozzer, C., Zanetti, G., Galliano, M., Sacchi, G. A., Minchiotti, L., and Curti, B. (1977). Molecular heterogeneity of ferredoxin-$NADP^+$ reductase from spinach leaves. *Biochim. Biophys. Acta* **485**, 278–290.

165. Gräber, P., Schlodder, E., and Witt, H. T. (1977). Conformational change of the chloroplast ATPase induced by a transmembrane electric field and its correlation to phosphorylation. *Biochim. Biophys. Acta* **461**, 426–440.

166. Gräber, P., and Witt, H. T. (1975). Direct measurement of the protons pumped into the inner phase of the functional membrane of photosynthesis per electron transfer. *FEBS Lett.* **59**, 184–189.

167. Gräber, P., and Witt, H. T. (1976). Relations between the electrical potential, pH gradient, proton flux and phosphorylation in the photosynthetic membrane. *Biochim. Biophys. Acta* **423**, 141–163.

167a. Gray, J. C. (1980). Maternal inheritance of cytochrome f in interspecific *Nicotiana* hybrids. *Eur. J. Biochem.* **112**, 39–46.

168. Green, D. E., and Hatefi, Y. (1961). The mitochondrion and biochemical machines: mitochondria or their equivalents are the principal energy transducers in all aerobic organisms. *Science* **133**, 13–19.

169. Green, D. E. and Wharton, D. C. (1963). Stoichiometry of the fixed oxidation–reduction components of the electron transfer chain of beef heart mitochondria. *Biochem. Z.* **338**, 335–348.

170. Griffiths, D. E. (1976). Studies of energy-linked reactions, net synthesis of adenosine triphosphate by isolated adenosine triphosphate synthase preparations: a role for lipoic acid and unsaturated fatty acids. *Biochem. J.* **160**, 809–812.

171. Griffiths, D. E. (1977). The role of lipoic acid in adenosine triphosphatase synthases. *Biochem. Soc. Trans.* **5**, 1283–1285.

172. Griffiths, D. E., Hyams, R. L. and Partis, M. D. (1977). Studies of energy linked reactions: a role for lipoic acid in the purple membrane of *Halobacterium halobium*. *FEBS Lett.* **78**, 155–160.

173. Grubmeyer, C., Duncan, I. and Spencer, M. (1977). Partial characterization of a soluble ATPase from pea cotyledon mitochondria. *Can. J. Biochem.* **55**, 812–818.

174. Grubmeyer, C., Melanson, D., Duncan, I., and Spencer, M. (1979). Oxidative phosphorylation in pea cotyledon submitochondrial particles. *Plant Physiol.* **64**, 757–762.

175. Grubmeyer, C., and Spencer, M. (1979). Effects of anions on a soluble ATPase from mitochondria of pea cotyledons. *Plant Cell Physiol.* **20**, 83–91.

176. Grubmeyer, C., and Spencer, M. (1980). ATPase activity of pea cotyledon submito-chondrial particles. Activation, substrate specificity, and anion effects. *Plant Physiol.* **65,** 281–285.

177. Gutman, M. (1980). Electron flux through the mitochondrial ubiquinone. *Biochim. Biophys. Acta* **594,** 53–84.

178. Haehnel, W. (1973). Electron transport between plastoquinone and chlorophyll a-1 in chloroplasts. *Biochim. Biophys. Acta* **305,** 618–631.

179. Haehnel, W. (1976). The reduction kinetics of chlorophyll a_1 as an indicator for proton uptake between the light reactions in chloroplasts. *Biochim. Biophys. Acta* **440,** 506–521.

180. Haehnel, W. (1977). Electron transport between plastoquinone and chlorophyll A_1 in chloroplasts. II. Reaction kinetics and the function of plastocyanin in situ. *Biochim. Biophys. Acta* **459,** 418–441.

181. Haehnel, W., Hesse, V., and Pröpper, A. (1980). Electron transfer from plasto-cyanin to P700. Function of a subunit of photosystem I reaction center. *FEBS Lett.* **111,** 79–82.

182. Haehnel, W., Pröpper, A., and Krause, H. (1980). Evidence for complexed plasto-cyanin as the immediate electron donor of P-700. *Biochim. Biophys. Acta* **593,** 384–399.

183. Hall, D. O. and Rao, K. K. (1977). Ferredoxin. *In* "Encyclopedia of Plant Physiology, New Series" (A. Pirson and M. H. Zimmermann, eds.), Vol. 5, Photosynthetic Elec-tron Transport and Photophosphorylation (A. Trebst and M. Avron, eds.), pp. 206–216. Springer-Verlag, Berlin.

184. Hanska, G., Reimer, S. and Trebst, A. (1974). Native and artificial energy-conserv-ing sites in cyclic photophosphorylation systems. *Biochim. Biophys. Acta* **357,** 1–13.

185. Harris, D. A. (1978). The interactions of coupling ATPases with nucleotides. *Bio-chim. Biophys. Acta* **463,** 245–273.

186. Harris, D. A., and Crofts, A. R. (1978). The initial stages of photophosphorylation. Studies using excitation by saturating, short flashes of light. *Biochim. Biophys. Acta* **502,** 87–102.

187. Harris, D. A., Rosing, J., Van de Stadt, R. J., and Slater, E. C. (1973). Tight binding of adenine nucleotides to beef-heart mitochondrial ATPase. *Biochim. Biophys. Acta* **314,** 149–153.

188. Harris, D. A., and Slater, E. C. (1975). Tightly bound nucleotides of the energy-transducing ATPase of chloroplasts and their role in photophosphorylation. *Bio-chim. Biophys. Acta* **387,** 335–348.

189. Hatefi, Y. (1966). The functional complexes of the mitochondrial electron-transfer system. *Compr. Biochem.* **14,** 199–231.

190. Hatefi, Y. (1978). Introduction-preparation and properties of the enzymes and enzyme complexes of the mitochondrial oxidative phosphorylation system. *Methods Enzymol.* **53,** 3–4.

191. Hatefi, Y. (1978). Preparation and properties of NADH: ubiquinone oxidoreduc-tase (Complex I), EC 1.6.53. *Methods Enzymol.* **53,** 11–14.

192. Hatefi, Y. (1978). Resolution of Complex II and isolation of succinate dehydro-genase (EC 1.3.99.1). *Methods Enzymol.* **53,** 27–35.

193. Hatefi, Y. (1978). Reconstitution of the electron-transport system of bovine heart mitochondria. *Methods Enzymol.* **53,** 48–54.

194. Hatefi, Y., Haavik, A. G., Fowler, L. R., and Griffiths, D. E. (1962). Studies on the electron transfer system XLII. Reconstitution of the electron transfer system. *J. Biol. Chem.* **237,** 2661–2669.

195. Hatefi, Y., and Stiggall, D. L. (1976). Metal-containing flavoprotein dehydrogenases. *In* "The Enzymes" (P. D. Boyer, ed.), Vol. XIII, pp. 175–297. Academic Press, New York.

196. Hatefi, Y., and Stiggall, D. L. (1978). Preparation and properties of succinate: ubiquinone oxidoreductase (Complex II). *Methods Enzymol.* **53**, 21–27.

197. Hauska, G. (1980). Transmembrane charge separation within the large subunit of photosystem I-reaction centers from chloroplasts. *FEBS Lett.* **119**, 232–234.

198. Hauska, G., Reimer, S., and Trebst, A. (1974). Native and artificial energy-conserving sites in cyclic photophosphorylation systems. *Biochim. Biophys. Acta* **357**, 1–13.

199. Hauska, G., Samoray, D., Orlich, G. and Nelson, N. (1980). Reconstitution of photosynthetic energy conservation. II. Photophosphorylation in liposomes containing photosystem-I reaction center and chloroplast coupling-factor complex. *Eur. J. Biochem.* **111**, 535–543.

200. Hauska, G. and Trebst, A. (1977). Proton translocation in chloroplasts. *Current Topics Bioenerg.* **6**, 151–220.

201. Heathcote, P., and Evans, M. C. W. (1980). Properties of the EPR spectrum of the intermediary electron acceptor (A_1), in several different photosystem I particle preparations. *FEBS Lett.* **111**, 381–385.

202. Heathcote, P., Timofeev, K. N., and Evans, M. C. W. (1979). Detection by EPR spectroscopy of a new intermediate in the primary photochemistry of photosystem I particles isolated using Triton X-100. *FEBS Lett.* **101**, 105–109.

203. Heathcote, P., Williams-Smith, D. L., and Evans, M. C. W. (1978). Quantitative electron-paramagnetic-resonance measurements of the electron-transfer components of the photosystem-I reaction centre. The reaction-centre chlorophyll (P700), the primary electron acceptor X and bound iron–sulphur centre. *Biochem. J.* **170**, 373–378.

204. Heathcote, P., Williams-Smith, D. L., Sihra, C. K., and Evans, M. C. W. (1978). The role of the membrane-bound iron–sulphur centres A and B in the photosystem I reaction centre of spinach chloroplasts. *Biochim. Biophys. Acta* **503**, 333–342.

205. Heber, U., Boardman, N. K. and Anderson, J. M. (1976). Cytochrome b-563 redox changes in intact CO_2-fixing spinach chloroplasts and in developing pea chloroplasts. *Biochim. Biophys. Acta* **423**, 275–292.

206. Heber, U., Egneus, H., Hanck, U., Jensen, M., and Koster, S. (1978). Regulation of photosynthetic electron transport and photophosphorylation in intact chloroplasts and leaves of *Spinacia oleracea* L. *Planta* **143**, 41–49.

207. Heber, U., Kirk, M. R., and Boardman, N. K. (1979). Photoreactions of cytochrome b_{559} and cyclic electron flow in photosystem II of intact chloroplasts. *Biochim. Biophys. Acta* **546**, 292–306.

208. Henningsen, K. W. and Boardman, N. K. (1973). Development of photochemical activity and the appearance of the high potential form of cytochrome b-559 in greening barley seedlings. *Plant Physiol.* **51**, 1117–1126.

209. Henry, M. F., and Nyns, E. J. (1975). Cyanide-insensitive respiration. An alternative mitochondrial pathway. *Subcell. Biochem.* **4**, 1–65.

210. Hiatt, A. J. (1961). Preparation and some properties of soluble succinic dehydrogenase from higher plants. *Plant Physiol.* **36**, 552–557.

211. Hildreth, W. W. (1968). Laser-induced kinetics of cytochrome oxidation and the 518 mμ absorption change in spinach leaves and chloroplasts. *Biochim. Biophys. Acta* **153**, 197–202.

212. Hildreth, W. W. (1968). Laser-activated electron transport in a *Chlamydomonas* mutant. *Plant Physiol.* **43**, 303–312.

213. Hill, R. (1954). The cytochrome b component of chloroplasts. *Nature (London)* **174**, 501–503.

214. Hill, R., and Bendall, F. (1960). Function of the two cytochrome components in chloroplasts: a working hypothesis. *Nature (London)* **186**, 136–137.

215. Hind, G. (1968). The site of action of plastocyanin in chloroplasts treated with detergent. *Biochim. Biophys. Acta* **153**, 235–240.

216. Hind, G., and Jagendorf, A. T. (1963). Separation of light and dark stages in photophosphorylation. *Proc. Natl. Acad. Sci. USA* **49**, 715–722.

217. Hind, G., and Jagendorf, A. T. (1965). Effect of uncouplers on the conformational and high energy states of chloroplasts. *J. Biol. Chem.* **240**, 3202–3209.

218. Hind, G., Nakatani, H. Y., and Izawa, S. (1969). The role of Cl^- in photosynthesis. I. The Cl^- requirement of electron transport. *Biochim. Biophys. Acta* **172**, 277–289.

219. Hiyama, T., and Ke, B. (1971). A further study of P430: a possible primary electron acceptor of photosystem I. *Arch. Biochem. Biophys.* **147**, 99–108.

220. Hiyama, T., and Ke, B. (1971). A new photosynthetic pigment, "P430": its possible role as the primary electron acceptor of photosystem I. *Proc. Natl. Acad. Sci. USA* **68**, 1010–1013.

221. Hiyama, T., Tsujimoto, H. Y., and Arnon, D. I. (1979). Photoreduction of membrane-bound paramagnetic component X by water as electron donor. *FEBS Lett.* **98**, 381–385.

222. Ho, K. K., and Krogmann, D. W. (1980). Cytochrome f from spinach and cyanobacteria. Purification and characterization. *J. Biol. Chem.* **255**, 3855–3861.

223. Homann, P. H. (1971). Actions of carbonylcyanide m-chlorophenylhydrazone on electron transport and fluorescence of isolated chloroplasts. *Biochim. Biophys. Acta* **245**, 129–143.

224. Horton, P. and Cramer, W. A. (1975). Acid–base induced redox changes of the chloroplast cytochrome b-559. *FEBS Lett.* **56**, 244–247.

225. Horton, P. and Cramer, W. A. (1976). Stimulation of photosystem I-induced oxidation of chloroplast cytochrome b-559 by pre-illumination and by low pH. *Biochim. Biophys. Acta* **430**, 122–134.

226. Horton, P., Croze, E., and Smutzer, G. (1978). Interactions between photosystem II components in chloroplast membranes. A correlation between the existence of a low potential species of cytochrome b-559 and low chlorophyll fluorescence in inhibited and developing chloroplasts. *Biochim. Biophys. Acta* **503**, 274–286.

227. Huber, S. C., and Edwards, G. E. (1977). The importance of reducing conditions for the inhibitory action of DBMIB, antimycin A and EDAC on cyclic photophosphorylation. *FEBS Lett.* **79**, 207–211.

228. Huchzermeyer, B., and Strotmann, H. (1977). Acid/base-induced exchange of adenine nucleotides on chloroplast coupling factors. *Z. Naturforsch.* **32C**, 803–809.

229. Huq, S., and Palmer, J. M. (1978). Isolation of a cyanide-resistant duroquinol oxidase from *Arum maculatum* mitochondria. *FEBS Lett.* **95**, 217–220.

230. Huq, S., and Palmer, J. M. (1978). Superoxide and hydrogen peroxide production in cyanide resistant *Arum maculatum* mitochondria. *Plant Sci. Lett.* **11**, 351–358.

231. Izawa, S., Heath, R. L., and Hind, G. (1969). The role of chloride ion in photosynthesis. III. The effect of artificial electron donors upon electron transport. *Biochim. Biophys. Acta* **180**, 388–398.

232. Izawa, S., Kraayenhof, R., Ruuge, E. K., and Devault, D. (1973). The site of KCN inhibition in the photosynthetic electron transport pathway. *Biochim. Biophys. Acta* **314**, 328–339.

233. Jagendorf, A. T. (1977). Photophosphorylation. *In* "Encyclopedia of Plant Physiol-

ogy, New Series" (A. Pirson and M. H. Zimmermann, eds.), Vol. 5, "Photosynthesis I. Photosynthetic Electron Transport and Photophosphorylation" (A. Trebst and M. Avron, eds.), pp. 307–337. Springer-Verlag, Berlin.

234. Jagendorf, A. T., and Hind, G. (1965). Relation between electron flow, phosphorylation, and a high energy state of chloroplasts. *Biochem. Biophys. Res. Commun.* **18**, 702–709.

235. Jagendorf, A. T., and Neumann, J. (1965). Effect of uncouplers on the light induced pH rise with spinach chloroplasts. *J. Biol. Chem.* **240**, 3210–3214.

236. Jagendorf, A. T., and Uribe, E. (1966). ATP formation caused by acid–base transition of spinach chloroplasts. *Proc. Nat. Acad. Sci. USA* **55**, 170–177.

237. Jaynes, J. M., Vernon, L. P., and Klein, S. M. (1975). Photophosphorylation and related properties of reaggregated vesicles from spinach photosystem I particles. *Biochim. Biophys. Acta* **408**, 240–251.

238. Johnston, R., and Criddle, R. S. (1977). F_1-ATPase-catalyzed synthesis of ATP from oleoylphosphate and ADP. *Proc. Natl. Acad. Sci. USA* **74**, 4919–4923.

239. Johnston, S. P., Møller, I. M., and Palmer, J. M. (1979). The stimulation of exogenous NADH oxidation in Jerusalem artichoke mitochondria by screening of charges on the membranes. *FEBS Lett.* **108**, 28–32.

240. Joliot, P. (1965). Études simutanées des cinétiques de fluorescence et d'émission d'oxygène photosynthétique. *Biochim. Biophys. Acta* **102**, 135–148.

241. Joliot, P., Barbieri, G., and Chabaud, R. (1969). Un Nouveau Modele des centres photochimique du systeme II. *Photochem. Photobiol.* **10**, 309–329.

242. Joliot, P. and Joliot, A. (1968). A polarographic method for detection of oxygen production and reduction of Hill reagent by isolated chloroplasts. *Biochim. Biophys. Acta* **153**, 625–634.

243. Joliot, P., and Joliot, A. (1979). Comparative study of the fluorescence yield and of the C550 absorption change at room temperature. *Biochim. Biophys. Acta* **546**, 93–105.

244. Jones, D. H., and Boyer, P. D. (1969). The apparent absolute requirement of adenosine diphosphate for the inorganic phosphate ⇌ water exchange of oxidative phosphorylation. *J. Biol. Chem.* **244**, 5767–5772.

245. Jung, D. W., and Hanson, J. B. (1973). Respiratory activation of 2,4-dinitrophenol-stimulated ATPase activity in plant mitochondria. *Arch. Biochem. Biophys.* **158**, 139–148.

246. Jung, D. W., and Laties, G. C. (1976). Trypsin-induced ATPase activity in potato mitochondria. *Plant Physiol.* **57**, 583–588.

247. Junge, W. (1977). Physical aspects of light harvesting electron transport and electrochemical potential generation in photosynthesis of green plants. *In* "Encyclopedia of Plant Physiology, New Series" (A. Pirson and M. H. Zimmermann, eds.), Vol. 5, Photosynthesis I. Photosynthetic Electron Transport and Photophosphorylation (A. Trebst and M. Avron, eds.), pp. 59–93. Springer-Verlag, Berlin.

248. Junge, W., Renger, G., and Ausländer, W. (1977). Proton release into the internal phase of the thylakoids due to photosynthetic water oxidation. *FEBS Lett.* **79**, 155–159.

249. Junge, W. B., Rumberg, B., and Schröder, H. (1970). The necessity of an electric potential difference and its use for photophosphorylation in short flash groups. *Eur. J. Biochem.* **14**, 575–581.

250. Junge, W., and Witt, H. T. (1968). On the ion transport system of photosynthesis—investigations on a molecular level. *Z. Naturforsch.* **23B**, 244–254.

251. Kagawa, Y. (1978). Reconstitution of the energy transformer, gate and channel

subunit reassembly, crystalline ATPase and ATP synthesis. *Biochim. Biophys. Acta* **505**, 45–93.

252. Kagawa, Y., Kandrach, A., and Racker, E. (1973). Partial resolution of the enzymes catalyzing oxidative phosphorylation. *J. Biol. Chem.* **248**, 676–684.

253. Kagawa, Y., and Racker, E. (1971). Partial resolution of the enzyme catalyzing oxidative phosphorylation. XXV. Reconstitution of vesicles catalyzing $^{32}P_i$-adenosine triphosphate exchange. *J. Biol. Chem.* **246**, 5477–5487.

254. Katoh, S. (1977). Plastocyanin. *In* "Encyclopedia of Plant Physiology, New Series" (A. Pirson and M. H. Zimmermann, eds.), Vol. 5, Photosynthesis I. Photosynthetic Electron Transport and Photophorylation (A. Trebst and M. Avron, eds.), pp. 247–252. Springer-Verlag, Berlin and New York.

255. Kayalar, C., Rosing, J., and Boyer, P. D. (1976). 2,4-Dinitrophenol causes a marked increase in the apparent K_m of P_i and of ADP for oxidative phosphorylation. *Biochem. Biophys. Res. Commun.* **72**, 1153–1159.

256. Kayalar, C., Rosing, J., and Boyer, P. D. (1977). An alternating site sequence for oxidative phosphorylation suggested by measurement of substrate binding patterns and exchange reaction inhibitions. *J. Biol. Chem.* **252**, 2486–2491.

257. Ke, B. (1978). The primary electron acceptors in green-plant photosystem I and photosynthetic bacteria. *Current Topics Bioenerg.* **7**, 75–138.

258. Ke, B. and Beinert, H. (1973). Evidence for the identity of P430 of photosystem I and chloroplast-bound iron–sulfur protein. *Biochim. Biophys. Acta* **305**, 689–693.

259. Ke, B., Hansen, R. E., and Beinert, H. (1973). Oxidation–reduction potentials of bound iron–sulfur proteins of photosystem I. *Proc. Natl. Acad. Sci. USA* **70**, 2941–2945.

260. Ke, B., Sugahara, K., and Shaw, E. R. (1975). Further purification of "Triton subchloroplast fraction I" (TSF-1 particles). Isolation of a cytochrome-free high-P-700 particle and a complex containing cytochromes f and b_6, plastocyanin and iron–sulphur protein(s). *Biochim. Biophys. Acta* **408**, 12–25.

261. Ke, B., Vernon, L. P., and Chaney, T. H. (1972). Photo-reduction of cytochrome b_{559} in a photosystem-II subchloroplast particle. *Biochim. Biophys. Acta* **256**, 345–357.

262. Kenney, W. C., Mowery, P. C., Seng, R. L., and Singer, T. P. (1976). Localization of the substrate and oxalacetate binding site of succinate dehydrogenase. *J. Biol. Chem.* **251**, 1369–2372.

263. Klimov, V. V., Dolan, E., and Ke, B. (1980). EPR properties of an intermediary electron acceptor (pheophytin) in photosystem-II reaction centers at cryogenic temperatures. *FEBS Lett.* **112**, 97–100.

264. Klimov, V. V., Dolan, E., Shaw, E. R., and Ke, B. (1980). Interaction between the intermediary electron acceptor (pheophytin) and a possible plastoquinone–iron complex in photosystem II reaction centers. *Proc. Nat. Acad. Sci. USA* **77**, 7227–7231.

265. Klimov, V. V., Klevanik, A. V., Shuvalov, V. A., and Krasnovsky, A. A. (1977). Reduction of pheophytin in the primary light reaction of photosystem II. *FEBS Lett.* **82**, 183–186.

266. Klingenberg, M. (1970). Localization of the glycerol phosphate dehydrogenase in the outer phase of the mitochondrial inner membrane. *Eur. J. Biochem.* **13**, 247–252.

267. Knaff, D. B. (1972). The photooxidation of chloroplast cytochrome b_6 by photosystem I. *FEBS Lett.* **23**, 92–94.

268. Knaff, D. B., and Arnon, D. I. (1969). Light-induced oxidation of a chloroplast b-type cytochrome at −189°C. *Proc. Natl. Acad. Sci. USA* **63**, 956–962.

269. Knaff, D. B., and Arnon, D. I. (1969). Spectral evidence for a new photoreactive

component of the oxygen-evolving system in photosynthesis. *Proc. Natl. Acad. Sci. USA* **63**, 963–969.

270. Knaff, D. B., and Arnon, D. I. (1969). A concept of three light reactions in photosynthesis by green plants, *Proc. Natl. Acad. Sci. USA* **64**, 715–722.

271. Knaff, D. B., and Malkin, R. (1978). The primary reaction of chloroplast photosystem II. *Current Topics Bioenerg.* **7**, 139–172.

272. Knaff, D. B., Malkin, R., Myron, J. C., and Stoller, M. (1977). The role of plastoquinone and beta-carotene in the primary reaction of plant photosystem II. *Biochim. Biophys. Acta* **459**, 402–411.

273. Koenig, F., Radunz, A., Schmid, G. H., and Menke, W. (1978). Antisera to the coupling factor of photophosphorylation and its subunits. *Z. Naturforsch.* **33C**, 529–536.

274. Kok, B., Forbush, B., and McGloin, M. (1970). Cooperation of charges in photosynthetic O₂ evolution. I. A linear four step mechanism. *Photochem. Photobiol.* **11**, 457–475.

275. Konishi, T., Packer, L., and Criddle, R. (1979). Purification and assay of a proteolipid ionophore from yeast mitochondrial ATP synthetase. *Methods Enzymol.* **55**, 414–421.

276. Kozlov, I. A., and Skulachev, V. P. (1977). H⁺-Adenosine triphosphatase and membrane energy coupling. *Biochim. Biophys. Acta* **463**, 29–89.

277. Krab, K., and Wikström, M. (1978). Proton-translocating cytochrome c oxidase in artificial phospholipid vesicles. *Biochim. Biophys. Acta* **504**, 200–214.

278. Krab, K., and Wikström, M. (1979). On the stoichiometry and thermodynamics of proton-pumping cytochrome c oxidase in mitochondria. *Biochim. Biophys. Acta* **548**, 1–15.

278a. Lambers, H. (1982). Cyanide-resistant respiration: a nonphosphorylation electron transport pathway acting as an energy overflow. *Physiol. Plant,* in press.

279. Lambowitz, A. M., and Bonner, W. D., Jr. (1974). The b-cytochromes of plant mitochondria. A spectrophotometric and potentiometric study. *J. Biol. Chem.* **249**, 2428–2440.

280. Lambowitz, A. M., Bonner, W. D., Jr., and Wikström, M. K. F. (1974). On the lack of ATP-induced midpoint potential shift for cytochrome b-566 in plant mitochondria. *Proc. Natl. Acad. Sci. USA* **71**, 1183–1187.

281. Lance, C., and Bonner, W. D., Jr. (1968). The respiratory chain components of higher plant mitochondria. *Plant Physiol.* **43**, 756–766.

282. Lappin, A. G., Segal, M. G., Weatherburn, D. C., and Sykes, A. G. (1979). Kinetic studies on 1 : 1 electron-transfer reactions involving blue copper proteins. 2. Protonation effects and different binding sites in the oxidation of parsley plastocyanin with Co(4,7-DPSphen)₃³⁻, Fe(CN)₆³⁻ and Co(Phen)₃³⁺. *J. Am. Chem. Soc.* **101**, 2287–2301.

282a. Laties, G. C. (1982). The cyanide-resistant alternative path in higher plant respiration. *Annu. Rev. Plant Physiol.* **33**, 519–555.

283. Lawford, H. G., and Garland, P. B. (1973). Proton translocation coupled to quinol oxidation in ox heart mitochondria. *Biochem. J.* **136**, 711–720.

284. Lee, C. P., Schatz, G., and Ernster, L. (eds.) (1979). "Membrane Bioenergetics" (in honour of E. Racker). Addison-Wesley, Reading, Massachusetts.

285. Lehninger, A. L., Reynafarje, B., Alexandre, A., and Villalobo, A. (1979). Proton stoichiometry and mechanisms in mitochondrial energy transduction. *In* "Membrane Bioenergetics" (C. P. Lee, G. Schatz, and L. Ernster, eds), pp. 393–404. Addison-Wesley, Reading, Massachusetts.

286. Lemberg, M. R. (1969). Cytochrome oxidase. *Physiol. Rev.* **49**, 48–121.
287. Leung, K. H., and Hinkle, P. C. (1975). Reconstitution of ion transport and respiratory control in vesicles formed from reduced coenzyme Q-cytochrome c reductase and phospholipids. *J. Biol. Chem.* **250**, 8467–8471.
288. Levine, R. P., and Gorman, D. S. (1966). Photosynthetic electron transport chain of *Chlamydomonas reinhardi*. III. Light-induced absorbance changes in chloroplast fragments of the wild type and mutant strains. *Plant Physiol.* **41**, 1293–1300.
289. Lindsay, J. G., and Wilson, D. F. (1972). Apparent adenosine triphosphate induced ligand change in cytochrome a_3 of pigeon heart mitochondria. *Biochemistry* **11**, 4613–4621.
290. Lipmann, F. (1941). Metabolic generation and utilization of phosphate bond energy. *Adv. Enzymol.* **1**, 99–162.
291. Lockau, W. (1979). The inhibition of photosynthetic electron transport in spinach chloroplasts by low osmolarity. *Eur. J. Biochem.* **94**, 365–373.
292. Lundegårdh, H. (1962). Quantitative relations between chlorophyll and cytochromes in chloroplasts. *Physiol. Plant.* **15**, 390–398.
293. Lundegårdh, H. (1965). The influence of diuron (3-(3,4-dichlorophenyl)-1,1-dimethylurea) on the respiratory and photosynthetic systems of plants. *Proc. Nat. Acad. Sci. USA* **53**, 703–710.
294. McCarty, R. E. (1978). The ATPase complex of chloroplasts and chromatophores. *Current Topics Bioenerg.* **7**, 245–278.
295. McCarty, R. E., and Fagan, J. (1973). Light-stimulated incorporation of N-ethylmaleimide into coupling factor 1 in spinach chloroplasts. *Biochemistry* **12**, 1503–1507.
296. McCarty, R. E., Pittmann, P. R., and Tsuchiya, Y. (1972). Light-dependent inhibition of photophosphorylation by N-ethylmaleimide. *J. Biol. Chem.* **247**, 3048–3051.
297. McIntosh, A. R., Chu, M., and Bolton, J. R. (1975). Flash photolysis electron spin resonance studies of the electron acceptor species at low temperatures in photosystem I of spinach subchloroplast particles. *Biochim. Biophys. Acta* **376**, 308–314.
298. Mackey, L. N., Kuwana, T., and Hartzell, C. R. (1973). Evaluation of the energetics of cytochrome c oxidase in the absence of cytochrome c. *FEBS Lett.* **36**, 326–329.
299. McKinney, D. W., Buchanan, B. B., and Wolosiuk, R. A. (1979). Association of a thioredoxin-like protein with chloroplast coupling factor (CF_1). *Biochem. Biophys. Res. Commun.* **86**, 1178–1184.
300. MacLennan, D. H., and Asai, J., (1968). Studies on the mitochondrial adenosine triphosphatase system. V. Localization of the oligomycin-sensitivity conferring protein. *Biochem. Biophys. Res. Commun.* **33**, 441–447.
301. Maeshima, M., and Asahi, T. (1978). Purification and characterization of sweet potato cytochrome c oxidase. *Arch. Biochem. Biophys.* **187**, 423–430.
302. Mahler, H. R., and Cordes, E. H. (1971). "Biological Chemistry," 2nd ed. Harper, New York.
303. Malhotra, S. S., and Spencer, M. (1974). Preparation and properties of purified (Na^+ plus K^+)-stimulated mitochondrial ATPase from germinating pea seeds. *Can. J. Biochem.* **52**, 491–499.
304. Malhotra, S. S., and Spencer, M. (1974). Effects of ethylene, carbon dioxide, and ethylene–carbon dioxide mixtures on the activities of "membrane-containing" and "highly purified" preparations of adenosine triphosphatase from pea-cotyledon mitochondria. *Can. J. Biochem.* **52**, 1091–1096.
305. Malkin, R., and Aparicio, P. J. (1975). Identification of a g equals 1.90 high-potential iron–sulfur protein in chloroplasts. *Biochem. Biophys. Res. Commun.* **63**, 1157–1160.

306. Malkin, R., and Bearden, A. J. (1971). Primary reactions of photosynthesis: photoreduction of a bound chloroplast ferredoxin at low temperature as detected by EPR spectroscopy. *Proc. Natl. Acad. Sci. USA* **68,** 16–19.

307. Malkin, R., and Bearden, A. J. (1976). The effect of alkaline pH on chloroplast photosystem I reactions at cryogenic temperature. *FEBS Lett.* **69,** 216–220.

308. Malkin, R., and Bearden, A. J. (1978). Membrane-bound iron–sulfur centers in photosynthetic systems. *Biochim. Biophys. Acta* **505,** 147–181.

309. Malkin, R., and Knaff, D. B. (1973). Effect of oxidizing treatment on chloroplast photosystem II reactions. *Biochim. Biophys. Acta* **325,** 336–340.

310. Malkin, R., Knaff, D. B., and Bearden, A. J. (1973). The oxidation-reduction potential of membrane-bound chloroplast plastocyanin and cytochrome f. *Biochim. Biophys. Acta* **305,** 675–678.

311. Malkin, R., and Posner, H. B. (1978). On the site of function of the Rieske iron–sulfur center in the chloroplast electron transport chain. *Biochim. Biophys. Acta* **501,** 552–554.

312. Malkin, S. (1966). Fluorescence induction studies in isolated chloroplasts. II. Kinetic analysis of the fluorescence intensity dependence on time. *Biochim. Biophys. Acta* **126,** 433–442.

313. Malkin, S., and Kok, B. (1966). Fluorescence induction in isolated chloroplasts. I. Number of components involved in the reaction and quantum yields. *Biochim. Biophys. Acta* **126,** 413–432.

314. Malmström, B. G. (1974). Cytochrome c oxidase: some current biochemical and biophysical problems. *Q. Rev. Biophys.* **6,** 389–431.

315. Malmström, B. G. (1979). Cytochrome c oxidase: structure and catalytic activity. *Biochim. Biophys. Acta* **549,** 281–303.

316. Marsho, T. V., and Kok, B. (1970). Interaction between electron transport components in chloroplasts. *Biochim. Biophys. Acta* **223,** 240–250.

317. Mathis, P., and Sauer, K. (1978). Rapidly reversible flash-induced electron transfer in P_{700} chlorophyll-protein complex isolated with SDS. *FEBS Lett.* **88,** 275–278.

317a. Matsuoka, M., Maeshima, M., and Asahi, T. (1981). The subunit composition of pea cytochrome c oxidase. *J. Biochem.* **90,** 649–655.

318. Meeuse, B. J. D. (1975). Thermogenic respiration in aroids. *Annu. Rev. Plant Physiol.* **26,** 117–126.

319. Mehler, A. H. (1951). Studies on reactions of illuminated chloroplasts. I. Mechanism of the reduction of oxygen and other Hill reagents. *Arch. Biochem. Biophys.* **33,** 65–77.

320. Merli, A., Capaldi, R. A., Ackrell, B. A. C., and Kearney, E. B. (1979). Arrangement of complex II (succinate-ubiquinone reductase) in the mitochondrial inner membrane. *Biochemistry* **18,** 1393–1400.

321. Metzler, D. E. (1977). *In* "Biochemistry: the Chemical Reactions of Living Cells," p. 233. Academic Press, New York.

322. Meunier, J. C., and Dalziel, K. (1978). Kinetic studies of glyceraldehyde 3-phosphate dehydrogenase from rabbit muscle. *Eur. J. Biochem.* **82,** 483–492.

323. Mills, J. D., Crowther, D., Slovacek, R. E., Hind, G., and McCarty, R. E. (1979). Electron transport pathways in spinach chloroplasts. Reduction of the primary acceptor of photosystem II by reduced nicotinamide adenine dinucleotide phosphate in the dark. *Biochim. Biophys. Acta* **547,** 127–137.

324. Mills, J. D., and Hind, G. (1979). Light-induced Mg^{2+} ATPase activity of coupling factor in intact chloroplasts. *Biochim. Biophys. Acta* **547,** 455–462.

274   D. T. DENNIS

325. Mills, J. D., Mitchell, P., and Schürmann, P. (1980). Modulation of coupling factor ATPase activity in intact chloroplasts. The role of the thioredoxin system. *FEBS Lett.* **112**, 173–177.
326. Mills, J. D., Slovacek, R. E. and Hind, G. (1978). Cyclic electron transport in isolated chloroplasts. Further studies with antimycin. *Biochim. Biophys. Acta* **504**, 298–309.
327. Mitchell, P. (1961). Coupling of phosphorylation to electron and hydrogen transfer by a chemi-osmotic type of mechanism. *Nature (London)* **191**, 144–148.
328. Mitchell, P. (1966). Chemiosmotic coupling in oxidative and photosynthetic phosphorylation. *Biol. Rev.* **41**, 445–502.
329. Mitchell, P. (1972). Structural and functional organization of energy-transducing membranes and their ion-conducting properties. *FEBS Symp.* **28**, 353–370.
330. Mitchell, P. (1973). Hypothesis: cation-translocating adenosine triphosphatase models: how direct is the participation of adenosine triphosphate and its hydrolysis products in cation translocation? *FEBS Lett.* **33**, 267–274.
331. Mitchell, P. (1975). Protonmotive redox mechanism of the cytochrome b–c, complex in the respiratory chain: proton motive ubiquinone cycle. *FEBS Lett.* **56**, 1–6.
332. Mitchell, P. (1975). The protonmotive Q cycle: a general formulation. *FEBS Lett.* **59**, 137–139.
333. Mitchell, P. (1976). Vectorial chemistry and the molecular mechanics of chemiosmotic coupling: power transmission by proticity. *Biochem. Soc. Trans.* **4**, 399–430.
334. Mitchell, P. (1976). Possible molecular mechanisms of the protonmotive function of cytochrome systems. *J. Theor. Biol.* **62**, 327–367.
335. Mitchell, P. (1977). Vectorial chemiosmotic processes. *Annu. Rev. Biochem.* **46**, 996–1005.
336. Mitchell, P. (1977). A commentary on alternative hypotheses of protonic coupling in the membrane systems catalysing oxidative and photosynthetic phosphorylation. *FEBS Lett.* **78**, 1–20.
337. Mitchell, P. (1979). Compartmentation and communication in living systems. Ligand conduction: a general catalytic principle in chemical, osmotic and chemiosmotic reaction systems. *Eur. J. Biochem.* **95**, 1–20.
338. Mitchell, P. and Moyle, J. (1965). Stoichiometry of proton translocation through the respiratory chain and adenosine triphosphatase systems of rat liver mitochondria. *Nature (London)* **208**, 147–151.
339. Mitchell, P., and Moyle, J. (1965). Evidence discriminating between the chemical and the chemiosmotic mechanisms of electron transport phosphorylation. *Nature (London)* **208**, 1205–1206.
340. Mitchell, P., and Moyle, J. (1967). Respiration-driven proton translocation in rat liver mitochondria. *Biochem. J.* **105**, 1147–1162.
341. Mitchell, P., and Moyle, J. (1968). Proton translocation coupled to ATP hydrolysis in rat liver mitochondria. *Eur. J. Biochem.* **4**, 530–539.
342. Moore, A. L., Bonner, W. D., Jr., and Rich, P. R. (1978). The determination of the proton-motive force during cyanide-insensitive respiration in plant mitochondria. *Arch. Biochem. Biophys.* **186**, 298–306.
342a. Moore, A. L., and Rich, P. R. (1980). The bioenergetics of plant mitochondria. *Trends Biochem. Sci.* **5**, 284–288.
343. Moore, A. L., Rich, P. R., Bonner, W. D. Jr., and Ingledew, W. J. (1976). A complex EPR signal in mung bean mitochondria and its possible relation to the alternate pathway. *Biochem. Biophys. Res. Commun.* **72**, 1099–1107.
344. Morowitz, H. (1976). Stored energy lies in the electrochemical potential of H^+. *Trends Biochem. Sci.* **1**, N222–224.

345. Mortenson, L. E., Valentine, R. C., and Carnahan, J. E. (1962). An electron transport factor from *Clostridium pasteurianum*. *Biochem. Biophys. Res. Commun.* **7,** 448–452.

346. Moyle, J., and Mitchell, P. (1978). Cytochrome *c* oxidase is not a proton pump. *FEBS Lett.* **88,** 268–272.

347. Mühlethaler, K. (1977). Introduction to structure and function of the photosynthetic apparatus. *In* "Encyclopedia of Plant Physiology, New Series" (A. Pirson and M. H. Zimmermann, eds.), Vol. 5, Photosynthesis I. Photosynthetic Electron Transport and Photophosphorylation (A. Trebst and M. Avron, eds.), pp. 503–521. Springer-Verlag, Berlin.

348. Nelson, N. (1976). Structure and function of chloroplast ATPase. *Biochim. Biophys. Acta* **456,** 314–338.

349. Nelson, N. (1977). Chloroplast coupling factor. *In* "Encyclopedia of Plant Physiology, New Series" (A. Pirson and M. H. Zimmermann, eds.), Vol. 5, Photosynthesis I. Photosynthetic Electron Transport and Photophorylation" (A. Trebst and M. Avron, eds.), pp. 393–404. Springer-Verlag, Berlin.

350. Nelson, N., Eytan, E., Notsani, B.-E., Signist, H., Nelson, K. S., and Gitler, C. (1977). Isolation of a chloroplast *N,N'*-dicyclohexylcarbodiimide-binding proteolipid, active in proton translocation. *Proc. Natl. Acad. Sci. USA.* **74,** 2375–2378.

351. Nelson, N., Nelson, H., and Racker, E. (1972). Partial resolution of the enzymes catalyzing photophosphorylation. XII. Purification and properties of an inhibitor isolated from chloroplast coupling factor. *J. Biol. Chem.* **247,** 7657–7662.

352. Nelson, N., Nelson, H., and Schatz, G. (1980). Biosynthesis and assembly of the proton-translocating adenosine triphosphatase complex from chloroplasts. *Proc. Natl. Acad. Sci. USA* **77,** 1361–1364.

353. Nelson, N. and Neumann, J. (1972). Isolation of a cytochrome b_6–f particle from chloroplasts. *J. Biol. Chem.* **247,** 1817–1824.

354. Neumann, J., and Jagendorf, A. T. (1964). Light-induced pH changes related to phosphorylation by chloroplasts. *Arch. Biochem. Biophys.* **107,** 109–119.

355. Nicholls, P., and Peterson, L. C. (1974). Haem–haem interactions in cytochrome aa_3 during the anaerobic–aerobic transition. *Biochim. Biophys. Acta* **357,** 462–467.

356. Oesterhelt, D., and Stoeckenius, W. (1973). Functions of a new photoreceptor membrane. *Proc. Natl. Acad. Sci. USA* **70,** 2853–2857.

357. Oesterhelt, D., and Stoeckenius, W. (1974). Isolation of the cell membrane of *Halobacterium halobium* and its fractionation into red and purple membrane. *Methods Enzymol.* **31,** 667–678.

358. Ohnishi, T. (1975). Thermodynamic and EPR characterization of iron–sulfur centers in the NADH-ubiquinone segment of the mitochondrial respiratory chain in pigeon heart. *Biochim. Biophys. Acta* **387,** 475–490.

359. Ohnishi, T. (1976). Studies on the mechanism of site I energy conservation. *Eur. J. Biochem.* **64,** 91–103.

360. Ohnishi, T., Lim, J., Winter, D. B., and King, T. E. (1976). Thermodynamic and EPR characteristics of a HiPIP-type iron–sulfur center in the succinate dehydrogenase of the respiratory chain. *J. Biol. Chem.* **251,** 2105–2109.

361. Ohnishi, T., and Salerno, J. C. (1976). Thermodynamic and EPR characteristics of two ferredoxin-type iron–sulfur centers in the succinate-ubiquinone reductase segment of the respiratory chain. *J. Biol. Chem.* **251,** 2094–2104.

362. Okayama, S. and Butler, W. L. (1972). Extraction and reconstitution of photosystem II. *Plant Physiol.* **49,** 769–774.

363. O'Keefe, D. P. and Dilley, R. A. (1977). The effect of chloroplast coupling factor removal on thylakoid membrane ion permeability. *Biochim. Biophys. Acta* **461,** 48–60.

364. Oliver, D., and Jagendorf, A. (1976). Exposure of free amino groups in the coupling factor of energized spinach chloroplasts. *J. Biol. Chem.* **251,** 7168–7175.

365. Olsen, L. F., Cox, R. P., and Barber, J. (1980). Flash-induced redox changes of P700 and plastocyanin in chloroplasts suspended in fluid media at sub-zero temperatures. *FEBS Lett.* **122,** 13–16.

366. Orlich, G., and Hauska, G. (1980). Reconstitution of photosynthetic energy conservation. I. Proton movements in liposomes containing reaction center of photosystem I from spinach chloroplasts. *Eur. J. Biochem.* **111,** 525–533.

367. Orme-Johnson, N. R., Hansen, R. E., and Beinert, H. (1974). Electron paramagnetic resonance-detectable electron acceptors in beef heart mitochondria. *J. Biol. Chem.* **249,** 1922–1927.

368. Ort, D. R., and Dilley, R. A. (1976). Photophosphorylation as a function of illumination time. I. Effects of permeant cations and permeant anions. *Biochim. Biophys. Acta* **449,** 95–107.

369. Palmer, G. (1975). Iron–sulfur proteins. *In* "The Enzymes" (P. D. Boyer, ed.), Vol. XII, pp. 2–56. Academic Press, New York.

370. Palmer, J. M. (1976). The organization and regulation of electron transport in plant mitochondria. *Annu. Rev. Plant Physiol.* **27,** 133–157.

371. Palmer, G., Babcock, G. T., and Vickery, L. E. (1976). A model for cytochrome oxidase. *Proc. Natl. Acad. Sci. USA* **73,** 2206–2210.

372. Palmer, J. M., and Coleman, J. O. D. (1974). Multiple pathways of NADH oxidation in the mitochondrion. *Horiz. Biochem. Biophys.* **1,** 220–260.

373. Papa, S. (1976). Proton translocation reactions in the respiratory chains. *Biochim. Biophys. Acta* **456,** 39–84.

374. Papa, S., Capuano, F., Markert, M., and Altamura, N. (1980). The H⁺/O stoichiometry of mitochondrial respiration. *FEBS Lett.* **111,** 243–248.

375. Papa, S., Guerrieri, F., and Izzo, G. (1979). Redox Bohr-effects in the cytochrome system of mitochondria. *FEBS Lett.* **105,** 213–216.

376. Papa, S., Lorusso, M., and Guerrieri, F. (1975). Mechanism of respiration-driven proton translocation in the inner mitochondrial membrane. Analysis of proton translocation associated with oxidation of endogenous ubiquinol. *Biochim. Biophys. Acta* **387,** 425–440.

377. Paradies, H. H., Zimmermann, J., and Schmidt, U. D. (1978). The conformation of chloroplast coupling factor I from spinach in solution. *J. Biol. Chem.* **253,** 8972–8979.

378. Passam, H. C., and Palmer, J. M. (1972). Electron transport and oxidative phosphorylation in *Arum* spadix mitochondria. *J. Exp. Bot.* **23,** 366–374.

379. Passam, H. C., and Palmer, J. M. (1973). The ATPase activity of Jerusalem-artichoke mitochondria and submitochondrial particles. *Biochim. Biophys. Acta* **305,** 80–87.

380. Penefsky, H. S. (1979). ATPases associated with electron transport. *Methods Enzymol.* **55,** 297–303.

381. Penefsky, H. S. (1979). Preparation of beef heart mitochondrial ATPase. *Methods Enzymol.* **55,** 304–308.

382. Peterson, T. G., and Heisler, C. R. (1963). Purification and properties of a soluble mitochondrial ATPase from cabbage. *Brassica oleracea. Biochem. Biophys. Res. Commun.* **12,** 492–497.

383. Pick, U., and Racker, E. (1979). Purification and reconstitution of the *N,N*′-dicyclo-

hexylcarbodiimide-sensitive ATPase complex from spinach chloroplasts. *J. Biol. Chem.* **254**, 2793–2799.

384. Plesnicar, M., and Bendall, D. S. (1973). Some evidence on the site of action of plastocyanin in the photosynthetic electron-transport chain between cytochromes f and P_{700}. *Eur. J. Biochem.* **34**, 483–488.

385. Portis, A. R., Jr., and McCarty, R. E. (1974). Effects of adenine nucleotides and of photophosphorylation on H^+ uptake and the magnitude of the H^+ gradient in illuminated chloroplasts. *J. Biol. Chem.* **249**, 6250–6254.

386. Portis, A. R., Jr., and McCarty, R. E. (1976). Quantitative relationships between phosphorylation, electron flow, and internal hydrogen ion concentrations in spinach chloroplasts. *J. Biol. Chem.* **251**, 1610–1617.

387. Powers, L., Blumberg, W. E., Chance, B., Barlow, C. H., Leigh, J. S., Jr., Smith, J., Yonetani, T., Vik, S., and Peisach, J. (1979). The nature of the copper atoms of cytochrome c oxidase as studied by optical and X-ray absorption edge spectroscopy. *Biochim. Biophys. Acta* **546**, 520–538.

388. Poyton, R. O., and Schatz, G. (1975). Cytochrome *c* oxidase from bakers' yeast. III. Physical characterization of isolated subunits and chemical evidence for two different classes of polypeptides. *J. Biol. Chem.* **250**, 752–761.

389. Pozzan, T., Miconi, V., Di Virgilio, F., and Azzone, G. F. (1979). H^+/Site, charge/ site, and ATP/site ratios at coupling sites I and II in mitochondrial e^- transport. *J. Biol. Chem.* **254**, 10200–10205.

390. Pulles, M. P. J., VanGorkam, H. J. and Willemsen, J. G. (1976). Absorbance changes due to the charge-accumulating species in system 2 of photosynthesis. *Biochim. Biophys. Acta* **449**, 536–540.

391. Racker, E. (1975). Reconstitution, mechanism of action and control of ion pumps. *Biochem. Soc. Trans.* **3**, 785–802.

392. Racker, E. (1977). Mechanisms of energy transformations. *Annu. Rev. Biochem.* **46**, 1006–1014.

393. Racker, E. (1979). Preparation of coupling factor 6 (F_6). *Methods Enzymol.* **55**, 398–399.

394. Racker, E., and Stoeckenius, W. (1974). Reconstitution of purple membrane vesicles catalyzing light-driven proton uptake and adenosine triphosphate formation. *J. Biol. Chem.* **249**, 662–663.

395. Racker, E., Violand, B., O'Neal, S., Alfonzo, M., and Telford, J. (1979). Reconstitution, a way of biochemical research: some new approaches to membrane-bound enzymes. *Arch. Biochem. Biophys.* **198**, 470–477.

396. Radmer, R. J., and Kok, B. (1976). Photoreduction of O_2 primes and replaces CO_2 assimilation. *Plant Physiol.* **58**, 336–340.

397. Ragan, C. I. (1976). The structure and subunit composition of the particulate NADH-ubiquinone reductase of bovine heart mitochondria. *Biochem. J.* **154**, 295–305.

398. Ragan, C. I. (1976). NADH-ubiquinone oxidoreductase. *Biochim. Biophys. Acta* **456**, 249–290.

399. Ragan, C. I., and Hinkle, P. C. (1975). Ion transport and respiratory control in vesicles formed from reduced nicotinamide adenine dinucleotide coenzyme Q reductase and phospholipids. *J. Biol. Chem.* **250**, 8472–8476.

400. Ragan, C. I., and Racker, E. (1973). Resolution and reconstitution of the mitochondrial electron transport system IV. The reconstitution of rotenone-sensitive reduced nicotinamide adenine dinucleotide-ubiquinone reductase from reduced ni-

cotinamide adenine dinucleotide dehydrogenase and phospholipids. *J. Biol. Chem.* **248**, 6876–6884.

401. Ramshaw, J. A. M., Brown, R. H., Scawen, M. D., and Boulter, D. (1973). Higher plant plastocyanin. *Biochim. Biophys. Acta* **303**, 269–273.

402. Reeves, S. G., and Hall, D. O. (1978). Photophosphorylation in chloroplasts. *Biochim. Biophys. Acta* **463**, 275–297.

403. Reich, R., Scheerer, R., Sewe, K.-U., and Witt, H. T. (1976). Effect of electric fields on the absorption spectrum of dye molecules in lipid layers. V. Refined analysis of the field-indicating absorption changes in photosynthetic membranes by comparison with electrochromic measurements in vitro. *Biochem. Biophys. Acta* **449**, 285–294.

404. Renger, G. (1972). The action of 2-anilinothiophenes as accelerators of the deactivation reactions in the watersplitting enzyme system of photosynthesis. *Biochim. Biophys. Acta* **256**, 428–439.

405. Renger, G. (1977). A model for the molecular mechanism of photosynthetic oxygen evolution. *FEBS Lett.* **81**, 223–228.

406. Renger, G., Bouges-Bocquet, B., and Delosme, R. (1973). Studies on the ADRY agent-induced mechanism of the discharge of the holes trapped in the photosynthetic watersplitting enzyme system Y. *Biochim. Biophys. Acta* **292**, 796–807.

407. Renger, G., and Wolff, Ch. (1976). The existence of a high photochemical turnover rate at the reaction centers of system II in Tris-washed chloroplasts. *Biochim. Biophys. Acta* **423**, 610–614.

408. Rich, P. R. (1978). Quinol oxidation in *Arum maculatum* mitochondria and its application to the assay, solubilization and partial purification of the alternate oxidase. *FEBS Lett.* **96**, 252–256.

409. Rich, P. R., and Bonner, W. D., Jr. (1978). EPR studies of higher plant mitochondria. II. Center S-3 of succinate dehydrogenase and its relation to alternative respiratory oxidations. *Biochim. Biophys. Acta* **501**, 381–395.

410. Rich, P. R., and Bonner, W. D., Jr. (1978). The nature and location of cyanide and antimycin resistant respiration in higher plants. *FEBS Symp.* **49**, 149–158.

411. Rich, P. R., Boveris, A., Bonner, W. D., Jr., and Moore, A. L. (1976). Hydrogen peroxide generation by the alternate oxidase of higher plants. *Biochem. Biophys. Res. Commun.* **71**, 695–703.

412. Rich, P. R., and Moore, A. L. (1976). The involvement of the protonmotive ubiquinone cycle in the respiratory chain of higher plants and its relation to the branchpoint of the alternate pathway. *FEBS Lett.* **65**, 339–344.

413. Rich, P. R., Moore, A. L., and Bonner, W. D. Jr. (1977). The effects of bathophenanthroline, bathophenanthrolinesulphonate and 2-thioyltrifluoroacetone on mung-bean mitochondria and submitochondrial particles. *Biochem. J.* **162**, 205–208.

414. Rich, P. R., Moore, A. L., Ingledew, W. J., and Bonner, W. D., Jr. (1977). EPR studies of higher plant mitochondria. I. Ubisemiquinone and its relation to alternative respiratory oxidations. *Biochim. Biophys. Acta* **462**, 501–514.

415. Rieske, J. S. (1976). Composition, structure, and function of complex III of the respiratory chain. *Biochim. Biophys. Acta* **456**, 195–247.

416. Rosa, L. (1979). The ATP/2e ratio during photosynthesis in intact spinach chloroplasts. *Bichem. Biophys. Res. Commun.* **88**, 154–162.

417. Rosing, J., Kayalar, C., and Boyer, P. D. (1977). Evidence for energy-dependent change in phosphate binding for mitochondrial oxidative phosphorylation based on measurements of medium and intermediate phosphate-water exchanges. *J. Biol. Chem.* **252**, 2478–2485.

418. Rottenberg, H. (1977). Proton and ion transport across the thylakoid membranes. *In* "Encyclopedia of Plant Physiology, New Series" (A. Pirson and M. H. Zimmermann, eds.), Vol. 5, Photosynthesis I. Photosynthetic Electron Transport and Photophosphorylation (A. Trebst and M. Avron, eds.), pp. 338–349. Springer-Verlag, Berlin/New York.

419. Roux, S. J., McEntire, K., Slocum, R. D., Cedel, T. E., and Hale, C. C. (1981). Phytochrome induces photoreversible calcium fluxes in a purified mitochondrial fraction from oats. *Proc. Natl. Acad. Sci. USA* **78**, 283–287.

420. Rumberg, B., Schmidt-Mende, P., and Witt, H. T. (1964). Different demonstrations of the coupling of two light reactions in photosynthesis. *Nature (London)* **201**, 466–468.

421. Rupp, H., and Moore, A. L. (1979). Characterization of iron–sulphur centres of plant mitochondria by microwave power saturation. *Biochim. Biophys. Acta* **548**, 16–29.

421a. Rustin, P., Moreau, F., and Lance, C. (1980). Malate oxidation in plant mitochondria via malic enzyme and the cyanide-insensitive electron transport pathway. *Plant Physiol.* **66**, 457–462.

422. Ryrie, I. J., and Jagendorf, A. T. (1971). An energy-linked conformational change in the coupling factor protein in chloroplasts. *J. Biol. Chem.* **246**, 3771–3774.

423. Ryrie, I. J., and Jagendorf, A. T. (1972). Correlation between a conformational change in the coupling factor protein and the high energy state in chloroplasts. *J. Biol. Chem.* **247**, 4453–4459.

424. Sadewasser, D. A., and Dilley, R. A. (1978). A dual requirement for plastoquinone in chloroplast electron transport. *Biochim. Biophys. Acta* **501**, 208–216.

425. Salerno, J. C., Harmon, H. J., Blum, H., Leigh, J. S., and Ohnishi, T. (1977). A transmembrane quinone pair in the succinate dehydrogenase–cytochrome b region. *FEBS Lett.* **82**, 179–182.

426. Sane, P. V. (1977). The topography of the thylakoid membrane of the chloroplast. *In* "Encyclopedia of Plant Physiology, New Series" (A. Pirson and M. H. Zimmermann, eds.), Vol. 5, Photosynthesis I. Photosynthetic electron transport and photophosphorylation (A. Trebst and M. Avron, eds.), pp. 522–542. Springer-Verlag, Berlin and New York.

427. Saphon, S., and Crofts, A. R. (1977). Protolytic reactions in photosystem II: a new model for the release of protons accompanying the photooxidation of water. *Z. Naturforsch.* **32C**, 617–626.

428. Saphon, S., and Crofts, A. R. (1977). The H^+/e ratio in chloroplasts is 2. Possible errors in its determination. *Z. Naturforsch.* **32C**, 810–816.

429. Sato, N., Wilson, D. F., and Chance, B. (1971). The spectral properties of the b cytochromes in intact mitochondria. *Biochim. Biophys. Acta* **253**, 88–97.

430. Sauer, K. (1980). A role for manganese in oxygen evolution in photosynthesis. *Accounts Chem. Res.* **13**, 249–256.

431. Schmidt, U. D., and Paradies, H. H. (1977). The structure of the delta-subunit from chloroplast coupling factor (CF_1) in solution. *Biochem. Biophys. Res. Commun.* **78**, 1043–1052.

432. Schneeman, R., and Krogmann, D. W. (1975). Polycation interactions with spinach ferredoxin-nicotinamide adenine dinucleotide phosphate reductase. *J. Biol. Chem.* **250**, 4965–4971.

433. Schneider, D. L., Kagawa, Y., and Racker, E. (1972). Chemical modification of the inner mitochondrial membrane. *J. Biol. Chem.* **247**, 4074–4079.

434. Schonbaum, G. R., Bonner, W. D., Jr., Storey, B. T., and Bahr, J. T. (1971). Specific

inhibition of the cyanide-insensitive respiratory pathway in plant mitochondria by hydroxamic acids. *Plant Physiol.* **47**, 124–128.

435. Schreiber, U. (1980). Light-activated ATPase and ATP-driven reverse electron transport in intact chloroplasts. *FEBS Lett.* **122**, 121–124.

436. Segal, H. L., and Boyer, P. D. (1953). The role of sulfhydryl groups in the activity of D-glyceraldehyde 3-phosphate dehydrogenase. *J. Biol. Chem.* **204**, 265–281.

437. Senior, A. E. (1973). The structure of mitochondrial ATPase. *Biochim. Biophys. Acta* **301**, 249–277.

438. Senior, A. E. (1979). Oligomycin-sensitivity-conferring protein. *Methods Enzymol.* **55**, 391–397.

439. Serrano, R., Kanner, B. I., and Racker, E. (1976). Purification and properties of the proton-translocating adenosine triphosphate complex of bovine heart mitochondria. *J. Biol. Chem.* **251**, 2453–2461.

440. Shahak, Y., Posner, H. B., and Avron, M. (1976). Evidence for a block between plastoquinone and cytochrome f in a photosynthetic mutant of *Lemna* with abnormal flowering behavior. *Plant Physiol.* **57**, 577–579.

441. Shavit, N. (1980). Energy transduction in chloroplasts: structure and function of the ATPase complex. *Annu. Rev. Biochem.* **49**, 111–138.

442. Shin, M., and Oshino, R. (1978). Ferredoxin–Sepharose 4B as a tool for the purification of ferredoxin-NADP+ reductase. *J. Biochem. (Tokyo)* **83**, 357–361.

443. Shoshan, V., and Selman, B. R. (1979). The relationship between light-induced adenine nucleotide exchange and ATPase activity in chloroplast thylakoid membranes. *J. Biol. Chem.* **254**, 8801–8807.

444. Shuvalov, V. A., Dolan, E., and Ke, B. (1979). Spectral and kinetic evidence for two early electron acceptors in photosystem I. *Proc. Natl. Acad. Sci. USA* **76**, 770–773.

445. Shuvalov, V. A., Ke, B., and Dolan, E. (1979). Kinetic and spectral properties of the intermediary electron acceptor A_1 in photosystem I. Subnanosecond spectroscopy. *FEBS Lett.* **100**, 5–8.

446. Shuvalov, V. A., Klevanik, A. V., Sharkov, A. V., Kryukov, P. G., and Ke, B. (1979). Picosecond spectroscopy of photosystem I reaction centers. *FEBS Lett.* **107**, 313–316.

447. Sigel, E., and Carafoli, F. (1978). The proton pump of cytochrome c oxidase and its stoichiometry. *Eur. J. Biochem.* **89**, 119–123.

448. Siggel, U., Khanna, R., Renger, G., and Govindjee (1977). Investigation of the absorption changes of the plastoquinone system in broken chloroplasts. *Biochim. Biophys. Acta* **462**, 196–207.

449. Siggel, U., Renger, G., Stiehl, H. H., and Rumberg, B. (1972). Evidence for electronic and ionic interaction between electron transport chains in chloroplasts. *Biochim. Biophys. Acta* **256**, 328–335.

450. Singer, T. P., Kearney, E. B., and Kenney, W. C. (1973). Succinate dehydrogenase. *Adv. Enzymol.* **37**, 189–272.

451. Singh, A. P. and Bragg, P. D. (1978). Lack of involvement of lipoic acid in membrane-associated energy transduction in *Escherichia coli. Biochem. Biophys. Res. Commun.* **81**, 161–167.

452. Singh, A. P., and Bragg, P. D. (1979). ATP synthesis driven by a pH gradient imposed across the cell membranes of lipoic acid and unsaturated fatty acid auxotrophs of *Escherichia coli. FEBS Lett.* **98**, 21–24.

453. Slater, E. C. (1953). Mechanism of phosphorylation in the respiratory chain. *Nature (London)* **172**, 975–978.

454. Slater, E. C. (1974). From cytochrome to adenosine triphosphate and back. *Biochem. Soc. Trans.* **2**, 1149–1163.
455. Slater, E. C., Kemp, A., van der Kraan, I., Muller, J. L., Roveri, O. A., Verschoor, G. J., Wagenvoord, R. J., and Wielders, J. P. M. (1979). The ATP- and ADP-binding sites in mitochondrial coupling factor F_1 and their possible role in oxidative phosphorylation. *FEBS Lett.* **103**, 7–11.
456. Slovacek, R. E., Crowther, D., and Hind, G. (1979). Cytochrome function in the cyclic electron transport pathway of chloroplasts. *Biochim. Biophys. Acta* **547**, 138–148.
457. Slovacek, R. E., Crowther, D., and Hind, G. (1980). Relative activities of linear and cyclic electron flows during chloroplast CO_2-fixation. *Biochim. Biophys. Acta* **592**, 495–505.
458. Slovacek, R. E., and Hind, G. (1978). Flash spectroscopic studies of cyclic electron flow in intact chloroplasts. *Biochim. Biophys. Res. Commun.* **84**, 901–906.
459. Slovacek, R. E., Mills, J. D., and Hind, G. (1978). The function of cyclic electron transport in photosynthesis. *FEBS Lett.* **87**, 73–76.
460. Smith, D. J., and Boyer, P. D. (1976). Demonstration of a transitory tight binding of ATP and of commited P_i and ADP during ATP synthesis by chloroplasts. *Proc. Natl. Acad. Sci. USA* **73**, 4314–4318.
461. Smith, D. J., Stokes, B. O., and Boyer, P. D. (1976). Probes of initial phosphorylation events in ATP synthesis by chloroplasts. *J. Biol. Chem.* **251**, 4165–4171.
462. Smith, S., and Ragan, C. I. (1980). The organization of NADH dehydrogenase polypeptides in the inner mitochondrial membrane. *Biochem. J.* **185**, 315–326.
463. Solomos, T. (1977). Cyanide-resistant respiration in higher plants. *Annu. Rev. Plant Physiol.* **28**, 279–297.
464. Soper, J. W., and Pedersen, P. L. (1979). Isolation of an oligomycin-sensitive ATPase complex from rat liver mitochondria. *Methods Enzymol.* **55**, 328–333.
465. Spector, M., and Winget, G. D. (1980). Purification of a manganese-containing protein involved in photosynthetic oxygen evolution and its use in reconstituting an active membrane. *Proc. Natl. Acad. Sci. USA* **77**, 957–959.
466. Stemler, A. (1977). The binding of bicarbonate ions to washed chloroplast grana. *Biochim. Biophys. Acta* **460**, 511–522.
467. Stemler, A. (1979). A dynamic interaction between the bicarbonate ligand and photosystem II reaction center complexes in chloroplasts. *Biochim. Biophys. Acta* **545**, 36–45.
468. Stemler, A. (1980). Inhibition of photosystem II by formate. Possible evidence for a direct role of bicarbonate in photosynthetic oxygen evolution. *Biochim. Biophys. Acta* **593**, 103–112.
469. Stemler, A. (1980). Forms of dissolved carbon dioxide required for photosystem II activity in chloroplast membranes. *Plant Physiol.* **65**, 1160–1165.
470. Stiehl, H. H., and Witt, H. T. (1968). Die kurzzeitigen ultravioletten Differenzspektren bei der Photosynthese. *Z. Naturforsch.* **23B**, 220–224.
471. Stiehl, H. H., and Witt, H. T. (1969). Quantitative treatment of the function of plastoquinone in photosynthesis. *Z. Naturforsch.* **24B**, 1588–1598.
472. Stiggal, D. L., Galante, Y. M., and Hatefi, Y. (1979). Preparation and properties of complex V. *Methods Enzymol.* **55**, 308–315.
473. Stoeckenius, W., Lozier, R. H., and Bogomolni, R. A. (1979). Bacteriorhodopsin and the purple membrane of halobacteria. *Biochim. Biophys. Acta* **505**, 215–278.
474. Storey, B. T. (1970). The respiratory chain of plant mitochondria. VI. Flavoprotein

components of the respiratory chain of mung bean mitochondria. *Plant Physiol.* **46**, 13–20.

475. Storey, B. T. (1971). The respiratory chain of plant mitochondria. X. Oxidation-reduction potentials of the flavoproteins of skunk cabbage mitochondria. *Plant Physiol.* **48**, 493–497.

476. Storey, B. T. (1971). The respiratory chain of plant mitochondria. XI. Electron transport from succinate to endogenous pyridine nucleotide in mung bean mitochondria. *Plant Physiol.* **48**, 694–701.

477. Storey, B. T. (1973). The respiratory chain of plant mitochondria. XV. Equilibration of cytochromes $c549$, $b553$, $b557$ and ubiquinone in mung bean mitochondria: placement of cytochrome $b557$ and estimation of the midpoint potential of ubiquinone. *Biochim. Biophys. Acta* **292**, 592–603.

478. Storey, B. T. (1974). The respiratory chain of plant mitochondria. XVI. Interaction of cytochrome b_{562} with the respiratory chain of coupled and uncoupled mung bean mitochondria: evidence for its exclusion from the main sequence of the chain. *Plant Physiol.* **53**, 840–845.

479. Storey, B. T. (1974). The respiratory chain of plant mitochondria. XVII. Flavoprotein–cytochrome b_{562} interaction in antimycin-treated skunk cabbage mitochondria. *Plant Physiol.* **53**, 846–850.

480. Storey, B. T. (1976). Respiratory chain of plant mitochondria. XVIII. Point of interaction of the alternate oxidase with the respiratory chain. *Plant Physiol.* **58**, 521–525.

481. Storey, B. T. (1980). Electron transport and energy coupling in plant mitochondria. *In* "The Biochemistry of Plants" (E. E. Conn and P. K. Stumpf, eds.), Vol. 2, pp. 125–195. Academic Press, New York.

482. Storey, B. T., and Bahr, J. T. (1972). The respiratory chain of plant mitochondria. XIV. Ordering of ubiquinone, flavoproteins and cytochromes in the respiratory chain. *Plant Physiol.* **50**, 95–102.

483. Stuart, A. L., and Wasserman, A. R. (1973). Purification of cytochrome b_6. A tightly bound protein in chloroplast membranes. *Biochim. Biophys. Acta* **314**, 284–297.

484. Stuart, A. L., and Wasserman, A. R. (1975). Chloroplast cytochrome b_6, molecular composition as a lipoprotein. *Biochim. Biophys. Acta* **376**, 561–572.

485. Tagawa, K., and Arnon, D. I. (1962). Ferredoxins as electron carriers in photosynthesis and in the biological production and consumption of hydrogen gas. *Nature (London)* **195**, 537–543.

486. Takeuchi, Y. (1975). Respiration-dependent uncoupler-stimulated ATPase activity in castor bean endosperm mitochondria and submitochondrial particles. *Biochim. Biophys. Acta* **376**, 505–518.

487. Takruri, I., and Boulter, D. (1979). The amino acid sequence of ferredoxin from *Triticum aestivum* (wheat). *Biochem. J.* **179**, 373–378.

488. Thayer, W. S., and Hinkle, P. C. (1973). Stoichiometry of adenosine triphosphate-driven proton translocation in bovine heart submitochondrial particles. *J. Biol. Chem.* **248**, 5395–5402.

489. Thayer, W. S., and Hinkle, P. C. (1975). Synthesis of adenosine triphosphate by an artificially imposed electrochemical proton gradient in bovine heart submitochondrial particles. *J. Biol. Chem.* **250**, 5330–5335.

490. Thayer, W. S., and Hinkle, P. C. (1975). Kinetics of adenosine triphosphate synthesis in bovine heart submitochondrial particles. *J. Biol. Chem.* **250**, 5336–5342.

491. Tiemann, R., Renger, G., Gräber, P., and Witt, H. T. (1979). The plastoquinone pool as a possible hydrogen pump in photosynthesis. *Biochim. Biopnys. Acta* **546**, 498–519.

492. Trebst, A. (1974). Energy conservation in photosynthetic electron transport of chloroplasts. *Annu. Rev. Plant Physiol.* **25**, 423–458.

493. Trebst, A. (1976). Coupling sites, native and artificial, in photophosphorylation by isolated chloroplasts. *Trends Biochem. Sci.* **1**, 60–62.

494. Trebst, A. (1978). Organization of the photosynthetic electron transport system of chloroplasts in the thylakoid membrane. *In* "Energy Conservation in Biological Membranes" (G. Schafer and M. Klingenberg, eds.), pp. 84–95. Springer-Verlag, Berlin.

495. Trebst, A. (1978). Plastoquinones in photosynthesis. *Philos. Trans. R. Soc. London Ser. B.* **284**, 591–599.

496. Trentham, D. R. (1971). Rate determining processes and the number of simultaneously active sites of D-glyceraldehyde 3-phosphate dehydrogenase. *Biochem. J.* **122**, 71–77.

496a. Trumpower, B. L. (1981). New concepts on the role of ubiquinone in the mitochondrial respiratory chain. *J. Bioenerg. Biomembr.* **13**, 1–24.

497. Trumpower, B. L., and Katki, A. (1975). Controlled digestion with trypsin as a structural probe for the *N*-terminal peptide of soluble and membranous cytochrome c. *Biochemistry* **14**, 3635–3642.

498. Tsujimoto, H. Y., Hiyama, T., and Arnon, D. I. (1980). Affinity of ferredoxin for electrons from water and the regulation of cyclic photophosphorylation. *Biochem. Biophys. Res. Commun.* **93**, 215–222.

499. Uribe, E. G., and Jagendorf, A. T. (1967). On the localization of organic acids in acid-induced ATP synthesis. *Plant Physiol.* **42**, 697–705.

500. Uribe, E. G., and Jagendorf, A. T. (1967). Organic acid specificity for acid-induced ATP synthesis by isolated chloroplasts. *Plant Physiol.* **42**, 706–711.

501. Uribe, E. G., and Jagendorf, A. T. (1968). Membrane permeability and internal volume as factors in ATP synthesis by spinach chloroplasts. *Arch. Biochem. Biophys.* **128**, 351–359.

502. Vadineanu, A., Berden, J. A., and Slater, E. C. (1976). Proteins required for the binding of mitochondrial ATPase to the mitochondrial inner membrane. *Biochim. Biophys. Acta* **449**, 468–479.

503. Vallejos, R. H., Viale, A., and Andreo, C. S. (1977). Essential role of an arginyl residue at the catalytic site(s) of chloroplast coupling factor. *FEBS Lett.* **84**, 304–308.

504. Van Gorkom, H. J. (1974). Identification of the reduced primary electron acceptor of photosystem II as a bound semiquinone anion. *Biochim. Biophys. Acta* **347**, 439–442.

505. Velthuys, B. R. (1979). Electron flow through plastoquinone and cytochromes b6 and f in chloroplasts. *Proc. Natl. Acad. Sci. USA* **76**, 2765–2769.

506. Velthuys, B. R. (1980). Mechanisms of electron flow in photosystem II and toward photosystem I. *Annu. Rev. Plant Physiol.* **31**, 545–567.

507. Velthuys, B. R., and Amesz, J. (1974). Charge accumulation at the reducing side of system 2 of photosynthesis. *Biochim. Biophys. Acta* **333**, 85–94.

508. Vermeglio, A., and Mathis, P. (1973). Photoreduction of C-550 and oxidation of cytochrome b_{559} in chloroplasts. Dependence on the state of photosystem II. *Biochim. Biophys. Acta* **314**, 57–65.

509. Villalobo, A., and Lehninger, A. L. (1979). The proton stoichiometry of electron transport in Ehrlich ascites tumor mitochondria. *J. Biol. Chem.* **254**, 4352–4358.

510. Villalobo, A., and Lehninger, A. L. (1980). Stoichiometry of H⁺ ejection coupled to electron transport through site 2 in ascites tumor mitochondria. *Arch. Biochem. Biophys.* **205**, 210–216.

511. von Jagow, G., and Engel, W. D. (1980). A model for the cytochrome b dimer of the ubiquinol : cytochrome c oxidoreductase as a proton translocator. *FEBS Lett.* **111**, 1–5.

512. Wada, K., and Arnon, D. I. (1971). Three forms of cytochrome b_{559} and their relation to the photosynthetic activity of chloroplasts. *Proc. Natl. Acad. Sci. USA* **68**, 3064–3068.

513. Walker, W. H., Singer, T. P., Ghisla, S., and Hemmerich, P. (1972). Studies on succinate dehydrogenase. 8-α-histidyl-FAD as the active center of succinate dehydrogenase. *Eur. J. Biochem.* **26**, 279–289.

514. Warburg, O., and Krippahl, G. (1960). Notwendigkeit der Kohlensäure für die Chinon-und Ferricyanid-Reaktionen in grünen Grana. *Z. Naturforsch.* **15B**, 367–369.

515. Weiss, H. (1976). Subunit composition and biogenesis of mitochondrial cytochrome b. *Biochim. Biophys. Acta* **456**, 291–313.

516. Whatley, F. R., Allen, M. B., and Arnon, D. I. (1955). Photosynthetic phosphorylation as an anaerobic process. *Biochim. Biophys. Acta* **16**, 605–606.

517. White, C. C., Chain, R. K., and Malkin, R. (1978). Duroquinol as an electron donor for chloroplast electron transfer reactions. *Biochim. Biophys. Acta* **502**, 127–137.

518. Whitmarsh, J., and Cramer, W. A. (1977). Kinetics of the photoreduction of cytochrome b_{559} by photosystem II in chloroplasts. *Biochim. Biophys. Acta* **460**, 280–289.

519. Whitmarsh, J., and Cramer, W. A. (1978). A pathway for the reduction of cytochrome b_{559} by photosystem II in chloroplasts. *Biochim. Biophys. Acta* **501**, 83–93.

520. Whitmarsh, J., and Cramer, W. A. (1979). Cytochrome f function in photosynthetic electron transport. *Biophys. J.* **26**, 223–234.

521. Whitmarsh, J., and Cramer, W. A. (1979). Photooxidation of the high-potential iron–sulfur center in chloroplasts. *Proc. Nat. Acad. Sci. USA* **76**, 4417–4420.

522. Wikström, M. K. F. (1973). The different cytochrome b components in the respiratory chain of animal mitochondria and their role in electron transport and energy conservation. *Biochim. Biophys. Acta* **301**, 155–193.

523. Wikström, M. K. F. (1977). Proton pump coupled to cytochrome c oxidase in mitochondria. *Nature (London)* **266**, 271–273.

524. Wikström, M. K. F., Harmon, H. J., Ingledew, W. J., and Chance, B. (1976). A reevaluation of the spectral, potentiometric and energy-linked properties of cytochrome c oxidase in mitochondria. *FEBS Lett.* **65**, 259–277.

525. Wikström, M., and Krab, K., (1979). Proton pumping cytochrome c oxidase. *Biochim. Biophys. Acta* **549**, 177–222.

526. Wikström, M. and Krab, K., (1980). Respiration-linked H⁺ translocation in mitochondria: stoichiometry and mechanism. *Current Topics Bioenerg.* **10**, 51–101.

527. Wikström, M. K. F., and Saari, H. T. (1977). The mechanism of energy conservation and transduction by mitochondrial cytochrome c oxidase. *Biochim. Biophys. Acta* **462**, 347–361.

528. Williams, R. J. P. (1961). Possible functions of chains of catalysts. *J. Theor. Biol.* **1**, 1–17.

529. Williams, R. J. P. (1976). Proton free energy differences in membrane drive ATP formation. *Trends Biochem. Sci.* **1**, N222–N224.
530. Williams, R. J. P. (1977). Fundamental features of proton coupled transport. *Biochem. Soc. Trans.* **5**, 29–32.
531. Williams, R. J. P. (1978). The multifarious couplings of energy transduction. *Biochim. Biophys. Acta* **505**, 1–44.
532. Williams, R. J. P. (1978). The history and the hypotheses concerning ATP-formation by energized protons. *FEBS Lett.* **85**, 9–19.
533. Williams-Smith, D. L., Heathcote, P., Sihra, C. K., and Evans, M. C. W. (1978). Quantitative electron-paramagnetic-resonance measurements of the electron-transfer components of the photosystem-I reaction centre. *Biochem. J.* **170**, 365–371.
534. Wilson, D. F., and Dutton, P. L. (1970). Energy dependent changes in the oxidation–reduction potential of cytochrome b. *Biochem. Biophys. Res. Commun.* **39**, 59–64.
535. Wilson, D. F., Lindsay, J. G., and Brocklehurst, E. S. (1972). Heme-heme interaction in cytochrome oxidase. *Biochim. Biophys. Acta* **256**, 277–286.
536. Wilson, S. B. (1980). Energy conservation by the plant mitochondrial cyanide-insensitive oxidase. Some additional evidence. *Biochem. J.* **190**, 349–360.
537. Winget, G. D., Kanner, N., and Racker, E. (1977). Formation of ATP by the adenosine triphosphatase complex from spinach chloroplasts reconstituted together with bacteriorhodopsin. *Biochim. Biophys. Acta* **460**, 490–499.
538. Witt, H. T. (1971). Coupling of quanta, electrons, fields, ions and phosphorylation in the functional membrane of photosynthesis. Results by pulse spectroscopic methods. *Q. Rev. Biophys.* **4**, 365–477.
539. Witt, H. T. (1979). Energy conversion in the functional membrane of photosynthesis. Analysis by light pulse and electric pulse methods. *Biochim. Biophys. Acta* **505**, 355–427.
540. Witt, K. (1973). Further evidence of X-320 as a primary acceptor of photosystem II in photosynthesis. *FEBS Lett.* **38**, 116–118.
541. Wood, P. M. (1974). Rate of electron transfer between plastocyanin, cytochrome f, related proteins and artificial redox reagents in solution. *Biochim. Biophys. Acta* **357**, 370–379.
542. Wood, P. M. (1977). The roles of c-type cytochromes in algal photosynthesis. Extraction from algae of a cytochrome similar to higher plant cytochrome f. *Eur. J. Biochem.* **72**, 605–612.
543. Wood, P. M., and Bendall, D. S. (1976). The reduction of plastocyanin by plastoquinol-1 in the presence of chloroplasts. A dark electron transfer reaction involving components between the two photosystems. *Eur. J. Biochem.* **61**, 337–344.
544. Yamashita, T. and Tomita, G. (1975). Comparative study of the reactivation of oxygen evolution in chloroplasts inhibited by various treatments. *Plant Cell Physiol.* **16**, 283–296.
545. Yoshida, K., and Takeuchi, Y. (1970). Properties of a soluble ATPase from castor bean endosperm mitochondria. *Plant Cell Physiol.* **11**, 403–409.
546. Younis, H. M., Winget, G. D., and Racker, E. (1977). Requirement of the delta subunit of chloroplast coupling factor I for photophosphorylation. *J. Biol. Chem.* **252**, 1814–1818.
547. Yu, C. A., Yu, L., and King, T. E. (1975). The presence of multiple cytochrome b proteins in succinate-cytochrome c reductase. *Biochem. Biophys. Res. Commun.* **66**, 1194–1200.

548. Yu, C. A., Yu, L., and King, T. E. (1977). The existence of an ubiquinone binding protein in the reconstitutively active cytochrome b–c_1 complex. *Biochem. Biophys. Res. Commun.* **78,** 259–265.

549. Zanetti, G., and Arosio, P. (1980). Solubilization from spinach thylakoids of a higher molecular weight form of ferredoxin-NADP$^+$ reductase. *FEBS Lett.* **111,** 373–376.

550. Zanetti, G., and Forti, G. (1966). Studies on the triphosphopyridine nucleotide-cytochrome f reductase of chloroplasts. *J. Biol. Chem.* **241,** 279–285.

CHAPTER THREE

Carbon Nutrition of Plants: Photosynthesis and Respiration[1]

R. G. S. BIDWELL

[1] Abbreviations used in this chapter: ADP, adenosine diphosphate; ATP, adenosine triphosphate; CAM, Crassulacean acid metabolism; CoA, coenzyme A; DCMU, 3-(3,4-dichlorophenol)-1,1-dimethylurea; EDTA, ethylenediaminetetraacetic acid; HPMS, α-hydroxy-2-pyridinemethanesulfonic acid; IAA, indole acetic acid; LEM, light-effect mediators; NAD(H), nicotine adenine dinucleotide (reduced); NADP(H), nicotine adenine dinucleotide phosphate (reduced); PEP, phosphoenolpyruvic acid; PEPcase, phosphoenolpyruvic acid carboxylase; PCK, phosphoenolpyruvate carboxykinase; PGA, 3-phosphoglyceric acid; P_i, inorganic phosphate; PP_i, pyrophosphate; R_m, mesophyll resistance; R_s, stomatal resistance; RuBP, ribulose 1,5-bisphosphate; RuBPcase, ribulose bisphosphate carboxylase; SAM, submersed aquatic macrophyte; Sm, mesophyll succulence index; Γ, CO_2 compensation point.

Plant Physiology
A Treatise
Vol. VII: Energy and Carbon Metabolism

ISBN 0-12-668607-6

I. Introduction

More than twenty years have passed since the earlier review of respiration and photosynthesis appeared in Volume I of this treatise, and over fifteen years since organic nutrition was covered in Volume IVA. Many aspects of carbon metabolism have been discovered or clarified during this time. Perhaps the most important concept to have emerged is that carbon metabolism is not merely a collection of isolated pathways conveniently packaged in specialized cells or organelles; it must be viewed as part of the integrated whole of the growth, metabolism, and behavior of the plant. This chapter presents a synthesis of the various aspects of carbon nutrition, covering developments in photosynthesis and the interactions of photosynthetic and respiratory metabolism (including photorespiration). The major metabolic pathways are described, but the evidence leading to their development is outlined only. Major emphasis is placed on the operation, integration, and control of photosynthetic and respiratory systems and their role in the nutrition of the growing, developing plant. These now appear to be the main areas of research that will lead most directly to broader understanding and more effective control of plant metabolism and development.

A. RELATIONSHIPS BETWEEN PHOTOSYNTHETIC AND RESPIRATORY METABOLISM

Until recently photosynthesis and respiration were considered as separate and different metabolic processes. Parts of the photosynthetic reductive cycle seemed to resemble reversed respiration, but with different enzymes. Furthermore, the processes of photosynthesis and respiration appeared to be neatly compartmented; each went at its own rate regardless of the other, and each process was regulated (if photosynthetic regulation was needed) in relation to itself and its own association with cellular needs. One thing seemed certain: photosynthesis occurred in chloroplasts, and the products (sugars and starch) were made there. Photosynthate was mobilized as required and exported to the cytoplasm for respiratory and synthetic metabolism.

This tidy, neatly compartmented framework no longer adequately explains these processes. Photosynthetic carbon metabolism is now known to involve numerous nitrogenous intermediates. It takes place throughout the cell and, as in C_4 photosynthesis, even requires the cooperative metabolism of cytoplasm and chloroplasts of cells in distant locations in the leaf. The newly discovered complexity of intermediary metabolism associated with photosynthesis requires precise regulation and, perhaps, hitherto unsuspected levels of organization in the association of organelles.

In the past, the question of respiration in the light was a theoretical one, a bugaboo for those who wished to measure precisely the rate of photosynthesis. The concept of photorespiration had been denied editorially, and only John P. Decker's enlightened analysis of light–dark respiratory transients, initially denied publication in the United States, suggested that a real phenomenon lurked beneath the nightmare. By 1965 everyone knew about photorespiration, and by 1970 they thought they knew how it worked. However, the situation is still cloudy, and there is much about the process we are not certain about, including its precise metabolic pathway, its relationship with photosynthetic metabolism, and the reason (if any) for its existence. Various and complex metabolic systems have been described that involve several organelles and metabolic shuttles throughout the cell.

Complexities have multiplied. Instead of one simple, invariable cycle of photosynthetic carbon metabolism, nearly a dozen are now known that involve new metabolites and, frequently, a substantial subsidiary nitrogen metabolism. Indeed, the complexity of photosynthetic metabolism is now such that it begins to resemble auxin physiology in the difficulty of extracting coherent concepts and principles. In one way the plant, in its metabolism, seems to have foreshadowed modern civilization: the job at hand is done as effectively as possible with whatever resources the organism (or society) has available to it, regardless of predetermined patterns or precedents. Thus most of the newly discovered photosynthetic and photorespiratory metabolic processes do not involve new or unknown enzymes. They were all, or nearly all, previously known in other associations such as respiration or cellular metabolism. What is new is the aggregation of enzymes, their association (in sequences or organelles), and the organization of metabolic sequences to perform previously unsuspected accomplishments. Thus the question whether a given plant is or is not, for example, a C_4 plant is irrelevant. What matters is the photosynthetic metabolism it uses under any given set of conditions, whether environmental, internal, or developmental.

B. PHOTOSYNTHETIC AND RESPIRATORY ENZYMES

As soon as it became evident that photosynthetic metabolism involved the same basic carbon pathways as respiration (but running backward in chloroplasts), it was proposed and later shown that photosynthetic enzymes are similar to their respiratory counterparts but sufficiently different for the necessarily different sorts of controls required. Now, however, many newly discovered photosynthetic enzymes associated with the C_2 and C_4 pathways are not so clearly distinguished, although the metabolic pathways they mediate are more precisely regulated than was formerly thought. Thus the carboxylase and various decarboxylating enzymes of the C_4 cycle and the enzymes handling the $C_2 \rightarrow C_3$ portion of the C_2 cycle are, in their origin, not photosynthetic enzymes at all; they are enzymes of cellular intermediary metabolism that have been pressed into service as photosynthetic metabolism has evolved and developed. In the process some of these enzymes have become specifically located and, perhaps, specifically regulated; they can now be considered truly part of the photosynthetic machinery of the plant. Others, however, apparently remain respiratory enzymes doing double duty. As such, their regulation and disposition, relative to substrates from photosynthetic or respiratory metabolism, is of prime importance to the integration of the metabolism of the cell. Furthermore, some of these enzymes, by their location in the main lines of widely divergent metabolic sequences, are of extreme importance in the overall regulation of the metabolic activity of the cell.

C. CELL AND WHOLE-PLANT METABOLISM

The relationship between the metabolism of the whole plant and that of its individual cells is necessarily precise, and many aspects of it have been well understood for some time. The parallel relationship between photosynthetic metabolism of leaf cells and the metabolic requirements of other parts of the plant is not as clear. The concept and understanding of the need for regulation of photosynthetic metabolism is a relatively new development. Earlier it was felt that the most beneficial situation for a plant would be to have photosynthesis running at full speed whenever conditions permitted, regulated only by supplies of energy and raw materials.

More recently, two concepts have been recognized. First, many plants have photosynthetic capacities that greatly exceed their needs under ideal conditions, and these high capacities need regulation to avoid damage from overindulgence in photosynthesis. Second, under limiting conditions, available light energy and reduced carbon must be partitioned among different cellular processes and among various parts of the whole plant. Thus the problem of regulation involves not only the control of the rate of photosynthesis, but also the regulation of the uses of light energy and the nature of photosynthetic end products and their distribution throughout the plant.

This further implies a much closer correlation of photosynthesis with other aspects of metabolism, including regulation. Evidently, many areas of plant physiology are involved in problems of how photosynthesis is affected by the need for photosynthate in other parts of the plant and how (and where) the required controls act. Clearly, respiration in the root (which may relate to salt uptake or growth) depends ultimately on photosynthesis in the leaf. Just how closely these seemingly distant and separate phenomena interrelate is only now beginning to be realized. In part, the realization has come from the development of computer models of plant and crop growth that highlight the interdependence of metabolic phenomena in different parts of the plant. But this realization has come primarily from the recognition that developmental correlative effects operate through a complex network of controls that interrelate every aspect of the metabolism of every part of the plant as well as its growth. Thus important aspects of photosynthesis are its relation to the need for photosynthate in other parts of the plant, the mechanisms whereby photosynthesis is controlled, and the control systems that regulate the distribution of its products.

Perhaps of even greater concern in the analysis of photosynthesis is consideration of the question, To what extent is photosynthetic chemical potential (reducing power or ATP) partitioned among various metabolic needs in the plant, and how and to what extent is it regulated? Photosynthesis has traditionally been thought of as the synthesis of carbohydrates. Recent data suggest that much light energy is channeled directly from the photosynthetic machinery to other, doubtless competing metabolic sequences including nitrate reduction, amino acid synthesis, some noncarbohydrate carbon metabolism, and some aspects of cellular maintenance and growth. This challenges the definition and

TABLE I

Activity of Spinach (*Spinacia oleracea*) Leaf Discs, Chloroplast Preparations, and
C_3-Cycle Enzymes[a]

	μmol/hr/mg Chlorophyll at 20°C
Leaf discs, air-level concentration CO_2	100
Reconstituted chloroplast preparation, 10 mM $NaHCO_3$	123
Phosphoriboisomerase	3217
Phosphoribulokinase	2422
Ribulosebisphosphate carboxylase	432
3-Phosphoglycerate kinase	5721
Triosephosphate dehydrogenase	1128

[a] From data of Slabas and Walker (530).

extends the scope of any analysis of photosynthesis. For the purposes of this chapter, particularly Sections II through V dealing with the photosynthetic carbon cycles, photosynthesis is discussed with reference to the associated carbon metabolism. The broader outlook is developed in Sections VI and VII, which examine the interrelationships of photosynthesis with total plant metabolism.

II. The C_3 Photosynthetic Cycle

A. Reactions of the Cycle

1. Outline of Reactions

The main outline of the cycle has changed little from the formulation proposed by Bassham *et al.* (25), and it is shown in Fig. 1. Over the years a number of controversies have arisen about details of the cycle, mainly resulting from inability to detect adequate amounts of required enzymes or from anomalous labeling patterns. These have largely subsided, and most of the reported anomalies had been resolved by 1970 (229). It can now be reasonably concluded that the scheme shown in Fig. 1 describes the reactions of the C_3 cycle. Questions currently under discussion concern the operation and regulation of the individual enzymes of the cycle, the nature of its end products, the control of the cycle and of the pattern

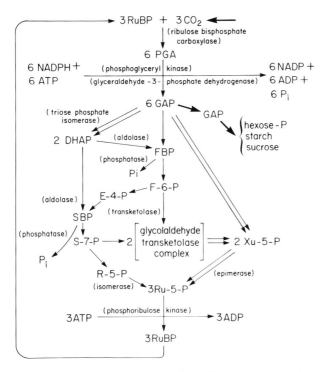

FIG. 1. The C_3 (Calvin) cycle of photosynthesis. Boldface arrows show input and output of carbon. DHAP, dihydroxyacetone phosphate; E-4-P, erythrose 4-phosphate; FBP, fructose bisphosphate; F-6-P, fructose 6-phosphate; GAP, glyceraldehyde 3-phosphate; PGA, 3-phosphoglyceric acid; R-5-P, ribose 5-phosphate; RuBP, ribulose bisphosphate; Ru-5-P, ribulose 5-phosphate; SBP, sedoheptulose bisphosphate; S-7-P, sedoheptulose 7-phosphate; Xu-5-P, xylulose 5-phosphate.

of its end products, and possible alternative reactions of some intermediates. These and other aspects (e.g., the dual function of the carboxylase—oxygenase) are considered in detail in the following subsections.

The reported activity of enzymes is generally affected by isolation or measurement procedures. Earlier measurements of activity (199, 447) were in some cases too low for cycle operation. Recent measurements (e.g., 593) show that rates range from more than adequate to tenfold excess, as shown in Table I.

2. Autocatalysis

One of the important characteristics of the C_3 cycle is its autocatalytic potential (589), that is, its ability to generate increasing amounts of the

CO_2-acceptor molecule, RuBP, instead of its normal product. In simplified form, the cycle can be shown to produce a molecule of hexose:

$$6\ CO_2 + 6\ RuBP \rightarrow 12\ C_3 \rightarrow 6\ Ru\text{-}5\text{-}P + 1\ Hexose$$

It is possible to rearrange the reactions in Fig. 1 so that the product of the cycle is RuBP instead of hexose:

$$5\ CO_2 + 5\ RuBP \rightarrow 10\ C_3 \rightarrow 5\ Ru\text{-}5\text{-}P + 1\ Ru\text{-}5\text{-}P$$

Thus the concentration of starting material is increased, and hence the rate of reaction may increase. The importance of this is that if the intermediates of the cycle have been reduced to a low level (e.g., during the night) their amounts can quickly be built up within the chloroplast to the concentration needed for rapid photosynthesis without the intervention of anaplerotic or secondary metabolism.

The C_3 cycle is the only sequence of photosynthetic reactions having this property. This may be an important reason why all photosynthetic systems now known, including CAM and C_4 photosynthesis, rely ultimately on the C_3 cycle for reductive carbon fixation (284). Other photosynthetic mechanisms are but adjuncts or supplementary systems that improve the CO_2-absorbing capability of the plant.

3. Photosynthesis of Chloroplasts

There is no question that the reactions of the carbon reduction cycle are located in chloroplasts, and isolated chloroplasts have been repeatedly shown to carry out the reduction cycle without perturbations attributable to isolation (e.g., 270, 525, 588). Furthermore, chloroplasts damaged in various ways during isolation, particularly with respect to the loss of the outer envelope, are capable of essentially normal operation of the reduction cycle if suitable cofactors (ferredoxin, stromal protein, Mg^{2+}, NADP, and ADP) are added (29, 591, 592). Self-sufficient reconstituted chloroplast systems have been developed by Walker and co-workers (593) that operate with partially purified enzymes (530). They permit close analysis of the precise conditions and cofactor requirements of each step in the cycle and detailed study of its regulation.

4. Cells and Tissue Cultures

Cells may be isolated by enzymatic maceration or by careful physical destruction of leaves. Enzymatic maceration usually produces cells the rates and products of photosynthesis of which differ substantially from those of the parent tissue (92, 272). Modified isolation procedures have produced cells of high photosynthetic rate that are viable for up to 80 hr (438). However, activity depends not only on carefully adjusted isolation media that include 0.2% methylcellulose and 10 mM cellobiose to protect against the action of cellulases, but also on a precise regime of dark and light periods following isolation.

Careful grinding enabled Gnanam and Kulandaivelu to produce photosynthetically active cells (202) that compare in many respects with algal cells or whole leaves (322). About 25% of the photosynthetic products leak out of such cells. No starch or other insoluble compounds were produced, suggesting that the preparations were uncontaminated by pieces of intact tissue. This also suggests that the photosynthetic metabolism of the cells may have been somewhat perturbed by the treatment. Not all plants tested yielded active cell suspensions; attempts with tobacco (*Nicotiana tabacum*), maize (*Zea mays*), rice (*Oryza sativa*), and cotton (*Gossypium* sp.) were unsuccessful, whereas *Canna indica, Crataeva religiosa, Crotalaria laburnifolia,* and *Thunburgia grandiflora* provided the best material.

Mechanically isolated mesophyll cells from leaves of *Ipomoea purpurea* (*Pharbitis purpurea*) were studied immediately after isolation and during division (155). Whereas the nondividing cells mainly fixed carbon by the C_3 cycle and synthesized a range of amino acids, dividing cells produced only malate, citrate, aspartate, and glutamate, indicating that only β-carboxylation took place. Cells isolated from different tissues, the palisade parenchyma and spongy parenchyma of *Vicia faba*, were found to contain similar amounts of photosynthetic enzymes and to produce identical patterns of labeled photosynthate from $NaH^{14}CO_3$ (432). Evidently these cell types do not differ in photosynthetic metabolism.

Photosynthetic activity has been observed in chlorophyllous callus from time to time, but only recently has continuous photoautotrophic growth of such cultures been achieved. Liquid suspension cultures of tobacco callus (38), *Chenopodium rubrum* (261), and rice (453) have been reported that double their mass in 12–20 days. All these systems require high concentrations of CO_2, saturating in the vicinity of 1% CO_2. This may be due to the diffusive resistance of tissue masses to carbon dioxide, but it could also be due to inherent restrictions on the activity of carbon metabolism in these tissues. Modification of the culture system permits growth of cultures in CO_2-enriched air on solid medium (39). Photorespiration is high in these cultures in relation to their rates of photosynthesis, in spite of the high CO_2 concentration (2–3%). It has been suggested that these systems may be useful for characterizing photorespiratory properties and in selecting varieties with improved ratios of photosynthesis to photorespiration.

5. Postillumination CO_2 Fixation

An early demonstration of the degree of independence of light trapping and carbon reduction was the finding that algae, when preilluminated in an atmosphere of N_2, could subsequently fix carbon in the dark and produce an approximately normal range of labeled photosynthetic intermediates as well as sugars (27, 37). Later, Togasaki and Gibbs (562) and Hogetsu and Miyachi (246) reinvestigated this phenomenon in algae and found that the C_3 cycle operated to but a limited extent, if at all, in the dark period following illumination. However, the latter workers concluded from comparable experiments that the carbon reduction cycle does operate under these conditions in leaves and chloroplasts of higher plants (379). They found that O_2 had no significant effect on either rates or products of postillumination fixation. Unfortunately, their analyses did not compare the magnitude of light-enhanced dark fixation with photosynthesis and certain anomalies, such as the reduction or elimination of sugar—phosphate production and the absence of an oxygen effect, were not explained.

This phenomenon was carefully reinvestigated in *Phaseolus vulgaris* and *Zea mays* by Leonard (339) and, more recently, in *Zea mays* by Samejima and Miyachi (509) and Creach (122). Leonard showed that fixation at low levels of CO_2 in air continued at an undiminished rate in darkness for up to 1 (bean) or 2 min (maize) and that the duration of postillumination fixation was doubled in the abscence of O_2. The main products of postillumination fixation in barley (*Hordeum vulgare*) were PGA and alanine, together with small amounts of aspartic and glyceric acids. In maize over 90% of postillumination-fixed carbon was in C_4 acids (most of it in aspartic acid). Evidently only the primary carboxylation reactions were operating. This experimental system may be useful for study of the primary carboxylation of C_3 and C_4 plants, in particular the effects of O_2 on these reactions.

6. CO_2 Uptake

The C_4 cycle (see Section IV) provides a powerful mechanism in plants that possess it for concentrating CO_2 at the site of RuBP carboxylation. Plants that do not have this cycle apparently rely on diffusion for the transport of CO_2 to chloroplasts. Certainly the conditions are poor for rapid diffusion: the gradient is shallow, cell walls and membranes present barriers, and changes of state for the diffusion of CO_2 or HCO_3^-

through cell fluids is required. The result is that the level of CO_2 at the chloroplast or at the site of CO_2 fixation has frequently been considered to be at or near zero for the purposes of calculating diffusion resistance in estimating gross CO_2 exchange. Evidently this is impossible; no CO_2 fixation could occur at zero CO_2 concentration. Furthermore, the best estimates of the K_{mCO_2} of RuBPcase place it in the range of 10 μM, which is roughly equivalent to 330 $\mu l/l$ CO_2 in air (see Section II,B,2). It would seem likely, considering the rapidity of CO_2 fixation, that some sort of CO_2-concentrating mechanism exists in plants lacking the C_4 cycle.

Many freshwater and all marine algae live in a medium containing about 2 mM HCO_3^-, which is in equilibrium with air concentrations of CO_2 (330 $\mu l/l$). Freshwater and marine aquatic plants are widely thought to absorb HCO_3^- (48a, 180, 460a, 558). Recent work by Lucas and others suggests that *Chara* actively absorbs HCO_3^- using a hydroxyl extrusion pump to provide the energy input (e.g., 361). Poincellot has shown that chloroplasts of higher plants absorb HCO_3^- (450) and that ATPase-mediated transport of bicarbonate through chloroplast membranes is proportional to whole plant photosynthesis (452). On entirely different experimental grounds, Champigny and co-workers proposed a mechanism for enhancing CO_2 or HCO_3^- levels in chloroplasts that is driven by a high-energy photosynthetic intermediate (96, 97).

Bicarbonate could be concentrated in chloroplasts as the result of an actively maintained pH gradient (596) associated, perhaps, with the activity of carbonic anhydrase (170, 180, 208). This enzyme has the necessary widespread distribution (14). It is found in much higher amounts in C_3 than in C_4 plants (172) and in much higher amounts in air-grown than in 5% CO_2-adapted *Chlorella* (e.g., 208), as would be expected if the enzyme were involved in a CO_2-concentrating mechanism. Various carbonic anhydrase inhibitors also inhibit photosynthesis (e.g., 24, 267). Carbonic anhydrase has been found to facilitate diffusion of CO_2 and HCO_3^- in solution (169) and across a membrane (76). Jacobson *et al.* have argued that carbonic anhydrase could not affect photosynthesis by facilitating the conversion of HCO_3^- to CO_2 (267). However, their argument is based on the amount of bicarbonate estimated to be present in chloroplasts and does not take into account the possibility of active HCO_3^- accumulation by transport across the chloroplast membrane.

A hypothetical model for CO_2 concentraton is possible in which CO_2 at substomatal concentration is hydrated at the cell surface (facilitated by carbonic anhydrase) and diffuses as bicarbonate to chloroplasts. The high pH of the chloroplast stroma relative to that of the cytoplasm (76, 170), perhaps together with active transport of bicarbonate into the chloroplast or diffusion down a proton-transport-generated gradient of the sort found in *Chara* cell surfaces, would result in the concentration of bicarbonate inside the chloroplast stroma. This would in turn be converted to CO_2 by the action of carbonic anhydrase. If the substomatal CO_2 concentration is about 100 μl/l and a pH gradient of 0.5 unit is maintained between chloroplast stroma and cytoplasm (597), the equilibrium concentration of bicarbonate inside the chloroplast will be as high as, or higher than, the K_{mCO_2} of RuBPcase. Furthermore, the action of carbonic anhydrase to facilitate diffusion might permit flux sufficient to maintain such a concentration even during active photosynthesis. It should be emphasized that this model is hypothetical; although supporting evidence can be adduced, no proof is possible at present. However, this model is compatible with the evolution of higher plants from aquatic ancestors, which presumably used some such mechanism for absorbing bicarbonate from the sea.

B. RIBULOSE BISPHOSPHATE CARBOXYLASE

The enzyme, RuBPcase (E.C. 4.1.1.39) has been the subject of intense study because it mediates the only pathway for the entry of CO_2 into the cycle of reducing reactions. Recently it has been recognized that RuBPcase also functions as an oxygenase, O_2 being a competitive substrate leading to the formation of alternative products. As a consequence, not only is O_2 a competitive inhibitor of CO_2 fixation, but a whole pathway of metabolism is necessary to convert the otherwise useless products of the oxygenase reactions into reusable substrates for carbon reduction. This aspect of the enzyme and the consequent or associated photorespiratory metabolism is discussed in Section III.

1. Synthesis of RuBPcase

RuBPcase has been shown to consist of eight large subunits, which are apparently homologous in all plants and contain the catalytic sites for RuBP and CO_2, and eight small subunits, which have a regulatory role and are structurally quite different among different species (551). The large subunits are made on the 70 S chloroplast ribosomes, whereas cytoplasmic 80 S ribosomes are the site of formation of the small subunits (269, 410). There is evidence that the synthesis of the two kinds of subunit is coupled in spinach (*Spinacia oleracea*) (410), pea (*Pisum sativum*) (91), and *Chlamydomonas reinhardtii* (265), but this view is disputed. For example, isolated chloroplasts can make large subunits although whole cells or protoplasts are needed for the formation of small subunits

(259). Apparently the synthesis of the two kinds of subunit can be uncoupled in barley (177), and evidence has been presented that they are not tightly coupled in *Euglena* (350). The regulation of RuBPcase synthesis and turnover evidently needs further study (see Volume VIII, Chapter 3 of this treatise for an in-depth discussion). Under normal circumstances it seems improbable that photosynthesis is regulated by the amount of RuBPcase present in the leaf (see Section II,D).

2. Substrate and K_{mCO_2}

Early reports suggested that the K_m of RuBPcase for bicarbonate, the supposed substrate, was too high for normal photosynthesis at air-level concentrations of CO_2. Then Cooper *et al.* showed that CO_2 rather than bicarbonate is the substrate of the carboxylase (119). Recalculation of the K_{mCO_2} from equilibrium concentrations of CO_2 at the pH used gives a value of approximately 45 μM, and this value was found experimentally in spinach using CO_2 (344). Then the K_{mCO_2} was found to be strongly pH dependent, and values as low as 7 μM were reported (68). Under appropriate conditions, values for K_{mCO_2} ranging from 7–15 μM have been found for several species including spinach, lettuce (*Lactuca sativa*), pea, and sunflower (*Helianthus annuus*) (144). Variations in K_{mCO_2} have been reported that show taxonomic patterns but that are unrelated to photosynthetic pathway (624a).

The values that have been reported are excessive for photosynthesis at air level concentrations of CO_2 (330 $\mu l/l$; ~8 μM), given V_{max} values of 800–1000 μmol CO_2/hr/mg chlorophyll (144), and they are sufficient for normal photosynthesis (~100 μmol CO_2/hr/mg chlorophyll) at a CO_2 concentration of approximately 80–100 $\mu l/l$. This compares well with estimates of leaf intracellular CO_2 concentrations of 100–150 $\mu l/l$ (219, 344), but, as noted earlier, it suggests the need of a mechanism for maintaining this concentration inside chloroplasts.

3. Activation and Regulation

Bassham and Jenson (26) first noted that the activity of RuBPcase decreases rapidly to zero when the light is turned off and that its activity does not return immediately upon reillumination. The kinetics of the reaction are such that CO_2 fixation should continue in the dark until all available RuBP is consumed, whereas, in fact, carboxylation stops while a substantial proportion of the steady-state concentration of the RuBP remains (271). This was interpreted as showing that the enzyme is light activated.

It is now apparent that the light effect is indirect. Robinson *et al.* have shown that wheat (*Triticum*) and barley protoplasts in an appropriate medium lack dark inactivation of RuBPcase (501). The light effect appears to be based on a complex series of events that hinge upon the effect of Mg^{2+} and CO_2 (as an activator rather than as substrate) on the enzyme and perhaps on chloroplast pH changes in light as well. It was noted some years ago that Mg^{2+} activates RuBPcase (346), and this has been confirmed and extended (9, 327, 357). Walter (589) suggested that Mg^{2+} moves to the chloroplast stroma from thylakoids or cytoplasm under the influence of the light-generated proton gradient (see 268 and discussion in Chapter 2). Mg^{2+} has been shown to move into chloroplasts upon illumination (346).

Elucidation of the Mg^{2+} effect hinges on the early observation that the activity of the enzyme can be increased by preincubation with bicarbonate (454). This observation has been confirmed and extended (9, 109, 327), and it is now evident that CO_2, rather than bicarbonate, is the molecular species that activates RuBPcase (357). It is also clear that the activating CO_2 has no chemical role in the Hill reaction, which it stimulates (543). Furthermore, the activator and substrate sites of CO_2 on the enzyme are separate and discrete (353, 382).

Lorimer *et al.* (42, 357) have suggested, on the basis of kinetic analysis, that the enzyme reacts first with CO_2 in a rate-determining, reversible reaction. A rapid reaction with Mg^{2+} follows, resulting in the formation of a ternary complex that is the enzymatically active form. These workers suggested that CO_2 reacts with an alkaline group, probably an uncharged amino group, because the activity of the preincubated enzyme increases as the preincubation pH is raised. They proposed that activation involves the formation of a carbamate that would in turn increase the effectiveness of metal binding. More data are required, however, before this (or any other) model of RuBPcase activation can be accepted without reservation.

Regulation of the carboxylase by several other chloroplast metabolites has been reported (108, 109). Reaction velocity of RuBPcase that has been activated by Mg^{2+} and bicarbonate preincubation is tripled on incubation with 1 mM NADPH, and the effect is saturated somewhere between 0.5 and 1.0 mM NADPH (108). Similar activation occurs with the addition of 0.05–0.1 mM 6-phosphogluconate, and a parallel but lesser activation (about 70%) results from addition of 0.1 mM fructose 1,6-bis-phosphate. Increasing the concentration of either of these metabolites causes less activation and, eventually, at concentrations over 1–1.5 mM, inhibition. Increasing the NADPH concentration over 2 mM has no further effect.

The addition of RuBP to an enzyme that has not been preincubated with bicarbonate or Mg^{2+} causes some inhibition (109). If the enzyme has been preincubated (and thus activated) by HCO_3^- and Mg^{2+}, RuBP carboxylation is competitively inhibited by 6-phosphogluconase or fructose 1,6-bisphosphate. Chu and Bassham (109) have suggested, on the basis of these data, that RuBP may react with an allosteric regulatory site that prevents binding at the reactive site; this concept has been supported by Vater and Salnikow on the basis of fluorometric studies of RuBP binding (582). However, Sicher *et al.* could find no effect of endogenous high levels of RuBP on RuBPcase activation, although activation of purified enzyme was irreversibly inhibited by incubation with added RuBP (527a). Binding of Mg^{2+} or HCO_3^- at separate allosteric sites is thought to change the enzyme conformation so that inhibitory binding of RuBP does not occur or is reduced. Additional allosteric activating sites are postulated for the binding of NADPH and 6-phosphogluconate or fructose bisphosphate. Again, the model presented to explain the experimental results requires further confirmation. The models presented by Chu and Bassham (109) and Lorimer *et al.* (357) are in disagreement with regard to the relationship of the CO_2 and Mg^{2+} binding sites, and Lorimer *et al.* regard the activation by NADPH and PGA or fructose bisphosphate as secondary reactions because prior activation by CO_2 and Mg^{2+} is necessary for their effect.

Benedetti *et al.* (36) have shown that 2 mM ammonium stimulates CO_2 fixation by increasing the rate of the RuBPcase reaction. They suggest that this is not due, as might be expected, to the pH rise that would be induced by NH_4^+ uptake into chloroplasts because NH_4^+ stimulates more strongly when chloroplasts are incubated in saturating concentrations of HCO_3^- (over 1–2 mM). However, the chloroplasts were not preincubated with HCO_3^-, and it is possible that the NH_4^+ effect resulted from a pH effect upon CO_2 activation of the RuBPcase (357).

A "light-activating factor" has been described by Wildner and Criddle that binds to RuBPcase and increases its activity when illuminated by long-wave UV light (300–400 nm) (601). Other research has failed to confirm this work, but a more recent paper from the same laboratory indicates that young (half-grown) tobacco leaves provide the best source of "activatable" RuBPcase and that other special precautions, including stabilization by dimethyl sulfoxide, are necessary (131). Further confirmation from other laboratories is necessary before a critical examination of this concept is possible.

It has been widely reported that RuBPcase quantity and activity change, particularly with development, in leaves (e.g., 541). Steer (542) measured the activity of the enzyme during leaf expansion, and the data clearly show that there is a strong net gain in enzyme activity during

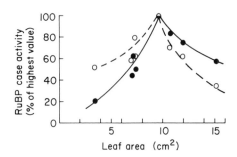

FIG. 2. Specific activity (units/cm^2; ○) and total activity (units per leaf; ●) of RuBPcase in leaves of *Capsicum frutescens*. Data recalculated from Steer (542).

early growth, followed by a substantial loss as the leaf matures (Fig. 2). These changes are paralleled by measured changes in fraction I protein. Attempts to perturb this pattern by removing other leaves during the development of the study leaf or by applying kinetin had little effect on the overall shape of the curve. It was concluded that intrachloroplast mechanisms, rather than effects originating in other organs, control carboxylase during the later stages of leaf growth. The inference is that rates of photosynthesis, which do seem to be affected by ontogenetic events (e.g., 191), are not closely controlled or limited by the activity of RuBPcase.

4. Inhibition

It was early postulated that the immediate product of the RuBPcase carboxylation of RuBP (A in Fig. 3) is an ephemeral, unstable intermediate, 2-carboxy-3-keto-D-ribitol 1,5-bisphosphate (B in Fig. 3) (87). This view was strengthened when 2-carboxy-D-ribitol 1,5-bisphosphate (C in Fig. 3) was found to bind tightly to the enzyme, apparently at the active site (528). More recently, xylitol 1,5-bisphosphate (D in Fig. 3) (503) and xyulose 1,5-bisphosphate (E in Fig. 3) (372) have been found to be potent inhibitors of RuBPcase, apparently binding at the RuBP active site. Similarities in the phosphate groups, the hydroxyl group on C-3 (D and E in Fig. 3), and the carboxyl group on C-2 (A and D in Fig. 3) have given rise to the suggestion that these groups are responsible for substrate or inhibitor binding (372).

The question whether a C_6 intermediate really occurs in the RuBPcase reaction has not been resolved. However, it seems to be relatively unimportant because the enzyme action is such that, if it is formed, it is immediately hydrolyzed to two molecules of PGA. Furthermore, the C_6

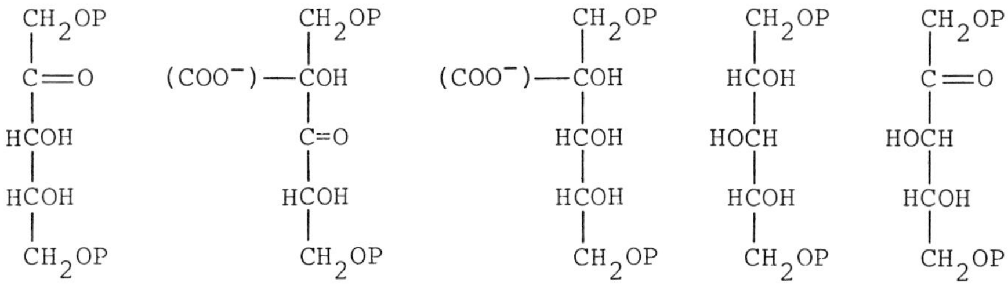

CH_2OP	CH_2OP	CH_2OP	CH_2OP	CH_2OP
$C=O$	$(COO^-)-COH$	$(COO^-)-COH$	$HCOH$	$C=O$
$HCOH$	$C=O$	$HCOH$	$HOCH$	$HOCH$
$HCOH$	$HCOH$	$HCOH$	$HCOH$	$HCOH$
CH_2OP	CH_2OP	CH_2OP	CH_2OP	CH_2OP
(A)	(B)	(C)	(D)	(E)

FIG. 3. Substrate (A, ribulose 1,5-bisphosphate), proposed unstable first product (B, 2-carboxy-3-ketoribitol 1,5-bisphosphate), and inhibitors (C, 2-carboxyribitol 1,5-bisphosphate; D, xylitol 1,5-bisphosphate; E, xylulose 1,5-bisphosphate) of RuBPcase action. P, $-PO_3^{2-}$.

inhibitor (C in Fig. 3) seems not to have provided any further insight into the enzyme mechanism than the C_5 analogs (D and E in Fig. 3). Studies with 2-carboxy-D-ribitol 1,5-bisphosphate (C in Fig. 3) showed independently that RuBPcase first reacts with Mg^{2+} before reacting with the inhibitor (or substrates) (528).

5. Oxygenase Function

During the 1960s a number of reports showed that O_2 acts as a competitive inhibitor of CO_2 fixation in low to medium concentrations in a way that closely parallels its effect on photorespiration (44). The reason for this became clear when Bowes *et al.* (69) showed that RuBPcase mediates an oxygenase reaction catalyzing the conversion of RuBP into one molecule of PGA and one molecule of phosphoglycolate (269, 355). The reaction mechanism is presumed to be that shown in Fig. 4. It has been suggested that the mechanism of reaction involves the superoxide anion, O_2^- (41). As would be expected, O_2 was found to competitively inhibit CO_2 fixation (67) with a K_i of about 0.6–0.8 mM (326), whereas CO_2 competitively inhibits oxygenation (18). This situation has been effectively reviewed by Lorimer (353a).

A number of early problems in the study of this enzyme system resulted from the difficulty of its assay. The oxygenase function of the enzyme, like that of the carboxylase, is activated by CO_2 (356). Stringent conditions are therefore necessary when determining oxygenase activity in the absence of CO_2 to prevent its inactivation.

There has been considerable argument about whether the two activities are really functions of the same enzyme. Certain data suggest they

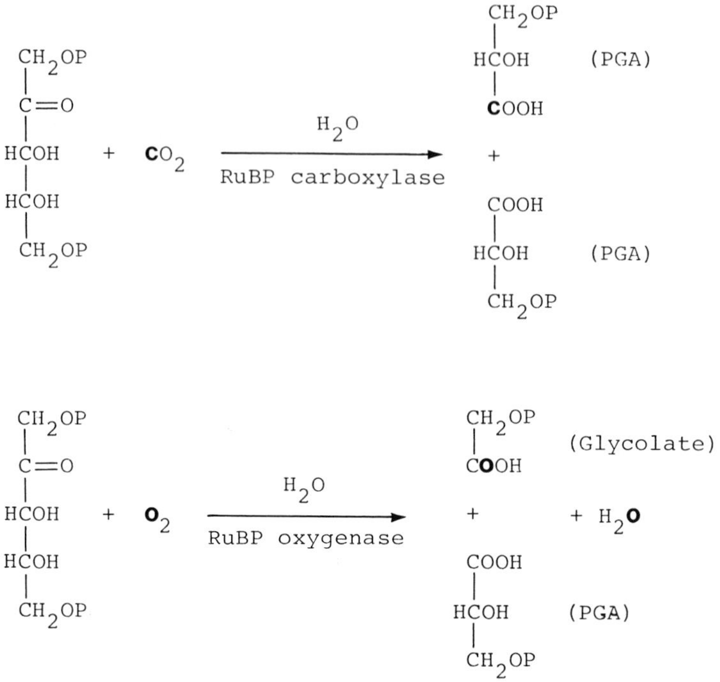

```
                                          CH2OP
                                          |
    CH2OP                                 HCOH    (PGA)
    |                                     |
    C=O                      H2O          COOH
    |               ------------------>
    HCOH   +  CO2            RuBP carboxylase     +
    |
    HCOH                                  COOH
    |                                     |
    CH2OP                                 HCOH    (PGA)
                                          |
                                          CH2OP
```

```
    CH2OP                                 CH2OP
    |                                     |         (Glycolate)
    C=O                      H2O          COOH
    |               ------------------>
    HCOH   +  O2             RuBP oxygenase    +        + H2O
    |
    HCOH                                  COOH
    |                                     |
    CH2OP                                 HCOH    (PGA)
                                          |
                                          CH2OP
```

FIG. 4. Comparison of RuBP carboxylase and RuBP oxygenase action. P, $-PO_3^{2-}$.

are not. Photorespiration, a consequence of oxygenase activity, does not always vary in strict proportion with photosynthesis. This, however, may be a function of other aspects of the photorespiratory metabolic sequence (see Section III). The pH optima of carboxylation and oxygenation differ: carboxylation has a narrow peak at or near pH 8.2, whereas oxygenase activity peaks broadly between pH 8.6 and 9 (18, 22). This difference is not necessarily related to the reaction, however, and may be dependent only on conditions affecting access of the substrates to the enzyme. Carboxylase and oxygenase activities have been shown to behave independently in certain preparations (218), but difficulties have been encountered in preserving adequate activity (cf. 356, 357). A stronger argument is the isolation of a mutant of *Chlamydomonas reinhardtii, ac i72*, that has greatly reduced RuBPcase activity but whose oxygenase activity is twice that of the wild type (406). However, the pH effect on oxygenase activity is also modified in the mutant, and it has been suggested that the enzyme is modified to have an increased affinity for O_2, which greatly increases its competitive effect. Finally, it has been suggested that the oxygenase and carboxylase functions can be separated by Sepharose gel filtration (72). It was further suggested that the oxidase is a copper enzyme. However, this work has been strongly criticized (see 73 and appended discussion) and will need confirmation before it can be further considered.

The other side of this question is strongly supported. Careful analysis of the kinetics and activation of carboxylation and oxygenation suggests that they are activities of the same enzyme (357). Moreover, it has been found that the K_{mCO_2} of the carboxylase is the same as the K_{iCO_2} of the oxygenase; the K_{mO_2} of the oxygenase is similar to the K_{iO_2} of the carboxylase; and the K_{mRuBP} is the same for both carboxylase and oxygenase reactions (326). This strongly suggests that the two functions are attributes of one active site on one enzyme. Data have been presented showing that the two functions have the same mode of action, are tightly coupled (20), and are expressions of a single reactive site on the enzyme (107). Oxygenase activity has been universally found to be associated with RuBPcase activity, even in *Rhodospirillum rubrum* (504), a photosynthetic bacterium that has the smallest known RuBPcase, consisting of only two large subunits (284). It appears safe to conclude that the RuBP oxygenase and RuBPcase are two competing activities of one enzyme. The inference is that the substrate recognition of RuBPcase is not sufficiently specific to discriminate strongly against molecular oxygen.

C. OTHER CYCLE ENZYMES

All of the required enzymes for the Calvin cycle (Fig. 1) have now been demonstrated in chloroplasts, and information regarding their behavior and control is slowly accumulating. Regulation undoubtedly occurs at several steps in the cycle (see Section II,D), particularly in the two strongly exergonic phosphatase reactions, so it is of interest to examine individual reactions (94).

The reduction step, involving glyceraldehyde-3-phosphate dehydrogenase (NADP) (E.C. 1.2.1.13), seems to have a variety of control mechanisms. It has been known for some time to be light activated (635). The enzyme tends to aggregate in the presence of NAD but to dissociate, and so be activated, on incubation with NADP or ADP (464). Cerff (93) has shown that binding with NADH and NADPH is mutually exclusive in the enzyme from *Brassica hirta* (*Sinapis alba*). Cerff suggested that this forms the basis for a light-mediated activation of the enzyme, possibly through the agency of phytochrome, resulting from the effect of light on the interconversion of NADH and NADPH. It is also possible that light-generated reducing power may activate the enzyme. Reduced thioredoxin, which may be reduced chemically by dithiothreitol or photochemically by ferredoxin, activates NADP-linked glyceraldehyde 3-phosphate dehydrogenase but not the NAD-linked enzyme in spinach chloroplasts (612). However, the system is not entirely clear; the enzyme activated by chemically reduced thioredoxin is inactivated by oxidized glutathione, but it is unaffected if photochemically reduced thioredoxin is the activating agent. Activation of the enzyme by various amino acids has been found in *Chlorella* (570), but no mechanism is apparent.

The reaction of 3-phosphoglyceric acid kinase (E.C. 2.7.2.3) is reversible, but it is closely controlled by energy charge (434) and inhibited by ADP (345). Both energy charge (376) and absolute ATP levels (367) of chloroplasts increase with illumination, and this doubtless promotes active forward operation of this kinase. Ribulose 5-phosphate kinase (E.C. 2.7.1.19) is also responsive to the energy charge level of chloroplasts (334), which may explain its activation by light (16). However, a portion of the chloroplast complement of phosphoribulokinase, like that of glyceraldehyde 3-phosphate dehydrogenase, has recently been shown to be activated by thioredoxin reduced by light-reduced ferredoxin (611). This fraction of the enzyme has been designated the *regulatory form* to distinguish it from the *nonregulatory form*, which is unaffected by thioredoxin.

The same factor, reduced thioredoxin, has also been implicated in the control of fructose-1,6-bisphosphate phosphatase (E.C. 3.1.3.11) (610). This reaction and that involving sedoheptulose bisphosphate phosphatase are strongly affected by the chloroplast Mg^{2+} concentration (e.g., 455), which, as we have seen, is affected by light. These enzymes, like the ones previously discussed, are subject to dual control measures triggered by light.

D. Regulation of the C_3 Cycle

Initially it was thought that the cycle did not need precise regulation because of the supposed sluggish nature of the carboxylase and the teleological view that plants should be opportunistic and use available conditions of light, temperature, and water to conduct photosynthesis always at maximum rate. However, the need for regulation has been pointed out by the discovery that the carboxylation reaction is rapid and effective and may thus compete for light energy with other cellular needs (such as nitrate reduction) and the recognition of the importance of the autocatalytic nature of the cycle. Examination of the cycle (Fig. 1) suggests that the main effective control mechanism to promote autocatalysis would be to increase the rate of triose phosphate isomerase, aldolase, and transketolase with reference to other cycle enzymes and to prevent reactions leading to end-product formation. However, little information has been gathered regarding the relative control of these steps in the cycle.

It was early recognized that protection of the levels of cycle intermediates during darkness requires light activation (or dark inactivation) of certain enzymes in order to prevent exhaustion of pools of key intermediates and a long induction phase on illumination. This aspect of the control of photosynthesis, involving at least two and perhaps three mechanisms of light activation of carboxylase and other cycle enzymes, has received considerable attention.

Finally, an obligatory and uncontrolled photosynthetic machinery is incompatible with orderly development and growth in plants. A long debate continues about how much and by what agency, if at all, the need for photosynthate in distant parts of a plant controls the rate of photosynthesis. The discussion centers around the possibility of direct or indirect control of photosynthesis by hormone signals, or feedback control of photosynthesis through product buildup related to translocation loading or unloading, or the possibility that no control mechanism exists that is directly related to needs at distant sites. The first two concepts have been strongly supported; the last appears unlikely because it suggested that growth of plants is dependent on and controlled by photosynthesis at all times.

1. Light Activation

Three main schemes for light activation of photosynthesis have been proposed: light activation of specific enzymes through light-mediated changes in hydrogen ion and magnesium ion concentrations; activation of enzymes through light reduction of a low-molecular-weight activating factor, now known to be thioredoxin; and by light-mediated changes in the energy charge level of chloroplasts.

a. Phosphate, Mg^{2+}, and pH. It was early noted that 5 mM Mg^{2+} would stimulate photosynthesis in chloroplasts that had been isolated from light-pretreated leaves (346). Mg^{2+} moves into the chloroplast on illumination, and the chloroplast Mg^{2+} concentration has been estimated to rise to 10 mM as a result of this and of light-stimulated loss of water from chloroplasts. Later work showed that light-mediated cation transport affects chloroplast pH, and this in turn drives Mg^{2+} transport into the chloroplast stroma (455, 465). This phenomenon has been confirmed by several laboratories. Light activation of fructose bisphosphate phosphatase is particularly pH dependent (285), and the two photosynthetic phosphatases are strongly regulated by Mg^{2+} through this mechanism (465). Thus, this may well be the mechanism, or one of the mechanisms, that controls the autocatalytic ratio of the cycle through the direction of triose phosphate carbon to the transketolase reactions, and hence away from end-product synthesis.

Magnesium has been reported to inhibit photosynthesis in chloroplasts (251) unless accompanied by a chelator such as EDTA (17). This may be explained by the capacity of Mg^{2+} to complex pyrophospate (162, 343), thus tying up the supply of inorganic phosphate. Excess P_i strongly inhibits photosynthesis, but in its absence chloroplast photosynthesis slows rapidly (e.g., 162). This is presumably because starch synthesis (which would release P_i for the kinase reactions) is too slow in chloroplasts to maintain the required P_i levels. Moreover, in intact tissue much photosynthate leaves the chloroplast as triosephosphate for sucrose synthesis, particularly in plants such as grasses, resulting in the release of both P_i and PP_i. Thus the availability of P_i would be affected by Mg^{2+} concentration in the chloroplast stroma as well as P_i concentration in the cytoplasm. The phosphate translocator (240) would permit the exchange of triose phosphate and P_i required to maintain chloroplast P_i, thus mediating the control system effected by light on Mg^{2+} concentration through the agency of light-driven proton transport (465).

b. Light-Generated Reducing Power. An assimilation regulation protein that affects the activity of several photosynthetic enzymes has been shown to be thioredoxin (162), a low-molecular-weight electron transport protein the redox function of which is a disulfide–sulfhydryl interconversion. It is a widespread factor, and the photosynthetic protein closely resembles thioredoxin extracted from various bacterial or animal sources (248). It appears to function in the same manner as nonbiological factors such as dithiothreitol, which is widely used (particularly with chloroplast preparations) to stimulate and activate many enzyme systems. Thioredoxin activates the main reducing step, triose phosphate dehydrogenase (612), the kinase that makes RuBP (611), and both photosynthetic phosphatases (455, 610). Thioredoxin may be reduced either by nonphysiological agents such as dithiothreitol or by light-generated reducing power through the action of ferredoxin (16, 611). Glutathione and glutathione reductase have been demonstrated in chloroplasts (190). It is possible that this reducing system may be involved in the mediation of thioredoxin reduction directly or through its role in the reduction of ascorbic acid.

It appears very probable that the thioredoxin system is actively involved in photosynthetic regulation. That it activates both the reducing step and the phosphatases more or less to the same extent suggests that it is responsible for controlling the autocatalytic ratio of the cycle. However, it is an obvious candidate for the overall regulation of photosynthetic activity within the framework of the requirements of the whole photosynthesizing plant.

An alternate approach comes from the work of Anderson *et al.* (e.g., 6). They suggest that light modulates the activity of several chloroplast enzymes through the reduction of light-effect mediators by the photosynthetic electron transport system. The LEMs are thought to be thylakoid-bound proteins that catalyze thiol–disulfide group exchange reactions required to activate certain enzymes. Chloroplast glucose-6-phosphate dehydrogenase is inactivated by this system (6).

c. Energy Charge. The direct control of photophosphorylation and of the phosphorylating or dephosphorylating cycle enzymes (e.g., kinases and phosphatases) is an obvious possibility and has been demonstrated (334, 376, 434). However, several enzymes show allosteric effects and react to the energy charge level. (e.g., 94, 95). Champigny has shown that 3-phosphoglyceric acid kinase, although it responds directly to ATP concentration from $0-10$ mM in the absence of ADP, is totally inactive at an energy charge level below 0.5 regardless of the ATP concentration (95). This prevents exhaustion of the cellular energy charge under conditions (e.g., limiting light) that might result in low ratios of ATP to P_i. Further, because the energy charge of the chloroplast stroma is about 0.5 in darkness and rises to about 0.9 in light (95, 376), this provides an effective light-activation–dark-inactivation control of the C_3 cycle.

d. Multiple Light Controls. There appears to be some advantage in the variety of different controls regulating the C_3 cycle. The cycle is of central importance to the growth and economy of plants and must be able to react rapidly and precisely to various kinds of adverse or favorable conditions. Moreover, it is more complex than it appears from its simplest representation (e.g., Fig. 1) because of its capacity to be reorganized into an autocatalytic cycle. Thus not only the amount of end produce (rate of operation of the cycle) and the relative amounts of various possible end products (including triose phosphate, sucrose, and starch), but also the proportion of carbon reentering the cycle in the autocatalytic reaction need to be regulated.

Direct activation by light reducing power, either by increase of the ratio of NADPH to NADP (94, 108) or via thioredoxin, would activate carboxylation and carbon reduction reactions. More specifically, it would strongly activate the phosphatases, leading to enhanced autocatalytic synthesis of RuBP. After a short time, as the charge level of the chloroplast rose, full activation of the reaction producing triose phosphate would occur through the energy charge activation of phosphoglyceric acid kinase, and an increasing proportion of carbon would be available for end product formation. The operation of the cycle would at all times be finely tuned to the amount of available light through the pH–Mg^{2+} effect and through carboxylase activation by NADPH (108).

2. CO_2 and O_2

Because RuBPcase is bifunctional and fixes CO_2 or O_2 competitively, it might be assumed that the enzyme has some regulating capacity to balance the operation of the C_3 cycle of photosynthesis and the C_2 cycle of photorespiration (see Section III). Laing et al. (326) have shown that the increased O_2 sensitivity of soybean (Glycine max) photosynthesis to temperature can be explained by the behavior of RuBPcase, which suffers a reduction in affinity for CO_2 and a slight increase in affinity for O_2 as temperature rises. They calculate that at air level concentration of CO_2 and O_2 approximately one molecule of O_2 is fixed for four of CO_2. Carboxylase has the interesting attribute that it is inactivated when exposed to RuBP in the absence of CO_2 (109). This characteristic doubtless protects the plant when CO_2 levels become low, due, for example, to stomatal closure under water stress. Under this circumstance, light, energy charge, and reduction-level controls would tend to activate the enzyme strongly leading to high levels of photorespiratory loss of carbon.

Several reports have suggested that O_2 may sometimes stimulate photosynthesis. The effect is usually small, sometimes temporary, and the reported environmental conditions for its manifestation vary. Viil et al. (583) and Canvin (88) have suggested that the effect at saturating CO_2 and light observed in bean (Phaseolus vulgaris) or sunflower (Helianthus annuus), respectively, may be due to an effect of O_2 on either the photosynthetic electron transport chain or the regeneration of RuBP. Miyachi and Okabe (381) found that O_2 stimulates photosynthesis of Anacystis nidulans, but only at saturating light and low CO_2 concentration. They suggested that oxygen may influence ATP production (perhaps via pseudocyclic electron transport), which in turn would stimuate CO_2 transport into cells. Apparently these effects are not directly related to the O_2 inhibition of photosynthesis or its stimulating effect on photorespiration. Their exact nature is still to be discovered.

3. Levels of Intermediates and Induction Phenomena

Walker has provided a very thorough and detailed discussion of induction phenomena (589). Numerous observations have shown that O_2 and CO_2 exchange start slowly on illumination and build steadily to a maximum rate in about 10 min. The effect is not caused by slow stomatal opening, and the induction phase does not occur if the preceeding dark interval is less than 15 min (577). Isolated chloroplasts also show an induction phase in CO_2 fixation, but not in the Hill reaction supported by artificial electron donors (589). However, little or no induction phase occurs when chloroplasts are provided with cycle intermediates (23). This suggests that the induction is caused by loss or export of cycle intermediates during darkness so that their concentration is reduced to a low level. The lag period would be the time required to raise concentration of intermediates, by autocatalytic action of the cycle, to a level sufficient to permit its rapid operation. It has been concluded that this lag, rather than the time required for light induction of enzymes, is the cause of induction phenomena in chloroplasts (590).

It is likely that the same explanation holds for intact leaves. Attached bean leaves show a pronounced lag phase on illumination, but detached leaves show little or no lag phase even after prolonged darkness (577). In attached leaves the translocation of photosynthate presumably stimulates the export from the chloroplasts of more photosynthate and, consequently, carbon cycle intermediates during periods of darkness. This would contribute to a lag phase. It is, of course, possible that light-activation phenomena also contribute to the induction phase of leaves. In addition, levels of certain photosynthetic intermediates may have allosteric effects on some cycle enzymes, particularly on the carboxylase (94). This would tend to exaggerate induction phenomena.

4. Control by End Products

Several possible mechanisms of end product control of photosynthesis have been evoked to explain the often observed relationship between the rate of carbon assimilation and apparent demand for photosynthate:

1. Direct feedback control of photosynthesis by its products
2. Physical effects of starch accumulation on chloroplasts or CO_2 diffusion
3. Remote or hormonal feedback effects on photosynthetic metabolism
4. Stomatal control by remote or hormone effects or by products of photosynthesis

A detailed consideration of this whole question by Neals and Incoll in 1968 resulted in the conclusion that although there is undoubtedly a relationship of some sort between photosynthesis and leaf carbohydrate content, there was at that time no proof that the two are casually related (404). Since then many research papers have contributed either positive or negative evidence for each of the four mechanisms. A brief analysis of representative data may bring us closer to a conclusion on this question, although proof is still elusive.

Physical factors (such as reduced light interception or increased resistance to CO_2 diffusion) presumably have some effect, but probably only when the leaf or chloroplast is glutted from long photosynthesis and reduced translocation (599). The relationship between starch content and photosynthesis of sugarbeet (*Beta vulgaris*) leaves may be due to mechanical effects of starch accumulation (378). However, the effect is not large, and it has been concluded that it is unlikely to be a determining factor in photosynthesis under field conditions.

No relationship between leaf carbohydrate content and rate of photosynthesis could be found during diurnal changes in *Vicia faba* (443), following ear removal in wheat (15), or after debudding in cotton (*Gossypium*) (397). Export rates are controlled by sucrose content in tomato (*Lycopersicon esculentum*), but this appears to be a consequence of photosynthesis and is not controlled by it (244). Neither sucrose nor starch regulates photosynthesis in eggplant (*Solanum melongena*) (110), and even massive additions of sugar have no effect on *Chlorella* (179). Net assimilation is reduced by sugar addition in cultured spinach cells (134), but this is probably due to increased respiration. One may conclude that direct feedback control of photosynthesis by products of photosynthesis does not occur, except in a minor way through the mechanical effects of a glut of photosynthate.

Nevertheless, relationships between photosynthesis and sugars or starch are frequently observed. In some studies the effects have been attributed to stomatal response to photosynthate concentrations (e.g., 491) or some other factors related to assimilate demand that regulate stomata (360). It seems likely that stomatal mechanisms are often responsible for regulating photosynthesis (316).

Examination of the effect of native or added carbohydrates on barley leaf segments (402) and *Brassica hirta* (*Sinapis alba*) cotyledons (386) reveal effects that cannot be related to specific causative factors. The effect of sink removal and senescence on photosynthesis in soybean leaves is unrelated to a number of physiological parameters (RuBPcase activity, leaf weight, carbohydrate, protein, or chlorophyll contents of leaves) (384). These and many other observations on the effect of sink demand

for photosynthate (e.g., 50, 421) have led to the general concept that
photosynthesis may be regulated by a sink signal, possibly of hormonal
nature (e.g., 197). This may operate through an effect on stomata
(317), but direct control of some aspect of the photosynthetic mechanism
is a possibility that needs to be more closely examined.

5. Hormonal Control of Photosynthesis

Numerous careful experiments with a variety of plant systems (see 35,
260, 338) seem to require the interpretation that assimilation is directly
related to or controlled by sink demand, perhaps through the agency of
sink-produced hormones. Bidwell et al. (46, 50) proposed that IAA produced in actively metabolizing tissue and translocated to the leaves might
serve as a sink signal to regulate photosynthesis. IAA stimulates photosynthesis in leaves when applied directly or to nearby parts of bean
(Phaseolus vulgaris) plants (50). Tamas et al. found that low concentrations of IAA stimulate CO_2 uptake and photophosphorylation in isolated chloroplasts by increasing coupling between ATP formation and
electron transport (552, 554). Applied IAA has been found to increase
the root sink effect (3). Besides stimulating photosynthesis, applied IAA
increases translocation of photosynthate from photosynthesizing leaves
to developing or darkened leaves (e.g., 49). Auxin is translocated rapidly
enough both acropetally and basipetally to accomodate this effect in
plants (48, 204). Carmi and Koller have reported that photosynthesis in
bean (P. vulgaris) leaves is regulated by unknown factors (possibly hormones) that are translocated in the water-conducting system from the
roots (90).

The capacity of IAA to direct translocation is well known, but opinions
differ regarding whether it affects the activity of the sink (396), the
source (65), or the translocation path (437, 506), and it is not possible to
draw a conclusion at present. The mechanism by which IAA might act is
not yet clear, though it might relate to its effect on hydrogen ion excretion (111), which affects translocation vein loading (580). It seems unlikely that IAA affects photosynthesis through an effect on sucrose concentration by increasing translocation because, as we have seen,
photosynthesis is normally unaffected by the carbohydrate content of
leaves. Therefore, its effect is more likely to be on chloroplast activity. If
it operates on membrane transport via proton transport, this effect
would be expected to be at the chloroplast level, as suggested by Tamas
et al. (554), rather than on vein loading.

However, the situation is not clear. Robinson et al. (502), using preparations that were much more photosynthetically active than those used

by Tamas *et al.* (552, 554), were unable to confirm the IAA effect on isolated chloroplasts. I. A. Tamas and R. G. S. Bidwell (unpublished data) have noted variations in the IAA effect that relate to the growth season or source of the leaf material, and it is possible that the effect is masked at times by high endogenous levels of hormones. Only very low concentrations of IAA have been found to be effective, and the effect peaks sharply at $0.1 \mu M$ (554). Robinson *et al.* used concentrations of $0.2 \mu M$ or higher in all but one of their experiments. This problem clearly cannot be resolved without further study.

Growth substances or hormones other than IAA have been implicated in the control of photosynthesis. Kriedemann and Lovies (318) have shown that phaseic acid, elaborated in grapes (*Vitis* sp.) in response to water stress, affects photosynthesis through some mechanism other than stomatal regulation. Applied gibberellic acid affects kikuyu grass (*Pennisetum clandestinum*) (340) and bean leaf (85) photosynthesis, and treatment of pea seeds during imbibition results in higher photosynthetic rates in the developing seedlings (457). These effects do not appear to be mediated by stomata. Miles (377) reported that chloroplasts from gibberellic acid-treated tomato leaves have enhanced rates of photosynthetic electron transport, but Robinson *et al.* (502) found no effect of added gibberellic acid on isolated pea chloroplasts. Abscisic acid directly inhibits photosynthesis at concentrations of $1-3$ mg/l (10^3 times higher than that required to effect stomatal closure) (457). Wort (621) has shown that the complex mixture of acids called naphthenic acid extracted from petroleum by alkali has a powerful growth-stimulating effect on a number of plants. The effect is due to increased photosynthetic CO_2 fixation and ATP synthesis resulting from greater photosynthetic electron transport. Paul *et al.* noted that 2,4-D stimulates photosynthetic CO_2 fixation and glutamine synthesis in cells from *Papaver somniferum* (439). They suggested that the response may be due to the capacity of 2,4-D to stimulate the reduction of disulfides, glutathione in particular, and so stimulate reduction of nitrate to ammonia.

No unifying concept seems possible at this time, but one is led to conclude that there are several possible mechanisms that permit regulation of photosynthesis to accomodate changing demands for reduced carbon nutrition in distant, heterotrophic parts of plants. Mechanisms that operate through the production by sinks of hormones or hormone-like substances are strongly indicated but other, more direct control systems are by no means excluded. Photosynthesis is probably finely tuned by a network of unrelated systems that provide an integrated response to environmental and physiological signals.

E. Products of Photosynthesis

1. Range of Photosynthetic Products

Immediate storage products of photosynthesis, aside from intermediates of various pathways associated with photosynthesis, are sucrose and starch in most higher plants. Little variation seems to occur, except under unusual or artificial conditions. However, lower plants, particularly the algae, produce a range of polysaccharides and a number of low-molecular-weight soluble carbohydrates that are often (but not always) analogous in their behavior to starch and sucrose of higher plants. These have been reviewed in detail by Craigie (121), and only a brief summary together with mention of some more recent observations are necessary here.

Several starch-like polysaccharides are found in red and blue-green algae. These are α-(1,4)-glucans with various degrees of branching. The role of floridean starch found in a red alga (*Serraticardia maxima*) is strikingly similar to that of starch in higher plants; it is formed rapidly in photosynthesis and utilized for respiration (399). However, in contrast to higher-plant starch, floridean starch is synthesized on particles outside chloroplasts not in chloroplasts (398). The enzymes synthesizing and metabolizing various algal amylopectins are usually similar to those of higher plants (121).

Soluble products of photosynthesis appear to be taxonomically diagnostic. Mannitol is produced mainly in brown algae and dinoflagellates, whereas red algae produce galactosides and green algae produce sucrose (42). Among the Rhodophyceae, Kremer (313) has shown that digeneaside (2-D-glycerate α-D-mannopyranoside) is produced in the order Ceramiales, whereas floridoside (2-O-D-glycerol α-D-galactopyranoside) is produced in all other orders. The red algae also produce two polyols, dulcitol and sorbitol (312). These compounds are available as substrates of respiration, but they are not converted to polysaccharides. In this respect they differ from the behavior of mannitol in brown algae, which acts as a substrate both of respiration and for laminarin synthesis in *Fucus vesiculosus* (43) and *Eisenia bicyclis* (622).

Evidently there is a high degree of taxonomic specificity for enzymes mediating the synthesis of final storage products in the marine algae (121), although the intermediary metabolism of the carbon reduction cycle is the same as in higher plants (284). This may reflect the much broader spread of taxonomic relationships among marine algae than among the groups of land plants. Alternatively, it may reflect adaptation of marine plants to environmental situations characteristic of the marine habitat, such as drying, wave action, or the need for osmoregulation (121).

2. Starch

A number of recent investigations have shown that ADPglucose α-1,4-glucan, α-4-glucosil transferase, or α-glucan synthetase, is primarily responsible for starch synthesis in plants (e.g., 158, 234). Starch breakdown is probably mediated by phosphorylase (444). Because chloroplasts normally make starch in light and degrade it in darkness, some control mechanism for these activities appears to be required. Preiss *et al.* showed that ADPglucose pyrophosphorylase, the enzyme that synthesizes the substrate of α-glucan synthetase, is activated by PGA and inhibited by P_i (460). Held *et al.* have shown that starch synthesis and breakdown is regulated through this mechanism by the activity of the phosphate translocator, which exchanges P_i and triose phosphate across the chloroplast membrane (239). Free sugars have been reported to regulate starch accumulation in rice (*Oryza sativa*) grains (529), but no effect of added sugar phosphates on starch metabolism of spinach chloroplasts was detected (444). The role of P_i and the phosphate translocator parallels very closely the mediation of photosynthetic cycle enzymes by the same system and provides an effective intergration of the rates of photosynthesis and starch synthesis.

Starch degradation appears to take place via amylolytic reactions followed by glycolytic reactions to yield PGA (444). A key enzyme previously undetected in chloroplasts, phosphofructokinase, has now been shown to be present and sufficiently active (283). The ATP requirement for this reaction may be balanced by the production of ATP in the oxidation of glyceraldehyde-3-phosphate to PGA (444).

3. Sucrose

Two questions arise regarding photosynthetically produced sucrose, What is the enzymatic pathway by which it is formed, and where is the site of its synthesis? Both questions now seem to have been settled, although the answer to the latter has raised new questions about the organizational relationship of organelles in photosynthetic cells.

Sucrose may be synthesized by sucrose synthetase

$$UDPGlucose + fructose \rightleftharpoons UDP + sucrose$$

or by sucrose phosphate synthetase and sucrose phosphate phosphorylase

$$UDPGlucose + fructose-6-P \rightleftharpoons UDP + sucrose-6-P$$
$$Sucrose-6-P \rightarrow sucrose + P_i$$

There seems to be no doubt that sucrose phosphate synthetase provides
the main pathway of sucrose synthesis in photosynthetic tissue (233).
Photosynthetic leaves that actively export sucrose contain sucrose phosphate synthetase and sucrose phosphate phosphorylase, whereas nonphotosynthetic, sucrose-importing leaves lack these enzymes (198). The
synthetic pathway is well regulated by its end product, as would be
expected in a pathway containing an essentially irreversible step. Both
sucrose phosphate synthetase (232) and sucrose phosphate phosphorylase (507) are regulated by sucrose. Sucrose synthetase is also present in
leaves, but its main function seems to be the formation of nucleoside
diphosphate glucose from sucrose, presumably for polysaccharide synthesis (146).

The second question, the site of sucrose synthesis, is not quite so clear.
Some enzymes of sucrose synthesis are located in the cytoplasm, not
in chloroplasts (53). Moreover, if sucrose were formed in chloroplasts, it would not readily diffuse to the cytoplasm (451). On the other
hand, sucrose synthesis has been reported in isolated spinach chloroplasts (171), although this is now thought to have been due to cytoplasmic impurities in the chloroplast preparation (331). *Acetabularia* chloroplasts have high rates of sucrose synthesis (525), but it has been
suggested that this is also due to cytoplasmic impurities (607). However,
highly purified preparations containing no electron microscope-visible
contamination produce sucrose at the same rate as contaminated ones
(45, 524). Furthermore, sucrose so produced is not liberated into the
medium. Possibly, *Acetabularia* chloroplasts, which are remarkably and
uncharacteristically stable (up to a week *in vitro*), are in some way different from chloroplasts of higher plants. With the exception of *Acetabularia*, the general consensus is that isolated chloroplasts do not make
sucrose.

In experiments by Fry and Bidwell (196) leaves were supplied $^{14}CO_2$
for a short time in light, then chloroplasts were separated in as pure a
state as possible, using nonaqueous techniques. Radioactive sucrose was
always detected first in the chloroplast fraction, but it rapidly spread to
the cytoplasm. However, intermediates of the glycolate pathway, which
are known to be metabolized in peroxisomes and mitochondria, also be-
haved similarly to sucrose and were found initially in chloroplasts and

later in cytoplasm. This suggests that the chloroplasts, no matter how carefully purified, contained cytoplasmic contamination that includes peroxisomes and mitochondria as well as the enzymes for making sucrose.

Other attempts to purify chloroplasts by nonaqueous techniques always resulted in an irreducible cytoplasmic contamination of about 15% (51), about the level of impurity encountered by Fry and Bidwell. This point may be considered in relation to the association of chloroplasts with other subcellular organelles described by Wildman *et al.* (e.g., 600) and the location of mitochondria in chloroplast invaginations (385). Some chloroplasts of *Oenothera* have tubular extensions into the cytoplasm and even into mitochondria (518). It has been suggested (196) that some form of close association of chloroplasts and other subcellular organelles facilitates the rapid export of photosynthetate from chloroplasts, most likely as triose phosphate or [as in *Dunaliella* (395)] as fructose 6-phosphate, and its prompt conversion to sucrose. This may also explain the frequent observations of sucrose compartmentation in photosynthetic leaf cells (e.g., 198).

A parallel situation seems to occur in the brown alga *Fucus serratus.* Willenbrink and Kremer showed that chloroplasts isolated nonaqueously after short periods of photosynthesis in $^{14}CO_2$ contain most of the labeled mannitol, which afterwards spread to the cytoplasm (603). Isolated chloroplasts could synthetize mannitol, in spite of the location of at least one of the enzymes of mannitol synthesis (mannose phosphate dehydrogenase) in the cytoplasm (412). Like chloroplasts of higher plants, *Fucus* chloroplasts could not be satisfactorily purified. Although Willenbrink and Kremer concluded that mannitol is formed inside chloroplasts, this view has been questioned (196). It seems possible that mannitol is formed in *Fucus,* like sucrose in higher plants, outside of but in close association with chloroplasts.

4. Alternative Energy Storage

Although most plant physiologists tend to think of photosynthetic energy storage largely, or exclusively, in terms of the formation of reduced carbon compounds, two groups in the Soviet Union have found that large discrepancies may occur between energy storage (measured calorimetrically) and gas exchange in photosynthesis. Petrov *et al.* (e.g., 448) have worked with leaves of higher plants, and they suggest that energy stored in leaves as a result of cyclic or pseudocyclic phosphorylation or other direct photochemical reactions may amount to a considerable proportion of photosynthetic energy storage. Bell *et al.* (e.g., 34) found up to 50% more energy was stored in *Chlorella* cultures than would be expected on the basis of O_2 exchange, assuming one mole of carbon (as carbohydrate) formed (112 kcal) per mole of O_2 released. They suggested the possibility of light-stimulated oxidative processes coupled with phosphorylation (34). It is known that *Chlorella* may continue dark-type respiration in light under appropriate culture conditions, but this phenomenon is not light stimulated (47). That red light failed to stimulate excess energy storage was interpreted to mean that cyclic phosphorylation by photosystem I was not possible (147), although red light may, in fact, inhibit cyclic phosphorylation (235). Although the extent of this phenomenon has not been sufficiently explored, the work of Petrov and associates suggests that it may be widespread and probably quite important in the photosynthetic economy of plants.

III. Photorespiration and the C_2 Cycle

A. THE CONCEPT OF PHOTORESPIRATION

1. Definitions and Terminology

Early experiments established that respiratory gas exchange takes place in photosynthesizing tissue in light, but this was initially considered to be a continuation of dark respiration. Then experiments with carbon isotopes showed both quantitative and qualitative differences between the processes of respiration as it occurs in light and in darkness, and the light process was called photorespiration. However, as there are several different metabolic pathways that lead to gas-exchange phenomena associated with dark respiration, so there are several pathways that result in respiratory gas exchange in light. These fall into two categories that are distinguished in the subsequent discussion. Normal dark-respiratory processes continuing in light will be called *light respiration* (i.e., respiration in light). The reoxidation of carbon just previously fixed in photosynthesis, and the resulting gas exchange, will be called *photorespiration* (i.e., oxidative processes associated with photosynthetic metabolism). This distinction has been made in the literature (47), and it seems preferable to the terminology proposed by Jackson and Volk (266), who lumped all respiratory activity in light under the term photorespiration and used the terms *mitochondrial respiration* and *peroxisomal respiration* to designate classical dark respiration and glycolate metabolism, respectively. These terms may lead to confusion because glycolate metabolism also involves mitochondria, and some of the metabolism of classical dark respiration does not. It should be emphasized that the term respiration normally implies metabolism associated with the useful transfer of chemical potential, for example, to ATP synthesis. The term light respiration implies this. Photorespiration is quite a different matter. It may, but usually does not, result in useful release or storage of energy, and the term *photorespiration* should not be thought to imply a normal respiratory process; only the gas-exchange pattern is similar to respiration.

It is essential to distinguish between such phenomena as photosynthesis, respiration, or photorespiration as detected by gas exchange and the metabolism that gives rise to them. Precise terminology is difficult because the terms photosynthesis and respiration have historically been used both for the processes and for their manifestations. Fortunately, the usage is nearly always clear from the context. In general, unless modified, the terms photosynthesis, respiration, and photorespiration are used here to mean the gas exchange associated with metabolism. Specific references to metabolism itself are usually made by referring to the metabolic pathways in question (photosynthetic cycle or C_3 cycle, glycolate pathway or C_2 cycle, etc.).

Because the gas exchange of photorespiration and photosynthesis are opposite, distinctive terminology is required to separate the observed gas exchange from that which would result independently from either process. The terms most often used are *true* or *gross* photosynthesis for the gas exchange of photosynthesis alone, and *apparent* or *net* photosynthesis for the resultant of photosynthesis *and* photorespiration. The terms true and apparent have confusing connotations in this context. Gross photosynthesis is only "true" with respect to the gas exchange detected outside the leaf, not to intercellular or substomatal gas exchange or to the metabolism supporting it, whereas gross and net mean just that—the amounts or values that can be measured. Thus *gross* photosynthesis is used here to mean the measured amount of photosynthetic gas exchange corrected for the measured rate of gas exchange due to all respiratory processes (light respiration and photorespiration), whereas *net* photosynthesis means the measured rate of gas exchange uncorrected for respiration.

It must be emphasized that gross photosynthesis, photorespiration, and light respiration are difficult to measure because they are affected by the measuring system. For example, CO_2 released by respiration may escape to the outside of a leaf and be detected, or it may be recycled internally (or even in the boundary layer surrounding the leaf) and so escape detection (Fig. 5). Oxygen may behave the same way. Gas ex-

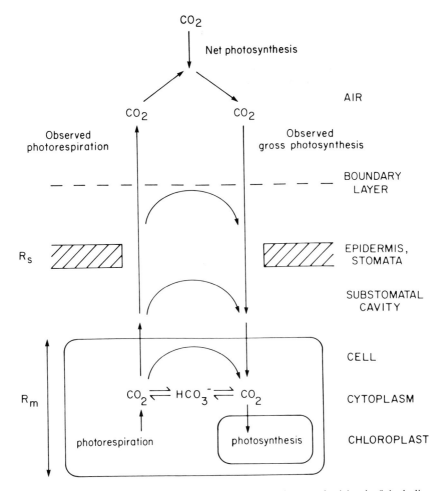

FIG. 5. Diagram showing diffusion path of CO_2 in a photosynthesizing leaf, including places at which recycling may occur. R_s and R_m are the so-called stomatal and mesophyll resistances, respectively. Each is a composite of several resistances; R_m includes state changes, solution, and chemistry of CO_2 fixation in addition to diffusion resistances.

change measurements of gross photosynthesis and photorespiration are reduced by the amount of recycling. Many parameters of the leaf environment affect the various resistances. For example, the characteristics of gas flow affect the boundary layer, and R_s varies with stomatal opening. R_m, which is a composite of many resistances, may be affected by the degree of cell wall hydration, the activity of RuBPcase as affected either

322   R. G. S. BIDWELL

by light intensity or by intrinsic factors, and the activity of carbonic anhydrase, to name only a few possibilities. The amount of recycling is affected by the flux of carbon, which is in turn affected by the intensity of metabolism, the concentration gradients of gas, and the resistances to gas exchange. Furthermore, various factors affect the exchange of O_2 and CO_2 in quite different ways; the two gases may behave differently because of the large differences in their concentrations and in physical factors such as their solubilities. Thus measured rates of gross photosynthesis and of photorespiration or light respiration must be considered in the context of the gas measured, the technique used to measure it, and the physiological as well as environmental factors surrounding the plant or tissue being measured.

As a consequence of these problems of measuring gas exchange (which have not always been clearly recognized), confusion has arisen in attempts to relate gas exchange and metabolism. For this reason much of the earlier work on photorespiration, while of the greatest importance in laying the ground work, is irrelevant to the present understanding of photorespiratory metabolism. Apart from a brief historical introduction, this discussion is restricted to the present concept of photorespiration and essential background material; earlier literature has been thoroughly and adequately reviewed by several authors, especially by Jackson and Volk (266) and Zelitch (628).

2. Early History of Photorespiration

Rabinowitch reviewed the early literature on photorespiration, a term coined for "a photochemical acceleration of normal respiration . . . a nightmare oppressing all who are concerned with the exact measurement of photosynthesis [468]," and Rabinowitch concluded that such a phenomenon was possible but unproven. Decker, in a series of experiments in which gas exchange was measured in leaves of various plants with an infrared gas analyzer, found high rates of CO_2 production from leaves immediately after darkening (called the *postillumination burst*), which soon decreased to the dark-respiratory rate (138). This was interpreted as the overshoot (in time) of rapid photorespiration. Experiments on the effects of temperature and CO_2 concentration on photosynthesis in *Mimulus* showed that the increase of respiration in light, caused by a rise in temperature, was 3.3× greater than that of dark respiration (140). Decker concluded that photorespiration was 3.3× greater than dark respiration. This was diametrically opposed to the work of Brown (77), who had shown, using isotopes of oxygen, that respiration did not increase in light, and the word photorespiration was editorially banned from the journal *Plant Physiology* in 1954 (141). However, Decker recognized a flaw in the oxygen work; photosynthesis decreases the internal concentration of isotopically labeled O_2, so respiration was underestimated by this technique (139, see also 266). By 1959 the term had struggled back into print (142), but it did not appear in *Plant Physiology* until 1966 (187).

The CO_2 burst after darkening was found by Krotkov and co-workers to be affected by O_2 concentration in a manner that differed from the effect of O_2 on dark respiration (187), and they suggested that photorespiration might be a process different from dark respiration. Many experiments in several laboratories linked the magnitude of the CO_2 burst with conditions affecting the rate of previous photosynthesis. Certain plants, like corn (*Zea mays*), have a CO_2-compensation point close to zero. The CO_2-compensation point (Γ) is defined as the CO_2 concentration at which photosynthetic CO_2 uptake exactly equals respiratory CO_2 output and at which net photosynthesis is zero. Evidently, because the compensation point of corn is zero, its rate of photorespiration is zero, and corn leaves have been found to have no postillumination burst of CO_2 (188). Furthermore, extrapolation of the rates of photosynthesis below the compensation point to zero CO_2 gave values of CO_2 output that parallel the magnitude and oxygen sensitivity of the CO_2 burst. This evidence strongly favors identification of the postillumination burst with photorespiration (628).

3. Measuring Photorespiration

a. Use of Isotopes. The problem of measuring the evolution of a gas occurring simultaneously with its uptake was solved by the use of isotopes of oxygen (433) or carbon (241). The stable isotope of oxygen, $^{18}O_2$, requires a mass spectrometer for detection and determination. The gas $^{14}CO_2$, being radioactive, is readily measured with Geiger-Muller tubes or with an ionization chamber. The theory of the technique, as applied to CO_2 exchange, is as follows: the leaf absorbs $^{14}CO_2$ and $^{12}CO_2$ at rates that relate directly to the proportional concentration of the two gases. However, prior to the experimental introduction of $^{14}CO_2$ the leaves are grown only in $^{12}CO_2$; they metabolize only $^{12}CO_2$, so they are made exclusively of ^{12}C. Consequently, any respiratory processes, because they draw on endogenous carbon, result in the liberation of $^{12}CO_2$ only (until photosynthetically fixed ^{14}C enters the metabolic pools). Thus, initially, the uptake of $^{14}CO_2$ measures gross photosynthesis, its rate being adjusted by the proportion of total CO_2 represented by $^{14}C_2$; the uptake of $^{12}CO_2$, similarly adjusted, represents net CO_2 uptake, or

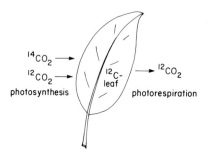

Gas-flow diagrams for measurement systems:

Open system (using differential infrared CO_2 analyzer):

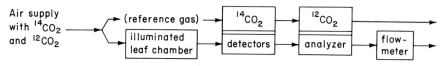

Closed system (using absolute infrared CO_2 analyzer):

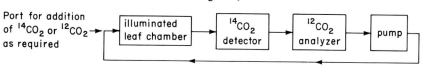

FIG. 6. The measurement of gross and net photosynthesis using $^{14}CO_2$ and $^{12}CO_2$. The leaf absorbs $^{14}CO_2$ and $^{12}CO_2$ during photosynthesis in proportion to their relative concentrations. For a short time after the introduction of $^{14}CO_2$, during which measurements are made, $^{12}CO_2$ is released by photorespiration. Rates of photosynthesis and photorespiration are calculated from the rates of $^{14}CO_2$ and $^{12}CO_2$ uptake as follows. Gross photosynthesis is equal to $^{14}CO_2$ uptake (estimated as units of radioactivity per unit time) times the proportional concentration of $^{14}CO_2$ (estimated as units of total CO_2 per unit radioactivity). Net photosynthesis is equal to $^{12}CO_2$ uptake times the proportional concentration of $^{12}CO_2$ (the value of which is so close to unity that no correction need be applied). Photorespiration is equal to gross photosynthesis less net photosynthesis.

the algebraic sum of CO_2 uptake (positive) and CO_2 output (negative). Photorespiration is measured by the difference between gross and net photosynthetic. These calculations are illustrated in Fig. 6.

The analytical system usually involves an infrared gas analyzer for $^{12}CO_2$ measurement. The gas $^{14}CO_2$ is not effectively detected by the gas analyzer although its concentration is usually very small in relation to that of $^{12}CO_2$, so no correction need be applied to measurements of $^{12}CO_2$. A differential analyzer, which measures the difference in CO_2

concentration between two gas streams, may be used to measure $^{12}CO_2$ uptake in the open system. The open system has the advantage of measuring photosynthesis under steady-state conditions, but it requires the use of premixed $^{12}CO_2$ and $^{14}CO_2$. The closed system can make more precise measurements because measurements are integrated over time, and $^{14}CO_2$ is added as required. However, the closed system cannot measure a steady state because the CO_2 concentration is continuously changing. It is useful for making measurements at or near the compensation concentration of CO_2 and for determining the compensation point.

Certain assumptions are inherent in using isotopes as a technique to measure photorespiration. The first is that $^{12}CO_2$ and $^{14}CO_2$ are taken up in direct proportion to their concentrations. Yemm and Bidwell, using corn leaves that had zero compensation and thus no CO_2 production in light, showed that relative uptake must be corrected for approximately 2% discrimination against $^{14}CO_2$ (624). Bikov (84; personal communication) claims discrimination of about 15%, as noted originally by Weigl and Calvin (595). This difference probably arises because Bikov apparently neglected to correct for the rapid production of $^{14}CO_2$ in photorespiration following saturation of the pools of intermediates with ^{14}C, whereas Weigl and Calvin supplied saturating levels of CO_2 over long intervals of time, permitting isotope discrimination at many chemical (enzymatic) levels during photosynthesis and photorespiration. Yemm and Bidwell used atmospheric CO_2 concentrations (as would normally be used in photosynthetic rate determinations), and their value of 2.1% discrimination against $^{14}CO_2$ agrees very well with the value of 2.2% calculated from Graham's law for diffusion of gases (624). Yemm and Bidwell concluded that at this CO_2 concentration the major resistance to CO_2 fixation is diffusion.

The second assumption is that no $^{14}CO_2$ is photorespired during the period of measurement. Krotkov showed that recently fixed carbon is a substrate of photorespiration by showing that the CO_2 burst after darkening was enriched with $^{14}CO_2$ following photosynthesis in $^{14}CO_2$ (319), and Ludwig et al. showed that the major substrate of photorespiration in light is indeed carbon that has just been fixed (363). This means that measurements must be made immediately after the introduction of $^{14}CO_2$ and before it has had time to pass through the metabolic pools of photosynthetic and photorespiratory intermediates. The time available for making measurements varies with the relative metabolic activity of the leaf and the size of pools of photosynthetic intermediates, and it is usually from 0.5 [sunflower (*Helianthus annuus*)] to 3–5 [bean (*Phaseolus vulgaris*)] min (544).

The third assumption is that $^{14}CO_2$ and $^{12}CO_2$ are continuously available to the CO_2-fixing enzyme in proportion to their measured concentrations in the gas stream. Because recycling undoubtedly occurs at many levels in the leaf and, with inadequate gas-flow rates, in the boundary layer, it is clear that this assumption cannot be correct. The result is that gross photosynthesis and photorespiration must be underestimated by dilution of incoming $^{14}CO_2$ by outgoing $^{12}CO_2$. Many attempts have been made to estimate the amount of such recycling using resistance models based on water-vapor loss of leaves, but none is satisfactory. Net photosynthesis can be an accurate measure of the rate of carbon accretion. Gross photosynthesis and hence photorespiration must inevitably be underestimated by an unknown and possibly large factor. The situation is alleviated somewhat by the extremely small size (or nonexistence) of the free exchange pool of CO_2 in leaves (275), so recycling does not involve complex equilibration phenomena.

The foregoing analysis dealt with CO_2; measurement of photosynthetic or photorespiratory O_2 exchange is subject to the same problems. The difference between mesophyll resistance to CO_2 uptake (which includes carboxylase activity) and O_2 release as well as differences in solubility and external concentrations of the gases may be partly responsible for differences in measurements of gross photosynthesis and photorespiration that have been made using isotopes of the two gases (e.g., 266). Care must be taken in relating the external gas exchange or photosynthesis and photorespiration to the fluxes of the underlying metabolism (444a).

b. Extrapolation or Measurement at Zero CO_2 Concentration. Net photosynthesis is an algebraic sum of gross photosynthesis and photorespiration; at the compensation point (where net photosynthesis is zero), the two processes are equal. In a closed system, photosynthesis continues at the expense of photorespired CO_2. A curve for photosynthesis as affected by CO_2 concentration can be constructed, and the curve may be extrapolated to zero CO_2 (Fig. 7). Provided that photorespiration is constant over all CO_2 concentrations, the point where the curve intercepts the vertical axis provides a measure of CO_2 output.

There has been much debate about the validity of extrapolation or measurement at zero CO_2 because the percentage of recycling might change as CO_2 concentration is changed, or both photosynthesis and photorespiration might change in a fixed ratio as CO_2 concentration is changed. The linearity of curve A (Fig. 7) requires that if one process change, the other must change proportionally. In fact, a nonlinear relationship is sometimes obtained, which throws further doubt on the validity of the technique. Another source of concern is that the substrate of

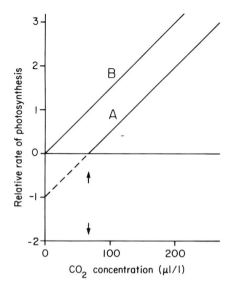

FIG. 7. Extrapolation of experimentally measured rate of net photosynthesis to zero CO_2 concentration (dotted line) to estimate photorespiration. Curve A represents data that might be obtained from a photorespiring leaf; the arrow indicates the CO_2 compensation point Γ. Curve B might be obtained from a leaf of a plant, such as maize, that does not photorespire.

photorespiration is recent photosynthate; if photosynthesis stops because of the absence of CO_2, it is likely that photorespiration will be affected. Another problem is that the substrate of photorespiration is formed from the oxygenase function of RuBPcase (see following discussion). Because CO_2 and O_2 are in competition for the active site of RuBPcase, reducing the CO_2 concentration will probably increase the proportion of carbon passing through the photorespiratory metabolic pathways.

c. Oxygen Effect on Photosynthesis. One of the early noted characteristics about photorespiration was its sensitivity to O_2 (e.g., 187). Whereas dark respiration saturates at low partial pressure of O_2 (~5%), photorespiration ceases at O_2 concentrations below 2% and does not saturate even at 100%, as shown in Fig. 8. Thus the difference between net photosynthesis in air (21% O_2) and in nitrogen containing the same CO_2 concentration has frequently been equated with photorespiration. Unfortunately, O_2 has many potential effects on photosynthesis (including reoxidation of members of the photosynthetic electron transport system) in addition to its competitive effect on carboxylation through the RuBP

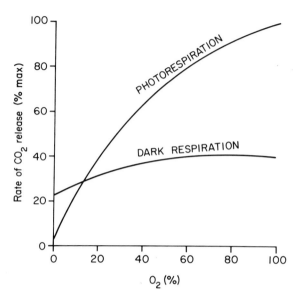

FIG. 8. Effect of O_2 on photorespiration and dark respiration of soybean leaves. Adapted from data of Forrester *et al.* (187).

oxygenase reaction. Consequently, measurements so made may overestimate the gas exchange of photorespiration.

d. Measurement of Postillumination Burst. This brief, rapid production of CO_2 from a photosynthesizing leaf after darkening led originally to the discovery of photorespiration (138). It was found to be affected by environmental factors in the same manner as photorespiration (187), and it is absent in plants lacking photorespiration (188). The postillumination burst is ephemeral and can be accurately measured only by a sensitive open-system CO_2 analyser. However, it has often been suggested that because it is a continuation into darkness of photorespiratory CO_2 release, its magnitude should be directly proportional to photorespiration in the preceding light period. In a carefully reasoned analysis, Doehlert *et al.* (152) demonstrated, using the mathematics for a nonsteady-state system, that the rates of the postillumination burst could be readily and accurately calculated from gas analyzer measurements. They argued effectively that the magnitude of the postillumination burst measured in this manner gives an accurate quantitative estimation of photorespiration over a wide range of environmental conditions. One decided advantage of this measurement is that it is not affected by internal recycling of CO_2. However, the possibility that the postillumination burst gives a low measure of photorespiration, because it is measured during its declining phase, cannot be discounted.

e. Photorespiration of Added $^{14}CO_2$. A more precise measurement of the metabolism of photorespiration was attempted by Zelitch (627), who labeled leaf discs with $^{14}CO_2$ and then measured $^{14}CO_2$ production in light and in darkness in a stream of CO_2-free air. This method requires the assumption that the specific radioactivity of respired CO_2 is the same in light and darkness, a condition that Ludwig and Canvin (362) found is not met, because substrates of light and dark respiration may be entirely different. Chollet (106) measured photorespiration by this and other techniques and found that the method of Zelitch invariably overestimates photorespiration by a large factor, (see Section III,E,2). The measurement of $^{14}CO_2$ production using added radioactive compounds supposed to be substrates of photorespiration (e.g., [^{14}C]glycine or [^{14}C]glycolate) suffers from this uncertainty as well as from the difficulty of establishing the degree of endogenous dilution of added tracer substrates.

f. Conclusion. The inescapable conclusion is that the metabolic processes leading to photorespiration are very difficult or impossible to determine accurately from measurements of photorespiratory gas exchange. The best that can be done is to measure the uptake of O_2 or the loss of carbon as CO_2 resulting from photorespiratory metabolism. This can be done most easily by measuring CO_2 release into a stream of CO_2-free air or by extrapolating the photosynthetic curve to zero CO_2. However, both measurements apply to low or zero CO_2, and it is inherently dangerous to extrapolate to higher levels of CO_2. Accurate measurements at physiological or higher CO_2 concentrations can be made using isotopes of carbon or oxygen, but measurements must be made within as few as 30–45 sec of introducing $^{14}CO_2$ in some species (241). The postillumination burst appears also to be a useful indication of the prior activity of photorespiration.

4. Substrates of Photorespiration

Tregunna *et al.* (572), using $^{14}CO_2$, noted that the photorespiratory CO_2 burst is derived from carbon recently fixed in photosynthesis. Zelitch discovered that α-hydroxypyridinemethanesulfonic acid, an inhibitor of glycolic acid oxidase, reduces photorespiration and correspondingly increases photosynthesis (626). Zelitch also proposed that oxidation of photosynthetically produced glycolate is the source of photorespiratory CO_2. Although it now appears that this inhibitor is much less specific in its action than was originally supposed, the conclusion derived from the work of Zelitch stands firm. Much experimental evidence, including comparison of the specific radioactivity of metabolic intermediates and of photorespired CO_2 (e.g., 1, 436), now leaves little doubt that the metabolism of glycolate derived from intermediates of the photosynthetic reduction cycle provides the main substrate of photorespiration. Some alternative or secondary pathways have been proposed, and questions have been raised regarding the source of glycolate and the pathway of its metabolism. These are discussed in Sections III,C and D.

B. Physiological Parameters
of Photorespiration

Before describing in detail the metabolism leading to or associated with photorespiration, it is necessary to consider briefly the characteristic attributes of photorespiration and its behavior under varying conditions and in different plants. Photorespiration will be discussed primarily in the context of CO_2 exchange, and the reactions of O_2 associated with inorganic compounds (e.g., nitrate reduction) will not be considered unless carbon metabolism is also involved.

1. Respiratory Quotient

Early measurements by Ozbun et al. suggested that the O_2 uptake of photorespiration greatly exceeds the output of CO_2 (433), but it now seems likely that the respiratory quotient (RQ = CO_2/O_2) of photorespiration is approximately 1.0. The measurements by Ozbun et al. were made with $^{13}CO_2$ and $^{18}O_2$; the exchange of $^{18}O_2$ is more or less unaffected by the duration of the experiment, but added isotopic carbon is rapidly converted to substrates of photorespiration and photorespired (as was quickly discovered when $^{14}CO_2$ began to be used). The result is that exposure to isotopic carbon for more than 30 sec–5 min (depending on the plant) will greatly underestimate CO_2 exchange.

In experiments by Fock et al. (184) it was found that the photosynthetic quotient is 1 at 2% O_2, but significantly higher (1.26) at 21% O_2. This suggests that greater proportions of oxidized carbon were produced in the presence of O_2 or that the photosynthetic quotient was significantly less than 1.0. But if this is so, then it is unlikely that the plants were in metabolic steady state because they must have been synthesizing large amounts of oxidized compounds (e.g., organic acids).

Alternatively, a systematic error in either O_2 or CO_2 exchange may have occurred. The latter explanation seems more plausible because O_2 consumption and CO_2 production exactly balance in the generally accepted reaction scheme of the C_2 cycle (see Section III,C,2). Kaplan and Björkman have confirmed a quotient of 1 for a C_3 and a C_4 species of higher plant (280a).

2. Factors Directly Affecting Photorespiration

a. Temperature. Photorespiration and the postillumination burst are directly affected by temperature. Decker (140) calculated that a 10° rise in temperature more than triples photorespiration, and Zelitch (628) indicated that Q_{10} values between 2 and 3 are probably representative.

b. O_2. Because the K_{mCO_2} of RuBPcase is quite high, photorespiration is strongly affected by O_2 concentration. No photorespiration occurs in the absence of O_2 (573), whereas photorespiration may increase threefold between air (21% O_2) and 100% O_2 (187). Because dark respiration is virtually unaffected between about 5–50% O_2 (187), this provides a useful diagnostic characteristic.

c. Light. The postillumination burst is affected by the intensity (571, 572) and quality (520) of previous illumination in the same way as photosynthesis. A strong inference is that photorespiration itself is affected in the same way, and various direct measurements of photorespiration confirm this.

d. CO_2. The effect of CO_2 on photorespiration is a result of several factors, including O_2 concentration, rate of photosynthesis, and photosynthetic capacity of the tissue. Holmgren and Jarvis (249) found that photorespiration decreases with decreasing CO_2 content between 40 and 5 µl/l, which suggests that photorespiration may be limited by substrate availability. On the other hand, it is widely held that photorespiration is reduced if the CO_2 concentration is sufficiently high to competitively inhibit RuBP oxygenase. It has been reported that photorespiration is virtually eliminated at CO_2 concentrations over 1000 µl/l (163).

Reports by Bravdo and Canvin (75) and Fock *et al.* (185) suggest that photorespiration in sunflower (*Helianthus annuus*) leaves, as measured by the $^{14}CO_2$–$^{12}CO_2$ technique, is not systematically affected by CO_2 concentration or the solubility ratio of O_2 to CO_2. Doehlert *et al.* (152), measuring photorespiration in wheat (*Triticum aestivum*) leaves by the size of the postillumination burst, showed that photorespiration decreases with increasing CO_2 concentration and is directly related to the solubility ratio of O_2 to CO_2. All workers agree that the ratio of photorespiration to photosynthesis decreases greatly with increasing concentration of CO_2 or ratio of CO_2 to O_2. The present lack of agreement on the actual rate of photorespiration probably reflects not only the complexity of photorespiratory metabolism but also the difficulty of measuring the resultant gas exchange and assessing the magnitude and effect of quantities such as internal recycling, stomatal apperture, and actual internal CO_2 concentration in the leaf (see discussion in 444a).

3. The CO_2 Compensation Concentration

Photosynthesis varies with CO_2 concentration and therefore approaches zero at zero CO_2 concentration. On the other hand, the rate of CO_2 evolution in light is not strongly affected by CO_2 concentration. Therefore, if a plant is placed in a closed container, CO_2 concentration will fall (or rise if the air in the container was initially CO_2-free) to a level at which its uptake in photosynthesis exactly balances its production. This level is called the CO_2 compensation concentration (or compensation point) and is frequently designated by Γ (see Fig. 7). Other factors being equal, Γ is proportional to the rate of CO_2 production in light, but it is not useful for measuring the rate of photorespiration because the rate of photosynthesis at low CO_2 concentrations varies considerably from plant to plant. However, plants that do not release CO_2 in light, primarily plants having the C_4 cycle of photosynthesis, are sharply distinguished because their Γ is very low or zero. Whether such plants truly lack photorespiration or whether it is merely masked by more efficient CO_2 uptake is considered in Section VI.

The value of Γ is very similar in those plants that produce CO_2 in light, being in the range 35–60 $\mu l/l$ CO_2 under normal conditions. However, Γ increases in approximately linear fashion with temperature (576) and may double with increase from 20 to 30°C (164, 274). This is because the Q_{10} of photorespiration, which may reach 3.0 or more (140), is substantially above that of photosynthesis, which is usually considered to be about 2.0 (628). Because O_2 is a required substrate for photorespiration but not for photosynthesis, Γ varies with O_2, becoming zero at 2% O_2 or lower and increasing at very high O_2 levels due to both increased photorespiration and O_2 inhibition of photosynthesis.

Any factor that affects photosynthesis also affects Γ. Thus drought strongly increases Γ through its differential effect on photosynthesis and stomatal resistance (576). Nitrogen nutrition also affects Γ. The supply of nitrate depresses Γ, whereas the supply of ammonia may raise it substantially, to the extent that even C_4 plants may show Γ of 10–25 $\mu l/l$ CO_2 (211). The reason for this appears to be that activities of enzymes associated with the metabolism of photorespiration are substantially increased by the supply of ammonia as compared to nitrate (175).

The effect of light on Γ is complex because the dependence of photorespiration on light is not proportional to that of photosynthesis on light and because, particularly at low light intensity, light respiration may vary inversely with light intensity. As would be expected, Γ is quite constant above photosynthetically saturating light intensities. It increases strongly as light intensity decreases until it becomes unmeasurably great at zero light with the cessation of photosynthesis (264). The increase with decreasing light is the result of decreasing photosynthesis, decreasing photorespiration, and increasing light respiration (dark respiration in light). As a consequence, it is not a particularly useful parameter in the study of any of these processes.

4. Magnitude of Photorespiration

As we have seen, it is extremely difficult to estimate the magnitude of photorespiration from measurement of its gas exchange, and widely diverging values have been published. Among the highest are those of Fock *et al.*, who showed that increasing the O_2 level from 2 to 50% decreased net photosynthesis of several species by 50% (186). They concluded that 80–90% of the decrease was due to photorespiration and that photorespiration could equal the rate of net photosynthesis, or one-half the rate of gross photosynthesis. However, Ludwig and Canvin (362), using a sensitive short-duration $^{14}CO_2$–$^{12}CO_2$ technique, showed that photorespiration in sunflower leaves is 5.7 mg $CO_2/dm^2/hr$, or about 20% that of net photosynthesis. They found that inhibition of gross photosynthesis by 21% O_2 is twice the stimulation of photorespiration, which casts doubt on the much higher values of Fock *et al.* Numerous subsequent measurements support the conclusion that under normal conditions of CO_2, light, and O_2, the rate of photorespiration is approximately 15–20% of the rate of gross photosynthesis. Thus net photosynthesis is normally reduced by approximately 20% as the result of photorespiration. However, if the rate of CO_2 recycling is considered, the true rate of photorespiratory metabolism may be somewhat higher, perhaps as much as 30% that of light-saturated photosynthesis (510).

Ludwig and Canvin commented on the stability of photorespiration and noted that it is not affected by changes in CO_2 concentrations that strongly affect photosynthesis (362). Various reports have suggested that the rate of photorespiration is not directly correlated with photosynthesis or with environmental factors that affect CO_2 uptake. Several studies have suggested that photorespiration in a variety of plants increases (303, 508) or decreases (174) with leaf age. The discrepancies may be related to the possibility that photorespiratory metabolism is linked in some way with other metabolic events that depend upon the overall activity of the plant. Bidwell et al. (48) found that applied hormones or breaking of bud dormancy stimulates photorespiration, and Steward et al. (544) showed that photoperiod and events during the development of various plants strongly affect photorespiration as well as the ratio of photorespiration to photosynthesis. Fraser and Bidwell (191) showed that photorespiration in individual bean leaves is related to ontogeny of the whole plant as well as to the leaf in question. Photorespiration rises sharply when new leaves or flower buds begin rapid growth. These ephemeral events are superimposed on a weak overall correlation of photorespiration with age; photorespiration increases slightly during leaf expansion, is stable during leaf maturity, and decreases again during senescence.

Correlation of photorespiration with internal and stomatal resistances of leaves has not been well studied. Data of Samish et al. (510) suggest correlation between leaf resistances and photorespiration. Fraser and Bidwell used measurements of relative humidity to calculate resistances of bean leaves and a sensitive (short exposure) $^{14}CO_2$–$^{12}CO_2$ technique for measuring photorespiration (191). They showed that photorespiration does not correlate with either stomatal or mesophyll resistances of bean leaves over a normal range of external conditions or during development of either individual leaves or the whole plant. That R_m is largely a resistance to photosynthesis and that stomatal aperture (R_s) would have little effect on CO_2 escape (because the rate of CO_2 production in the leaf is unaffected by CO_2 concentration) probably explains this lack of correlation. Thus if R_s increased, the substomatal concentration of CO_2 would also increase, counteracting the tendency of stomatal closure to decrease CO_2 outward diffusion.

5. The Effect of O_2

Early experiments demonstrated that the photorespiratory burst is affected by O_2 (187, 573). Oxygen has long been known to affect photosynthesis (the Warburg Effect), and it was a natural conclusion that this effect is due, at least in part, to an O_2 effect on photorespiration. The demonstration that glycolic acid oxidase is an important component of photorespiration (626) provided a partial explanation for the role of O_2 in photorespiration. However, an oxidation step is also needed to make glycolic acid, and it was not until the discovery of the oxygenase function of RuBPcase (69) that the nature of this step began to be understood.

Although there was considerable controversy during the development of our present understanding (and alternative mechanisms are still under discussion; see Section III,D), it is now widely agreed that oxidation of RuBP to produce phosphoglycolate (shown in Fig. 4) is the ultimate source of the substrate of photorespiration and the site of the major component of the Warburg Effect.

The O_2 effect has been widely used to separate photorespiration from dark respiration and light respiration because of the great difference in sensitivity of the two processes to O_2 concentration, as shown in Fig. 8. A problem with this approach (as with the use of the O_2 effect on photosynthesis as a measure of photorespiration) is that O_2 has several possible effects on photosynthesis as well as on photorespiration (320), and it is impossible to distinguish these effects merely by measuring gas exchange.

6. Taxonomy of Photorespiration

a. C_4 *Plants.* It was noted in 1962 that some plants, particularly maize (*Zea mays*), have Γ values in the range 0–5 μl/l CO_2 (375, 389), and Tregunna *et al.* in 1964 related this to the absence of photorespiration (572). By 1967, Downton and Tregunna (159) were able to relate low Γ to the photosynthetic and anatomical characteristics of C_4 plants and to the systematics of Gramineae. Since then several genera have been found with species that have both high Γ (>40 μl/l) and low Γ (<10 μl/l). Low-Γ C_4 species have been found in several dicotyledonous families as well as among the monocots (315). The question of photorespiration among C_4 plants is considered in Section IV,C.

b. C_3 *Plants.* Several authors have sought photorespiratory variations that are not associated with C_4 photosynthesis but that are inherent in the relationship of photorespiration with the C_3 photosynthetic mechanism. This has been done because it has frequently been asserted that substantial improvement in crop productivity should be possible by eliminating photorespiration or by selecting varieties that have low photorespiration (e.g., 628). Zelitch and Day (634) presented results suggesting that varieties of tobacco (*Nicotiana tabacum*) selected for diminished photorespiration would result in large increases in net photosynthesis and plant productivity, but this hopeful prognosis does not seem to have been realized. Wilson (606) found some variation in photorespiration within a population of plants of a *Lolium* hybrid, but careful analysis of 114 maize genotypes by Moss *et al.* (390) and 107 sunflower varieties or selections by Lloyd and Canvin (347) failed to reveal any significant variation that might form a useful basis for a breeding program. Many thousands of C_3 plants have been screened by an imaginative technique proposed by Widholm and Ogren (598) in which plants are tested for their capacity to survive in an illuminated, closed chamber in which C_4 plants (such as maize) are growing. Only plants with much-reduced photorespiration and low Γ can survive. So far, none have been found. A few species, such as *Panicum milioides,* appear to be intermediate between C_3 and C_4 and have somewhat-reduced photorespiration. However, this appears to be due to a weakly active C_4 cycle (478); see Section IV,C,4.

  c. Aquatic Plants and Algae.   Controversy still exists about photorespiration (or its manifestation in gas exchange) in aquatic plants and algae. Green algae have been found to have a glycolate pathway (see Section III,C) that differs from that of higher plants only in having NADH-linked glycolate dehydrogenase instead of glycolic acid oxidase (83). Peroxisomes have been found in several algae, including *Chlorella* (113). However, it has been widely noted that algae excrete glycolic acid, suggesting that glycolate metabolism leading to photorespiratory CO_2 production is weak or absent. Glidewell and Raven (200) noted the O_2 exchange associated with glycolate synthesis. However, Bidwell (47) and Lloyd *et al.* (348) in parallel studies were unable to demonstrate photorespiration in *Chlorella* or other unicellular algae. Similarly, although several indirect measurements suggest CO_2 exchange and glycolate metabolism exist in marine algae (567), Bidwell *et al.* were unable to detect photorespiration in some 30 species of marine algae (48a). These experiments were all done with cells or tissues suspended in moist air, which avoids the difficulty of measuring CO_2 exchange in the presence of a large volume of HCO_3^--containing medium. Extensive tests have shown that neither microalgae (47) nor seaweeds (48a) are affected by this treatment, which permits precise and sensitive measurement of CO_2 exchange using $^{14}CO_2$–$^{12}CO_2$ techniques.

  Although photorespiratory CO_2 exchange does not appear to occur in algae or seaweeds, many organisms, including *Chlorella* (47) and red, brown, and green marine algae exhibit O_2-insensitive light respiration under appropriate culture conditions. Furthermore, brown algae have been found to exhibit wound respiration that mimics photorespiration in being O_2 sensitive but that is not derived from recent photosynthate (48a). It appears probable that if photorespiratory CO_2 is produced in aquatic plants, it must be produced within the cells and largely or entirely reassimilated before it can leave them. Wound respiration and light respiration, when they occur, presumably result in CO_2 liberation at or near the cell surface, permitting the escape of CO_2. Some claims have been made that marine algae are C_4 plants, but extensive experiments by Kremer and Küppers (314) and Bidwell *et al.* (48a) indicate clearly that these claims are not true.

C. THE GLYCOLATE PATHWAY

1. The C_2 Cycle

The evidence leading to the present formulation of the glycolate pathway or C_2 cycle of photosynthesis, shown in Fig. 9, has been effectively reviewed by Tolbert (564, 568) and need not be detailed here. This scheme rests on the earlier recognition of the metabolic implications of the interconversion of serine and glycine by Rabson *et al.* (469) and Wang and Waygood (594). Zelitch (626) and Goldsworthy (201) initially recognized the possibility that glycolate formed in photosynthesis might be a photorespiratory substrate. Tolbert *et al.* (565) demonstrated the involvement of peroxisomes and Kisaki *et al.* (304) showed the involvement of mitochondria. Finally, Bowes *et al.* (69) discovered the reaction of RuBP with O_2 which appears to have resolved the problem of the source of photosynthetic glycolate.

The scheme shown in Fig. 9 coordinates the several components of the C_2 cycle and its complementary metabolic pathways, and it will be discussed in more detail later. In addition to the carbon pathway, the cycles of oxidation and reduction reactions, the physical location of the various reactions within the cell, and the associated nitrogen cycles will also be considered. In addition, questions about the source of glycolate and the quantity and direction of carbon flux through the cycle will be discussed.

The C_2 cycle itself starts with the production of phosphoglycolate in chloroplasts and continues with the oxidation of glycolate and the formation of glycine in peroxisomes. Two molecules of glycine are converted in mitochondria to serine and photorespired CO_2, and serine is converted to glyceric acid in peroxisomes. Glyceric acid may be a substrate for cytoplasmic sugar synthesis (see Section III,C,8) or, completing the C_2 cycle, it may reenter chloroplasts and be reassimilated into the C_3 cycle as PGA. The net effect of this cycle is a scavenging reaction that converts three-quarters of the carbon lost as glycolate back into PGA; the other quarter is liberated as photorespired CO_2. Alternatively, the cycle could supply various intermediates (glycine, serine, glyceric acid) that might be needed for cellular syntheses.

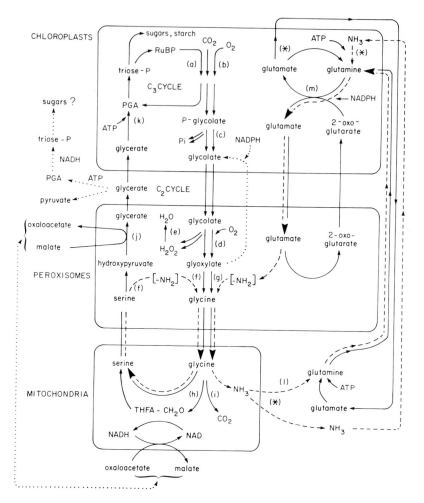

FIG. 9. The C_2 cycle of glycolate metabolism. The path of nitrogen is shown by dashed arrows; possible ancillary carbon pathways are shown by dotted arrows. Although all arrows are shown in a forward direction, all of the reactions, except those converting RuBP via glycolate to glycine (reactions b–g), are reversible. That two molecules of glycolate are needed to make one of serine is indicated by the double arrows. NH_3 released from glycine decarboxylation in mitochondria may react with glutamate to make glutamine (reaction l), either in the cytoplasm or in the chloroplast, or both; the alternative shuttles are marked (*). Reactions: (a), RuBPcase; (b), RuBP oxygenase (same enzyme); (c), P-glycolate phosphatase; (d), glycolic acid oxidase; (e), catalase; (f), serine–glyoxylate aminotransferase; (g), glutamate–glyoxylate aminotransferase; (h), hydroxymethyltransferase; (i), glycine decarboxylase; (j), hydroxypyruvate reductase; (k), glycerate kinase; (l), glutamine synthetase; (m), glutamate synthetase (GOGAT).

2. Oxidation and Reduction Reactions

Other considerations are the pattern and cofactors of oxidation and reduction reactions of the cycle. There are several distinct sites of reduction and oxidation, and their separation requires subsidiary cycles or transfers of redox potential. The cycle shown in Fig. 9 requires the reduction of a cofactor (presumably NAD) in the mitochondrial hydroxymethyltransferase (Fig. 9, reaction h), and the peroxisomal oxidation of NAD in the conversion of hydroxypyruvate to glyceric acid (Fig. 9, reaction j). The peroxisomal reduction is coupled with the oxidation of malate to oxaloacetic acid. This may possibly be converted to aspartic acid, which in turn could be reconverted to oxaloacetate and reduced to malate in chloroplasts, presumably by light-generated NADPH. The net effect would be a transfer of one molecule of photosynthetic reducing equivalent to the peroxisomes for each turn of the C_2 cycle. This set of reactions is not shown in Fig. 9. A more likely alternative is that the oxaloacetate produced in the peroxisomal reduction of hydroxypyruvate is transferred to mitochondria, where it is reduced to malate by the reducing power generated in the hydroxymethyltransferase reaction. The malate would then return to peroxisomes for the reduction of hydroxypyruvate. Some support for this shuttle has been presented by Woo (617).

The oxidation of glycolate produces H_2O_2, which is converted to water and O_2 by catalase, an invariable and prominent constituent of peroxisomes. Some of the glyoxylate so reduced could be returned to the chloroplasts where it might be reduced to glycolate at the expense of photosynthetic NADPH. The required reductase is not present in large amounts of chloroplasts (568). However, these reactions do provide a possible shuttle that would result in light-generated reducing power being degraded by glycolic acid oxidase in peroxisomes. Such "wasteful" reactions may be useful in preventing damage to the photosynthetic machinery when CO_2 is in short supply.

It is sometimes suggested that mitochondrial NADH produced in the hydroxymethyl transferase reaction (Fig. 9, reaction h) may be reduced by the cytochrome electron transport chain producing ATP (e.g., 294, 515). However, Woo (617) points out that the respiratory electron transport chain may be inhibited or shut down in light. Mitochondrial ATP is not required in light; it is more rational to suppose that NADH is used to reoxidize oxaloacetate produced in the peroxisomal reduction of hydroxypyruvate.

The balance of O_2 uptake and CO_2 production may be derived from Fig. 9. In each turn of the C_2 cycle, two molecules of O_2 are absorbed for the oxidation of two molecules of glycolate, and one is produced by the catalase reaction. The O_2 absorbed by the oxygenase reaction is balanced by O_2 produced in the light reaction required to generate RuBP. Because one molecule of CO_2 is produced in each turn of the cycle, the theoretical balance is one molecule of O_2 absorbed per CO_2 produced, giving an expected RQ of 1.0. This expectation has not always been realized (see Section III,B,1), but the discrepancy is as likely to be due to difficulties in measuring photorespiration as to anomalies in the C_2 cycle or light reactions.

3. Reaction Sites

One of the key points about the C_2 cycle is the degree to which it involves the metabolism of the whole cell. The grouping of the reactions in the peroxisomes would seem to satisfy a requirement to isolate the potentially dangerous H_2O_2-producing glycolic acid oxidase together with the catalase necessary to neutralize this poisonous by-product. It seems probable that some of the accessory metabolism may have resulted from the evolutionary trend that allowed the development of peroxisomes from glyoxysomes (which are concerned with fat metabolism in seedlings) or other precursor particles (178). The involvement of mitochondria is essential to the cyclic regeneration of glyceric acid or PGA from glycolate because they house the required C_1 metabolism. Mitochondria may also be the ultimate source of reducing power needed to reduce hydroxypyruvate to glyceric acid.

Plant peroxisomes were reported by Mollenhauer et al. (383) in 1966 and demonstrated in leaves by electron microscopy in 1967 by Frederick and Newcomb (193). They were shown by Tolbert et al. in 1968 (565) to contain the enzymes of glycolate metabolism. By 1971 the overall scheme involving peroxisomes, mitochondria, and chloroplasts had been developed (304). It is now widely accepted that this interorganelle carbon traffic is primarily responsible for photorespiration. The rather large fluxes of carbon and nitrogen required to maintain photorespiration (see Section III,C,4) have raised questions about the degree of metabolic "traffic control" required. The compartmentation of the various reaction sequences also implies rather direct channels of access between the organelles. It has been speculated that loose associations of organelles occur in intact cells, bound or held sufficiently close together to permit the ready and rapid movement of all the required metabolites to maintain the integrated operation of the various carbon and nitrogen cycles. The arguments for this idea were presented earlier (Section II,E,3); however, little direct evidence supports them.

4. Formation of Glycolic Acid

Controversy has surrounded the source of photosynthetic glycolic acid since 1958. Tolbert (563) suggested that it might be derived from RuBP, which was further supported by Pritchard et al. (461) in experiments on the excretion of photosynthetic glycolate by *Chlorella*. Bassham and Kirk (28) noted that phosphoglycolate is formed first, but that at least one-half of the glycolate formed when *Chlorella* cells are gassed with 100% O_2 apparently comes from other sources. A specific mechanism for the formation of glycolate from RuBP by the RuBP oxygenase reaction (Fig. 4) was offered by Ogren and Bowes (416) and Bowes et al. (69), and convincing evidence that this is a major pathway of glycolate production in photosynthesizing leaves has been offered by Tolbert's group (11, 355).

Alternative pathways have been proposed. Zelitch (625) found that the specific radioactivity of photosynthetic glycolate in tobacco leaf discs was higher than that of PGA or phosphoglycolate, and Zelitch proposed a direct carboxylation or condensation of two molecules of CO_2 to explain these results. However, no enzymatic confirmation of this proposed pathway has been advanced, and it seems likely that compartmentation of intermediary metabolites may have caused these results. Zelitch has also proposed alternative pathways for glycolate production in maize on the basis of ^{14}C-labeling patterns (630). It appears that the majority of glycolate formed in these low-photorespiratory leaves is derived from organic metabolites via pyruvic acid.

A second alternative has been developed by Shain and Gibbs using isolated spinach (*Spinacia oleracea*) chloroplasts (527). They proposed that the glycolaldehyde transketolase addition product (dihydroxyethyl-thiamine pyrophosphate or "active glycolaldehyde") derived from fructose 6-phosphate is oxidized by a photosystem-II oxidant or by H_2O_2 produced by the reoxidation of NADPH by ferredoxin in chloroplasts. Gibbs et al. showed that added intermediates of the C_3 cycle would eliminate the O_2 inhibition of photosynthesis (the Warburg Effect) without affecting glycolate formation (499). However, further investigations with limiting or zero CO_2 and ^{14}C-labeled substrates failed to produce data that showed unequivocally which intermediate is the source of glycolate or by what pathway it is formed (309, 500). Furthermore, the mechanism of the oxidation reaction (whether enzymatic or not) has not been elucidated. Recent evidence shows that H_2O_2 or free oxygen radicals inhibit rather than stimulate glycolate synthesis (302). The oxidants that best mediate the oxidation of the transketolase C_2 addition compound (ferrycianide or dichlorophenolindophenol) do not result in the incorporation of labeled oxygen into glycolate, as occurs *in vivo*. Finally, it should be noted that this pathway produces glycolic acid. Phosphoglycolate, which is the product of RuBP oxygenase, is widely considered to be the precursor of glycolic acid in chloroplasts.

Two different lines of evidence have appeared that seem to resolve this issue. Lorimer *et al.* (358) quantitatively estimated the enrichment of ^{18}O in glycolate produced by intact chloroplasts and found it to be in excess of 80% of that of the supplied $^{18}O_2$, regardless of carbon source (CO_2 or photosynthetic intermediates) and O_2 concentration. This indicates that at least 80% or more (allowing for dilution of the $^{18}O_2$ supplied) must have come from a reaction inserting molecular oxygen into the carboxyl group of glycolate. RuBP oxygenase is the only well-authenticated reaction that does this (see Fig. 4).

A second approach, by Kirk and Heber (302) and Krause *et al.* (311), is based upon the need for ATP for RuBP synthesis, whereas substrates of the transketolase reaction do not need ATP. Glycolate synthesis has been found to be sensitive to uncouplers of photosynthetic phosphorylation (302) even when the activity of enzymes of the photosynthetic cycle is unaffected (311). Both of these investigations show that glycolate is produced by ATP-requiring reactions (presumably via RuBP) at rates sufficient to satisfy the demand for photorespiration.

There is still room for the possibility that yet other pathways may provide glycolate (see Section III,D). However, no alternative mechanism has been sufficiently characterized to be strongly considered. Direct carboxylation and oxidation of the transketolase reaction are not consistent with much of the recent evidence. Direct carboxylation and oxidation are much less well authenticated and are not susceptible to quantitative estimation. The evidence supporting the RuBP oxygenase reaction as the major source of photosynthetic glycolate, though it falls short of absolute proof, is very strong.

5. Carbon Flux in the Glycolate Pathway

The uncertainties in the pathway of glycolate metabolism and its overall relationship to photorespiration have led several research groups to investigate the flux of carbon through the pathway in order to determine if a relationship exists with the rate of photorespiration. In addition, the specific radioactivity of glycolate-path intermediates has been compared with that of $^{14}CO_2$ supplied to leaves or produced in photorespiration. Such measurements should show the proportion of photorespired CO_2 derived from the glycolate path or from other reaction sequences that result in the oxidation of recent photosynthate.

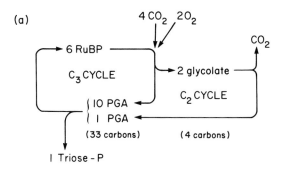

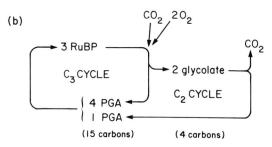

FIG. 10. Flux of carbon through C_2 and C_3 cycles during photosynthesis. (a) Gross photosynthesis is four times photorespiration. (b) Gross photosynthesis equals photorespiration (compensation point).

The formulation in Fig. 9 requires a rather precise relationship between the metabolic rates of the C_2 and C_3 cycles. If the rate of photorespiratory CO_2 production is 20–25% of the rate of gross photosynthesis, then the rate of turnover of the C_2 cycle must be 80–100% of the rate of the C_3 cycle. This sounds large, but if it is expressed in terms of the relative flux of total carbon through the C_2 and C_3 cycles, a more realistic picture emerges. As shown in Fig. 10a, 4 carbons enter the C_2 cycle for every one released as CO_2, whereas 33 carbons pass through the C_3 cycle for every one fixed. Thus the relative carbon flux through the C_2 cycle is about one-eighth that flowing through the C_3 cycle during normal photosynthesis. However, at compensation point this ratio rises to 4/15, or about one-quarter (Fig. 10b). Data of Lorimer *et al.* (359), who used $^{18}O_2$, support this balance of stoichiometry of the two cycles at compensation point (shown in Fig. 10b). The question then is, Can sufficient carbon flux through the C_2 cycle be demonstrated to satisfy the requirement just set forth?

TABLE II

RATES OF CO_2 EXCHANGE AND CALCULATIONS OF THE FLUX OF CARBON THROUGH THE C_2
CYCLE IN SUNFLOWER (*Helianthus annuus*) LEAVES[a]

Property	μmol C/dm^2/min
Gross photosynthesis	9.70
Net photosynthesis	8.00
Photorespiration (by difference)	1.70
Carbon flux from glycine entering serine	6.02
Carbon flux from glycine to CO_2 ($\frac{1}{3}$ × previous figure)	2.01
Total carbon flux through C_2 cycle (through glycine by sum of two previous figures)	8.03
Total carbon flux into PGA from sources other than CO_2	7.22
Total carbon flux through C_2 cycle (sum of previous figure and photorespiration)	8.92

[a] Photosynthesizing in 400 μl/l CO_2 in air, 2800 ft-candle. Data from Fock *et al.* (183).

A number of earlier measurements are in doubt because of difficulty in measuring rates of photorespiration or specific radioactivity of intermediates. However, Fock *et al.* (183) used a combination of kinetic analysis of carbon flux and careful determination of specific radioactivities of intermediates to develop precise measurements of carbon fluxes in sunflower leaves. A summary of their data is presented in Table II. Photorespiration was found to be about 20% that of gross photosynthesis. Because each carbon released by CO_2 in photorespiration requires four carbons as glycolate to enter the C_2 cycle, the total flux of carbon through the C_2 cycle should be about 80% of the rate of gross photosynthesis. Fock *et al.* found this to be true either whèn measuring the flux of carbon (other than as CO_2) into PGA or when summing the measured flux of carbon from glycolate to CO_2 and from glycine to serine. The agreement with the required figures, if all photorespiratory CO_2 is derived from operation of the C_2 cycle, is remarkable.

More recently, Kumarasinghe *et al.* (323) used a $^{14}CO_2$–$^{12}CO_2$ pulse-chase technique to follow the flush kinetics of C_2 intermediates in wheat (*Triticum*), and Berry *et al.* (40) followed the kinetics of incorporation of $^{18}O_2$ into these compounds in leaves of spinach, sunflower, and *Atriplex hastata*. Both groups derived data that agree very well with the data of Fock *et al.* (183) and the formulation of the C_2 cycle shown in Fig. 9. Some concern has been expressed by Heber *et al.* (236), who could not detect sufficient levels of glycerate kinase to accommodate the required conversion of glycerate to PGA in chloroplasts. However, photosynthetic enzymes have been notoriously hard to assay accurately in the past, and cytoplasmic glycerate kinase may also be involved.

Canvin *et al.* (89) have published more data supporting their earlier measurements (183) for sunflower that suggests that the flux of carbon through the C_2 cycle is much reduced in maize, a C_4 plant, and in C_3 plants held under 1% O_2. Their observations are consistent with the facts that neither maize nor C_3 plants exhibit photorespiration at low O_2. It seems safe to conclude that the C_2 cycle, as presently formulated, adequately describes the intermediary metabolism leading to photorespiration.

6. Nitrogen Cycles Associated with the C₂ Cycle

Prominent features of Fig. 9 are the nitrogen cycles associated with glycolate path metabolism. Twice as many amino groups are required in peroxisomes for the synthesis of glycine as are available from the conversion of serine to hydroxypyruvate, whereas an amount equal to that needed is released during the mitochondrial decarboxylation of glycine. Ammonia released during glycine decarboxylation needs to be scavenged for glycine synthesis to prevent a serious shortage of amino groups in peroxisomes and to prevent toxic quantities of ammonia from building up in the cytoplasm. It has been observed by Keyes *et al.* (294) that the flux of amino nitrogen through the photorespiratory cycle must be of an order of magnitude greater than that resulting from primary nitrogen assimilation.

There seems to be no question about the location and operation of serine–glyoxylate aminotransferase (Fig. 9, reaction f) in peroxisomes (493). However, the fate of ammonia released from the decarboxylation of glycine (Fig. 9, reaction i) is unclear. Two possibilities exist; the alternatives are shown in Fig. 9 by (*). Woo *et al.* (619) have suggested that ammonia moves to chloroplasts, where it is refixed by glutamine synthetase. This conclusion is based on the capability of illuminated spinach chloroplasts to fix $^{15}NH_4Cl$ into glutamine and glutamic acid and the inability of mitochondria to perform this reaction. Keys *et al.* (294) have suggested that because there is an active cytoplasmic glutamine synthetase present, the released ammonia encounters it and is converted to glutamine before reaching chloroplasts. However, they do not rule out the possibility of chloroplast refixation of ammonia. If, indeed, as has been suggested, peroxisomes are closely associated with chloroplasts, the diffusion path may not be so great. Furthermore, light-generated ATP would be available for the glutamine synthetase reaction, which might be more advantageous to the cell than ATP generated or released in the cytoplasm.

The nitrogen cycles are shown in Fig. 9 as though they were "tight" (i.e., as though no exchange of intermediates occurs between the cycle and cellular pools). This is unlikely to be true. Although compartmentation or close association of organelles may reduce leakage, some may occur (particularly in compounds such as ammonia). Presumably, ammonia lost from the nitrogen cycle of photorespiration is replaced by nitrate reduction in chloroplasts and fixation of ammonia by glutamine synthetase. This may be a primary function of chloroplast glutamine synthetase.

The nitrogen cycles of photorespiration are clearly a most important part of photosynthesis. Therefore, safeguards undoubtedly exist to prevent the loss of nitrogen from the cycle and to prevent the accidental poisoning of cells by excess ammonia. Doubtless when all the carbon and nitrogen reactions associated with photorespiration and photosynthesis are known, the picture will be even more complex than that shown in Fig. 9. Inevitably, however, all reaction sequences and cycles will be in perfect and dynamic balance.

7. Glycolate Pathway in Algae

It seems to be well established that algae have a glycolate pathway (83), but there has been much controversy about its nature, extent, and distribution. Many algae lack glycolic acid oxidase and have glycolic acid dehydrogenase instead (405). Glycolic acid dehydrogenase reacts directly with O_2, but its natural electron acceptor is unknown. Beezley et al., using electron microhistochemistry, showed that it is located in mitochondria in Chlamydomonas and green algae generally (33), unlike glycolic acid oxidase, which is located in peroxisomes. Paul et al. (440) have shown, using cell disruption and differential centrifugation, that this is also true for diatoms. However, there is some disagreement. Collins and Merrett, after mechanically disrupting Euglena cells and isolating organelles by isopycnic centrifugation on sucrose gradients, found glycolic acid dehydrogenase to be distributed between mitochondria and peroxisomes (114). In a study using osmotic shock followed by centrifugation on a discontinuous sucrose gradient, Grodzinski and Colman found that glycolic acid dehydrogenase and its natural electron acceptor were bound to the photosynthetic lamellae in Oscillatoria, a blue-green alga

(210). These organisms are taxonomically widely separated, which may explain the discrepancy among the results. Even within the green algae, however, the distribution of glycolic acid dehydrogenase and glycolic acid oxidase appears somewhat haphazard, although it may have phylogenetic significance (192).

Having the glycolate pathway, algae might be expected to photorespire, and some studies have been interpreted to show photorespiration in algae (567). However, careful direct measurements of a wide range of unicellular (47, 348) and marine algae (48a) have failed to detect any photorespiratory gas exchange. This may be because the flux of the glycolate pathway is limited by the relatively low activity of glutamic dehydrogenase in most algae in which it has been detected, as well as by the possibility that an efficient HCO_3^--concentrating mechanism exists in algae that would recover any CO_2 lost in photorespiration. Reports of photorespiratory metabolism in algae may relate to the presence of light respiration or to the presence in some algae of C_2 metabolism and glycolic acid oxidase like those in higher plants (e.g., 192).

This problem may also relate to the question of glycolate excretion, long known to accompany algal photosynthesis under appropriate conditions (448, 569). Glycolate excretion is increased at low CO_2 and high O_2 levels (e.g., 66), and excretion may be a consequence of the accumulation of glycolate, caused by its rapid formation via RuBP oxygenase and the relatively low activity of glycolate dehydrogenase. However, Fock *et al.* (182) found that *Chlorella* makes glycolate only when rates of photosynthesis are low, and then largely from carbon sources other than recent photosynthate. The oxidation of such nonphotosynthetic glycolate could give rise to the light respiration sometimes observed in *Chlorella* (47).

The work of Colman and associates suggests that glycolate excretion is directly related to the ability of cells to absorb CO_2 (116, 263). Cells that are adapted to high CO_2 levels and that lack carbonic anhydrase and the ability to absorb HCO_3^- (181) immediately begin to excrete glycolic acid when placed in low concentrations of CO_2. In time they acquire carbonic anhydrase and a HCO_3^--uptake mechanism and stop excreting glycolate. It is tempting to suggest that absence of glycolate excretion or photorespiration in air-adapted (i.e., naturally grown) algae is the result of a HCO_3^--uptake mechanism that increases the CO_2 concentration at the site of RuBPcase. This suggestion needs more hard evidence; other explanations are possible, and the situation is still unclear. It is evident, however, that one must be extremely cautious about drawing conclusions from experiments on algae grown in laboratory culture under elevated levels of CO_2.

8. The Glycolate Pathway to Sugars

When the glycolate pathway was first established on the basis of intra-molecular labeling patterns, it was envisaged as a metabolic sequence leading to sugars (469, 594). The supply of serine 3-^{14}C or glycolate 1- or 2-^{14}C produced sucrose in wheat leaves in which hexose was labeled in the 1-6, 1-2-5-6, or 3-4 positions, indicating synthesis of sucrose from trioses derived from glycolate-path metabolism (273). Data of Bird et al. (54) indicate that carbon from the glycolate path is indeed incorporated into sugars, as suggested in Fig. 9. However, although this type of experiment shows that some sucrose is doubtless made by the glycolate pathway, it does not indicate how much. Neither does it demonstrate clearly if the three-carbon product of the C_2 cycle needs to pass through the C_3 cycle prior to sucrose synthesis or if it can be reduced to triose phosphate in the cytoplasm and converted to sucrose directly.

Data of Poskuta et al. (458), Tamas and Bidwell (553), and Lawler and Fock (333) show that glycolate conversion to sugars is increased in leaves of various plants under stress conditions of high O_2, low CO_2, or drought, respectively. Bidwell et al. showed that inhibiting glycolic acid oxidase in bean leaves with HPMS decreases photosynthetic sucrose production in proportion to the pileup of glycolic acid (48). However, Fry and Bidwell (196) could find no evidence of cytoplasmic conversion of glycolate-path intermediates to sucrose in bean (Phaseolus vulgaris) leaves. Evidently, the synthesis of sugars from C_2-cycle intermediates occurs as the result of C_3-cycle activity following reentry of glyceric acid as PGA into the C_3 cycle.

Alternative fates of C_2-cycle intermediates are possible. Undoubtedly, some serine and glycine enter the protein metabolism of photosynthesizing cells, and glycine may also be a substrate of chlorophyll synthesis. In Chlorella pyrenoidosa, Lord and Merrett have shown that glycolate carbon may be converted via the glycolate pathway to pyruvate and acetyl CoA and thus enter the citric acid cycle (351). However, glycolate is also produced from sources other than recent photosynthate in Chlorella (182). It appears that the metabolism of glycolate in algae may be more complex than is presently thought.

D. OTHER POSSIBLE PHOTORESPIRATORY METABOLISM

In the preceding section it was concluded that operation of the glycolate cycle as the major or only pathway of photorespiration in leaves is consistent with most available data. However, the data do not rule out the possibility of other minor or even, in some plants, major pathways of metabolism leading to photorespiration. Furthermore, certain disturbing data have been published that seem to require alternative pathways. More than one alternative metabolic sequence has been suggested, although usually based on inadequate evidence, that would lead to photorespiratory gas exchange.

The most obvious problem with the pathway of photorespiration, as shown in Fig. 9, is the required high flux of carbon through the C_2 cycle. Data reviewed in Section III,C,5 appear to satisfy this requirement, but the question is nevertheless frequently raised when alternative paths of photorespiratory metabolism are proposed. Typical objections are those raised by Heber et al. (236), who were unable to find sufficient glyceryl kinase to account for the required carbon flux, and those by Thomas et al. (559), who found that the ratio of photosynthesis to photorespiration changes substantially as wheat leaves age whereas the ratio of RuBP carboxylase to RuBP oxygenase does not. In either case, alternative oxidative metabolism appears to be required. Proposed alternatives include oxidative reactions producing CO_2 from partial or total oxidation of glycolate via glyoxylate, reoxidation of the CO_2-addition product of RuBPcase to yield CO_2 (and usually glycolate), and an alternative pathway for the conversion of two molecules of glycolate to glycerate and CO_2 that bypasses the glycine–serine conversion.

1. Oxidation of Glyoxylate and Formate

Zelitch proposed that decarboxylation of glyoxylate is a source of photorespiratory CO_2 in chloroplasts (629). The residual formate presumably would be oxidized (213) or would enter folate–C_1 metabolism (352). Canvin et al. have cautiously supported the possibility that glyoxylate, rather than glycine, is the molecular species decarboxylated (89), but the evidence is rather weak. Leck et al. have shown that peroxisomal oxidation of formate or oxalate can proceed, but the activities of the enzymes involved are too low for them to play a major part in photorespiration (336). However, Grodzinski (209) has shown that under normal conditions the majority of glycolate is converted to glycine in peroxisomes, but if conditions arise that increase the production of H_2O_2, rapid oxidation of glyoxylate takes place. Illuminated chloroplasts under somewhat elevated temperature and low CO_2 concentration tend to produce larger amounts of H_2O_2. These are conditions quite likely to be met in brilliantly illuminated leaves, particularly those suffering water stress. It therefore seems quite likely that glyoxylate oxidation may be important as a source of photorespiratory CO_2 under stress conditions, which are known to stimulate photorespiration.

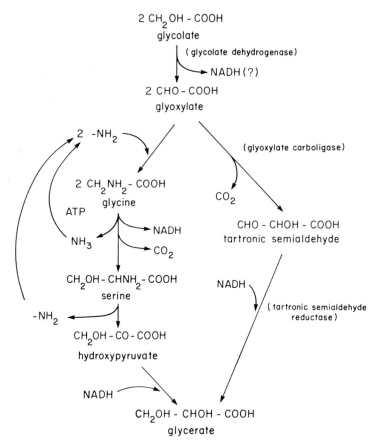

Fig. 11. Possible pathways of glycolate metabolism in diatoms as suggested by Paul and Volcani (441).

2. Glyoxylate Carboligase Pathway

Badour and Waygood (21) demonstrated glyoxylate carboligase in a blue-green alga (*Gloeomonas*) that converts two molecules of glyoxylate to CO_2 and tartronic semialdehyde. Paul and Volcani (441) have suggested an alternative pathway of glycolate metabolism in diatoms that involves this enzyme and tartronic semialdehyde reductase, producing glycerate. This pathway, shown in Fig. 11, has the advantage of not involving nitrogen, which is often in very short supply to pelagic phytoplankton. Energetically it is less satisfactory. Although no ATP is required to refix ammonia liberated in glycine decarboxylation, no reducing equivalent is generated in the decarboxylation step. So far, the enzymes have not been demonstrated in higher plants, and the relative importance of this pathway compared with the glycine decarboxylase pathway (which also occurs in diatoms) has not been established. There is certainly the possibility that higher plants may also have the glyoxylate carboligase pathway, which would rationalize the persistent concern about glyoxylate oxidation. It seems quite possible that if tartronic semialdehyde is sought in this connection, it will be found.

3. Phosphohydroxypyruvate Pathway

Data from $^{14}CO_2$–$^{12}CO_2$ pulse-chase experiments by Daley and Bidwell (129) suggest that phosphohydroxypyruvate and phosphoserine are early products of a photosynthetic carboxylation reaction in bean leaves. Phosphoserine has also been recorded as an early intermediate in maize chloroplast photosynthesis (98). Daley et al. (133) have proposed that nonenzymatic oxidation of hydroxypyruvate derived (presumably by phosphatase action) from phosphohydroxypyruvate might contribute substantially to photorespiration. Their data do not distinguish clearly between the possibilities that phosphohydroxypyruvate may arise from an oxidative carboxylation reaction or from the oxidation of PGA produced by normal RuBPcase activity. As much as 35% of ^{14}C fixed in 1-min photosynthesis was found in phosphohydroxypyruvate and phosphoserine, which suggests that the pathway may be more than a minor one in bean leaves. The probable reason that these intermediates have not been detected previously in short-term photosynthesis experiments is the extreme difficulty of their chromatographic separation (98, 130).

The distribution among plants and the quantitative importance of this pathway have not been established. However, it does provide a solution to two problems. First, it is possible that some photorespiratory O_2 uptake takes place in an oxidative carboxylation or in the oxidation of PGA to produce phosphohydroxypyruvate. This would explain the situation in maize, in which photorespiratory O_2 exchange occurs without CO_2 production (266). Second, the formation of phosphoserine from phosphohydroxypyruvate by transaminase activity would provide an alternative source for high-specific-radioactivity serine. Serine is sometimes observed to have higher specific radioactivity than the glycine from which it is supposedly derived (89). This would surmount the problem that has led Canvin et al. to suggest that glyoxylate decarboxylation, rather than glycine decarboxylation, leads to serine synthesis in sunflower photorespiration (89). It must be stressed, however, that this pathway is hypothetical and requires substantiation.

E. ROLE AND CONTROL OF PHOTORESPIRATION

Frequent and very strong claims have been made in the scientific and semipopular press to the effect that productivity could be greatly increased by inhibiting, controlling, or genetically eliminating photorespiration (e.g., 626, 628, 632, 634). As a result of this viewpoint, much effort has been spent on the control of photorespiration by chemical agents and on the reduction or elimination of photorespiration by selective breeding, so far without success. This approach implies that photorespiration is susceptible to control and that its control will be beneficial to plants, that is, that photorespiration is an undesirable biochemical behavior that is deleterious both to the plant and to the goals of agriculture. This viewpoint and the scientific methodology used to develop it have come under attack, and it appears worthwhile to consider alternative possibilities.

First, if the formulation of the C_2 cycle shown in Fig. 9 is (at least approximately) correct, any attempt to inhibit photorespiration by inhibiting the C_2 cycle will reduce photosynthesis, not increase it, because large amounts of reduced carbon will be uselessly diverted from the C_3 cycle. Second, if the claim is correct that the oxygenase function of RuBPcase is an inevitable consequence of the nature of its carboxylase function, then improvement is only possible by devising a strategy to increase the ratio of CO_2 to O_2 at the carboxylase or by modifying the enzyme itself to discriminate more effectively against O_2. Third, if the metabolism associated with photorespiration is beneficial or essential to the plant, eliminating it will do no good. This is an extension of the naive idea that reducing carbon losses in dark respiration will automatically result in bigger or better plants, a view that overlooks the need for respiratory activity to drive synthetic or developmental reactions that may proceed in the dark.

1. Why Photorespiration?

The question frequently asked is, Why should plants indulge in such an apparently wasteful process as reoxidizing a substantial proportion of recently fixed carbon? By now it should be apparent that this is not the right question. Based on the present formulation of the C_2 cycle, two questions might be asked. The first is whether the oxygenase function of RuBPcase is inescapable. The second concerns the extent to which plants have capitalized on the consequences of the oxygenase reaction or the degree to which photorespiration may have become a useful or even essential, as well as inescapable, reaction sequence.

a. Oxygenase Reaction. Lorimer and Andrews suggested in 1973 that oxygenation of RuBP is chemically unavoidable because an intermediate generated during the RuBPcase reaction is susceptible to attack by O_2 (354). A more recent reexamination indicates that this suggestion is probably correct (10), but the active-site chemistry of RuBPcase is still insufficiently known to be certain. It appears that all situations in which the manifestations of photorespiration are reduced or absent result from reassimilation of photorespired CO_2 or from operation of a CO_2-concentrating device that increases the ratio of CO_2 to O_2 at the chloroplast to the point at which O_2 ceases to be strongly inhibitory. In fact, if one considers the relative abundance of O_2 and CO_2 (nearly three orders of magnitude different), the specificity of RuBPcase is amazingly high. The K_{mO_2} and K_{mCO_2} of the enzyme differ by only two orders of magnitude (~700 and ~7 μM, respectively), but the difference in relative velocities of the two reactions permits carboxylation to predominate.

Numerous authors have suggested that carboxylase evolved at a time in the history of the earth when the atmosphere contained much larger amounts of CO_2 and less O_2, when the possibility of oxygenase function was unimportant. By the time the global atmosphere changed toward its present composition, to a considerable extent because of photosynthetic activity by plants, the reaction sequence of photosynthesis was established beyond the possibility of recall. If this is true, the questions are how effective has subsequent evolution been in utilizing this accidental deficiency in the photosynthetic apparatus, and whether it can be controlled by chemical or genetic means to increase agricultural yields.

b. C_2 Cycle. Several suggestions have been advanced concerning the usefulness of the C_2 cycle. As has been observed by many, and more recently has been clarified by Servaites and Ogren (521), the C_2 cycle is an essential and integral part of photosynthesis because it returns to the C_3 cycle three-quarters of the carbon that would otherwise have been lost as glycolate. This is undoubtedly its primary function. Somerville and Ogren (538) showed that photosynthesis is greatly reduced except at low O_2 or high CO_2 levels in mutants of *Arabidopsis thaliana* deficient in glycolate pathway enzymes. This strongly supports the necessity of photorespiration as a normal component of photosynthetic metabolism.

Certain derivative functions have been suggested. It is possible that the glycine-to-serine reaction may be coupled via NAD to the mitochondrial electron transport chain, thus producing ATP (52). Undoubtedly, some of the intermediates, particularly serine and glycine, are of use in cellular syntheses (173).

Another consequence of photorespiratory metabolism is that is may

354 R. G. S. BIDWELL

supply CO_2 when it is vitally needed to prevent O_2 damage, namely, when CO_2 is in short supply (perhaps because of stomatal closure) and the light intensity is high. Providing a CO_2 compensation concentration above zero is thus a valuable side effect of photorespiration. This viewpoint was first noted by Osmond and Björkman (429), and it has been provided with supporting evidence. Spinach chloroplasts or cells (310) and bean (459) or mustard [*Brassica hirta* (*Sinapis alba*)] (120) leaves illuminated in the absence of CO_2 and O_2 suffer a major loss of photosynthetic capacity that does not occur if compensation levels of CO_2 are present during illumination (either by adding CO_2 or by adding sufficient O_2 to permit photorespiratory generation of CO_2). Clearly, the presence of CO_2, or possibly the capacity of the system to react with O_2, is necessary to protect the photochemical apparatus from light damage. Presumably, it does this by the dissipation or degradation of photochemical energy through the operation of a "futile" or "wasteful" cycle of reactions.

2. Natural Control of Photorespiration

All the regulatory functions of RuBPcase affect both oxygenase and carboxylase functions of the enzyme. However, Chu and Bassham have pointed out that in the absence of CO_2 the enzyme becomes less active and, hence, the oxygenase function and glycolate synthesis are also reduced (109). This would serve to protect chloroplasts from complete drainage of C_3 intermediates to glycolate under conditions of CO_2 deprivation. Laing *et al.* (326) and Servaites and Ogren (522) have pointed out that regulation of net photosynthesis is mediated by the interaction of CO_2 and O_2 at the carboxylase reaction, and it is further affected by the differential effects of temperature on photorespiration and photosynthesis. However, no clear picture has yet emerged of a regulation site or mechanism that controls the C_2 cycle, except as it interacts with the C_3 cycle.

3. Chemical and Genetic Control of Photorespiration

The foregoing discussion strongly suggests that photorespiration cannot be controlled except by modification of the RuBPcase reaction to increase its discrimination against O_2, and further, that inhibiting any part of the C_2 cycle will inhibit photosynthesis by preventing the return of large amounts of glycolate carbon to the C_3 cycle. Nevertheless, there have been persistent and often-repeated claims that specific inhibitors of C_2 enzymes increase net photosynthesis by as much as 300% (e.g., 626) and that strain selection can produce offspring with higher photosynthetic and lower photorespiratory rates. It has been pointed out by Chollet (105, 106) that most of the claims for chemical or genetic control of photorespiration have been based on techniques for measuring photosynthesis and photorespiration that may not give accurate results. A review of this controversy seems to be in order.

Zelitch showed that HPMS, an inhibitor of glycolic acid oxidase, suppresses photorespiration and greatly increases net photosynthesis in tobacco leaf discs (626). Subsequently, many observers found that HPMS, in addition to inhibiting glycolate oxidation, also inhibits photosynthesis and raises Γ in many plants, including leaves of soybean, tobacco, bean, spinach, and isolated spinach chloroplasts (e.g., 521, 536). Substantial or prolonged stimulation of net photosynthesis by HPMS has not been reported by other workers.

Later, Zelitch reported that 2,3-epoxypropionic acid (glycidate), a putative analog of glycolic acid, inhibits glycolate formation and photorespiration and increases net photosynthesis by 40–50% (631). Zelitch also found that glycidate has no effect on RuBPcase or other glycolate pathway enzymes assayed (633) but that it inhibits glutamate : glyoxylate aminotransferase (335). Oliver and Zelitch had previously found that added glutamate raises photosynthesis and decreases photorespiration of tobacco leaf discs that have been grown under precisely defined conditions (419) and that added glyoxylate doubles net photosynthesis by inhibiting photorespiration (420) by an unknown mechanism (418). This led them to suggest that glycidate acts through its effect on the interconversion of glyoxylate and glutamate.

Poskuta and Kochánska (456), on the other hand, found that glycidate stimulates photorespiration and the formation of glycolate, serine, and glycine. Chollet found that photosynthesis rates are stimulated by glycidate in chloroplast preparations of low activity, but not in active preparations (105). Chollet also found that this inhibitor does not inhibit photorespiration or raise photosynthesis in tobacco leaf discs, as had been claimed. Servaites and Ogren (521) also tested the butyl ester of 2-hydroxy-3-butyonate, another glycolic acid oxidase inhibitor, and isonicotinyl hydrazide, an inhibitor of the serine–glycine conversion. They found that, whereas these chemicals inhibit glycolate-path metabolism, they do not stimulate net photosynthesis, and that they increase rather than decrease the O_2 sensitivity of photosynthesis. Dorovari and Canvin found that 2-hydroxy-3-butyonate in fact inhibits many aspects of plant metabolism (155a).

It appears probable that the source of disagreement lies in the assays used by Zelitch and co-workers for photosynthesis and photorespiration. Their assay for photosynthesis (626) is based on the measurement of ^{14}C in leaf discs after exposure to $^{14}CO_2$ of known specific activity for a short time in closed Warburg vessels. Leaf discs are shaken in illuminated vessels and attached to monometers but "open to the air" for 1 hr prior to $^{14}CO_2$ addition. Chollet (106) has noted that reported rates of photosynthesis using this technique are extremely low compared with rates measured by conventional techniques, especially when the elevated CO_2 levels used by Zelitch and co-workers are considered. Chollet pointed out that stimulation of chloroplast photosynthesis by glycidate requires one important precondition—the chloroplasts must have a low initial photosynthesis rate. It seems likely that the failure of other workers to repeat Zelitch's observations may be because leaf discs used by Zelitch were somewhat inhibited or reduced in photosynthetic capacity, and the effect of the added compounds was merely restorative, not stimulatory. The reason for this may be the long preexperimental period during which leaf discs are illuminated in "open" Warburg vessels. If the CO_2 concentration is considerably depleted in the vessels, as seems likely, the leaf discs could be severely photoinhibited by the time the ^{14}C assay is conducted.

Several workers have criticized the assay for photorespiration used by Zelitch and co-workers (627). This position has been well summarized by Chollet (106). In Zelitch's method, preilluminated leaf discs are allowed to absorb $^{14}CO_2$ to the compensation point. Then, after about 30 min of illumination, $^{14}CO_2$ released into flowing CO_2-free air is measured alternatively in light and darkness. The ratio of $^{14}CO_2$ produced in light and in darkness is considered to be proportional to photorespiration. However, this method does not account for the much higher specific radioactivity of photorespired CO_2 than that of dark-respired CO_2, nor that specific radioactivities of light- and dark-respired CO_2 change rapidly with time. Furthermore, because the leaf discs are preilluminated in attached (but open) Warburg vessels and then are held at compensation point for some 30 min following $^{14}CO_2$ absorption, the distribution of ^{14}C between light- and dark-respiratory substrates may be quite different from what it would be under normal conditions. Whatever the causes, careful measurements with direct methods have not substantiated the claims made by Zelitch and co-workers.

Whereas inhibition of the C_2 cycle seems to inhibit rather than stimulate photosynthesis, differential chemical inhibition of the oxygenase function should be beneficial. However, this goal has proved elusive, strengthening the view that the same catalytic site is involved in both reactions. Workers in several laboratories (e.g., 435) have been unable to substantiate the claim (e.g., 602) that glycidate inhibits oxygenase activity of RuBPcase without affecting its carboxylase function. It has been reported that hydroxylamine stimulates the carboxylase while inhibiting the oxygenase function of *Anabaena cylindrica* RuBPcase (417). However, the reported extract and methods of assay are not sufficiently specific for evaluation of this claim without further support.

The search for varieties or mutations with altered carboxylase : oxygenase activity has been unsuccessful (414). McCashin and Canvin were unable to find varieties of barley (*Hordeum vulgare*) with relatively higher carboxylase activity (369). Using bacteriological techniques, Andersen examined several mutants of *Alcaligines eutrophus*, a hydrogen bacterium, having altered RuBP carboxylase–oxygenase (8). Andersen found that in every case oxygenase activity was affected to the same degree and in the same way as carboxylase activity.

4. Conclusion

It seems most reasonable to conclude, with Serviates and Ogren (522), Chollet (106), and many others, that inhibiting photorespiration by inhibiting the C_2 cycle cannot increase net photosynthesis or productivity of plants. The C_2 cycle and resulting photorespiration are beneficial to the plant, given the unavoidability of the oxygenase reaction. Even growing plants in an artificial atmosphere without O_2 is counterproductive because, although the plants thrive vegetatively, their reproduction and seed development is greatly reduced at low O_2 (466). The possibilities for improving plant production by increasing photosynthesis seem to lie more in the direction of increasing CO_2 uptake (CO_2-concentrating mechanisms, decreased resistance, improved ratio of CO_2 fixation to water loss, increased rate or efficiency of carboxylation, etc.) than in the search for ways to eliminate photorespiration or the O_2 sensitivity of photosynthesis.

IV. The C_4 Cycle of Photosynthesis

A. CHARACTERISTICS OF THE C_4 CYCLE

1. Theme and Variations

a. General Outline. In 1965 Kortschack *et al.* (307) noted that the major early products of photosynthesis by sugarcane leaves in $^{14}CO_2$ are the C_4 acids malate and aspartate. In 1966, Hatch and Slack (227) elaborated on this observation and proposed a cyclic reaction mechanism in which a C_3 acid is carboxylated to yield a C_4 acid that subsequently transfers its β-carboxyl carbon as CO_2 to the reductive photosynthetic

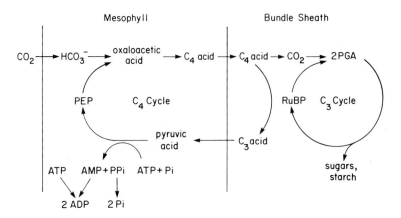

FIG. 12. Simple outline of the C$_4$ cycle.

cycle. A general outline of the C$_4$ cycle as it is now understood and its relationship with the C$_3$ cycle of photosynthesis is shown in Fig. 12. Evidence for this formulation and for variations on the basic theme, which will be described subsequently, have been summarized in several excellent reviews (e.g., 58, 161, 219, 226).

In simple outline, as shown in Fig. 12, the C$_4$ cycle consists of β-carboxylation of PEP in mesophyll parenchyma cells to produce oxaloacetic acid. This unstable first product of the carboxylation is converted to a more stable C$_4$ acid (either malate or aspartate) that diffuses to the bundle-sheath cells of the leaf. The C$_4$ acid is then decarboxylated, and the CO$_2$ thus produced is available for fixation by RuBPcase in a conventional C$_3$-cycle reaction. The C$_3$ acid remaining after the decarboxylation (either pyruvic acid or alanine formed from it) diffuses back to the mesophyll cells where it is converted through the action of pyruvate, phosphate dikinase to PEP, completing the cycle. The C$_4$ cycle is in essence a device for absorbing CO$_2$ at the mesophyll cells and transferring it, as the β-carboxyl of a C$_4$ acid, to the inner bundle-sheath cells of the leaf, where it is made available for C$_3$ photosynthetic fixation and reduction.

The C$_4$ cycle is almost always accompanied by characteristic, anatomical specialization called *Kranz anatomy*, which is discussed in greater detail in Section IV,B. Briefly, Kranz leaves have smaller air spaces, dense mesophyll parenchyma, and many bundles surrounded by conspicuous bundle-sheath cells having many chloroplasts, as shown in Fig. 13.

Leaves possessing the Kranz syndrome can often be recognized by eye becuase of their dark green veins, whereas the veins of C_3 leaves are usually pale. Junctions between mesophyll and bundle-sheath cells are characterized by many conspicuous plasmodesmata. Plants or leaves having the C_4 cycle (hereafter called C_4 plants) have other characteristics:

1. O_2 has little or no effect on photosynthesis up to and often beyond air-level concentrations.
2. They do not release photorespiratory CO_2.
3. Γ is usually very low (0–5 µl/l).
4. They usually have higher rates of photosynthesis than C_3 plants, particularly at low CO_2 levels and high light intensity.
5. They are often (but not always) acclimated to high temperature.
6. They usually withstand drought and maintain photosynthesis at levels of soil moisture that would cause stomatal closure and consequent reduction of photosynthesis in C_3 plants.
7. C_4 plants tend to discriminate less strongly against the naturally occurring heavy isotope of carbon, ^{13}C, than do C_3 plants. They are characterized by a higher ratio of ^{13}C to ^{12}C than C_3 plants.

These characteristics are almost all immediate consequences of the operation of the C_4 cycle, which seems to absorb CO_2 into the leaf more powerfully than the C_3 cycle is able to do and transfers and concentrates it deeply within the leaf so that it becomes more readily available to the RuBPcase located there. This has the double effect of increasing the rate of C_3 cycle operation and reducing the O_2 effect and photorespiration. The motive power for the C_4 cycle is derived from excess illumination (over that required to drive the C_3 cycle) to which most plants (particularly tropical plants) are subject. The significance of these and various secondary characteristics of the C_4 cycle are examined in Section IV,D. First, however, it is appropriate to consider in detail the various types of C_4 metabolism that have been distinguished, the details of the regulation of C_4 photosynthesis, and the taxonomic distribution and evolution of the C_4 syndrome.

b. Variations on the Theme. The outline of C_4 photosynthesis given in Fig. 12 cannot convey the beauty and balance of the various reaction mechanisms that have become adapted or have evolved to "pump" CO_2 into the bundle sheath, the site of C_3-cycle reductive carbon assimilation. The three main pathways are shown in Fig. 14; they are named from the

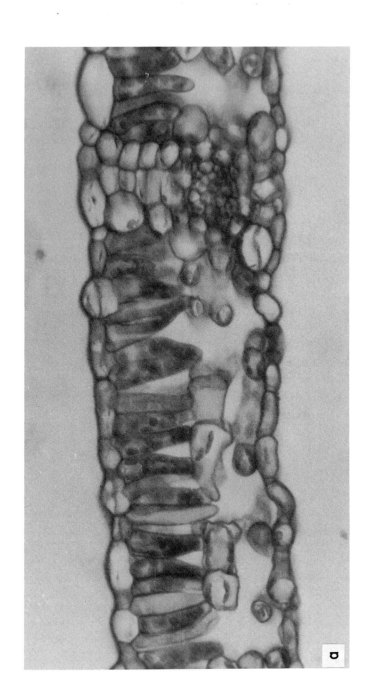

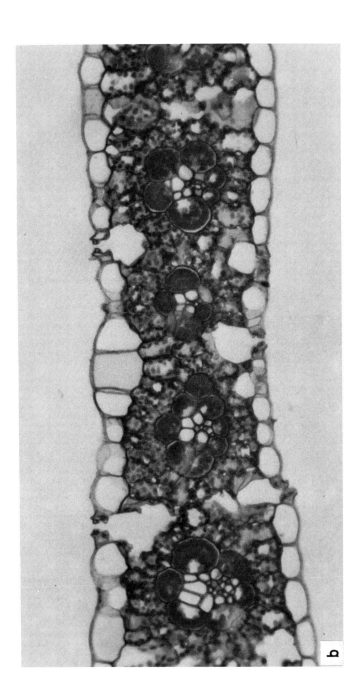

Fig. 13. Photomicrographs showing details of the internal anatomy of a C₃ leaf (a, *Acer rubrum*, maple) and a C₄ leaf (b, *Zea mays*, maize). Reprinted by permission of Macmillan Publishing Co., Inc. from "Plant Physiology," second ed., by R. G. S. Bidwell. Copyright 1979 by R. G. S. Bidwell.

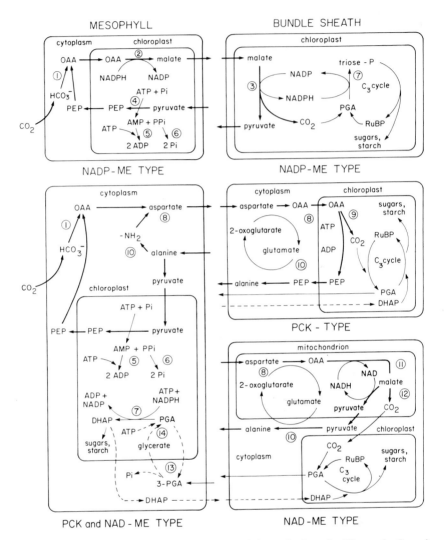

MESOPHYLL

BUNDLE SHEATH

NADP-ME TYPE   NADP-ME TYPE

PCK - TYPE

PCK and NAD - ME TYPE   NAD -ME TYPE

Fig. 14. Reactions of the three types of C_4 photosynthetic cycle. The main C_4-cycle reactions are shown with bold-face arrows; reactions not well established are shown with dashed arrows. The PGA–DHAP shuttle may also function in NADP-ME plants. The cytoplasmic location of alanine formation in NAD-ME bundle-sheath cells is not well established. The chloroplast location of PEP carboxykinase has also been questioned (321a). Enzymes: 1, PEP carboxylase; 2, NADP-malate dehydrogenase; 3, NADP-malic enzyme; 4, pyruvate, P_i dikinase; 5, adenylate kinase; 6, pyrophosphatase; 7, triose-P dehydrogenase, 3-PGA kinase and triose-P isomerase; 8, aspartate aminotransferase; 9, PEP carboxykinase; 10, alanine aminotransferase; 11, NAD-malate dehydrogenase; 12, NAD-malic enzyme; 13, 3-PGA phosphatase; 14, glycerate kinase.

characteristic mechanisms by which the C_4 β-decarboxylation takes place in bundle-sheath cells (223):

1. NADP-malic enzyme (NADP–ME) type. The major C_4 and C_3 acids transported are malate and pyruvate. Malate is decarboxylated in bundle-sheath chloroplasts to yield pyruvate and CO_2.
2. PEP-carboxykinase (PCK) type. Aspartate and alanine are the characteristic C_4 and C_3 acids transported. Aspartate is converted by amino transferase to oxaloacetic acid and decarboxylated to yield PEP (an ATP-requiring step) in the cytoplasm of bundle-sheath cells. PEP is presumably converted to pyruvate and then to alanine for return to the mesophyll cells. The details of the PCK type are less clearly defined than those of the NAD- or NADP-ME types. Indeed, it has been suggested that malate and PEP may be the C_3 and C_4 acids transported in this type (221), either instead of, or as well as, aspartate and alanine. If so, malate would be converted to oxaloacetate in bundle-sheath cells prior to decarboxylation, and the PEP produced in the decarboxylation might be transported directly back to participate in the mesophyll C_4 carboxylation. This alternative is speculative, however, and other reactions in addition to the ones shown in Fig. 14 are possible.
3. NAD-malic enzyme (NAD-ME) type. Aspartic acid and alanine are transported. The conversion of aspartate to oxaloacetate and its decarboxylation to yield pyruvate are mitochondrial reactions.

c. Experimental Identification of C_4 Types and Compartments. It was early noted that C_4 plants may be identified by their low Γ, green veins, and insensitivity to O_2. Furthermore, if supplied with $^{14}CO_2$ in light for a brief period (usually less than 1 min) the majority of fixed ^{14}C is present in C_4 acids. The NADP–ME type generally produces mainly malate, whereas the PCK and NAD-ME types tend to produce more aspartate.

Specific metabolic inhibitors of the various decarboxylation systems have been used to distinguish among the C_4 types (485). Oxalic acid, 100 μM, specifically blocks NAD-ME decarboxylation, whereas 200-μM 3-mercaptopicolinic acid inhibits PCK decarboxylation. Neither inhibitor affects NADP-ME decarboxylation. These inhibitors may thus serve to identify the main decarboxylation systems of a plant or the combinations and relative importance of types that may occur simultaneously (e.g., PCK and NAD-ME) in one plant.

The location of various metabolic events in the C_4 cycle has depended upon the possibility of separating bundle-sheath cells or chloroplasts from mesophyll cells or chloroplasts. Initially, mesophyll and bundle-sheath chloroplasts were separated on the basis of their differential density by centrifuging in a nonaqueous medium (531). However, this technique only works well for plants with dimorphic chloroplasts (NADP-ME type; see Section IV,B,3). Differential grinding, done because mesophyll cells are generally more easily separated and disrupted than bundle-sheath cells, proved somewhat more useful (57, 160). A more-effective technique, now widely adopted, is the treatment of tissue with enzymes (cellulase and pectinase) to dissolve cell walls, followed by gentle mechanical separation of cell types (279). This technique permits the essentially complete separation of mesophyll cells or protoplasts from strands of vascular bundles and their associated bundle-sheath cells. The factors required for successful use of the technique have been evaluated by Huber and Edwards (253).

Identification of the intracellular location of enzymes is more difficult, and the site of some enzymes (e.g., PEP carboxylase) is still not known with certainty. However, a combination of brief enzymatic disruption of cells and graded, mechanical blending has been useful in separating cell contents. Various standard techniques of subcellular fractionation have been used to separate chloroplasts, mitochondria, and cytoplasm. However, the well-known difficulty of separating the fractions, completely free of contamination, has made it very difficult to be absolutely certain of the exact location of some activities. Furthermore, certain enzymes may be present that have the function of a cycle enzyme but that do not take part in the cycle. It is therefore necessary to view with reservation schemes such as those in Fig. 14, which show the presently supposed location of cycle functions.

    d. Overlap of Types. Since the discovery of C_4 photosynthesis it has been recognized that both aspartate and malate may be transported in the same plant, and it has become evident that more than one decarboxylation mechanism may operate simultaneously in the same plant. For example, Rathnam and Edwards (484) have shown that *Eriochloa borumensis*, although predominantly a PCK type, may also transport malate (like the NADP-ME type), which is decarboxylated by the NAD-ME system. Furthermore, a single genus (e.g., *Panicum*) may contain species exhibiting each of the three types (212), suggesting that taxonomic or evolutionary distinction among the types is not very great. Créach *et al.* (123) have shown that *Zea mays*, although normally considered a NADP-ME type, may transport aspartate. However, Hatch and Mau (224) have suggested that NAD-ME activity found in NADP-ME plants such as *Zea mays* may be merely an associated activity of the NADP-ME enzyme and not normally functional in photosynthesis.

    In view of the widespread distribution and basic nature of the enzymes in question, it is probably unwise to stress too heavily the precise classification of C_4 plants. Although it is clear that the majority of C_4 plants emphasize one of the three distinct decarboxylation mechanisms and the other associated characteristics of the type, many of them may have subsidiary capabilities similar to or identical with other types. So the classification system, although useful, is not absolute. Doubtless other variants or combinations await discovery.

e. Cofactors and Subsidiary Cycles. Other important characteristics of the C_4 cycle include the cyclic metabolism of various cofactors necessary to balance its energetic, redox, and nitrogenous components. Malate is transported in the simplest type, the NADP-ME type. Malate is formed by reduction (NADP-malate dehydrogenase) in mesophyll chloroplasts, and NADPH is regenerated in its decarboxylation. NADPH is used again in the reduction of PGA to triose phosphate, and half the required reducing power of the bundle-sheath cells in NADP-ME type leaves is thus supplied from mesophyll chloroplast photosynthesis. It was noted early (156) that chloroplasts in the bundle sheath of malate-transporting C_4 leaves are agranal, whereas those in the mesophyll cells are normal. This may relate to the reduced need for the coupled photosystems I and II required for noncyclic electron flow and NADPH formation in these chloroplasts.

Both PCK- and NAD-ME-type leaves transport aspartate. They must therefore transport alanine in the opposite direction in order to maintain nitrogen balance between mesophyll and bundle-sheath cells. In both types deamination of aspartic acid and conversion of pyruvate to alanine are coupled by a glutamate-2-oxoglutarate shuttle. In PCK-type leaves the entire shuttle apparently takes place in the bundle-sheath cytoplasm, whereas in NAD-ME-type leaves the decarboxylation reaction takes place in mitochondria. It has been suggested that deamination of aspartate also occurs in mitochondria (226). This means that the glutamate-2-oxoglutarate shuttle in NAD-ME leaves must operate across the mitochondrial membrane. The aspartate-transferring PCK and NAD-ME types also require an aminotransferase system that operates between alanine and oxaloacetic acid in the mesophyll cells. The location of the aminotransferase system (whether cytoplasmic or mitochondrial) and its requirement of an intermediate transfer system (e.g., glutamate–oxoglutarate) are not known at present.

An additional subsidiary shuttle that may operate, particularly in the PCK and NAD-ME types, is the transport of PGA arising from C_3 carboxylation in bundle-sheath cells to mesophyll cells, where it may be reduced to dihydroxyacetone phosphate. This reaction is supported by the presence of the required C_3-cycle enzymes in mesophyll chloroplasts

(226). It would be useful in NADP-ME plants with agranal bundle-sheath chloroplasts, which have reduced capacity for the reduction of NADP and in which malate decarboxylation only supplies half of the required NADPH. It would be an equally effective way in PCK and NADP-ME types for utilizing the photosynthetic capacities of mesophyll chloroplasts. Supporting data for these reactions have been presented (189, 254), but their quantitative importance has not been established.

2. Details of the Reactions

a. Enzymatic Activities. The required enzymes have all been found with the required levels of activity in the appropriate locations to support the schemes shown in Fig. 14. A summary of major C_4 enzyme activities is presented in Table III. The levels of activities cited are all well in excess of that required to catalyze maximum photosynthesis, which is usually in the range 3–5 μmol CO_2/min/mg chlorophyll (226). Not all C_4 plants have been shown to have the required enzymes for reducing PGA in mesophyll cells, and the participation of this sequence of reactions in mesophyll chloroplasts of all C_4 plants has not been demonstrated.

b. Carboxylation Reaction. The primary carboxylation of the C_4 cycle is accomplished by PEP carboxylase (E.C. 4.1.1.31), an enzyme that catalyzes the β-carboxylation of PEP, using HCO_3^- for its substrate (e.g., 118), to yield oxaloacetic acid (58, 561). PEPcase is probably located in the cytoplasm of mesophyll cells (276), although occasional reports (e.g., 190a) suggest that it may be found in chloroplasts. The enzyme from mesophyll cells of C_4 plants has characteristics different from those of the similar enzyme isolated in smaller amounts from C_3 leaves, as shown in Table IV. Both enzymes may be present in the same plant (392, 561). The C_4-type enzyme has much higher thermostability (optimum at 40°C) (392). Care needs to be taken in order to ensure maximum activity of the extracted enzyme. Hatch and Oliver (225) have described precise conditions for its incubation and activation. The enzyme is inhibited by bisulfite compounds such as HPMS and glyoxal bisulfite (427) and sulfite ions (394). Sulfite inhibition may be caused by competitive inhibition of HCO_3^- fixation.

As might be expected, PEPcase is closely regulated. Mukerji has found that a complex relationship exists between the enzyme and Mg^{2+}, such that its K_m varies widely (2–20 μM PEP) as a function of the Mg^{2+} concentration (393). Mukerji suggested that a PEP–Mg^{2+} complex is the true substrate and that Mg^{2+} also activates the enzyme. The enzyme from *Zea mays* (but not from other plants) has been reported to be

TABLE III

Activities and Locations of Enzymes of the C_4 Cycle[a]

Enzyme	Activity (μmol/min/mg chlorophyll)	Location	C_4 type
Pyruvate, P_i dikinase	3–10	Mesophyll chloroplast	All
Adenylate kinase	17–45	Mesophyll chloroplast (mainly)	All
Phosphatase	20–60	Mesophyll chloroplast (mainly)	All
PEPcase	15–40	Mesophyll cytoplasm	All
NADP-malate dehydrogenase	10–17	Mesophyll chloroplast	NADP-ME[b]
NADP-ME	9–14	Bundle-sheath chloroplast	NADP-ME[b]
PEP carboxykinase	10–14	Bundle-sheath chloroplast	PCK
NAD-ME	5–9	Bundle-sheath mitochondria	NAD-ME[b]
Aspartate aminotransferase	45–60	Mesophyll and Bundle-sheath cytoplasm?	PCK
	25–45	Bundle-sheath mitochondria	NAD-ME[b]
Alanine aminotransferase	30–60	Mesophyll and Bundle-sheath cytoplasm?	PCK, NAD-ME[b]
3-PGA kinase Triose phosphate dehydrogenase Triose phosphate isomerase	—	Mesophyll and Bundle-sheath chloroplast	All
RuBPcase	—	Bundle-sheath chloroplast	All

[a] Data revised from Hatch and Osmond (226).
[b] These enzymes are also found in small amounts in other locations and other C_4 types.

activated by glycine (409), and the enzyme from various plants is inhibited by malate and aspartate (255), providing feedback control.

The product of carboxylation, oxaloacetic acid, is reduced to malate in mesophyll chloroplasts of NADP-ME leaves by NADP malate dehydrogenase (E.C. 1.1.1.82) (230). This enzyme is closely and proportionately regulated by light, presumably through a membrane-bound sulfhydril light-effect mediator, and by a low-molecular-weight, regulatory protein

TABLE IV

CHARACTERISTICS OF PEPCASE ALLOENZYMES FROM C_3 AND C_4 PLANTS[a]

Source (leaves)	Activity (μM/min/mg Chlorophyll)	K_{mPEP} (mM)	K_{mMg} (mM)	Role: formation of photosynthetic
C_3	1.5 ± 0.57	0.14 ± 0.07	0.09 ± 0.057	End product
C_4	29.0 ± 13.2	0.59 ± 0.35	0.5 ± 0.3	Intermediate

[a] Data from Ting and Osmond (561).

(220). In PCK or NAD-ME types, oxaloacetic acid is converted to aspartate by aspartate aminotransferase in the cytoplasm of mesophyll cells.

  c. Formation of PEP. This reaction, now well established as a major chloroplast activity of mesophyll cells in C_4 plants, was first described by Hatch and Slack in 1968 (228). It requires the intervention of 3 enzymes

$$\text{Pyruvate} + \text{ATP} + \text{P}_i \xrightarrow[\substack{\text{Pyruvate, P}_i \\ \text{dikinase}}]{\text{Mg}^{2+}} \text{PEP} + \text{AMP} + \text{PP}_i$$

$$\text{AMP} + \text{ATP} \xrightarrow{\text{Adenylate kinase}} 2\text{ADP}$$

$$\text{PP}_i \xrightarrow{\text{Pyrophosphatase}} 2\text{P}_i$$

In summary:

$$\text{Pyruvate} + 2\text{ATP} + \text{P}_i \longrightarrow \text{PEP} + 2\text{ADP} + 2\text{P}_i$$

The orthophosphate consumed in the pyruvate, phosphate dikinase (E.C. 2.7.9.1) reaction appears in the pyrophosphate (228). The Mg^{2+} requirement is specific.

  The formation of PEP by pyruvate kinase, for example, or even by phosphopyruvate synthetase,

$$\text{Pyruvate} + \text{ATP} \xrightarrow{\text{Pyruvate kinase}} \text{PEP} + \text{ADP}$$

$$\text{Pyruvate} + \text{ATP} \xrightarrow{\text{PEP synthetase}} \text{PEP} + \text{AMP} + \text{P}_i,$$

would be energetically very difficult because the equilibrium would strongly favor PEP breakdown. However, when pyruvate, phosphate dikinase is coupled with adenylate kinase and pyrophosphatase, the equilibrium is shifted toward the synthesis of PEP by the value of the hydrolysis of an additional ATP. This provides a rapid and sufficient synthesis of PEP, the substrate of C_4 carboxylation, at the expense of photosynthetically produced ATP.

Pyruvate, phosphate dikinase is inhibited by all its products: PEP, AMP, and PP_i. The levels of the other two enzymes and ATP seem to regulate the rate of PEP production. A very complex interaction with light has been discussed in detail by Hatch (220). The enzyme is virtually inactive in darkness and is activated in light to a level proportional to that of light intensity. Activation is closely correlated with the activity of the photosynthetic electron transport system (623). However, light activation also requires pyruvate or phosphate and is inhibited by AMP, and dark inactivation requires the presence of ATP. These interactions indicate the possibility for extremely precise regulation of the reactions generating PEP in relation to the environmental situation and the immediate needs of the plant. However, all the details of the control system have not yet been worked out.

d. Decarboxylation Mechanisms. The three major decarboxylation mechanisms located in bundle-sheath cells are NADP-specific malic enzyme, pyruvate carboxykinase, and NAD-specific malic enzyme. These enzymes provide the basis for the classification of C_4 plants according to their decarboxylation mechanism as NADP-ME, PCK, or NAD-ME types.

NADP-ME (E.C. 1.1.1.40; Fig. 14, reaction 3) is located in bundle-sheath chloroplasts. It is specific for NADP at physiological pH, although it will react with NAD at low (6.5–7.5) pH (224). Unlike the nonphotosynthetic enzyme from C_3 plants, NADP-ME from C_4 plants does not seem to be closely regulated by metabolites (220), although a substrate–pH–Mg^{2+} interaction with regulatory effect has been reported (13). This enzyme is responsible for the generation of NADPH required for C_3-cycle operation of the bundle-sheath chloroplasts of NADP-ME plants.

PEP carboxykinase (E.C. 4.1.1.49; Fig. 14, reaction 14) catalyzes the energetically economical decarboxylation of oxaloacetic acid

$$\text{Oxaloacetate} + \text{ATP} \xrightarrow{Mn^{2+}} \text{PEP} + CO_2 + \text{ADP}$$

The reaction has been shown to be located in the chloroplasts of PCK-type bundle-sheath cells (481) and in CAM plants (517a). The claim that it may, in fact, be cytoplasmic (321a) requires an additional shuttle system to provide cytoplasmic ATP for the reaction and to remove resulting ADP (see Section VI,C,3,b and Fig.19). PEP carboxykinase also catalyzes the carboxylation of PEP by reverse reaction and exchanges the β-carboxyl of oxaloacetic acid, but under physiological conditions and in

C_4 plants it acts primarily as a decarboxylase (224, 492). It is competitively inhibited by C_3 intermediates and ADP, providing negative feedback that may integrate C_3 and C_4 cycle operation (220).

NAD-ME (E.C. 1.1.1.39) differs from the other decarboxylases in that it is located in the mitochondria of bundle-sheath cells of NAD-ME-type leaves rather than in chloroplasts (278). It forms part of a cycle using and regenerating NAD

$$\text{NADH + oxaloacetate} \xrightarrow{\substack{\text{NAD-malate}\\\text{dehydrogenase}}} \text{NAD + malate} \xrightarrow{\substack{\text{NAD-}\\\text{malic}\\\text{enzyme}}} \text{NADH + pyruvate + CO}_2$$

The required oxaloacetic acid is derived from aspartate by aspartate aminotransferase. NAD-ME is regulated in various ways by several intermediates and factors. The more precise control of this enzyme than that of the other decarboxylases may relate to its mitochondrial location. The enzyme is competitively inhibited by HCO_3^-, which also inhibits the action of the activators fructose 1,6-bisphosphate and CoA (101). Acetyl-CoA is also an activator. The effect of HCO_3^- is consistent with a regulatory role of the CO_2-generating enzyme, but the reason for the effect of CoA and acetyl-CoA is not apparent.

Direct decarboxylation of oxaloacetate by the activity of oxaloacetate (OAA) decarboxylase has been reported in *Eleusine coracana* (480). The plant is reported to transfer aspartate, but no malic enzyme and insufficient PEP carboxykinase were detected. The OAA decarboxylase activity was found in the mitochondria of bundle-sheath cells. This report does not seem to have been substantiated, however, and it is possible that the plant is really a NAD-ME type like other members of the genus (212). Preparations from NAD-ME plants actively decarboxylate oxaloacetate or aspartate very rapidly (101) but only after conversion to malate. Because the formation of malate and its decarboxylation are tightly coupled by the NAD–NADH shuttle, malate concentrations may be quite low.

e. Integration with the C_3 Cycle. That the majority of CO_2 fixed by the C_3 cycle has been transported in the C_4 cycle of C_4 plants was shown by Kagawa and Hatch (277). Usuda and Miyachi (579) have shown that C_4 decarboxylation and C_3 carboxylation are, in fact, coupled in NADP-ME plants by the NADP–NADPH cycle in bundle-sheath chloroplasts, which links malic decarboxylation with PGA reduction.

It has been shown that mesophyll cells of C_4 plants do not possess a C_3 photosynthetic cycle because they lack RuBPcase. Hatch and Kagawa (222) separated mesophyll and bundle-sheath chloroplasts and found mesophyll chloroplasts deficient in RuBPcase. More recently, using an elegant immunofluorescent histochemical labeling technique, Hatterslie *et al.* (231) showed microscopically that RuBPcase is confined almost exclusively to bundle-sheath cells in C_4 plants of various types. Matsumoto *et al.* (368) have also shown, using a more general immunochemical technique, that the mesophyll cells of maize contain neither the enzyme itself nor any immunologically related protein. This point has been independently supported by Ray and Black using the PEPcase inhibitor 3-mercaptopicolinic acid in the PCK-type leaves of *Panicum maximum* (493). They found that the rate of C_3 photosynthesis was inhibited to the same extent as C_4 cycle turnover, indicating that little or no CO_2 is fixed directly by the C_3 cycle.

Mesophyll chloroplasts of C_4 plants do, however, contain the enzymatic machinery for reducing PGA. A number of studies have indicated that PGA formed in bundle-sheath chloroplasts may move to mesophyll chloroplasts, where it would be reduced to triose phosphate by the available photosynthetically generated NADPH. This would be most reasonable in PCK and NAD-ME plants that transport aspartate (e.g., 189). Their mesophyll chloroplasts are fully competent for the synthesis of NADPH, which would otherwise not be utilized. C_4 leaves contain substantial amounts of phosphoglycerate phosphatase in the cytoplasm of their mesophyll cells (475). This presumably facilitates the transfer of glycerate to mesophyll chloroplasts, where it could be reconverted to PGA at the expense of photosynthetic ATP. Triose phosphate resulting from PGA reduction might be used for starch or sucrose synthesis, either in the mesophyll or bundle-sheath cells. Most enzymes for sucrose synthesis are located in mesophyll cells, the major site of sucrose synthesis in *Zea mays* (158). It seems probable that much of the PGA exported from bundle-sheath chloroplasts of C_4 plants is used ultimately for sucrose or starch synthesis in mesophyll cells.

Some reports have suggested that all cells in C_4 plants may fix CO_2 directly by RuBPcase action without the necessary intervention of the C_4 cycle. Laber *et al.* (508, 324) have interpreted their data on ^{14}C kinetics through photosynthetic intermediates to suggest that C_4 acids produced in corn serve primarily for carbon storage and not as cycle intermediates. Data of most other workers do not support this view, however. Tissues cultured from sugarcane stalk parenchyma are weakly photosynthetic and fix carbon by the C_3 pathway (308). However, such cultures lack the anatomical characteristics of C_4 leaves, which seem to be a requisite for normal C_4 cycle operation (see Section IV,B). Whereas there is no reason to believe that C_4 plants are unable to fix CO_2 directly, at least in the bundle-sheath cells and other unspecialized photosynthetic tissue, the weight of the evidence indicates that most of their CO_2 is normally fixed via the C_4 pathway.

f. Regulation. Internal regulation of the C_4 cycle is possible by several mechanisms. For example, Rathnam and Edwards (483) have shown that C_4 decarboxylation in all three C_4 types is affected by CO_2 concentration, a typical feedback control mechanism. Another interesting example is the work of Hatch, which indicates that aspartate greatly stimulates malate decarboxylation in *Zea mays* (102). Because aspartate would presumably be produced in mesophyll cells if the malate concentration became very high, it could act to normalize the flow of C_4-cycle intermediates if it diffused to bundle-sheath cells, where it would speed up the decarboxylation of malate and hence the diffusion of malate from mesophyll to bundle-sheath cells. In this case, aspartate appears to act on the transport of malate or the product of its decarboxylation, pyruvate, rather than on the malate-decarboxylating enzyme itself.

The regulation of C_4 photosynthesis is very precise. Close internal controls based upon the mechanisms described prevent imbalance in the operation of the various, distant components of the C_4 and C_3 cycles. The proportional light activation of pyruvate, phosphate dikinase, and NADP-malate dehydrogenase ensures balanced operation of the C_4 cycle with respect to the capacity of the C_3 cycle as affected by light (220). Triose phosphate and PGA from the C_3 cycle do not have a regulatory role on the C_4 cycle (117). However, glucose 6-phosphate, which could be produced in mesophyll chloroplasts from bundle-sheath-derived PGA, does fill this role and may serve to integrate the activity of the two cycles.

An interesting peculiarity of C_4 plants is that they require sodium ions at the level of about 100 μM for normal growth (81). The requirement is highly specific to plants possessing the C_4 cycle. Related C_3 plants, even within the same genus, have no detectable sodium requirement. The biochemical site of sodium requirement is not known; presumably, it is in some metabolic sequence that is not essential to C_3 photosynthesis. CAM plants, which also have an active β-carboxylation reaction, also require sodium.

B. ANATOMICAL CHARACTERISTICS OF C_4 PLANTS

1. Kranz Anatomy

In 1967 Downton and Tregunna (159) noted that low-compensation plants, characterized by the C_4 cycle, also have a specialized leaf anatomy. In these plants, but not in high-compensation C_3 plants, the ring (*Kranz*) of sheath cells surrounding the bundles are thick walled, highly developed, and contain many chloroplasts that are often highly specialized (498, and see Section IV,B,3). They are usually prominently loaded with starch (125). In Kranz leaves, stomata are often more numerous, bundles are usually closer together (127), intercellular air spaces are much reduced, and mesophyll is much denser and more compact than in C_3 leaves (see Fig. 13).

These anatomical modifications are evidently consistent with the function of the C_4 cycle, which is to pump or concentrate CO_2 into the bundle-sheath cells where C_3 photosynthesis takes place. This has led to the often-repeated statement that Kranz anatomy is essential for a functional C_4 photosynthetic cycle. This view has been further bolstered by the negative argument that in experimental crosses between C_3 and C_4 species progeny may exhibit either partial or apparently fully developed Kranz anatomy, but they do not have a functional C_4 cycle (56). However, in this circumstance the segregation of genes conferring biochemical and anatomical characteristics would need to be tightly coupled, which is improbable. The real question seems to be, Is it possible for a partly or fully functional C_4 cycle to exist without specialized anatomical characteristics that are apparently necessary for its operation?

At least one C_4 species, *Arundinella hirta*, has aberrant Kranz anatomy (126). Its leaves have strands of parenchyma cells that are morphologically (126) and biochemically (494) like Kranz cells, but which are not associated with bundles. *Arundinella hirta* also has normal Kranz cells around its bundles. However, Shomer-Ilan *et al.* have shown that *Suaeda monoica* (Chenopodiaceae) is a C_4 plant without Kranz anatomy (537). Although this plant is succulent in appearance, the authors carefully tested for the presence of CAM and demonstrated quite clearly that *S. monoica* is indeed a C_4 plant. Although this plant lacks the typical Kranz bundle sheath, it does have an internal layer of chlorenchyma that appears morphologically similar to normal Kranz cells and that is distinct from the thinner-walled outer layer of chlorenchyma. A somewhat similar abberant Kranz anatomy has been described for *Calligonum persicum*, a C_4 member of Polygonaceae, by Winter *et al.* (608). It appears probable from these reports that as long as there is spatial separation of the two cell types the operation of a C_4 cycle is possible.

A few species have been reported that are biochemically or physiologically intermediate between C_3 and C_4 plants, that is, their levels of PEP-case, decarboxylases, photorespiration, and CO_2 compensation are intermediate, and $^{14}CO_2$ pulse-chase experiments show some turnover of C_4 acids. These include *Mollugo verticillata*, a member of Aizoaceae, a family that contains C_3, C_4, and CAM species, (290) and several species of *Panicum* (*P. milioides, P. hians, P. laxum*) (78, 205). However, all these plants have typical Kranz anatomy (78, 159, 290). Furthermore, a number of C_3 dicots have a well-developed chlorophyllous bundle sheath (125). It would seem, therefore, that biochemical adaptations may be as important for C_4 photosynthesis as anatomical ones.

Attempts to analyze the photosynthetic capacity of tissue cultures or green, nonleafy tissues of C_4 plants have not shed much light on the subject. Cultures of *Portulaca oleracea* (288) and cotyledons, stems, or pith callus of *Amaranthus retroflexus* (578) were found to form C_4 acids and exhibit low photorespiration. However, the rates of photosynthesis were very low, and it is difficult to assess whether normal photosynthetic processes were occurring at all in these cultures. Sugarcane (*Saccharum officinarum*) stem-parenchyma cultures were found by Kortschack and Nickell (308) to lack both Kranz anatomy and C_4 photosynthesis. Most plant tissues exhibit some degree of nonphotosynthetic β-carboxylation that is not associated with C_4-cycle activity. Hence it is difficult to draw any conclusion about the operation of a C_4 photosynthetic cycle in the absence of Kranz anatomy from these experiments.

From time to time reports are published that suggest algae, including unicellular forms, have C_4 photosynthesis. However, these claims have not been substantiated, and as discussed earlier it is most unlikely that algae have a functional C_4 cycle.

It seems safe to conclude that the fully effective operation of a C_4 cycle requires some degree of anatomical specialization. At least, spatial separation of carboxylation (near the leaf surface) and decarboxylation (internal) reactions would appear to be necessary for successful operation of a CO_2-pumping mechanism. The association of the cells containing the carbon reduction mechanism and that of sucrose synthesis with vascular tissue seems to be frequent but not invariable, which suggests that this arrangement is advantageous but not essential. Clearly, different degrees of modification would suit plants to various environmental situations in which aspects of the C_4 photosynthetic syndrome are advantageous.

2. Sites and Organelles

Aside from the tissue-level specialization of Kranz anatomy, certain intercellular biochemical specializations are characteristic of C_4 plants (see Fig. 14). These have mostly been located with some degree of certainty, as shown in Fig. 14 and as discussed in Section IV,A. Difficulties in locating some enzymatic activities associated with the C_4 cycle may arise because two to several isoenzymes, often with different intracellular locations, may exist. In addition, the same enzyme may be associated with different metabolic sequences in different subcellular compartments. However, with the usual caveat about subcellular associations of organelles, and with the exception of the location of aminotransferase activities in mesophyll cells of PCK and, possibly, NAD-ME bundle-sheath cells, the location of enzymes in organelles seems reasonably certain.

Black and Mollenhauer (59) have presented data showing that photosynthetic capacity and intermediary metabolism are closely related to the distribution pattern and density of organelles in various tissues. This highlights the concept of organelle-packaged enzyme sequences. This point together with the preceding discussion of the interrelationship of anatomy and biochemical pathways emphasize the importance of structural or architectural organization of biochemical systems for the functioning of complex photosynthetic pathways.

3. Chloroplast Dimorphism[2]

Long before the discovery of C_4 photosynthesis, Rhoades and Carvalho (498) noted that chloroplasts in the bundle sheath of some tropical grasses, particularly maize, are agranal. Laetsch (e.g., 325) examined the chloroplasts of C_4 plants and showed that although mesophyll chloroplasts are small and lack starch, bundle-sheath chloroplasts are larger and agranal in some species. Chloroplasts in both types of cells have a well-developed reticulum of anastomosing tubules that are contiguous with the inner plastid membrane. Laetsch interpreted these morphological factors as adaptations for rapid transport of intermediates. Slack et al. (532) noted that malate transporters specifically lack grana and interpreted this in terms of the reducing power generated in bundle-sheath chloroplasts on decarboxylation of malate.

It is now clear that NADP-ME leaves usually contain agranal bundle-sheath chloroplasts, whereas leaves of other types do not. It has been suggested that agranal chloroplasts do not contain photosystem II (618), or that PSII is not linked with photosystem I (55), and that the reducing power for C_3-cycle operation is supplied by the transport of malate and its oxidative decarboxylation by NADP malic enzyme. However, this would supply only one reducing equivalent per CO_2, and the C_3 cycle requires two. More-careful investigation has demonstrated that agranal bundle-sheath chloroplasts do indeed have photosystem-II activity, although in lesser amounts than granal chloroplasts (e.g., 30, 176). Plastocyanin is an essential factor linking photosystems I and II, and its loss by

[2] See discussion in Chapter 1.

leaching from chloroplasts during preparation may have led to the earlier conclusion (533). Some more-recent data, based on chloroplast fluorescence yields, indicate that the relative proportion of photosystems I and II may in fact be the same in the mesophyll and bundle-sheath chloroplasts (166). It is therefore not immediately certain if the reduction or absence of grana stacks in NADP-ME leaves really reflects their lesser need for reducing power. On the other hand, it is possible that the experimentally determined high values may not reflect true natural conditions. Certainly, the coincidence of agranal chloroplasts and the decarboxylation step generating NADPH is suggestive.

It has been reported that agranal chloroplasts have higher ratios of chlorophyll a to b (620). Holden (247) analyzed the mesophyll and bundle-sheath chloroplasts of a wide range of C_4 plants and compared them with the values of C_3 plants. Some of these data are summarized in Table V. Holden found generally higher ratios for C_4 plants than for C_3. Bundle-sheath chloroplasts from monocots often (but not always) have higher ratios than mesophyll chloroplasts. But the difference between granal and agranal bundle-sheath chloroplasts is not significant. In most dicotyledonous C_4 plants there is little difference between bundle-sheath and mesophyll chloroplasts. It seems that C_4 plants tend to have higher ratios of chlorophyll a to b, but the relationship is neither precise nor usefully diagnostic.

It has often been noted that bundle-sheath chloroplasts of C_4 plants are the major sites of starch formation and that they can be separated from mesophyll chloroplasts because of their high density (due to the presence of starch granules; eg., 532). The enzymes of starch synthesis appear to be concentrated in bundle-sheath cells, although the mesophyll appears to be the main site of sucrose synthesis (158). This distinction is not absolute, however, After prolonged illumination, mesophyll cells also acquire starch, and the density difference between mesophyll and bundle-sheath chloroplasts then tends to disappear (532).

4. Transport

An obvious characteristic of C_4 photosynthesis, and one which has disturbed many plant physiologists, is the high level of inter- and intracellular traffic. For every molecule of hexose produced, at least six (more if photorespiration occurs) molecules of C_4 acid move from mesophyll to bundle-sheath, and six of C_3 acids return, in addition to cycling of cofactors and nitrogenous compounds. The question has often been raised, Is energy input through active transport necessary to maintain all the transport, or does it depend only on concentration gradients?

TABLE V

Chlorophyll Ratios (A/C) in Leaves and Tissues of C_3 and C_4 Plants[a]

	Whole-leaf chlorophyll (a/b)		Ratio of bundle sheath to mesophyll (a/b)	
	Average[b]	Range	Average[c]	Range
C_4 Dicotyledons	4.4	3.1–5.6	1.1	0.95–1.5
C_3 Dicotyledons	3.3	2.6–3.7	—	—
C_4 Monocotyledons	3.9	2.7–5.2	1.3	0.83–2.0
C_3 Monocotyledons	3.0	2.7–3.6	—	—

[a] Data from Holden (247).
[b] 10–25 species or varieties.
[c] 7–17 species.

Clearly, C_4 leaves are modified in ways that facilitate metabolite transport. The veins are closer and the distance between any mesophyll cell and the nearest bundle-sheath cell is usually no more than one cell width (e.g., 127). Osmond (422) calculated diffusion coefficients through plasmodesmata of the C_4 plant *Atriplex spongiosa* and concluded that the rapid movement of metabolities between cells could take place by diffusion and, therefore, that no special transport mechanism should be required.

On the other hand, Raghavendra and Das have shown that ouabain, which inhibits Na^+–K^+-dependent ATPase activity in C_4 leaves, also blocks the transfer of label from C_4 acids to photosynthetic end products in C_4 leaves, but not in C_3 leaves (472). They also found that the production of alanine in the NAD-ME plant *Amaranthus paniculatus* is light dependent and that the light requirement can be replaced by ATP in the presence of sodium ions (471). These observations suggest active transport mechanisms that might assist in the intercellular movement of C_4 metabolites, but the details are by no means clear.

A number of intracellular metabolite transport systems have been reported. Spinach (C_3) chloroplasts actively transport C_4 dicarboxylic acid across the inner membrane (337). Huber and Edwards have characterized translocators that move ATP (256), PEP and inorganic phosphates (258), and pyruvate (257) across the boundary of mesophyll chloroplasts of the C_4 plant *Digitaria sanguinalis*. It is possible that the energy-dependent transport of acids observed by Raghavendra and Das (471, 472) may be the result of this sort of chloroplast-membrane transport system. Although active transport of metabolites into and out of organelles seems likely by analogy with other systems, mechanism that might accomplish long-range active transport of metabolites are lacking.

C. DISTRIBUTION OF C_4 PHOTOSYNTHESIS

1. Systematics

The first clear analysis of the phylogenetic relationship of C_4 photosynthesis was made by Downton and Tregunna (159). They followed Brown's (80) analysis of the phylogenetic distribution of Kranz anatomy, studying various phylogenetic characteristics of C_4 photosynthesis as well as leaf anatomy and greatly extending the taxonomic range of the study. Since then a number of checklists of C_4 plants have appeared. Space does not permit their reproduction here. However, data to 1974 is provided by Downton (157) and Krenzer et al., (315), and that for 1974–1977 is provided in a carefully and critically documented list by Raghavendra and Das (473). So far, C_4 photosynthesis has been reported in 18 families (3 monocots and 15 dicots), 196 genera, and 943 species. A list of families containing C_4 plants is given in Table VI.

The most interesting systematic relationship is in Poaceae, the family that also contains the majority of C_4 species. The C_4 syndrome appears in many, but not all, of the species in the subfamilies Panicoideae and Eragrostoideae, but in none of the Festucoideae, Arundinoideae, Oryzoideae, or Bambusoideae. Gutierrez et al. (212) found that Panicoideae contained all three types of C_4-cycle plants, whereas Eragrostoideae were either NAD-ME or PCK.

The same workers found no PCK plants among dicots (212). They found that all C_4 Chenopodiaceae and Euphorbiaceae were NADP-ME, but other families contained representatives of both NAD- and NADP-ME types. It has been found repeatedly that C_3 and C_4 plants exist in the same family or genus, and it has been reported that both C_3 and C_4 individuals may occur within a single taxonomic species (56).

2. Evolution

All of the previous information suggests that the C_4 cycle has arisen in plants with C_3 photosynthesis on many occasions and in many diverse taxa at different periods during the evolution of angiosperms (56). Fossil

TABLE VI

FAMILIES REPORTED TO HAVE C$_4$ REPRESENTATIVES

Monocotyledonae		Dictoyledonae	
Cyperaceae	Acanthaceae	Boraginaceae	Nyctaginaceae
Liliaceae	Aizoaceae	Capparaceae	Polygalaceae
Poaceae	Amaranthaceae	Caryophyllaceae	Portulacaceae
	Asclepiadaceae	Chenopodiaceae	Scrophulariaceae
	Asteraceae	Euphorbiaceae	Zygophyllaceae

plants that have Kranz anatomy and the ^{13}C levels characteristic of C$_4$ plants have been found from the Pliocene epoch (401). However, this is fairly recent ($2–5 \times 10^6$ yr) in the evolutionary history of angiosperms. It does not seem possible at present to fix a time or period for the evolution of C$_4$ photosynthesis, except that it must have been quite recent, probably largely in the late Tertiary and early Quaternary Periods, following the major development and spread of angiosperms.

3. Genetics of C$_4$ Photosynthesis

Björkman has analyzed the genetic relationship among several species of *Atriplex*, some C$_3$ and some C$_4$ (56). His analysis suggests that the C$_4$ syndrome has arisen at least twice within this genus. Several crosses between related species of *Atriplex* have been possible, and one between *A. rosea* (C$_4$) and *A. patula* (*A. triangularis*) (C$_3$) has been followed for several generations. A number of hybrids having various C$_3$–C$_4$-intermediate morphologies have been produced, but they are all C$_3$ with respect to the normal criteria (photorespiration, O$_2$ effect, isotope discrimination, C$_4$ acid production, etc.). Some have more PEPcase than the C$_3$ parent or other enzymatic characteristics that are intermediate. None, however, is C$_4$.

Björkman and co-workers concluded that the genetic differences between C$_3$ and C$_4$ species of *Atriplex* are not great, perhaps amounting to only a few genes. Apparently, however, the requirements for functional C$_4$ photosynthesis are very precise, and the individual characteristics that together comprise the genetic basis for the C$_4$ cycle are, to a large degree, inherited independently of each other. This means that the chances of a hybrid inheriting a fully functional C$_4$ cycle are very low, and the chance of transferring the C$_4$ syndrome to economically important C$_3$ plants seems vanishingly small.

This conclusion is further supported by the fact that unorganized tissue from C_4 plants such as callus or nonleafy green tissues do not exhibit normal C_4 photosynthesis (288, 308, 578). Clearly, the C_4 cycle is not genetically coherent, but it is extremely sensitive to the absence of any of a number of essential but independently inherited anatomical and biochemical characteristics.

4. Intermediate Forms

In apparent contradiction to the conclusion reached by Björkman and colleagues, certain plants have been reported that appear to be intermediate in character between C_3 and C_4, that is, they possess an "imperfect" C_4 cycle. *Panicum milioides* has been found to be intermediate with respect to anatomical (478) and biochemical (205) characteristics. Its main fixation pathway is C_3 (205), but an operational PEPcase is apparently able to increase internal CO_2 to a limited degree, thus reducing photorespiration (479). Kennedy *et al.* (290, 516) have shown that *Mollugo verticillata* is a C_3–C_4 intermediate with ecotypic variations in the degree of C_4 metabolism. Similarly, species of *Moricandia* show various C_3–C_4 - intermediate anatomical and biochemical characteristics (12). The inference is that the syndrome may first appear imperfectly or partially in an evolutionary series. It has been reported that Japanese larch [*Larix kaempferi* (*L. leptolepis*)] has certain characteristics of a C_3–C_4 intermediate (195), but the evidence is not yet sufficient to extend the taxonomic range of C_4 photosynthesis to the gymnosperms. The small number of intermediate forms and their absence in nearly all genera having both C_3 and C_4 species argue against the likelihood that partial C_4 capability offers much evolutionary advantage.

5. Submersed Aquatic Macrophytes

Holaday and Bowes (246a) have published data for the aquatic macrophyte *Hydrilla verticillata* that indicates that this plant has partial C_4 capability under certain conditions of growth. In summer, or on long days, the plant has low Γ (10–25 μl/l) and fixes CO_2 by β-carboxylation into C_4 acids. These are later decarboxylated, and the resulting CO_2 is fixed by C_3 photosynthesis. *Hydrilla verticillata* has high levels of pyruvate, phosphate dikinase, which is uncharacteristic of C_3 plants. The primary decarboxylation mechanism is NAD-ME.

This type of photosynthetic metabolism, which has been called SAM photosynthesis, appears to be intermediate between C_4 and CAM (see Section V). It appears to be advantageous in a summer lake environment, where CO_2 or HCO_3^- levels may become very low during the day, but are usually normal at night.

6. C_4 Fixation in C_3 Plants

All plants show some capacity for C_4 carboxylation, but in C_3 plants this is usually a dark reaction and unconnected with photosynthesis. However, under certain conditions C_4 carboxylation in C_3 plants may be light enhanced, and this has led to much confusion about the extent of C_4 cycle operation in C_3 plants.

There seems to be no question that β-carboxylation frequently occurs in light in many plants and may even be stimulated or driven by light. The phenomenon has been noted, for example, in greening barley leaves (555), barley pericarp (413), *Chlorella* during photosynthetic induction (153), and wheat leaves in which C_3 photosynthesis is greatly reduced by abscisic-acid treatment (511). Stomatal photosynthesis has been shown to produce mainly C_4 acids (486), and photosynthesizing *Elodea canadensis* may produce C_4 acids that serve to maintain pH (143) (although this last example may in fact represent SAM photosynthesis). Many marine algae have active β-carboxylation (314). However, in none of these examples is there any hint of the operation of a C_4 cycle as such.

The operation of a "C_4 pathway" in spinach chloroplasts has been claimed (63). Certainly, there is no queston that spinach chloroplasts (like most other photosynthesizing systems) have a minor β-carboxylation fixation of CO_2 (517) and may transfer carbon from malate to the C_3 cycle (63) under conditions of malate breakdown. However, this does not constitute a C_4 cycle of carbon fixation. It has been suggested several times that blue-green algae have C_4 photosynthesis (e.g., 115, 154). However, the evidence for a C_4 cycle is equivocal, and cells exhibit varying rates of C_3–C_4 fixation depending on culture conditions (86). Blue-greens, like other algae (314), undoubtedly have high levels of β-carboxylation and active PEPcase. It may be that development of the C_4 syndrome in higher plants was possible because of the existence of a β-carboxylation ability in the earliest plant forms. Furthermore, there is no reason to doubt that this primitive β-carboxylation capacity is photosynthetic in some organisms. However, this simple type of light-driven β-carboxylation must be clearly differentiated from the elaborate cycle of photosynthetic reactions implicit in the term C_4 photosynthesis.

7. C_4 Photosynthesis during Development

It has been widely reported that juvenile as well as senescent leaves of corn (*Zea mays*) and other C_4 plants tend to emphasize C_3 photosynthesis. Senescing leaves of corn and *Portulaca oleracea* show this tendency due to a more extensive loss of C_4 enzymes (particularly PEPcase) than of RuBPcase (286, 604). In *Portulaca*, Kennedy and Laetsch found that not only the intensity of the C_4 cycle but also the ratio of aspartate to malate were markedly affected by the stage of leaf development (289). The C_3–C_4-intermediate plant *Mollugo nudicaulis* privides an interesting developmental situation. Its young leaves appear to be C_3 and lack Kranz anatomy, but Kranz anatomy and characteristics of C_4 photosynthesis develop as the leaves approach maturity (474). These varying characteristics are doubtless responsible for some of the confusion that has arisen regarding types of C_4 metabolism exhibited by different plants.

The development of C_4 photosynthesis in greening sugarcane leaves appears to involve the conversion of dark CO_2-fixation-type PEPcase into the enzyme form associated with the C_4 cycle, rather than *de novo* synthesis of new enzyme (203). This is in direct contrast to the pattern of *de novo* biosynthesis of RuBPcase in greening leaves of C_3 plants (see Volume VIII, Chapter 3). The concomitant development of the characteristics of Kranz anatomy and of the enzymatic machinery of photosynthesis seems not to have been studied in detail, so it is not possible to examine the developmental relationship between these two aspects of C_4 photosynthesis.

D. Significance of the C_4 Cycle

1. Ecological Significance

It has become generally accepted that the function of the C_4 cycle is to transfer CO_2 to bundle-sheath cells in the interior of the leaf (220). This increases the CO_2 concentration at the site of C_3 photosynthesis, permitting more rapid reductive fixation, less O_2 toxicity, and lower photorespiration. Because C_4 CO_2 absorption is more powerful, high rates of fixation are maintained with reduced stomatal aperture, permitting active photosynthesis under drought conditions. Characteristics of the carboxylation allow higher temperature optima in many C_4 species. These factors are largely interrelated, but they will be examined individually for the sake of clarity.

a. Temperature. Many C_4 species have much higher temperature optima than most C_3 plants, in the range 30–45°C (56). This is partly due to the low-O_2 effect on C_4 photosynthesis and photorespiration. In C_3 plants, O_2 toxicity and photorespiration increase with rising temperature, lowering net photosynthesis above 25–30°C. This is further exacerbated by the increased ratio of the solubility of O_2 to CO_2 at higher temperatures (321).

TABLE VII

QUANTUM YIELDS OF C_3 AND C_4 PHOTOSYNTHESIS AS
AFFECTED BY TEMPERATURE

Plant	Quantum yield (mol CO_2/einstein)		
	15°C	25°C	35°C
Zea mays (C_4)[a]	0.057	0.059	0.058
Atriplex rosea (C_4)[b]	0.054	0.054	0.054
Triticum aestivum (C_3)[a]	0.055	0.050	0.041
Encelia californica (C_3)[b]	0.067	0.059	0.046

[a] Data from Ku and Edwards (321).
[b] Data from Ehleringer and Björkman (165).

The quantum yield of C_4 plants is lower than that of C_3 plants at low temperature, but it is maintained at high temperature in C_4 plants, whereas in C_3 plants it drops substantially (Table VII). This indicates that C_4 photosynthesis is more efficient at high temperature, in addition to being more rapid. However, not all C_4 plants have high-temperature optima. Caldwell *et al.* (86) showed that cold-tolerant species of *Atriplex* undergo normal C_4 photosynthesis with only minor alterations in pathway at temperatures as low as 5°C. At this temperature many other C_4 species would suffer extensive chilling damage or death. Evidently, the C_4 syndrome does not require high temperature, though it confers specific benefits in this circumstance. Similarly, Long and Woolhouse showed that the temperate C_4 grass *Spartina townsendii*, although it has high rates of photosynthesis at elevated temperature, equals or exceeds the rates of C_3 grasses even at low temperature (349).

The high-temperature tolerance of C_4 plants appears to be due in part, at least, to the high-temperature tolerance of some of the C_4-cycle enzymes. PEPcase from C_4 plants has maximal rates at 30–35°C, whereas RuBPcase peaks at about 20–25°C (574). That PEPcase from C_4 leaves is protected from heat inactivation by the C_4 acids aspartate and malate is considered an adaptive protective mechanism by Rathnam (476). Conversely, the cold lability of maize pyruvate, phosphate dikinase (526) and PEPcase (392), which show a sharp break in activation energy at 11–12°C (the critical low temperature for maize growth), probably sets the low-temperature tolerance of this plant. In general, the cold lability of pyruvate, phosphate dikinase correlates with the cold senstivity of C_4 plants (547). It seems probable, therefore, that high-temperature tolerance is conferred by the enzymatic characteristics of C_4 plants. However, it is necessary that all photosynthetic enzymes exhibit adequate if not maximal activity at elevated temperatures, which has not yet been established.

b. O_2. Oxygen at atmospheric concentration inhibits C_3 photosynthesis by up to 25%, but it may actually stimulate C_4 photosynthesis, possibly by increasing pseudocyclic electron flow and hence ATP supply (252). The absence of air inhibiton of C_4 photosynthesis is almost certainly due to the insensitivity of PEPcase to O_2 (103) and to the absence of photorespiration in C_4 plants. Ehleringer and Björkman (165) have pointed out that absence of the O_2 effect on the quantum yield of C_4 photosynthesis almost precisely offsets the additional ATP requirement of the C_4 pathway, so C_3 and C_4 plants have nearly identical quantum yields (see Table VII).

c. CO_2 and H_2O. The capacity for rapid photosynthesis at reduced levels of CO_2 permits C_4 plants to maintain high rates of photosynthesis with stomata nearly closed, thus conserving water. In this respect the C_4 cycle, although it may not necessarily confer higher rates of photosynthesis, may produce much higher growth rates because photosynthesis can continue under conditions of water stress that would severely or totally inhibit it in C_3 plants. Powles *et al.* (458a) have suggested that this capability may also be beneficial by providing a sink for dissipating excess light energy under drought conditons, thus avoiding damage to the photosynthetic apparatus.

d. Light. The C_4 syndrome is most commonly found in plants of tropical or subtropical origin. The limiting factors for photosynthesis under full sunlight are CO_2 and water; light is usually superabundant. Under these conditions, the requirement for additional ATP to drive the reactions of the C_4 cycle appears to be an effective device for utilizing an otherwise wasted resource. Even when light is not limiting, the increased CO_2-absorbing capacity of the C_4 cycle offsets its lower efficiency (i.e., its higher quantum requirement per CO_2). Whereas the ratio (C_4/C_3) of growth rates for maize and rye grass may be 1.5 at high light intensity, it does not fall below 1.0 even at low light (387).

2. Photorespiration in C_4 Plants

The virtual absence of detectable CO_2 production in light from C_4 leaves might be due (*a*) to a genuine absence of the enzymes or organelles of photorespiration; (*b*) to reassimilation of photorespired CO_2; (*c*) to a reduction or inhibition of RuBP oxygenase by the high level of CO_2 developed in bundle-sheath cells by the C_4 cycle; or (d) to some other accessory mechanism that inhibits photorespiration in C_4 leaves. So far, no naturally occurring inhibitors of photorespiration have been found, and the last possibility remains no more than an unlikely speculation. However, a large number of papers have been published that support one or another of the first three alternatives.

Certain C_4 plants have been reported to have very low levels of photorespiratory enzymes or organelles. Osmond and Harris found that the activity of the glycolate pathway is so low in *Sorghum bicolor* leaves that little CO_2 production could occur (431), and Huang and Beevers found that the microbodies of *S. sudanense* are unspecialized and enzymologically unable to support photorespiration (250). However, many authors have reported the presence of peroxisomes (e.g., 194) and enzymes of photorespiration (e.g., 495) in C_4 leaves, although often at lower levels than in C_3 leaves (566). Numerous reports have also appeared showing that C_4 plants are capable of oxidizing supplied glycolate at rates approaching those of C_3 plants (e.g., 306), suggesting that they have the capacity for photorespiration. In this context, it has been found that certain C_4 plants (e.g., *Atriplex rosea, Amaranthus edulis*) produce a postillumination burst of CO_2 (e.g., 167), which is normally considered to be a carryover of photorespiration. However, the postillumination burst of these plants is O_2 insensitive and appears to result from decarboxylation of the pool of C_4 acid in aspartate-transporting types, not from photorespiration (156).

El-Sharkaway *et al.* (167) early proposed that C_4 plants do photorespire, but that much or all of the photorespired CO_2 is trapped and photosynthetically refixed before it can escape from the leaf. Numerous analyses have confirmed this conclusion. Rathnam (477) demonstrated photorespiration in the bundle-sheath cells of C_4 leaves by using inhibitors of C_4 carboxylation and decarboxylation, and Rathnam and Chollet (479) showed that the reduced photorespiration of the C_3–C_4-intermediate *Panicum milioides* is due to refixation of CO_2 by PEPcase. Chollet (104) showed that bundle-sheath cells of maize are capable of apparently normal photorespiratory metabolism. Using mass isotopes of CO_2, Volk and Jackson (584) also demonstrated its photosynthetic reassimilation in maize. Finally, Kennedy (287) showed that tissue cultures of the C_4 plant *Portulaca oleracea* do photorespire, although at a lower rate than comparable cultures from a C_3 plant, *Streptanthus tortuosus*.

The weight of evidence favors the conclusion that the majority of C_4 leaves do photorespire, though at a reduced rate, and that the photorespired CO_2 is reassimilated by PEPcase located in the mesophyll cells.

Certain C_4 leaves (e.g., *Sorghum* spp.) appear to have very little or no photorespiratory metabolism. However, for the majority of C_4 plants it is the presence and operation of the C_4 cycle that prevents external loss of photorespired CO_2.

3. Carbon Isotope Fractionation by C_4 Plants

It has been known for many years that plants discriminate against heavier isotopes of carbon during photosynthesis and, consequently, contain a smaller proportion of ^{13}C than that normally present in air. The difference is expressed as $\delta^{13}C‰$, which is given by the equation

$$\delta^{13}C‰ = \left(\frac{^{13}C/^{12}C \text{ in sample}}{^{13}C/^{12}C \text{ in standard}} - 1 \right) \times 1000.$$

Values of $\delta^{13}C$ are negative for plants because they contain less ^{13}C than the standard. Atmospheric air has a $\delta^{13}C$ value of $-7‰$ because the standard used, fossil skeleton of *Belemnitella americana* from the Peedee formation of South Carolina (PDB), is richer in ^{13}C than air. For this reason, $\delta^{13}C$ values for plants, as commonly expressed, need to be increased by $7‰$ to give true levels of discrimination. Several analyses of C_3 and C_4 plants (e.g., 534, 575) show that the $\delta^{13}C$ values for C_4 plants range from -10 to $-19‰$, whereas the range for C_3 plants is $-(21–33)‰$. As shown in Fig. 15, the values for C_3 and C_4 plants are well separated, and the differences appear to be maintained over a wide range of environmental conditions (535).

Various reasons for discrimination have been advanced. Ultimately, it relates to Graham's Law that diffusion rates of molecules are proportional to the square root of their mass. This would give a discrimination based on diffusion alone of $22‰$ for $^{14}CO_2$ and $11‰$ for $^{13}CO_2$. Very high values of about $150‰$ discrimination against $^{14}CO_2$ have been reported (581, 595) and are frequently quoted. However, these involve the assumption, now known to be incorrect, that $^{12}CO_2$ production continues at undiminished rate in light during $^{14}CO_2$ supply. Yemm and Bidwell found discrimination of 20 and $21‰$ against $^{14}CO_2$ during CO_2 uptake by NaOH solutions and photosynthesizing maize leaves, respectively, suggesting that the limiting factor in maize CO_2 absorption is diffusion (624). This would give an equivalent value for $\delta^{13}CO_2$, corrected for the PDB standard, of $-17‰$, which is low for maize but within the limits for C_4 plants (Fig. 15). However, plants grown in closed systems (greenhouses or growth chambers) tend to have $\delta^{13}C$ values that are $5–9‰$ more negative than those for field-grown plants (535), which would account for the low value found by Yemm and Bidwell.

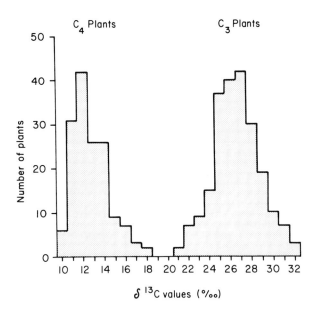

FIG. 15. Ratios of ^{13}C to ^{12}C in C_3 and C_4 plants. Data from Troughton (575) and Smith and Brown (534).

It has been shown in careful experiments by Wong *et al.* that RuBP-case discriminates against $^{14}CO_2$ by 27.1‰ (616), whereas the fractionation by PEPcase is only 2.03‰ (496). This strongly suggests that the difference in the primary carboxylating enzymes is responsible for the observed difference in $\delta^{13}C$ between C_3 and C_4 plants. As Wong *et al.* have pointed out in their discussion (616), the participation of RuBPcase in C_4 plants does not lead to high discrimination because the bundle-sheath RuBPcase fixes all available CO_2 and is thus unable to show discrimination.

Other components in the photosynthetic system (diffusion and hydration of CO_2, other enzymatic steps) may also contribute to a lesser degree to the $\delta^{13}C$ values (e.g., 496, 616). However, it seems clear that the major factor affecting $\delta^{13}C$ values is the presence or absence of PEPcase, indicating that $\delta^{13}C$ values are indeed diagnostic of C_4 photosynthesis. It should be noted that CAM plants fix carbon via PEPcase and that many CAM plants have $\delta^{13}C$ values in the range occupied by C_4 plants (424). Thus an appropriate $\delta^{13}C$ value alone cannot be considered sufficient to identify a plant as C_4 unless CAM can be ruled out.

4. Nitrogenous Metabolism in C_4 Photosynthesis

Aside from the obvious requirement for aminotransferase activity in the aspartate–alanine-transferring C_4 types (PCK and NAD-ME), C_4 plants need the capacity for very rapid assimilation of nitrogen to form the nitrogenous transfer compounds as required. Rathnam and Edwards (482) found adequate amounts of all nitrogen-assimilating enzymes in examples of all three types of C_4 plants. They also found that most of the required enzymes (nitrate reductase, nitrite reductase, glutamine synthetase, glutamic acid synthetase) are located mainly in mesophyll cells. NADH-glutamate dehydrogenase is present mainly in bundle-sheath cells, whereas NADPH-glutamate dehydrogenase is in both bundle-sheath and mesophyll cells. Based on experiments with *Panicum miliaceum*, Rathnam and Edwards concluded that mesophyll chloroplasts are the major site of nitrate assimilation.

Blackwood and Miflin (61) found that supplying excess nitrogen fertilizer to maize plants causes an increase in C_4 acid formation and a decrease in sugar and starch. Neyra and Hageman (408) showed, however, that malate (plus other organic acids) and nitrate content of C_4 plants are negatively correlated. They proposed that oxidation of malate can supply NADH for cytoplasmically located nitrate reductase but that it is not specifically required. They also found that nitrogen levels are negatively correlated with the level of the CO_2 supply.

It seems evident that sufficient enzymatic machinery and energy sources exist for the active nitrogen metabolism associated with C_4 photosynthesis. However, it is also clear that the level of C_4-cycle activity may affect the distribution of nitrogenous compounds and the pattern of their metabolism in C_4 plants. Conversely, the level of nitrogen nutrition also affects the pattern of C_4 metabolism. These factors may be important in the study of relative metabolic patterns of C_4 plants, and they need to be carefully considered in the analysis of experimental data.

An interesting point has been made by Brown (79), who observes that C_4 plants have a greater nitrogen-use efficiency (i.e., biomass production per unit of N in the plant) than C_3 plants. Brown suggests that this difference is due to the much smaller photosynthetic protein requirement of C_4 plants. Whereas C_3 plants may have up to 60% of their soluble protein as RuBPcase, it is widely reported to be present in 3–10-fold lower concentrations in C_4 plants. This means that although C_4 crops may have lower levels of leaf nitrogen and thus make less-nutritious fodder, they can produce larger amounts of carbohydrate on a limited nitrogen supply.

5. Summary

Very briefly, current evidence strongly favors the interpretation that the C_4 cycle improves photosynthetic rates under conditions of high temperature and light and intermediate or moderate water stress. The advantage of C_4 plants is thus not only that high rates of photosynthesis occur under optimal conditions, but also that photosynthesis may continue under conditions that would prevent it in C_3 plants. On the other hand, C_4 plants may be less efficient than C_3 plants under reduced light, intermediate temperature, or adequate water supply. Some C_4 plants suffer a distinct disadvantage if the temperature falls below a temperate level (about 11°C for maize).

Analysis of the distribution of C_3 and C_4 plants over a wide range of environment and habitats strongly reflects the selective advantages of these two photosynthetic types (548).

V. Crassulacean Acid Metabolism

A. Introduction

1. Definition

Crassulacean acid metabolism is a specialized pattern of photosynthesis in which CO_2 is absorbed and stored at night and released during the day inside the tissue, where it is fixed photosynthetically. This permits a remarkable degree of water conservation, because stomata can remain shut during the day. The nocturnal absorption of CO_2 is via β-carboxylation of PEP, which is catalyzed by PEPcase to produce malic acid. Decarboxylation of malic acid during the day yields CO_2 inside the photosynthetic tissue, which is fixed by RuBPcase and the C_3 cycle. Criteria for CAM given by Kluge and Ting (301) are that

1. Malate accumulates at night and is depleted by day.
2. Carbohydrates in storage fluctuate inversely with malate content.
3. CO_2 uptake usually takes place mainly at night.
4. Stomata are open at night and usually are closed during most of the day.

CAM has been the subject of numerous reviews, including discussions by Osmond (424) and Kluge and Ting (301) that cover the subject in great depth. It is apparent, however, that the details of intermediary CAM have not been worked out as well as those of the other photosynthetic cycles. CAM is a remarkably unusual and successful modification of the photosynthetic process for effective survival under extreme conditions and therefore warrants careful consideration.

2. Outline of CAM Reactions

The CAM sequence of reaction is, simply, a β-carboxylation reaction at night producing C_4 acids that are stored and that are subsequently decarboxylated in daytime to release CO_2, which is fixed by the C_3 photosynthetic cycle. A diagrammatic representation of this sequence is shown in Fig. 16. The key points are (a) nocturnal CO_2 fixation and acidification resulting from C_4 acid (malate) accumulation; (b) use of stored carbohydrates to sustain the nocturnal carboxylation reaction; and (c) release of the stored β-carboxyl carbon and its refixation by C_3 photosynthesis during the day behind closed stomata.

CAM photosynthesis has often been considered to be a variation of C_4 photosynthesis. In fact, CAM appears much earlier in the evolutionary sequence (ferns and gymnosperms), and, although it may possibly have been a forerunner of C_4 photosynthesis, there are important distinctions between the two photosynthetic systems. The first difference is that the C_4 carboxylation in CAM is separated from C_3-cycle operation in time rather than in space (as in C_4 plants). The second, a consequence of this temporal separation, is that no closed cycle of intermediates occurs. Instead, stored carbohydrates are used to provide the C_3 acid substrate for β-carboxylation, and the C_3 acid remaining after diurnal decarboxylation is disposed along various metabolic pathways (including the resynthesis of polysaccharides). Hence, although the reaction is cyclic in time, it is not cyclic in terms of the flow of carbon intermediates involved.

As might be expected, there are many biochemical and physiological variations on the pattern outlined in Fig. 16. Furthermore, the reaction mechanisms of different CAM types are complex, and the patterns of gas exchange in CAM plants are many. Evidently, precise control mechanisms are required to maintain a balance of the various temporal aspects of CAM. In addition, because such controls (e.g., over malate accumulation) must work in opposite directions at different times, some periodic or rhythmic control mechanism is required for integration of the temporal processes of CAM. These points, together with some discussion of the carbon nutritional significance of CAM, are considered in the following subsections.

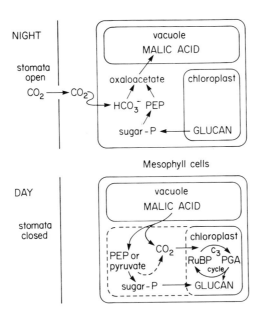

Fig. 16. Outline of Crassulacean Acid Metabolism. Intracellular location of malate decarboxylation and the subsequent metabolism of C₃ acids are not certain.

3. Succulence

The significant advantage of CAM is water economy, because CAM plants are able to maintain closed stomata during the day, when water loss is potentially greatest, and under extreme conditions of drought they can refix and retain essentially all the carbon that might be lost in respiration. In order to realize most fully the advantages from this sort of metabolism, a water-storage system is essential. Furthermore, large water-filled vacuoles are required in the photosynthetic cells for the storage of malic acid produced during nocturnal carboxylation. Thus CAM is associated in an obligatory manner with succulence, that is, the presence of water-storage tissue in either leaves or stems (301). The inverse relationship is not absolute; some succulent plants do not possess CAM. The requirement is that the water-storage cells are also photosynthetic.

The definition of succulence is not an easy one, and the concept may be misleading. Succulents are normally considered to have a saturating water content of about 5–15 g/dm² (S value), but a number of CAM plants fall within the range or close to the range of nonsucculents (S =

0.1–1.2) (299, 373). In some CAM plants much of the succulent tissue is unrelated to CAM because only photosynthetic tissue exhibits CAM. Kluge and Ting have proposed mesophyll succulence,

$$S_m = \frac{\text{water content of mesophyll (g)}}{\text{chlorophyll content of mesophyll (mg)}},$$

as an index of succulence more closely related to CAM (301). Apart from the experimental difficulty of accurately measuring S_m, the reported range of values for nonCAM species (0.33–1.25) comes rather close to overlapping the range for CAM species (1.34–13.0). Perhaps the concern is irrelevant; all that is necessary is the understanding that CAM requires the presence of some degree of succulence (i.e., large vacuoles) in the photosynthesizing cells.

Other than succulence, obvious structural or anatomical modifications do not appear requisite for the operation of CAM. Agranal chloroplasts are found in some CAM plants, but not in all (301). Apparently their presence is not linked with specific metabolic pathways, as it is in C_4 plants.

B. Pathways of CAM

1. Details of Reactions

As with C_4 photosynthesis, several possible pathways of CAM have been discovered. Like C_4, they primarily involve differences in the mechanisms of decarboxylation. Unlike C_4, they are not as easily categorized. Schemes showing the patterns of CAM, as they are presently understood, are shown in Fig. 17.

a. *CO_2 Fixation.* The fixation pathway (Fig. 17a) seems to be common and requires the breakdown of starch and other polyglucans to produce PEP, the substrate of carboxylation. PEPcase probably reacts with HCO_3^- (by analogy with the same enzyme in C_4 plants) to produce oxaloacetate, which is reduced by NAD-malate dehydrogenase to malate (587). At one time it was thought that a double carboxylation involving RuBPcase as well as PEPcase was required (71), based on the labeling pattern of malate after fixation of $^{14}CO_2$. Reexamination of the situation by Sutton and Osmond (548) and the development of more-precise degradation methods for malate have shown that only β-carboxylation takes place. In addition, extensive equilibration of the α- and β-carboxyls of malate occurs rapidly by fumarase equilibration (148). It now seems quite clear that PEPcase is the primary dark-carboxylating enzyme in CAM.

b. Malate Decarboxylation. Decarboxylation of malate may occur by one of three pathways, as in C_4 photosynthesis. Two of the pathways involve direct decarboxylation of malate, either by NAD-malic enzyme, as occurs in Crassulaceae, or, more commonly, by NADP-malic enzyme. The NAD enzyme is mitochondrial (149, 539), whereas NADP-malic enzyme is nonparticulate (539). Pyruvate resulting from malate decarboxylation is converted to PEP by pyruvate, phosphate dikinase located, as in C_4 plants, in chloroplasts (539).

CAM plants in certain families, including Liliaceae, Asclepiadaceae, Bromeliaceae, and some others, have low levels of NAD-malic enzyme and high levels of PEP carboxykinase (150). In these plants (Fig. 17c) malic acid is oxidized to oxaloacetic acid by NAD-malic dehydrogenase (mitochondrial) or, possibly, by NADP-malic dehydrogenase (probably cytoplasmic) (301). Oxaloacetate is decarboxylated by PEP carboxykinase, an enzyme that has variously been reported to be in the cytoplasm (150, 321a) or, as seems more likely, in chloroplasts (481, 517a). Kinetic analysis of the enzyme from pineapple (*Ananas comosus*) leaves by Daley *et al.* (132) suggests that its primary function is indeed decarboxylation in light, and it probably does not function as a carboxylase.

c. Refixation of CO_2. CO_2 produced by either decarboxylation step is refixed by RuBPcase in what appears to be a conventional C_3 cycle (301, 424).

d. Fate of the C_3 Product of Decarboxylation. It has been suggested that pyruvate or PEP resulting from malic decarboxylation might be oxidized to CO_2 by mitochondrial respiration and refixed by C_3 photosynthesis. However, several lines of evidence indicate that the major pathway is reductive conversion to triose phosphate and gluconeogenesis, as shown in Fig. 17b,c. Isotope experiments with specifically labeled pyruvate and analysis of respiratory behavior indicate that the major pathway is directly to sugars without intervening oxidation to CO_2 (see 424). However, it is not certain where the conversion of PEP to triose phosphate takes place. Spalding *et al.* (539) have shown that enolase and phosphoglyceryl mutase (Fig. 17, reaction 5) are cytoplasmic in the CAM plant *Sedum dendroideum* subsp. *praealtum* (*S. praealtum*), showing that the formation of PGA is a cytoplasmic reaction. However, the reduction of PGA to triose phosphate could take place in cytoplasm (NAD linked) or chloroplasts (NADP linked). Because chloroplast membranes are not very permeable to hexose phosphates, the formation of glucan precursors presumably takes place in plastids.

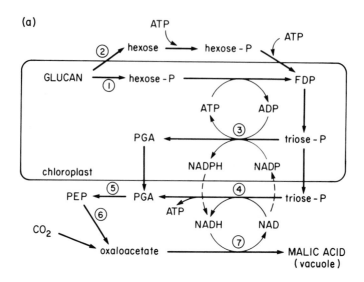

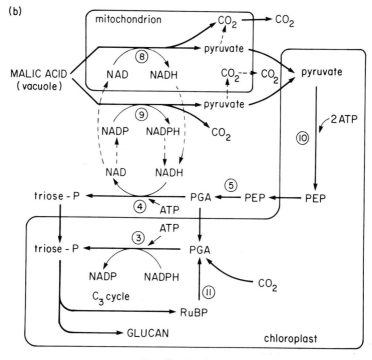

FIG. 17. (*Continues*)

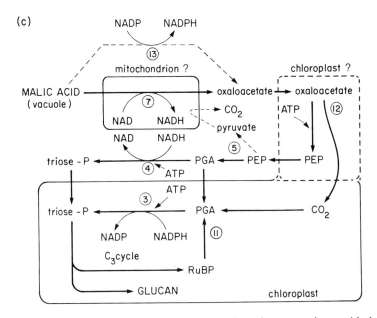

FIG. 17. Details of CAM patterns. Minor or uncertain pathways are shown with dashed arrows. Where the location or enzymatic machinery of major carbon pathways is uncertain, both alternatives are shown. (a) Nocturnal CO_2 fixation and malic acid synthesis. (b) Malic acid decarboxylation: NAD- or NADP-ME type. (c) Malic acid decarboxylation: PEP carboxykinase type. Enzymes: 1, phosphorylase; 2, amylase; 3, NADP-triose-P dehydrogenase; 3-PGA kinase; triose-P isomerase; 4, NAD-triose-P dehydrogenase; 3-PGA kinase; triose-P isomerase; 5, P-glyceryl mutase; enolase; 6, PEP carboxylase; 7, NAD-malate dehydrogenase; 8, NAD-malic enzyme; 9, NADP-malic enzyme; 10, pyruvate, phosphate dikinase; 11, RuBP carboxylase; 12, PEP carboxykinase; 13, NADP-malate dehydrogenase.

2. Transport

The location of enzymatic steps is not as certain in CAM as in C_4 plants. Inevitably, a certain amount of cofactor transport or reducing shuttle is required, but the nature or extent of such shuttles is not clear. As may be seen in Fig. 17, there are several points where oxidation and reduction steps may be coordinated by redox cofactors, but it is usually unclear whether NAD or NADP is the operating cofactor and if transmembrane transport is required. Osmond has suggested that all processes associated with decarboxylation and refixation in CAM plants may be located in chloroplasts (424). However, in light of the demonstration by Spalding *et al.* (539) that NAD malic enzyme is mitochondrial and that NADP malic enzyme, enolase, and phosphoglycerate mutase are nonparticulate, it seems probable that Fig. 17b,c is a somewhat more realistic, if less tidy, approximation of the situation. Clearly, much remains to be done to develop an understanding of the operation and integration of CAM.

C. CAM Patterns and Controls

1. CO_2 Exchange

The general pattern of CAM, in which CO_2 is fixed at night but is not fixed (or is weakly fixed) in daytime, is not invariable. Neales (403) recognizes various degrees of CAM, depending on the capacity for nighttime fixation and daytime retention of CO_2 and on the inverse behavior of water vapor. It is well known that the degree or intensity of CAM in some plants may be altered by changed environmental conditions, particularly water supply (428). In addition, the overall diurnal pattern of CAM has been shown (e.g., 423) to exhibit as many as four clearly defined phases, which are illustrated in Fig. 18.

Phase 1 (Fig. 18) is the CO_2 dark-uptake–acidification stage that is most characteristic of CAM plants. Variations in night temperature affect the rate of CO_2 uptake (549), partly as a direct result of the temperature effect on enzymes, but such variations also perturb the temporal pattern of CO_2 fixation by shifting the timing of peak uptake in an as yet unexplained way (299).

Phases 2–4 occur in light. In Phase 2, a brief surge of CO_2 fixation occurs that is not accompanied by further acidification. This is probably the result of RuBPcase activity before stomatal closure takes place. Because deacidification does not begin at once (295), this represents a lag phase before the biochemical reactions of deacidification begin.

Phase 3 represents the operation of the deacidification process and the fixation of CO_2 released by the C_3 cycle. Stomata are normally closed during this phase, even when plants are watered well.

Following deacidification, particularly if plants are watered well (423), there is a period (phase 4) during which stomata open and CO_2 fixation continues by operation of the C_3 cycle. Some labeled malate is formed when $^{14}CO_2$ is supplied during phase 4, but its labeling pattern indicates that it arises from double carboxylation involving both RuBPcase and PEPcase (425). However, malate accumulation does not occur during phase 4. Nalborczyk et al. (400) obtained $\delta^{13}C$ values of -10.6% in darkness (phase 1) and -25.9% in light (phase 4) (see Section IV,D,4). This strongly supports the concept that fixation is primarily by PEPcase in phase 1 and by RuBPcase in phase 4.

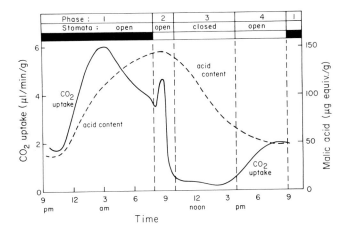

FIG. 18. Phases of gas exchange in a CAM plant, *Kalanchoë daigremontiana*. Adapted from Osmond (423, 424).

The pattern of CO_2 fixation may differ widely among CAM species and with time or environmental variations, and the outline shown in Fig. 18 probably represents an approximation, at best. Phase 2 may be reduced or absent in either weak or very strong CAM plants, and phase 4 may be reduced or absent in strong CAM plants or under extreme conditions of drought (403).

2. Environmental Factors

In general, the stomatal behavior of CAM plants (shown in Fig. 18) is modified in the expected way by environmental factors that affect the moisture content of plants. Decreased water content results in a longer and more-complete daytime stomatal closure (e.g., 297). However, drought sometimes causes greater stomatal opening at night (e.g., 299), with the result that net CO_2 assimilation may actually be increased. On the other hand, as Kluge and Ting point out (301), extreme drought might reduce the malic acid storage capacity of vacuoles, thus reducing the CO_2 storage capacity.

Other factors also affect CAM directly. Kluge and Ting (301) have suggested that the rate of deacidification and glucan synthesis during phase 3 is limited by the activity of the C_3 photosynthetic cycle. However, the duration of phase 3 would also be limited by factors affecting the amount of malate present and the rate of CO_2 fixation. Thus, more rapid CO_2 fixation at night (resulting in a high initial malate content) and factors reducing photosynthesis by day (low light, high temperature) prolong phase 3 at the expense of phase 4. Correspondingly, daytime and nighttime fixation of CO_2 are reciprocally related. Illumination intensity during the day directly affects phase-1 CO_2 fixation, perhaps by increasing the supply of C_3 precursors. CAM is also somewhat influenced by photoperiod, but this seems to relate primarily to the influence of photoperiod on stomatal behavior and the effects of total light received.

3. Endogenous Rhythms

Many workers have shown that circadian rhythms of CO_2 uptake or output occur in CAM plants placed in constant light or darkness (see 301). However, it is not yet possible to determine if the regulation of CAM requires biorhythmic phenomena. Endogenous stomatal rhythms are well known, and these may account for the rhythmic patterns of CO_2 exchange under constant conditions without specific effects on the underlying metabolic events. Lüttge and Ball (365) have shown that undamped diurnal oscillations of CO_2 uptake continue in constant light or dark but that malate oscillation continues for only one cycle, and then it is substantially damped. Circadian rhythms of photosynthetic and respiratory enzymes are known (128). Oscillation of PEPcase and malic enzyme (PEPcase high at night; malic enzyme high in daytime) have been reported in CAM plants (e.g., 467). Thus it is possible that the free-running oscillation of CO_2 uptake–release observed under constant conditions is a manifestation of the periodic behavior of some CAM enzymes or metabolic systems, but not of the operation of the whole integrated CAM. It appears probable, however, that elucidation of the nature and operation of biological clocks will shed some light on the periodic behavior of CAM.

4. Control Mechanisms

It is evident that precise and effective control mechanisms are necessary to integrate the temporal aspects of CAM. The known light-regulating mechanisms of the C_3 cycle (Section II,D) are adequate to account for the operation of this cycle in CAM. However, additional controls are required to regulate the dark operation of PEPcase, the enzymes of glucan conversion to PEP, and the alternating accumulation, storage, and remobilization of malic acid as well as the integration of the stomatal behavior of CAM plants with their metabolic needs. Several hypotheses or models of CAM behavior have been proposed, but, unfortunately, they all rely on assumptions or hypothetical propositions, and much further work is required before it will be possible to choose among them.

The four main hypotheses involve control by (*a*) diurnal temperature fluctuations; (*b*) competition between carboxylases for available CO_2 or C_3 intermediates; (*c*) endogenous rhythms; and (*d*) feedback inhibition of PEPcase by malate, perhaps with malate-transport regulation. In addition, PEPcase is reported to be controlled by light (585a).

a. Thermoperiodic Regulation. It has been noted that the enzymes of malate formation, PEPcase and malate dehydrogenase, have a lower temperature optimum (35°C) than malic enzyme (>53°C). Brandon (74) suggested that this differential would allow net malate synthesis during cooler periods, at night, and malate breakdown during the hotter day. However, diurnal acid variation has been shown to continue under constant temperatures (297), and CAM operates equally well at high or low average temperatures. Although this mechanism might be involved in CAM control, it could not be solely responsible for it.

b. Competition. If PEPcase and RuBPcase are not otherwise regulated, they must compete for CO_2 and a source of C_3 acid. However, such competition would occur only in daytime because the C_3 cycle is inactive at night. Beevers *et al.* (32) suggested that PEPcase activity is reduced in daytime due to competition with RuBPcase for CO_2, allowing deacidification to take place. At night, however, PEPcase activity is maximal, causing acidification. Osmond and Alloway (425) proposed that competition is rather for the C_3 substrate, PGA, which is preferentially channeled into the C_3 cycle by day, permitting deacidification. The strongest evidence against these models is that malic acid accumulation continues if the plant is illuminated during the night period and that diurnal oscillation of malate content continues in continuous light (365), although for one cycle only. Furthermore, from what is known of the $K_{mCO_2-HCO_3^-}$ of RuBPcase and PEPcase (see Sections II and IV), the uptake of CO_2 by PEPcase even in light might be expected to be as vigorous as that by RuBPcase, other factors being equal.

c. Endogenous Rhythms. As mentioned earlier, many enzymes are known to fluctuate in activity in a circadian rhythm. The strong circadian pattern of CAM suggests that some endogenous pacemaker may well be involved in its regulation. Unfortunately, not enough is known about the nature of such a pacemaker or the possible site of its action to allow useful speculation here about the way in which it might control CAM.

d. Feedback Regulation. Kluge and co-workers (296, 300) investigated the inhibition of PEPcase by malate and formed the hypothesis that malate carboxylation is controlled through feedback regulation of PEPcase by malate. Later they reported that PEPcase becomes much less sensitive to malate control at night (296a). That malate is compartmented from cytoplasmic PEPcase by storage in the vacuole led Kluge and Heininger (298) to postulate that control of malate transport across the tonoplast might also be involved, and a model based on a hypothetical turgor mechanism has been developed (366). In essence, it is assumed that increased turgor increases malate export from the vacuole and that vacuolar storage of malate increases at low turgor. Thus in phase 1 malate is transported into the vacuole as fast as it is formed, allowing continuous rapid operation of PEPcase. As the malate concentration in the vacuole increases, the osmotic potential in the vacuole falls, and turgor pressure increases until the point is reached at which malate efflux commences, corresponding with the onset of phase 2 or 3. Deacidification follows because PEPcase becomes inhibited. An additional point of feedback control is that phosphofructokinase activity is inhibited by PEP (281). This provides a means of controlling glycolysis, thus regulating the supply of C_3 precursors of PEP and the fate of C_3 residues of decarboxylation.

Some data support this model. Tissue acidification or deacidification is affected in the expected way by lowering or raising the water potential of the bathing fluid (366). However, the calculated internal pressure potential of *Kalanchoë daigremontiana* leaves appears to be essentially unchanged during the day (364). It is possible that light, perhaps through the influence of reducing equivalents from chloroplasts, may modulate tonoplast transport of malate (400). This would explain the lag period during phase 2 before deacidification begins.

However, this phenomenon could also be explained by the slow (~90 min to completion) inactivation of PEPcase by light (585a) or by the fact that PEPcase inhibition by malate appears to be greatest by day (296a). Investigations by Pierre and Queiroz (449) show that there is great flexibility in CAM patterns relating to culture conditions. These workers stress that there may be alternative regulating mechanisms that could operate at different times or under different environmental conditions to regulate the overall integration of the various metabolic sequences that are part of CAM.

e. Stomatal Regulation. In spite of the paradoxical behavior of stomata in CAM plants, the mechanisms underlying their stomatal responses are probably identical in most respects with those of conventional C_3 plants. Potassium-ion accumulation has been reported in the guard cells of a CAM plant, *Crassula argentea,* during night opening (137). Stomata of CAM plants are responsive to internal CO_2 concentration (112). Thus the CO_2 concentration mechanism probably controls CAM stomata; internal concentration is reduced at night (phase 1) during acidification, stimulating opening, and is greatly increased during deacidification (phase 3), causing closure. Stomata are normally open during phases 2 and 4, when deacidification is not taking place.

D. SIGNIFICANCE OF CAM

1. Environmental Significance

It has been clearly demonstrated that CAM is an adaptation to extremely xerophytic environments (e.g., 301). The carbon-fixation rates in CAM plants are generally somewhat lower than those of C_3 or C_4 plants, as shown in Table VIII, and the net daily assimilation of CAM leaves is probably only about half that of comparable C_3 leaves. However, photosynthesis by C_3 leaves may cease entirely during severe water stress, whereas CO_2 uptake continues in CAM leaves, particularly in the dark phase. Finally, under extreme stress, C_3 and C_4 leaves experience net carbon loss due to dark respiration, whereas CAM leaves or tissues are able to retain and reassimilate respired CO_2, essentially eliminating respiratory loss. Certain cacti can maintain almost constant dry weight over periods of weeks under virtually absolute drought conditions.

That CO_2 assimilation by the C_3 pathway can function at times (e.g., phase 4) without the intervention of β-carboxylation has led to the suggestion, supported by $\delta^{13}C$ values and acid fluctuation data, that CAM is

TABLE VIII

ESTIMATED RATES OF CO_2 UPTAKE AND CARBON ASSIMILATION OF CAM, C_3, AND C_4 LEAVES[a]

Leaves	Net CO_2 uptake (mg/dm²/hr)	Net daily carbon gain (mg/dm²)[b]
CAM (dark, phase 1)	15	57
CAM (light, phases 2 and 4)	10	
C_3	30	110
C_4	45	165

[a] Based on averages from numerous literature reports.
[b] Carbon-assimilation values assume a 14-hr day and include dark-respiratory loss.

an option rather than an obligate metabolic system in many succulent plants (426). Certain plants, such as *Mesembryanthemum* spp. (609), *Portulacaria afra*, and *Peperomia obtusifolia* (215), appear able to switch to C_3 photosynthesis when adequately watered. Others, mostly cacti such as *Opuntia basilaris* (214) and *Zygocactus truncatus* (215), appear to be obligately CAM regardless of the moisture available. Apparently cacti are not unique; *Agave deserti* continues CAM even when watered (411).

By contrast, *Mesembryanthemum crystallinum*, a facultative CAM species, is apparently unable to exhibit CAM in the absence of high levels (100 mM) of inorganic salts in the soil (62). *Kalanchoë tubiflora* (*Bryophyllum tubiflorum*), also a facultative CAM species, requires small amounts of NaCl before CAM can takes place (82). This may be analogous to the Na^+ requirement of C_4 plants (81). At present, the nature of this requirement and the mechanism of Na^+ action are unknown.

2. Taxonomy, Distribution, and Development of CAM

a. Taxonomy and Evolution. The occurrence of CAM has been widely documented, and several check lists have been published (e.g., 60, 548a, 550). Twenty-three families of flowering plants as well as two more primitive families, listed in Table IX, have been reported to have CAM representatives. The CAM ferns are epiphytic, and their metabolism has been carefully examined by Hew and co-workers (242, 615). The inclusion of *Welwitschia mirabilis* is based on substantial investigation by Dittrich and Heber (151) and Schulze *et al.* (519). A report by Santakumari *et al.* (512) provides data on the diurnal cycles of malic acid and CO_2 fixation of chickpea (*Cicer arietinum*), suggesting that this important leguminous crop plant has CAM. However, Winter has been unsuccessful in his attempts to repeat this work and has found no signs of CAM in chickpea even under water stress (607a). Evidently it would be inappropriate to add chickpea (and Leguminosae) to the list of CAM plants at present.

Several interesting points arise from a comparison of Tables IX and VI (which lists families having C_4 representatives). Like the C_4 syndrome, CAM seems to be represented in widely scattered families and is found in a fern and a primitive gymnosperm. CAM and C_4 photosynthesis evidently appeared independently in many families or genera. This suggests that an evolutionary development resulting in β-carboxylation photosynthesis is relatively "close to the surface" in all groups of plants. Presumably, it occurs often but confers an advantage and is thus retained only under specific environmental conditions. This argument is supported by the identification of SAM photosynthesis, a condition in-

TABLE IX

FAMILIES REPORTED TO HAVE CAM REPRESENTATIVES

Filicophyta	Gymnospermae	Monocoty-ledonae	Dicotyledonae	
Polypodiaceae	Welwitschiaceae	Agavaceae	Aizoaceae[a]	Didiereaceae
		Bromeliaceae	Asclepiadaceae[a]	Euphorbiaceae[a]
		Liliaceae[a]	Asteraceae[a]	Geraniaceae
		Orchidaceae	Bataceae	Lamiaceae
			Cactaceae	Oxalidaceae
			Caryophyllaceae[a]	Piperaceae
			Convolvulaceae	Plantaginaceae
			Crassulaceae	Portulacaceae[a]
			Cucurbitaceae	Vitaceae

[a] Family also has C_4 representatives (see Table VI).

termediate between CAM and C_4 photosynthesis, which confers an advantage in summer lakes that may have very low daytime CO_2 content. However, the long and fruitless search for C_4 capabilities in C_3 plants and the failure to produce functional C_4 plants from $C_4 \times C_3$ crosses suggest that some unknown but very specific prerequisite is necessary for the development of CAM or C_4 photosynthesis. That only a few families contain representatives of both types provides support for their independent evolution.

Comparison of the occurrence of CAM and C_4 photosynthesis is interesting. Assuming that they would appear with roughly equal frequency under equal evolutionary pressure (although this assumption may not be justified), one would conclude from the somewhat wider distribution of CAM that the environmental pressure leading to it is somewhat greater than that resulting in C_4 photosynthesis. In other words, the survival value of CAM may be somewhat greater than that of the C_4 cycle, in spite of the greater productivity associated with C_4 photosynthesis.

b. Distribution. As might be expected, CAM plants are distributed largely in the semiarid or arid areas of the tropical and subtropical climates (301). The major desert areas of the world, physiologically dry habitats that include saline and mountainous areas, and epiphytic habitats contain the richest diversity of CAM plants.

c. Development. Two separate studies show that young leaves of the CAM plants *Mesembryanthemum crystallinum* (585) and *Kalanchoë pinnata* (605) are C_3 and that they progressively acquire CAM potential as they mature. Because the guard cells of *Tulipa gesneriana* and *Commelina communis* fix CO_2 by β-carboxylation at night and appear to metabolize small amounts of the fixed carbon via the C_3 cycle in light, Willmer and Dittrich have suggested that these cells exhibit CAM (605). However, the data are also consistent with normal dark fixation and limited conventional C_3-cycle activity (cf. 486). The concept of CAM in guard cells, although it has interesting possibilities in connection with stomatal operation, needs further experimental work.

3. Photorespiration

Because CAM plants fix CO_2 by the C_3 cycle, it appears very likely that they do photorespire. There is not much opportunity for its manifestation because they absorb CO_2 only during brief intervals in light. However, several lines of evidence support the existence of photorespiration. Crews *et al.* (124) have shown that characteristic postillumination bursts occur. Badger *et al.* (19) have shown that RuBPcase from *Kalanchoë daigremontiana* has the same oxygenase activity as that of C_3 plants, and Kapil *et al.* (280) have reported that CAM leaves have peroxisomes. Finally, the O_2 sensitivity of CO_2 assimilation in light, as determined by Osmond and Björkman (430), and the compensation point and its sensitivity to O_2, as measured by Allaway *et al.* (2), are all characteristic of C_3 plants. Evidently, CAM plants reabsorb photorespired CO_2 during phase 3 when stomata are closed and release it normally when they are open.

4. Carbon Isotope Discrimination

The carbon isotope discrimination of CAM plants falls in the range of both C_3 and C_4 plants (cf. Fig. 15). Their $\delta^{13}C$ values range between -10 and $-27‰$. This is hardly surprising, however, for two reasons. First, PEPcase discriminates weakly against ^{13}C, whereas RuBPcase discriminates strongly (see Section IV,D,3), and the two enzymes may act as the primary carboxylase in different ratios, depending on environmental conditions. Second, phase 3 CO_2 fixation in light by RuBPcase utilizes a closed pool of CO_2, so there is little chance for selective uptake of isotopes. Thus $\delta^{13}C$ values of CAM plants indicate the major metabolic pathways used to fix CO_2. Nalborczyk *et al.* (400) showed that $\delta^{13}C$ values for *Kalanchoë daigremontiana* were -10.6, -25.9, and $-15.0‰$ when CO_2 was supplied only in dark, only in light, or continuously, respectively. Osmond *et al.* (428) found good correlation between $\delta^{13}C$ values and the level of water stress during growth of the same plant. Deleens *et al.* (145) showed that ^{13}C content of photosynthetic intermediates in *Kalanchoë blossfeldiana* varies with day length. During short days CAM is active and

[13]C levels of intermediates are higher than during long days, when the C3 pathway predominates. Thus $\delta^{13}C$ values may be useful in determining the environmental conditions under which CAM plants grew, insofar as it might affect the proportion of PEPcase and open-stomata RuBPcase incorporation. However, they cannot be used as a diagnostic characteristic for CAM plants.

5. Summary

Much less is known about CAM than the C_4 pathway of photosynthesis, partly because CAM does not carry with it the same striking increase in productivity. However, there are good reasons why much more study should be devoted to this distinctive and complex metabolic system. First, an understanding of its regulation will undoubtedly throw much light on the whole picture of the regulation of complex biological systems. This is particularly so because of the evidence that several unrelated but interlocking regulatory systems cooperate to control CAM. Second, an understanding of the evolution of CAM may be essential for the understanding and use of genetic principles in the development of CAM or C_4 crop plants. Third, there may be unexpected economic returns from a better understanding of CAM. CAM appears to be more flexible and adaptive than C_4 photosynthesis. At present, there is only one major economic CAM plant—the pineapple (*Ananas comosus*). However, a number of minor economic plants such as *Yucca*, *Agave*, and *Vanilla* exhibit CAM. With the looming world crisis in energy and food, the need will surely arise to make economic use of the huge amounts of solar energy that fall on desert or semidesert lands, and new CAM crop plants will undoubtedly be an effective way to meet this need.

Finally, the demonstration that the C_4 succulent *Portulaca oleracea* can show CAM activity under drought stress (306a) and the suggestion that an important crop plant such as chickpea may survive and grow under quite severe drought conditions (512) strongly recommend facultative CAM as an important future goal in the genetic engineering of crops.

VI. Nonphotosynthetic Reactions

A. Introduction

1. Scope

The major reaction sequence involved in carbon nutrition, apart from the photosynthetic cycles, is respiration. However, the energy relations of respiration and the overall picture of respiration as an integrated system of plant metabolism are covered elsewhere in this volume (Chapters 2 and 4). This section is confined, therefore, to an analysis of the roles of respiration in the context of the carbon nutrition of the plant and the relationship between the assimilation, storage, mobilization, and use of carbon for plant growth and development. The role of CO_2 dark fixation in the carbon economy of the plant is also briefly discussed.

2. Roles of Respiration

One of the major developments of the past has been the attempt to account for respiration in terms of the carbon and energy economy of plants. Intense interest in the improvement of crop production led initially to the concept that respiration, like photorespiration, should be controlled to eliminate wasteful carbon loss. However, a more liberal view began to emerge in 1969 (523; summarized in 371), which recognizes the dual role of respiration in maintenance and growth.

Part of the respiratory process is used to provide energy for the conversion of the photosynthate and mineral nutrients into new plant substance and has been called growth respiration. The remainder provides energy for all the chemical (e.g., turnover) and physical (e.g., transport) facets of plant maintenance. Because growth depends on the rate of photosynthesis, and maintenance on the mass (dry weight) of the plant, the concept has been expressed by McCree (371, also in 523)

$$R = kP + cW,$$

where R and P are the rates of respiration and gross photosynthesis, respectively, W is dry weight, and k and c are constants. Preliminary data reported by McCree (371) indicate that values of 0.25–0.3 for k and 0.01–0.04 for c may be expected for growing crop plants. This means that 25–30% of photosynthetically fixed carbon is lost as CO_2 in the course of its conversion into plant substance and that a plant loses 1–4% of its weight per day for maintenance processes.

The value for k is consistent with the known efficiencies of biochemical pathways that convert primary photosynthate to structural materials in the plant. It suggests that substantial gains in productivity might be made only by reducing undesirable or unproductive growth, because it is unlikely that the efficiency of growth respiration could be improved. The lower value of maintenance respiration makes it a less worthwhile target for studies of crop improvement, except late in the life of the plant when growth ceases and maintenance becomes the major component of respiratory activity. Even then, it is difficult to see which components of maintenance could be restricted while maintaining healthy operation of the plant. Penning de Vries has made an extensive analysis of this problem (445) and has concluded that the most important processes supported by maintenance respiration are active transport and protein turnover. It appears unlikely that active transport could be reduced without deleterious effects on the plant or that protein turnover is susceptible to manipulation. It seems unlikely that substantial gains in growth or productivity can be realized.

3. Alternative Sources of Energy and Materials

The simple approach just outlined served to clarify the confusion about the roles of respiration and the degree to which it might be manipulated. However, the term describing growth respiration, kP, is only applicable under steady-state conditions because of the existence of large pools of stored photosynthate, which may be tapped for respiratory substrates during periods of low photosynthesis. Thornely (560) and Hansen (216) have attacked this problem by considering the dry weight of the plant to be of two categories, nondegradable and degradable. Mathematical analyses based on this approach have been developed (e.g., 560).

A second problem arises because the model requires the assumption that dark respiration continues unabated in light. Certainly, growth and maintenance continue in light, perhaps at accelerated rates. However, the assumption that the required energy is derived only from respiration is unlikely to be correct. It is probable that a substantial amount of photochemical potential becomes directly available to drive functions that would, in darkness, use respiratory (i.e., stored photochemical) potential. Thus measurement even of the total carbon flux of a plant does not necessarily give an estimate of energy flux. Furthermore, certain processes (e.g., photorespiration) may result in carbon flux with little or no energy transfer. Various attempts have been made to analyze the extent to which light-generated reducing power or ATP replaces or supplements respiration. These ideas and the inverse proposition (the amount and mechanism by which light controls respiration) are considered in Section VI,C,3 and 4.

B. RESPIRATORY EFFICIENCY[3]

1. Estimation of Maintenance and Growth Respiration

It is necessary to know approximately for what the energy derived from respiration is being used in order to determine the efficiency of

[3] For further discussion of respiratory efficiency and the physiology of respiration, see Chapter 4, Section VII,A. (Ed.)

respiration in a plant. As mentioned earlier, it is convenient to divide the main uses of respiratory energy into two components, maintenance and growth. The independent estimation of these two components requires certain assumptions. Growth respiration can be estimated by making a complete analysis of the biochemical requirements and pathways of growth, as illustrated by Penning de Vries (446). Alternatively, plants may be supplied briefly with $^{14}CO_2$ in light, and the proportion of assimilated ^{14}C remaining in the plant after 24 hr gives a measure of growth conversion (505). Analysis of the rates of respiration of clover plants (*Trifolium repens*) of different weights at varying light intensities (which caused varying rates of photosynthesis) showed that respiration fits the equation $R = kP + cW$ given earlier (370) and permits calculation of k and c. Maintenance respiration has also been measured in plants starved for 48 hr, by which time growth has ceased (217, 371). A number of values for the two constants (k and c) have been determined by these techniques for several crop plant species. They agree reasonably well, giving values of 0.2–0.3 for k and 0.015–0.040 for c (217, 370).

Some proportion of the maintenance respiration estimated in this manner may well be the result of wasteful respiration, a useless hydrolysis of ATP or oxidation of NADH or NADPH by pathways uncoupled with ATP formation such as ascorbic acid oxidase, glycolic acid oxidase, or phenol oxidase, as discussed by Beevers (31). The exact levels of such useless respiration are not known, but they doubtlessly vary with age and condition of the plant. The increased respiration at high temperature or during respiration climacteric may fall largely within this category of metabolism. Possibly the term "useless" implies a value judgment based on ignorance, and such respiratory processes may be an essential part of plant metabolism. This question cannot be resolved until more information is at hand.

2. Variations in Respiratory Efficiency

The productivity of a plant is affected by the intensity and efficiency of its respiratory processes, so factors that affect these parameters in crop plants are of great interest. Lambers and Steingröver (329) have provided data that shows that respiratory efficiency of roots varies widely among species. Their data suggest that wasteful processes may occur at two levels; the efficiency of ATP formation during electron transport may be low because of the activity of alternative oxidases, and the efficiency of ATP use may vary, perhaps because of ATPase activity. Radin *et al.* (470) have shown that cotton (*Gossypium hirsutum*) roots contain a mechanism for partitioning respiratory energy output between growth and nitrate reduction so that growth continues in preference to nitrate reduction if sugar is limiting. No information is presently available about the nature of mechanisms affecting efficiency or partitioning respiratory output, but these observations suggest possible avenues for future studies of the control of wasteful or unnecessary respiratory processes.

3. Respiration and Demand

Respiratory control by cofactors [NAD(H), energy charge level] is well established. It follows that respiratory rates should vary with the demand for respiratory energy, perhaps modified by the requirement for metabolic intermediates. Numerous reports indicate that this is approximately so. For example, Mousseau showed that respiration of *Chenopodium polyspermum* (a short-day plant) is greater during short days but that the carbon balance of the plant is modified so that the relationship between CO_2 input and output remains constant (391). However, the data suggest that respiratory control is a complex phenomenon. If growing parts of the plants are excised, early-night respiration is reduced, as would be expected if the rate of respiration is controlled by the demands of growth. However, a late-night respiratory peak, which is strongly affected by photoperiod, was not affected by excision, suggesting an endogenous rhythmic control of this phase of respiration.

The effects of temperature on respiration are complex and difficult to resolve. Respiration rates increase greatly as temperature increases, but respiratory efficiency does not usually do so. Above the characteristic night-temperature optimum, growth and maintenance activities of a plant do not normally increase as rapidly with temperature as does respiration. The inference is that wasteful or uncoupled respiration increases at elevated temperatures. Evidently, plants that tolerate high night temperatures must possess mechanisms that prevent inefficiency or maintain respiratory control at high temperatures.

This situation has been examined by Pearcy using *Atriplex lentiformis* grown under conditions of either high (43°/30° day/night) or low (23°/8°) temperatures (442). The temperature dependence of plants grown at high temperature is dramatically less than that of plants grown at low temperature. High-temperature plants have a broad plateau in respiratory rate between 30 and 40°, whereas respiration of low-temperature plants has a Q_{10} of 2, and above 40°C the rate of low-temperature plants is two to three times higher than that of high-temperature plants. In addition, Pearcy observed that plants from different (hot desert or

cooler coastal) habitats have different capabilities of acclimating to high-
or low-temperature culture conditions.

Unfortunately, no information is presently available about the mechanisms underlying the plasticity of the response capability of respiration
to environmental variation or stress. However, it seems clear that experimental systems of the type just described will be most useful in the
search for ways to improve the performance of crop plants, particularly
those grown under extreme conditions.

4. Pathways of Respiration

Some thought has been given to the fact that different pathways of
respiration have different efficiencies of energy conversion. The ex-
tremes are probably represented by citric-acid-cycle metabolism coupled
as closely as possible with the electron transport chain on one hand, and
various oxidases (e.g., glycolate oxidase or ascorbic acid oxidase) on the
other. As conventionally represented, pentose-phosphate-pathway respiration is only slightly less efficient, in terms of ATP produced per
molecule of glucose oxidized, than the citric acid cycle. Pentose phosphate respiration is associated with the cellular NADPH requirement for
synthesis of lignin (462) or fat, for example. Well-known changes in
respiratory pathways during development doubtlessly relate to this phenomenon.

Alternate or cyanide-insensitive pathways appear to be somewhat less
efficient than cyanide-sensitive electron transport. However, present evidence does not link either electron transport path with specific sequences of carbon metabolism. Cyanide-insensitive electron transport is
closely concerned with thermogenic respiration (374), and its presence
in roots (e.g., 330) may partly explain the lower efficiency of root maintenance respiration noted in the previous section (cf. 328). Theologis
and Laties (556) concluded from studies of potato (*Solanum tuberosum*)
and sweet potato (*Ipomoea batatas*) slices that the operation of the alternate path is controlled by substrate fluxes. They suggested that the alternate path is engaged when the rate of substrate oxidation exceeds the
electron transport capacity of the cytochrome path. They also reported
that the climacteric rise in banana (*Musa*) and avocado (*Persea*) fruit
respiration is primarily mediated by the cytochrome path unless the
respiratory flux exceeds its capacity, when the alternate pathway becomes engaged (557).[4] It may be inferred that major improvements in
productivity by controlling the respiratory pathway, either carbon or
electron transport, are not likely.

---

[4] For a comprehensive analysis of alternate respiration, see Chapter 4. (Ed.)

C. RESPIRATION AND PHOTOSYNTHESIS

1. Net Assimilation

Net assimilation means the gain in carbon and other elements resulting from photosynthetic conversion of light energy, less the losses due to
respiration. It is clear that photosynthesis is normally the major term in
the definition of net assimilation, and as such it has received the largest
share of recent research in crop improvement. However, respiratory
losses, both in light and in darkness, may be substantial. It now seems
improbable that losses due to photorespiration will easily be reduced
because they are an integral part of the photosynthetic process. On the
other hand, losses due to dark respiration may be as large or larger and
perhaps more amenable to control at times. Glover has estimated that in
sugarcane (*Saccharum officinarum*) 20% of photosynthate is lost in respiration of young, actively growing plants and that losses may exceed 50%
per day in mature plants (201).

Ogren wrote that no evidence now exists that indicates if any of the
dark respiration can be eliminated without detriment to the plant and
questioned what the approach should be to this problem (415). The
questions that need to be answered are, How much of dark respiration is
essential, and can unnecessary respiration be eliminated? Answers to
these questions are not available yet. However, it is becoming evident
that respiration is perhaps not an inevitable consequence of active metabolism. Analyses suggest that photosynthetic reactions may under
some conditions directly supply much of the intermediates and energy
needed for growth and synthesis that are conventionally thought to
come from respiration. Work on the capacity of photosynthesis to substitute for respiration and the consequent extent and degree of respiratory
suppression by light is considered in the following section. It is hoped
that this research will provide leads to answer the questions raised about
control of respiration.

2. Light Respiration

The respiratory exchange resulting from the continuation of dark-
respiratory metabolism in light was defined as *light respiration* (47) to
distinguish it from photorespiration (see Section III). It is necessary to
distinguish between the operation of a respiratory pathway associated
with the production of reduced cofactors or ATP and a limited or partial
operation of sequences associated with the production of intermediates,
(e.g., from the citric acid cycle). Each of these metabolic situations might
be partially or wholly integrated with photosynthesis, either through use

of or control by nucleotides or through use of photosynthetic intermediates as substrates. In other words, it is necessary to consider the metabolism of the whole cell, tissue, or organism rather than the operation of one or another metabolic pathway, compartmented as it may be. In fact, the cross-links between compartments, provided by shuttle mechanisms that transport reducing equivalents and by direct transport of intermediates, are as important in their regulatory behavior as the main links in the individual metabolic chains. In a sense, the concept of inter- and intracellular metabolic coordination developed for C_4 photosynthesis, photorespiration, and CAM is now being extended to the relationship between dark respiration, particularly as it operates in light, and photosynthesis.

3. Respiration and Photosynthesis: Alternate Energy Sources

a. ATP. The alternate use of photosynthetically generated reducing power and ATP to drive processes normally coupled with respiration requires that amounts of cofactors and reducing equivalents in excess of those required for CO_2 fixation need to be generated and that mechanisms exist for exporting them from chloroplasts. Raven has presented calculations for several green algae (*Chlorella, Scenedesmus, Ankistrodesmus,* and *Hydrodictyon africanum*) showing that cyclic photosynthetic phosphorylation produces ATP at about the same rate as oxidative phosphorylation (489). Using inhibitors, he showed that a number of cellular functions, including solute transfer and sugar metabolism, are supported about equally by a cyclic ATP synthesis or oxidative phosphorylation. In addition, numerous experiments on the control of respiration by light (see the following discussion) show that photosynthetically produced cofactors influence and therefore have access to mitochondrial metabolism.

b. ATP Transport. The transport of photosynthetically generated ATP from chloroplasts to cytoplasm presents certain problems. Heldt pointed out that the maximum rate obtained for the ATP translocator of spinach (*Spinacia oleracea*) chloroplasts is only about 5μmol/hr/mg chlorophyll (238). Somewhat higher rates of ATP export, approximately 15 μmol/hr/mg chlorophyll, have been reported by Mangat *et al.* (367). These rates appear low when compared with reported rates of cyclic phosphorylation of 20–200 μmol/hr/mg chlorophyll (489) or with rates of oxidative phosphorylation in dark-respiring leaves (usually in the range 20–100 μmol/hr/mg chlorophyll). It is possible that better experimental conditions will increase the maximum attainable rate of ATP export. However, it appears that direct transport of ATP is very low. An alternative, much more rapid mechanism for exporting ATP from chloroplasts is the triose phosphate–PGA shuttle described by Stocking and Larson (545). In this shuttle, shown in Figure 19a, dihydroxyacetone-phosphate export and PGA import are mediated by the phosphate translocator. PGA is reduced to dihydroxyacetone phosphate in chloroplasts, and the triose is oxidized in the cytoplasm, generating NADH and ATP. The reaction has been experimentally demonstrated in chloroplasts (237, 545). The rate of operation of the shuttle does not seem to have been estimated under natural conditions, but the phosphate translocator has adequate capacity, up to 500 μmol/hr/mg chlorophyll (238). There seems to be no reason why photosynthetically generated ATP should not be available to the cytoplasm, either by direct transport or by shuttle.

c. *Reducing-Equivalent Shuttle.* Over the years several shuttle systems have been proposed that would transfer reducing power from chloroplasts to cytoplasm, and some of these have been experimentally shown to operate. C_4 photosynthesis exemplifies the operation of one such shuttle, in which malate transfers reducing equivalents from parenchyma mitochondria to bundle-sheath chloroplasts by reactions 2 and 3 in Fig. 14. Similarly, PGA from bundle-sheath chloroplasts may enter mesophyll chloroplasts and there be reduced; if the resulting dihydroxyacetone phosphate is returned to the bundle-sheath chloroplasts (as may happen), this also constitutes the transfer of reducing equivalents from mesophyll to bundle-sheath chloroplasts. CAM also involves more or less shuttle activity, as shown in Fig. 17, depending on the intercellular location of various reactions. Unlike the C_4 shuttles, the nature and extent of CAM shuttles have not yet been clarified.

Several proposed shuttle systems are shown in Fig. 19, all of which are capable of generating NADH or NADPH outside the chloroplast at the expense of reducing power generated by light in the chloroplast. Shuttle systems shown in Fig. 19a–c have been shown to operate (237, 282, 545). The glyoxylate shuttle (Fig. 19d) is hypothetical; glycolate oxidation does not appear to generate NADH in higher plants, but this may occur in microalgae (see Section III).

d. *Extent of Photosynthetic Contribution to Respiratory Processes.* Raven has demonstrated that cyclic phosphorylation in *Hydrodictyon* is quantitatively capable of supplying ATP levels generated in dark respiration by oxidative phosphorylation (489). In a subsequent paper Raven examined data from his own and a number of other laboratories to determine

CYTOPLASM   MEMBRANE   CHLOROPLAST

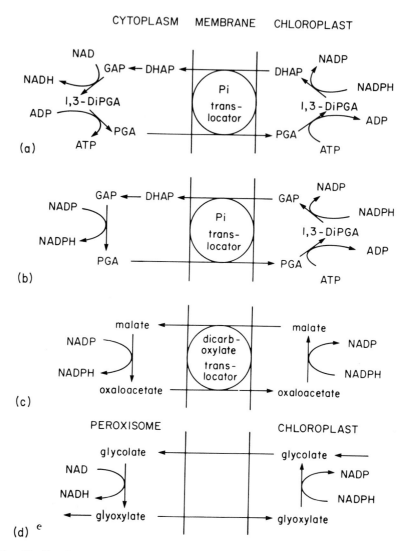

FIG. 19. Shuttle mechanisms transporting reducing equivalents or ATP from chloroplasts. (a) after Stocking and Larson (545); (b) after Kelly and Gibbs (282); (c) after Heldt (238); and (d) after Kisaki and Tolbert (305).

the extent to which organisms use light-generated reducing power and ATP for growth and other reactions that are normally considered to be driven by respiration (490). Raven calculated the ratio of CO_2 produced per C fixed (CO_2/C) for phototrophic or heterotrophic growth of a

TABLE X

Ratios of CO_2 Produced to Carbon Assimilated for
Heterotrophic and Phototrophic Growth of Plants[a]

Plant	CO₂/C	
	Heterotrophic	Phototrophic
Chlorella spp.	0.55–1.20	0.15–0.25
Euglena gracilis	0.4 –0.5	0.2 –0.35
Average of several algae	0.77	0.28
Helianthus annuus	2.0	0.21–0.59
Nicotiana tabacum	0.5	0.2
Oryza sativa	0.67–1.1	0.43–0.67
Solanum tuberosum	1.0	0.2 –0.55
Zea mays	0.53–1.0	0.29–1.0
Average of several higher plants	0.87	0.42

[a] Data summarized from Raven (490).

number of algae and leaves of higher plants. This ratio provides a measure of the respiratory metabolism required to sustain growth. A summary of Raven's data is presented in Table X.

In all cases the measurements made on algae are comparable, because the cells grow normally either heterotrophically (glucose) or phototrophically (light and CO_2). In heterotrophic growth, all the energy for growth must have been derived from respiration. In phototrophic growth, dark respiration is calculated for the light period as if dark respiration continued unabated in light, although this may produce a value that is higher than the true one. Nevertheless, the amount of respiration required to support phototrophic growth is always much lower, often in the range of one-half of the amount required to support heterotrophic growth.

The situation for higher plants is less clear than that for algae. Much of the heterotrophic data is for seedlings or for callus tissue that characteristically make more-expensive growth (more protein and lipid, less carbohydrate) than the leaves or whole mature plants used for phototrophic analysis. On the other hand, the maintenance respiration of mature plants growing phototrophically may be higher than that of seedlings or callus tissue growing heterotrophically, which in part balances the high cost of heterotrophic growth. In any event, Raven (490) concluded that in algae, and probably also in higher plants, a substantial part of the energy for growth in light is derived directly from photosynthetically produced ATP and reducing power.

Raven's analysis agrees with the analyses of Petrov (448) and Bell *et al.* (34) (see Section II), who found that energy storage (measured calorimetrically) exceeds by a substantial margin the energy equivalence of photosynthetic gas exchange. Thus it appears probable that the light-energy conversion mechanism of photosynthesis is coupled directly, presumably through shuttle mechanisms such as those shown in Fig. 19, with the reactions of maintenance and growth. It is well known that photosynthetic reducing power is used in chloroplasts for the reduction of nitrate, assimilation of ammonium, reduction and assimilation of inorganic sulfate, and other energy demanding processes (7). Exactly how much photosynthetic energy is used in this way, and what mechanisms exist to govern the fractionation of photosynthetic reducing power between CO_2 reduction, nitrate reduction, maintenance, and growth, remain to be discovered.

4. Light Control of Respiration

a. Magnitude of Light Respiration. The concept that light can substitute for respiratory activity as a source of energy for maintenance and growth is consistent with the idea that light controls or inhibits respiration. With the development of double-isotope techniques for simultaneously measuring the gas exchange of photosynthesis and respiration in light, it became evident, after some initial controversy, that rapid respiratory gas exchange continues in light. The clarification of the nature of photorespiration, however, left a new question, What proportion of the measured respiratory gas exchange in light is photorespiration, and what proportion is due to light respiration?

Ludwig and Canvin (362) concluded that the dark-respiratory pathway continues to function in light in sunflower (*Helianthus annuus*) leaves, noting that dark respiration appears to be unaffected in light when photosynthesis is inhibited. However, Mangat *et al.*, using $^{12}CO_2$ and $^{14}CO_2$, showed that CO_2 production by dark respiration is reduced by about 75% by light in bean (*Phaseolus vulgaris*) leaves, contributing only about one-quarter of the total CO_2 produced in light (367). Przybylla and Fock, measuring the specific activity of CO_2 and metabolic intermediates after the supply of $^{14}CO_2$, also concluded that dark respiration of sunflower leaves is strongly inhibited in light (463). Lambers *et al.* (328) reviewed a number of studies suggesting that light respiration is affected or controlled by products of photosynthetic reducing reactions (see Section VI,C,4,c) and concluded that dark respiration is probably considerably reduced in light. This would explain why Ludwig and Canvin (362) found high light respiration in leaves in which photosynthesis had been inhibited.

In an alternative approach, Raven (487, 488) calculated the carbon flux through various pathways on the basis of levels and requirements of intermediates and cofactors. He concluded that the pentose pathway of respiration was inhibited in light, and that the tricarboxylic acid cycle was partially inhibited. The general conclusion seems to be that light inhibits dark respiration to a considerable degree and that light respiration, the continuation of dark respiration in light, takes place at a much-reduced rate in those tissues that have been investigated.

b. Operation of the Tricarboxylic Acid Cycle in Light. Chapman and Graham (99, 100), using labeled tricarboxylic acid cycle intermediates, showed that the flux of carbon through the cycle is about the same in light as in dark, following a brief initial inhibition on transition from dark to light that is mediated by transient light-generated changes in the concentration of nucleotides. On the other hand, data of Fock and co-workers (332, 463) on the specific activity of metabolic intermediates in sunflower leaves after $^{14}CO_2$ was supplied indicate that operation of the tricarboxylic acid cycle is much reduced in light, although higher rates were observed in stressed leaves.

This problem may be resolved in light of work by Kent *et al.* (291, 293), whose degradation data for citric acid and 3-carbon compounds, obtained after labeled intermediates were supplied to *Vicia faba* leaves, show that, in light, serine rather than PGA is the carbon source of the tricarboxylic acid cycle. Kent's recent analysis suggests that the cycle functions in light primarily as a mechanism for the synthesis of glutamic acid. The data suggest that the C_2 and C_4 precursors of citric acid are derived from malate formed via a serine–formate condensation (292). Thus it appears possible that the activity of portions of the tricarboxylic acid cycle, diverted from its function as the main metabolic system driving respiratory electron transport, may be quite high in light when the requirement for nitrogen metabolism demands this. On the other hand, the total operation of the cycle, emphasizing its primary function in energy metabolism, may well be reduced in light.

c. Control by Photosynthetic Pyridine Nucleotides. Work by Sargent and Taylor (513) and Lewenstein and Bachofen (342) on *Chlorella* suggests that respiration in this organism may be controlled in light by photosynthetically produced ATP. The same conclusion was reached by Mangat *et al.* (367), who noted that photosynthetically produced ATP rapidly leaves bean chloroplasts. More recently, Hoffmann showed that increased light respiration serves to maintain ATP content and energy charge in illuminated wheat (*Triticum aestivum*) leaves in which photosynthesis has been poisoned by carbonylcyanide *m*-chlorophenylhydrazone (CCC) (245). Imafuku and Katoh suggested that direct interaction between photosynthesis and respiration occurs in the prokaryotic algae at some common point in the electron transport chain prior to ATP formation (262). They concluded that such interaction in eukaryotic organisms may be much more indirect, depending on cofactor concentration and common intermediates.

In view of the availability of shuttle mechanisms that could transport photosynthetic ATP or reducing equivalents to the cytoplasm (Fig. 19), and the well-known control of respiration by energy charge, it is reasonable to conclude that light does indeed influence dark respiration through the agency of photosynthetic pyridine nucleotides. The degree of such regulation may well be affected by the requirement for respiratory carbon metabolism to produce needed intermediates for cellular synthesis, nitrogen metabolism, and the activities supported by photosynthesis.

d. Control by Photosynthetic Carbon Intermediates. Some enzymes of glycolysis are under allosteric control by metabolic intermediates of respiratory or associated pathways. For example, phosphofructokinase is inhibited by PEP (281), and this fact has been invoked in an attempt to explain respiratory control in CAM plants by Pierre and Queiroz (449). However, their data show that other, unknown factors are also involved in maintaining the strict correlation that is observed between the carboxylation and glycolytic activities of CAM plants. A second possibility is the regulation of the tricarboxylic acid cycle by PGA. However, the chloroplast compartment of PGA appears to be tight in light, and little PGA is lost to the cytoplasm. On the whole, the compartments of photosynthetic cycles are relatively tight, so it does not seem likely that photosynthetic carbon intermediates play a major, direct role in controlling respiration.

e. Light Modulation of Respiratory Enzymes. Anderson *et al.* (4, 5) have extensively examined the mediation of glucose 6-phosphate dehydrogenase by light-effect mediators in pea (*Pisum sativum*). This enzyme exists both in chloroplasts and cytoplasm; the latter is important as the starting point of the oxidative pentose phosphate pathway. Light-effect mediators are also implicated in the regulation of photosynthetic enzymes. That DCMU inhibits light activation indicates that the photochemical apparatus is directly involved, through reduction of the light-effect mediator, in the control of pentose-phosphate-pathway respiration (4). The nature of the intermediary step that would seem to be required between the light-effect mediator, which is bound to the chloroplast membrane, and the cytoplasmic enzyme is not known. It appears likely that further experimentation will uncover other examples of mechanisms in which light, acting through the photochemical system of the chloroplasts, controls or regulates the activity of respiratory energy metabolism.

Other, less direct light effects have been reported. Blue light enhances respiration of *Chlorella*, as first reported by Emerson and Lewis (168). Sargent and Taylor (514) concluded that this effect is caused by feedback control of the electron-transport chain or tricarboxylic acid cycle following blue-light stimulation of RNA or protein synthesis. However, experiments by Ries and Gauss (497) strongly suggest that blue light modulates a pathway along which carbon is transported from the C_3 cycle to the tricarboxylic acid cycle following β-carboxylation of C_3 intermediates.

The well-known photoperiodic behavior of respiration in *Lemna* has been the subject of much investigation. Goto (207) has shown that the K_m of cytoplasmic (as well as chloroplast) glyceraldehyde 3-phosphate dehydrogenase is controlled by both the circadian oscillator and the hourglass mechanism in *L. gibba*. The control appears to operate through the intervention of reducing compounds, but how it is related to the circadian oscillator or hourglass mechanism is unclear. It would appear likely that a mechanism similar or analogous to the photoperiodically regulated respiratory system (e.g., in CAM; see 449) remains to be discovered.

f. Summary. The widely held assumption that light regulates respiration is supported by most of the evidence, although much of it is admittedly indirect. The overriding concern seems to be whether light energy may be used, presumably via light-generated reducing power and ATP, directly for cytoplasmic reactions that would otherwise deplete prior reserves. This concern could be extended, via effects of light on morphogenesis, to more-distant cells and organs. However, mechanisms must exist that protect those metabolic sequences that are essential for the production of required intermediates from indescriminate shutdown by light. The balance and operation of the required mechanisms represents an important area for future research.

D. Dark Fixation of CO_2

A treatment of the carbon nutrition of plants would be incomplete without some consideration of the role of dark fixation of CO_2. Aside from the dark reactions of CAM, this subject does not seem to have been reviewed in detail recently. The reactions of dark fixation have long been known, and the possibility was early considered that dark fixation might be of importance in the carbon balance of plants. However, Stolwijk and Thimann found that enrichment of roots of various plants with CO_2 does not stimulate growth itself except marginally at low concentrations and that it is strongly inhibitory at about 1% CO_2 (546). It seems unlikely, therefore, that dark CO_2 fixation is important in terms of the carbon nutrition of plants.

Nevertheless, a low level of dark fixation appears to be beneficial or necessary for roots. Splittstoesser (540) repeated the experiments of Stolwijk and Thimann and reviewed the available literature, finding that growth and protein synthesis are stimulated by added CO_2 or HCO_3^-. Splittstoesser's experiments with labeled HCO_3^- suggest that dark-fixed CO_2 is used in anaplerotic carboxylation reactions (PEPcase, pyruvate carboxylase), leading to replenishment of oxaloacetic acid lost by amino acid synthesis. This view is supported by Nesius and Fletcher for 'Paul's Scarlet' rose tissue in culture. Their data indicate synthesis of the aspartate family of amino acids directly from oxaloacetic acid produced from dark fixation of CO_2 (407). Bown et al. have shown that malate can substitute for dark-fixed CO_2 in the growth of Avena sativa coleoptiles (70). The effects of malate and CO_2 are not synergistic, supporting the idea that CO_2 functions to produce a C_4 acid. Data of Levi et al. (341) indicate similar metabolic pathways in leaves of C_3 and C_4 plants. Unfortunately, none of these studies provides evidence on which the rates of carboxylation can be determined, as compared with other metabolic functions. Levi et al. noted that the dark-fixation rate of leaves is low, and continues for only about 3 hr (excepting CAM plants) (341). Thus it seems that the requirement for dark fixation, though important in maintaining maximal growth and protein synthesis, is limited in extent.

A different proposal has been made by Davies, who suggested that CO_2 dark fixation mediated by malic enzyme and PEPcase is related to a metabolic pH-stat (135). This conclusion is based upon the regulating properties of the enzymes from potato tubers. Malic enzyme is activated by dicarboxylic acids and is inhibited by inorganic phosphate, triose phosphate, and AMP (136). PEPcase, on the other hand, is activated by sugar phosphates and glycolate and is inhibited by dicarboxylic acids (64). The pH optima of these functions provide a possible mechanism for the metabolic maintenance of cellular pH. This conclusion is sustained by Hill and Bown (243) in work on Avena coleoptile tissue, in which the regulatory effect of PEPcase is integrated with IAA-stimulated proton movement.

For some years it has been recognized that blue-light effects are related to the dark operation of PEPcase (586). Miyachi et al. have proposed that blue light stimulates the synthesis or activation of the enzyme

PEPcase in a *Chlorella vulgaris* mutant (380). They showed that the effect of blue light is to increase the anaplerotic formation of oxaloacetic acid by PEPcase while inhibiting the glyoxylate cycle (another possible anaplerotic reaction). This tends to prevent cyclic operation of the tricarboxylic acid cycle and promote utilization of 2-oxoglutarate and oxaloacetic acid for synthetic reactions (e.g., of amino acids) instead. Although this reaction sequence has been considered primarily in photosynthetic cells, it is actually dark fixation; the role of blue light is to regulate the metabolic balance of the cell.

Dark fixation of CO_2 in *Euglena gracilis* has been extensively investigated by Wolpert and Ernst-Fonberg (613, 614). They envisage CO_2-scavenging mechanisms that assimilate CO_2 for introduction into various metabolic pathways. They found that a highly efficient multienzyme complex is present in *Euglena*. It contains PEPcase, malate dehydrogenase, and acetyl-CoA carboxylase, which channels carbon from dark-fixed CO_2 into fatty acid synthesis. The complex appears to have a marked effect on the catalytic efficiencies of the individual enzymes; for example, free PEPcase has $K_{mHCO_3^-}$ of 7.3–9.8 mM, whereas the same enzyme in the complex has a value of 0.7–1.3 mM. This interesting development does not seem to have been investigated in higher plants.

The overall conclusion is that CO_2 dark fixation may be important in certain metabolic and cellular regulating functions but that it does not contribute in a major way to the carbon nutrition of plants. Attempts to stimulate plant or root growth by CO_2 or HCO_3^- fertilization have been largely unsuccessful, although it has been shown that optimum growth requires at least air-level concentrations of CO_2 and may benefit by slight enrichment of soil CO_2 levels. However, cost-effective CO_2 fertilization does not seem to be a workable possibility.

VII. Conclusion

A. THE RELATIONSHIP OF PHOTOSYNTHESIS AND CARBON ASSIMILATION

It is now clear that CO_2 assimilation is not the sole use of photosynthetic light energy by green plants. Photosynthetic potential is used directly in many ways: for nitrogen assimilation and nitrate reduction, for the production of ATP and reducing equivalents that are used in growth

422 R. G. S. Bidwell

and maintenance, and, possibly, for other photochemical reactions that store light energy in the plant. These reactions just as clearly represent the photosynthetic storage of light energy as do the reactions of reductive CO_2 fixation. This means that rates of CO_2 fixation, or even of O_2 production, cannot be equated with photosynthesis. It seems unlikely, although no hard data are available at present, that any fixed relationship exists between the various forms of energy storage at any given time. It is more likely that the relative intensities of carbon or nitrogen reduction or ATP synthesis are governed by the immediate external and internal environments of the photosynthesizing cells, which in turn depend on the environmental relationships and developmental stage of the whole plant.

The interest in the physiology of productivity and the great need to understand it, particularly in relation to crop plants, requires some more-effective measure of photosynthesis than CO_2 uptake or O_2 evolution. The problem has been analyzed in some depth by Raven (e.g., 490), but the solution seems to lie in the approach of Petrov and co-workers (448, and unpublished data), who use calorimetry and direct measurement of light absorption to determine the gross storage of photosynthetic energy. Unfortunately, all techniques devised so far are cumbersome and destructive, so repetitive or sequential measurements are not easy. However, greater use will have to be made of techniques of this sort if a complete understanding of the photosynthetic behavior of plants is to be reached.

B. Unsolved Problems

The pathway of C_3 photosynthesis is well established. The C_4 pathways still lack firmness and detail, and much work is required to determine the details of operation and intracellular location of many of the sequences in the C_4 cycle. Details of CAM are even less clear, although the general outline is now well established. The C_2 cycle is no longer a mystery, although its ramifications and interactions in cellular metabolism are by no means well understood. Nevertheless, the overall picture of photosynthetic carbon metabolism is now beginning to clear.

Less clear, and much in need of research effort, are the degree, function, and control of cooperative interaction in photosynthesis and associated metabolism. In particular, the ways in which the nucleotide cofactors in the C_4 and CAM cycles are regenerated or turned over, the coupling of reducing and oxidizing reactions, and the transport mechanisms or reactions that permit this coupling are all poorly understood.

Much of the current interest in photosynthesis is directed toward improving its efficiency. This, in turn, will depend to a considerable extent on effective cycling of cofactors.

Problems of the regulation of photosynthetic metabolism are currently under attack, and it is hoped that the new information gained will permit much closer modeling of crop growth and behavior. The problem of regulation or control of photosynthesis is bound up with the concept of photosynthate distribution. The natural pattern of plant growth is consistent with the survival of the species in natural conditions. The ways in which this pattern needs to be perturbed in order to achieve optimum production of usable crop (as opposed to total plant mass) under agricultural conditions can be derived only from a total knowledge of the control and operation of all physiological and metabolic levels in the plant. Not only the rates of assimilation, but the patterns of nutrient distribution and the whole pattern of correlative growth need to be considered. For example, under natural conditions, substantial reduced carbon and nitrogen may be directed to produce a large root system that will protect the plant against possible drought damage. In irrigated agriculture no more roots need to be formed than those necessary for optimum nutrient absorption; the extra carbon and nitrogen would be converted more usefully to fruit or produce. A complete understanding of metabolic integration and intrinsic control mechanisms is essential in order to achieve the required degree of control over the growth pattern of crop plants.

C. PHOTOSYNTHESIS AND CARBON NUTRITION

The problems of carbon nutrition of plants have tended to be obscured by the approach to photosynthesis. The tendency has been to view photosynthesis as the production of reduced-carbon compounds and to view all other metabolic nutrition as driven or powered by the oxidation of photosynthate. It is now clear that photosynthesis is more than merely the production of reduced carbon, yet less than the totality of these reactions. It is more in the sense that photosynthesis (as the conversion of light energy into chemical potential) directly drives many of the synthetic, metabolic, growth, and maintenance reactions of plants (7). Included in this are nitrate reduction and the assimilation of reduced nitrogen into organic form, sulfate reduction, amino acid interconversion and synthesis, and various intermediary carbon reactions (as in lipid or nucleotide synthesis) that may run at the expense of light-generated reducing power. It is less than the totality of reactions leading to sugar or starch in that the dark reactions of the C_3 cycle need to be considered not only as part of photosynthesis, but also as a part of carbon nutrition that competes with other metabolic demands for the available supplies of light-generated chemical potential. Thus not only does the distribution pattern of photosynthate and its control deserve attention, but an understanding of the distribution and control of energy among the metabolic requirements of the plant is of paramount importance in understanding and optimizing the growth of plants.

References

1. Agrawal, P. K., and Fock, H. (1978). The specific radioactivity of glycolic acid in relation to the specific activity of carbon dioxide evolved in light in photosynthesizing sunflower leaves. *Planta* **138**, 257–261.
2. Allaway, W. G., Austin B., and Slatyer, R. O. (1974). Carbon dioxide and water vapour exchange parameters of photosynthesis in a Crassulacean plant *Kalanchoë daigremontiana. Aust. J. Plant Physiol.* **1**, 399–405.
3. Altman, A., and Wareing, P. F. (1975). The effect of IAA on sugar accumulation and basipetal transport of ^{14}C-labelled assimilates in relation to root formation in *Phaseolus vulgaris* cuttings. *Physiol. Plant.* **33**, 32–38.
4. Anderson, L. E. (1977). Photosynthetic electron-transport system controls cytoplasmic glucose-6-phosphate dehydrogenase activity in pea leaves. *FEBS Lett.* **76**, 64–66.
5. Anderson, L. E., and Duggan, J. X. (1976). Light modulation of glucose-6-phosphate dehydrogenase. Partial characterization of the light activation system and its effects on the properties of the chloroplastic and cytoplasmic forms of the enzyme. *Plant Physiol.* **58**, 135–139.
6. Anderson, L. E., Nehrlich, S. C., and Champigny, M.-L. (1978). Light modulation of enzyme activity. Activation of the light-effect mediators by reduction and modulation of enzyme activity by thiol-disulfide exchange? *Plant Physiol.* **61**, 601–605. See also Anderson *et al.* (1980). *What's New Plant Physiol.* **11**(10).
7. Anderson, J. M. (1978). Light coupled metabolic reactions of plants. *What's New in Plant Physiol.* **9**, 17–20.
8. Andersen, K. (1979). Mutations altering the catalytic activity of a plant-type ribulose bisphosphate carboxylase/oxygenase in *Alcaligines eutrophus. Biochim. Biophys. Acta* **585**, 1–11.
9. Andrews, T. J., Badger, M. R., and Lorimer, G. H. (1975). Factors affecting interconversion between kinetic forms of ribulose diphosphate carboxylase-oxygenase from spinach. *Arch. Biochem. Biophys.* **171**, 93–103.
10. Andrews, T. J., and Lorimer, G. H. (1978). Photorespiration—still unavoidable? *FEBS Lett.* **90**, 1–9.
11. Andrews, T. J., Lorimer, G. H., and Tolbert, N. E. (1973). Ribulose diphosphate oxygenase. I. Synthesis of phosphoglycolate by Fraction-I protein of leaves. *Biochemistry* **12**, 11–18.
12. Apel, P., and Ohle, H. (1979). CO_2-Kompensationspunkt und Blattanatomie bie Arten der Gattung *Moricandia* DC (Cruciferae). *Biochem. Physiol. Pflanz.* **174**, 68–75.
13. Asami, S., Inoue, K., and Akazawa, T. (1979). NADP-malic enzyme from maize leaf: regulatory properties. *Arch. Biochem. Biophys.* **196**, 581–587.

14. Atkins, C. A., Patterson, B. D., and Graham D. (1972). Plant carbonic anhydreses. I. Distribution of types among species. *Plant Physiol.* **50,** 214–217.
15. Austin, R. B., and Edrich, J. (1975). Effects of ear removal on photosynthesis, carbohydrate accumulation and on the distribution of assimilated ^{14}C in wheat. *Ann. Bot. (London)* **39,** 141–152.
16. Avron, M., and Gibbs, M. (1974). Properties of phosphoribulokinase of whole chloroplasts. *Plant Physiol.* **53,** 136–139.
17. Avron, M., and Gibbs, M. (1974). Carbon dioxide fixation in the light and in the dark by isolated spinach chloroplasts. *Plant Physiol.* **53,** 140–143.
18. Badger, M. R., and Andrews, T. J. (1974). Effects of CO_2, O_2 and temperature on a high affinity form of ribulose diphosphate carboxylase-oxygenase from spinach. *Biochem. Biophys. Res. Commun.* **60,** 204–210.
19. Badger, M. R., Andrews, T. J., and Osmond, C. B. (1974). Detection in C_3, C_4 and CAM plant leaves of a low-K_m (CO_2) form of RuDP carboxylase, having a high RuDP oxygenase activity at physiological pH. *Proc. Int. Congr. Photosynth. 3rd* **2,** 1421–1429.
20. Badger, M. R., and Lorimer, G. H. (1976). Activation of ribulose-1,5-bisphosphate oxygenase. The role of Mg^{2+}, CO_2 and pH. *Arch. Biochem. Biophys.* **175,** 723–729.
21. Badour, S. S., and Waygood, E. R. (1971). Glyoxylate carboxy-lyase activity in the unicellular green alga *Gloeomonas* sp. *Biochim. Biophys. Acta* **242,** 493–499.
22. Bahr, J. T., and Jensen, R. G. (1974). Ribulose bisphosphate oxygenase activity from freshly ruptured spinach chloroplasts. *Arch. Biochem. Biophys.* **164,** 408–413.
23. Baldry, C. W., Walker, D. A., and Bucke, C. (1966). Calvin-cycle intermediates in relation to induction phenomena in photosynthetic carbon dioxide fixation by isolated chloroplasts. *Biochem. J.* **101,** 642–646.
24. Bamberger, E. S., and Avron, M. (1975). Site of action of inhibitors of carbon dioxide assimilation by whole lettuce chloroplasts. *Plant Physiol.* **56,** 481–485.
25. Bassham, J. A., Benson, A. A., Kay, L. D., Harris, A. Z., Wilson, A. T., and Calvin, M. (1954). The path of carbon in photosynthesis. XXI. The cyclic regeneration of carbon dioxide acceptor. *J. Am. Chem. Soc.* **76,** 1760–1770.
26. Bassham, J. A., and Jensen, R. G. (1967). Photosynthesis of carbon compounds. In "Harvesting the Sun" (A. San Pietro, F. A. Green and T. J. Army, eds.), pp. 79–110. Academic Press, New York.
27. Bassham, J. A., and Kirk, M. (1963). Synthesis of compounds from [14]CO_2 by *Chlorella* in the dark following preillumination. *Plant Cell Physiol.* (special issue: "Studies on Microalgae and Photosynthetic Bacteria") pp. 493–503.
28. Bassham, J. A., and Kirk, M. (1973). Sequence of formation of phosphoglycolate and glycolate in photosynthesizing *Chlorella pyrenoidosa*. *Plant Physiol.* **52,** 407–411.
29. Bassham, J. A., Levine, A. G., and Forger, J. (1974). Photosynthesis *in vitro*. I. Achievement of high rates. *Plant Sci. Lett.* **2,** 15–21.
30. Bazzaz, M. B., and Govindjee (1973). Photochemical properties of mesophyll and bundle sheath chloroplasts of maize. *Plant Physiol.* **52,** 257–262.
31. Beevers, H. (1970). Respiration in plants and its regulation. In "Prediction and Measurement of Photosynthetic Productivity" (I. Setlik, ed.), pp. 209–220. Centre for Agricultural Publishing and Documentation, Wageningen, The Netherlands.
32. Beevers, H., Stiller, M. L., and Butt, V. S. (1966). Metabolism of the organic acids. In "Plant Physiology, a Treatise" (F. C. Steward, ed.), pp. 119–262. Academic Press, New York.
33. Beezley, B., Gruber, P. J., and Frederick, S. E. (1976). Cytochemical localization of glycolate dehydrogenase in mitochondria of *Chlamydomonas*. *Plant Physiol.* **58,** 315–319.

34. Bell, L. N., Shuvalova, N. P., Volkova, T. V., and Krupenko, A. N. (1976). Discrepancy between gas exchange and energy storage rates in *Chlorella*. *Photosynthetica* **10**, 147–154.

35. Benjamin, L. R., and Wren, M. J. (1978). Root development and source–sink relations in carrot, *Daucus carota* L. *J. Exp. Bot.* **29**, 425–433.

36. Benedetti, E. D., Forti, G., Garlaschi, F. M., and Rosa, L. (1976). On the mechanism of ammonium stimulation of photosynthesis in isolated chloroplasts. *Plant Sci. Lett.* **7**, 85–90.

37. Benson, A. A., and Calvin, M. (1947). The dark reduction of photosynthesis. *Science* **105**, 648–649.

38. Berlyn, M. B., and Zelitch. I. (1975). Photoautotrophic growth and photosynthesis in tobacco callus cells. *Plant Physiol.* **56**, 752–756.

39. Berlyn, M. B., Zelitch, I., and Bandette, P. D. (1978). Photosynthetic characteristics of photoautrophically grown tobacco callus cells. *Plant Physiol.* **61**, 606–610.

40. Berry, J. A., Osmond, C. B., and Lorimer, G. H. (1978). Fixation of $^{18}O_2$ during photorespiration. Kinetic and steady-state studies of the photorespiratory carbon oxidation cycle with intact leaves and isolated chloroplasts of C_3 plants. *Plant Physiol.* **62**, 954–967.

41. Bhagwat, A. S., and Sane, P. V. (1978). Evidence for the involvement of superoxide anions in the oxygenase reaction of ribulose-1,5-diphosphate carboxylase. *Biochem. Biophys. Res. Commun.* **84**, 865–873.

42. Bidwell, R. G. S. (1958). Photosynthesis and metabolism of marine algae. II. A survey of rates and products of photosynthesis in $C^{14}O_2$. *Can. J. Bot.* **36**, 337–349.

43. Bidwell, R. G. S. (1967). Photosynthesis and metabolism in marine algae. VII. Products of photosynthesis in fronds of *Fucus vesiculosus* and their use in respiration. *Can. J. Bot.* **45**, 1557–1565.

44. Bidwell, R. G. S. (1968). Photorespirator. *Science* **161**, 79–80.

45. Bidwell, R. G. S. (1972). Impurities in preparations of *Acetabularia* chloroplasts. *Nature (London)* **237**, 169.

46. Bidwell, R. G. S. (1973). A possible mechanism for the control of photoassimilate translocation. *Proc. Res. Inst. Pomol. Skierniewice Ser. E.* **3**, 77–89.

47. Bidwell, R. G. S. (1977). Photosynthesis and light and dark respiration in freshwater algae. *Can. J. Bot.* **55**, 809–818.

48. Bidwell, R. G. S., Levin, W. B., and Tamas, I. A. (1968). The effect of auxin on photosynthesis and respiration. *In* "Biochemistry and Physiology of Plant Growth Substances" (F. Wightman, ed.), pp. 361–376. Runge Press, Ottawa.

48a. Bidwell, R. G. S., McLachlan, J. L., and Lloyd, N. D. H. (1980). Rapid infra-red analysis of carbon dioxide exchange in marine algae. *Proc. Can. Soc. Plant Physiol.* **20**, 13.

49. Bidwell, R. G. S., and Quong, E. C.-K. (1975). Indoleacetic acid effect on the distribution of photosynthetically fixed carbon in the bean plant. *Biochem. Physiol. Pflanz.* **168**, 361–370.

50. Bidwell, R. G. S., and Turner, W. B. (1966). The effect of growth regulators on CO_2 assimilation in leaves, and its correlation with the "bud break" response in photosynthesis. *Plant Physiol.* **41**, 267–270.

51. Bird, I. F., Cornelius, M. J., Dyer, T. A., and Kemp, A. J. (1973). The purity of chloroplasts isolated in nonaqueous media. *J. Exp. Bot.* **24**, 211–215.

52. Bird, I. F., Cornelius, M. J., Keys, A. J., and Whittingham, C. P. (1972). Oxidation and phosphorylation associated with the conversion of glycine to serine. *Phytochemistry* **11**, 1587–1594.

53. Bird, I. F., Cornelius, M. J., Keys, A. J., and Whittingham, C. P. (1974). Intracellular sites of sucrose synthesis in leaves. *Phytochemistry* **13**, 59–64.
54. Bird, I. F., Cornelius, M. J., Keys, A. J., and Whittingham, C. P. (1978). Intramolecular labelling of sucrose made by leaves from [^{14}C]carbon dioxide or [3-^{14}C]serine. *Biochem. J.* **172**, 23–27.
55. Bishop, D. G., Andersen, K. S., and Smillie, R. M. (1971). Incomplete membrane-bound photosynthetic electron transfer pathway in agranal chloroplasts. *Biochem. Biophys. Res. Commun.* **42**, 74–81.
56. Björkman, O. (1976). Adaptive and genetic aspects of C_4 photosynthesis. *In* "CO_2 Metabolism and Plant Productivity" (R. H. Burris and C. C. Black, eds.), pp. 287–309. Univ. Park Press, Baltimore.
57. Björkman, O., and Gauhl, E. (1969). Carboxydismutase activity in plants with and without β-carboxylation photosynthesis. *Planta* **88**, 197–203.
58. Black, C. C. (1973). Photosynthetic carbon fixation in relation to net CO_2 uptake. *Annu. Rev. Plant Physiol.* **24**, 253–286.
59. Black. C. C., and Mollenhauer, H. H. (1971). Structure and distribution of chloroplasts and other organelles in leaves with various rates of photosynthesis. *Plant Physiol.* **47**, 15–23.
60. Black, C. C., and Williams, S. (1976). Plants exhibiting characteristics common to Crassulacean acid metabolism. *In* "CO_2 Metabolism and Plant Productivity" (R. H. Burris and C. C. Black, eds.), pp. 407–424. University Park Press, Baltimore.
61. Blackwood, G. C., and Miflin, B. J. (1976). The effect of nitrate and ammonium feeding on carbon dioxide assimilation in maize. *J. Exp. Bot.* **27**, 735–747.
62. Bloom, A. J. (1979). Salt requirement for Crassulacean acid metabolism in the annual succulent *Mesembryanthemum crystallinum*. *Plant Physiol.* **63**, 749–753.
63. Bocher, M., and Kluge, M. (1978). Der C_4-Weg der C-Fixierung bie *Spinacea oleracea*. II. Pulse-chase Experimente mit suspendierten Blattsreifen. *Z. Pflanzenphysiol.* **86**, 405–421.
64. Bonugli, K. J., and Davies, D. D. (1977). The regulation of potato phosphoenolpyruvate carboxylase in relation to a metabolic pH-stat. *Planta* **133**, 281–287.
65. Bowen, M. R., and Wareing, P. F. (1971). Further investigations into hormone-directed transport in stems. *Planta* **99**, 120–132.
66. Bowes, G., and Berry, J. A. (1972). The effect of oxygen on photosynthesis and glycolate excretion in *Chlamydomonas reinhardtii. Carnegie Inst. Washington, Year Book* **71**, 148–158.
67. Bowes, G., and Ogren, W. L. (1972). Oxygen inhibition and other properties of soybean ribulose 1,5-diphosphate carboxylase. *J. Biol Chem.* **247**, 2171–2176.
68. Bowes, G., Ogren, W. L., and Hageman, R. H. (1975). pH dependence of the K_m (CO_2) of ribulose 1,5-diphosphate carboxylase. *Plant Physiol.* **56**, 630–633.
69. Bowes, G., Ogren, W. L., and Hageman, R. H. (1971). Phosphoglycolate production catalysed by ribulose diphosphate carboxylase. *Biochem. Biophys. Res. Commun.* **45**, 716–722.
70. Bown, A. W., Dymock, I. J., and Aung, T. (1974). A synergistic stimulation of *Avena sativa* coleoptile elongation by IAA and carbon dioxide. *Plant Physiol.* **54**, 15–18.
71. Bradbeer, J. W., Ranson, S. L., and Stiller, M. L. (1958). Malate synthesis in Crassulacean leaves. I. The distribution of ^{14}C in malate of leaves exposed to $^{14}CO_2$ in the dark. *Plant Physiol.* **33**, 66–70.
72. Brandén, R. (1978). Ribulose-1,5-diphosphate carboxylase and oxygenase from green plants are two different enzymes. *Biochem. Biophys. Res. Commun.* **81**, 539–546.
73. Brandén, R., and Brandén, C. I. (1978). Separation of ribulose-1,5-biphosphate

carboxylase and oxygenase activities. *In* "Photosynthetic Carbon Assimilation" (H. W. Siegelman and G. Hind, eds.), pp. 391–397. Plenum, New York.

74. Brandon, P. C. (1967). Temperature features of enzymes affecting Crassulacean acid metabolism. *Plant Physiol.* **42,** 977–984.

75. Bravdo, B-A., and Canvin, D. T. (1979). Effect of carbon dioxide on photorespiration. *Plant Physiol.* **63,** 399–401.

76. Brown, G., Selegny, E., Tran Minh, C., and Thomas, D. (1970). Facilitated transport of CO_2 across a membrane bearing carbonic anhydrase. *FEBS Letters* **7,** 223–226.

77. Brown, A. H. (1954). The effects of light on respiration using isotopically enriched oxygen. *Am. J. Bot.* **40,** 719–729.

78. Brown, R. H. (1976). Characteristics related to photosynthesis and photorespiration of *Panicum milioides. In* "CO_2 Metabolism and Plant Productivity" (R. H. Burris and C. C. Black, eds.), pp. 311–325. University Park Press, Baltimore.

79. Brown, R. H. (1978). A difference in N use efficiency in C_3 and C_4 plants and its implications in adaptation and evolution. *Crop Sci.* **18,** 93–98.

80. Brown, W. V. (1958). Leaf anatomy in grass systematics. *Bot. Gaz. (Chicago)* **119,** 170–178.

81. Brownell, P. F., and Crossland, C. J. (1972). The requirement for sodium as a micronutrient by species having the C_4 dicarboxylic acid photosynthetic pathway. *Plant Physiol.* **49,** 794–797.

82. Brownell, P. F., and Crossland, C. J. (1974). Sodium requirement in relation to photosynthesis of a CAM plant. *Plant Physiol.* **54,** 416–417.

83. Bruin, W. J., Nelson, E. B., and Tolbert, N. E. (1970). Glycolate pathway in green algae. *Plant Physiol.* **46,** 386–391.

84. Bykov, O. D. (1970). The method of calculation of isotopic effect in photosynthesis and its value. *Photosynthetica* **4,** 195–201.

85. Bystrzejewska, G., Maleszewski, S., and Poskuta, J. (1971). Photosynthesis, ^{14}C-photosynthetic products and transpiration of bean leaves as influenced by gibberellic acid. *Bull. Acad. Pol. Sci.* **B19,** 533–536.

86. Caldwell, M. M., Osmond, C. B., and Nott, D. L. (1977). C_4 pathway photosynthesis at low temperature in cold-tolerant *Atriplex* species. *Plant Physiol.* **60,** 157–164.

87. Calvin, M. (1954). Chemical and photochemical reactions of thioctic acid and related disulfides. *Fed. Proc. Fed. Am. Soc. Exp. Biol.* **13,** 697–714.

88. Canvin, D. T. (1978). Photorespiration and the effect of oxygen on photosynthesis. *In* "Photosynthetic Carbon Assimilation" (H. W. Siegelman and G. Hind, eds.), pp. 61–76. Plenum, New York.

89. Canvin, D. T., Lloyd, N. D. H., Fock, H., and Przybylla, K. (1976). Glycine and serine metabolism and photorespiration. *In* "CO_2 Metabolism and Plant Productivity" (R. H. Burris and C. C. Black, eds.), pp. 161–176. University Park Press, Baltimore.

90. Carmi, A., and Koller, D. (1979). Regulation of photosynthetic activity in the primary leaves of bean (*Phaseolus vulgaris* L.) by materials moving in the water-conducting system. *Plant Physiol.* **64,** 285–288.

91. Cashmore, A. R. (1976). Protein synthesis in plant leaf tissue. The site of synthesis of the major proteins. *J. Biol. Chem.* **251,** 2848–2853.

92. Cataldo, D. A., and Berlyn, G. P. (1974). An evaluation of selected physical characteristics and metabolism of enzymically separated mesophyll cells and minor veins of tobacco. *Am. J. Bot.* **61,** 957–963.

93. Cerff, R. (1978). Glyceraldehyde-3-phosphate dehydrogenase (NADP) from *Sinapis alba* L. NAD(P)-induced conformation changes of the enzyme. *Eur. J. Biochem.* **82,** 45–53.

94. Champigny, M.-L. (1976). La régulation du cycle de Calvin. *Physiol. Veg.* **14**, 607–628.
95. Champigny, M.-L. (1978). Adenine nucleotides and the control of photosynthetic activities. *Proc. Int. Congr. Photosynth., 4th* 479–488.
96. Champigny, M.-L., and Bismuth, E. (1977). Inorganic carbon transport across the spinach chloroplast envelope. *Plant Cell Physiol.*, special issue "Photosynthetic Organelles," pp. 365–375.
97. Champigny, M.-L., and Miginiac-Maslow, M. (1971). Relations entre l'assimilation photosynthétique de CO_2 et la phosphorylation des chloroplasts isolés. I. Stimulation de la fixation de CO_2 par l'antimycine A, antagoniste de son inhibition par le phosphate. *Biochim. Biophys Acta* **234**, 335–343.
98. Chapman, D. J., and Leech, R. M. (1976). Phosphoserine as an early product of photosynthesis in isolated chloroplasts and in leaves of *Zea mays* seedlings. *FEBS Letters* **68**, 160–164.
99. Chapman, E. A., and Graham, D. (1974). The effect of light on the tricarboxylic acid cycle in green leaves. I. Relative rates of the cycle in the dark, and the light. *Plant Physiol.* **53**, 879–885.
100. Chapman, E. A., and Graham, D. (1974). The effect of light on the tricarboxylic acid cycle in green leaves. II. Intermediary metabolism and the location of control points. *Plant Physiol.* **53**, 886–892.
101. Chapman, K. S. R., and Hatch, M. D. (1977). Regulation of mitochondrial NAD-malic enzyme involved in C_4 pathway photosynthesis. *Arch. Biochem. Biophys.* **184**, 298–306.
102. Chapman, K. S. R., and Hatch, M. D. (1979). Aspartate stimulation of malate decarboxylation in *Zea mays* bundle sheath cells: possible role in regulation of C_4 photosynthesis. *Biochem. Biophys. Res. Commun.* **86**, 1274–1280.
103. Chollet, R. (1973). The effect of oxygen on CO_2 fixation in mesophyll cells isolated from *Digitaria sanguinalis* (L.) Scop. leaves. *Biochem. Biophys. Res. Commun.* **55**, 850–856.
104. Chollet, R. (1974). $^{14}CO_2$ fixation and glycolate metabolism in the dark in isolated maize (*Zea mays* L.) bundle sheath strands. *Arch. Biochem. Biophys.* **163**, 521–529.
105. Chollet, R. (1976). Effect of glycidate on glycolate formation and photosynthesis in isolated spinach chloroplasts. *Plant Physiol.* **57**, 237–240.
106. Chollet, R. (1978). Evaluation of the light/dark ^{14}C assay of photorespiration. *Plant Physiol.* **61**, 929–932.
107. Chollet, R., and Anderson, L. L. (1976). Regulation of ribulose 1,5-bisphosphate carboxylase–oxygenase activities by temperature pretreatment and chloroplast metabolites. *Arch. Biochem. Biophys.* **176**, 344–351.
108. Chu, D. K., and Bassham, J. A. (1974). Activation of ribulose 1,5-diphosphate carboxylase by nicotinamide adenine dinucleotide phosphate and other chloroplast metabolites. *Plant Physiol.* **54**, 556–559.
109. Chu, D. K., and Bassham, J. A. (1975). Regulation of ribulose-1,5-diphosphate carboxylase by substrate and other metabolites. *Plant Physiol.* **55**, 720–726.
110. Claussen, W., and Biller, E. (1977). Die Bedeutung der Saccharose- und Stärkegehalte der Blatter fur die Regulierung der Netto-Photosyntheseraten. *Z. Pflanzenphysiol.* **81**, 189–198.
111. Cleland, R. E. (1975). Auxin-induced hydrogen ion excretion: correlation with growth, and control by external pH and water stress. *Planta* **127**, 233–242.
112. Cockburn, W., Ting, I. P., and Sternberg, O. (1979). Relationship between stomatal behavior and internal carbon dioxide concentration in Crassulacean acid metabolism plants. *Plant Physiol.* **63**, 1029–1032.

113. Codd, G. A., and Schmid, G. H. (1972). Enzymic evidence for peroxisomes in a mutant of *Chlorella vulgaris. Arch. Mikrobiol.* **81,** 264–272.

114. Collins, N., and Merrett, M. J. (1975). The localization of glycolate-pathway enzymes in *Euglena. Biochem. J.* **148,** 321–328.

115. Colman, B., Cheng, K. H., and Ingle, R. K. (1976). The relative activities of PEP carboxylase and RuDP carboxylase in blue-green algae. *Plant Sci. Lett.* **6,** 123–127.

116. Colman, B., Miller, A. G., and Grodzinski, B. (1974). A study of the control of glycolate excretion in *Chlorella. Plant Physiol.* **53,** 395–397.

117. Coombs, J., and Baldry, C. W. (1975). Metabolic regulation in C_4 photosynthesis: phosphoenol pyruvate carboxylase and 3C intermediates of the photosynthetic carbon reduction cycle. *Planta* **124,** 153–158.

118. Coombs, J., Maw, S. L., and Baldry, C. W. (1975). Metabolic regulation in C_4 photosynthesis: the inorganic carbon substrate for PEP carboxylase. *Plant Sci. Lett.* **4,** 97–102.

119. Cooper, T. G., Filmer, D., Wishnik, M., and Lane, M. D. (1969). The active species of "CO_2" utilized by ribulose-diphosphate carboxylase. *J. Biol. Chem.* **244,** 1081–1083.

120. Cornic, G. (1978). La photorespiration se déroulant dans un air sans CO_2 a-t-elle une fonction? *Can. J. Bot.* **56,** 2128–2137.

121. Craigie, J. S. (1974). Storage products. *In* "Algal Physiology and Biochemistry" (W. D. P. Stewart, ed.), pp. 206–235. University of California Press, Berkeley.

122. Créach, E. (1979). Enhanced dark carbon dioxide fixation in maize. *Plant Physiol.* **64,** 435–438.

123. Créach, E., Michel, J. P., and Thibault, P. (1974). Aspartic acid as an internal CO_2 reservoir in *Zea mays:* effect of oxygen concentration and of far-red illumination. *Planta* **118,** 91–100.

124. Crews, C. E., Vines, H. M., and Black, C. C. (1975). Postillumination burst of carbon dioxide in Crassulacean acid metabolism plants. *Plant Physiol.* **55,** 652–657.

125. Crookston, R. K., and Moss, D. N. (1970). The relation of carbon dioxide compensation and chlorenchymatous vascular bundle sheaths in leaves of dicots. *Plant Physiol.* **46,** 564–567.

126. Crookston, R. K., and Moss, D. N. (1973). A variation of C_4 leaf anatomy in *Arundinella hirta* (Gramineae). *Plant Physiol.* **52,** 397–402.

127. Crookston, R. K., and Moss, D. N. (1974). Interveinal distance for carbohydrate transport in leaves of C_3 and C_4 grasses. *Crop Sci.* **14,** 123–125.

128. Cumming, B. G., and Wagner, E. (1968). Rhythmic processes in plants. *Annu. Rev. Plant Physiol.* **19,** 381–416.

129. Daley, L. S., and Bidwell, R. G. S. (1977). Phosphoserine and phosphohydroxypyruvic acid: evidence for their role as early intermediates in photosynthesis. *Plant Physiol.* **60,** 109–114.

130. Daley, L. S., and Bidwell, R. G. S. (1978). Separation of phosphohydroxypyruvate, 3-phosphoglyceric acid and O-phosphoserine by paper chromatology and chemical derivatization. *J. Chromatogr.* **147,** 233–241.

131. Daley, L. S., Dailey, F., and Criddle, R. S. (1978). Light activation of ribulose bisphosphate carboxylase. Purification and properties of the enzyme in tobacco. *Plant Physiol.* **62,** 718–722.

132. Daley, L. S., Ray, T. B., Vines, H. M., and Black, C. C. (1977). Characterization of phosphoenolpyruvate carboxykinase from pineapple leaves *Ananas comosus* (L.) Merr. *Plant Physiol.* **59,** 618–622.

133. Daley, L. S., Vines, H. M., and Bidwell, R. G. S. (1979). Oxidation of phosphohy-

droxypyruvate and hydroxypyruvate: physiological implications in plants. *Can. J. Bot.* **57**, 1–3.

134. Dalton, C. C., and Street, H. E. (1977). The influence of applied carbohydrates on the growth and greening of cultured spinach (*Spinacia oleracea* L.) cells. *Plant Sci. Lett.* **10**, 157–164.

135. Davies, D. D. (1973). Control of and by pH. *In* "Symposium 27: Rate Control of Biological Processes" (D. D. Davies, ed.), pp. 513–529. Cambridge Univ. Press, London/New York, 1973.

136. Davies, D. D., and Patil, K. D. (1974). Regulation of "malic" enzyme of *Solanum tuberosum* by metabolites. *Biochem. J.* **137**, 45–53.

137. Dayanandam, P., and Kaufman, P. B. (1975). Stomatal movements associated with potassium fluxes. *Am. J. Bot.* **62**, 221–231.

138. Decker, J. P. (1955). A rapid, postillumination deceleration of respiration in green leaves. *Plant Physiol.* **30**, 82–84.

139. Decker, J. P. (1958). The effects of light on respiration using isotopically enriched oxygen: an objection and alternate interpretation. *Plant Sci. Bull.* **4**, 3–4.

140. Decker, J. P. (1959). Some effects of temperature and carbon dioxide concentration on photosynthesis of *Mimulus*. *Plant Physiol.* **34**, 103–106.

141. Decker, J. P. (1970). Early history of photorespiration. *Bioeng. Bull.* 10.

142. Decker, J. P., and Tio, M. A. (1959). Photosynthetic surges in coffee seedlings. *J. Agric. Univ. P. R.* **43**, 50–55.

143. DeGroot, D., and Kennedy, R. A. (1977). Photosynthesis in *Elodea canadensis* Michx.: four-carbon acid synthesis. *Plant Physiol.* **59**, 1133–1135.

144. Delaney, M. E., and Walker, D. A. (1978). Comparison of the kinetic properties of ribulose bisphosphate carboxylase in chloroplast extracts of spinach, sunflower, and four other reductive pentose phosphate-pathway species. *Biochem. J.* **171**, 477–482.

145. Deleens, E., Garnier-Dardart, J., and Querioz, O. (1979). Carbon isotope composition of intermediates of the starch-malate sequence and the level of the crassulacean acid metabolism in leaves of *Kalanchoë blossfeldiana* Tom Thumb. *Planta* **146**, 441–449.

146. Delmer, D. P., and Albersheim, P. (1970). Biosynthesis of sucrose and nucleoside diphosphate glucoses in *Phaseolus aureus*. *Plant Physiol.* **45**, 782–786.

147. Demidov, E. D., and Bell, L. N. (1973). Effect of blue and red light on photophosphorylation in isolated chloroplasts of pea. *Fiziol. Rast.* (*Moscow*) **20**, 292–299.

148. Dittrich, P. (1976). Equilibration of label in malate during dark fixation of CO_2 in *Kalanchoë fedtschenkoi*. *Plant Physiol.* **58**, 288–291.

149. Dittrich, P. (1976). Nicotinamide adenine dinucleotide-specific "malic enzyme in *Kalanchoë daigremontiana* and other plants exhibiting Crassulacean acid metabolism. *Plant Physiol.* **57**, 310–314.

150. Dittrich, P., Campbell, W. H., and Black, C. C. (1973). Phosphoenolpyruvate carboxykinase in plants exhibiting Crassulacean acid metabolism. *Plant Physiol.* **52**, 357–361.

151. Dittrich, P., and Huber, W. (1974). Carbon dioxide metabolism in members of the Chlamydospermas. *Proc. Int. Congr. Photosynth. 3rd* **2**, 1573–1578.

152. Doehlert, D. C., Ku, M. S. B., and Edwards, G. E. (1979). Dependence of the postillumination burst of CO_2 on temperature, light, CO_2 and O_2 concentration in Wheat (*Triticum aestivum*). *Physiol. Plant.* **46**, 299–306.

153. Döhler, G. (1972). Untersuchung des Einflusses der Temperatur auf die Photosynthese-Induktion bie *Chlorella vulgaris* mit radioactivem CO_2. *Planta* 107, 33–42.

154. Döhler, G. (1976). Photosynthetische Carboxy-lierungsreaktionen verschiedenen pigmentierte *Anacystis*-Zellen. *Planta* **131**, 129–133.
155. Döhler, G., and Kohlenbach, H. W. (1976). CO_2-Fixierung isolierter und sich teilender Mesophyllzellen von *Pharbitis purpurea*. *Z. Pflanzenphysiol.* **80**, 81–86.
155a. Dorovari, S., and Canvin, D. T. (1980). Effect of butyl 2-hydroxy 3-butyonate on sunflower leaf photosynthesis and photorespiration. *Plant Physiol.* **66**, 628–631.
156. Downton, W. J. S. (1970). Preferential C_4-dicarboxylic acid synthesis, the postillumination burst, carboxyl transfer step, and grana configurations in plants with C_4-photosynthesis. *Can. J. Bot.* **48**, 1795–1800.
157. Downton, W. J. S. (1975). The occurrence of C_4 photosynthesis among plants. *Photosynthetica* **9**, 96–105.
158. Downton, W. J. S., and Hawker, J. S. (1973). Enzymes of starch and sucrose metabolism in *Zea mays* leaves. *Phytochemistry* **12**, 1551–1556.
159. Downton, W. J. S., and Tregunna, E. B. (1968). Carbon dioxide compensation—its relation to photosynthetic carboxylation reactions, systematics of the Gramineae, and leaf anatomy. *Can. J. Bot.* **46**, 207–215.
160. Edwards, G. E., and Black, C. C. (1971). Photosynthesis in mesophyll cells and bundle sheath cells isolated from *Digitaria sanguinalis* leaves. *In* "Photosynthesis and Photorespiration" (M. D. Hatch, C. B. Osmond, and R. O. Slatyer, eds.), pp. 153–168. Wiley–Interscience, New York.
161. Edwards, G. E., Huber, S. C., Ku, S. B., Rathnam, C. K. M., Gutierrez, M., and Mayne, B. C. (1976). Variation in photochemical activities of C_4 plants in relation to CO_2 fixation. *In* "CO_2 Metabolism and Plant Productivity" (R. H. Burris and C. C. Black, eds.), pp. 83–112. Univ. Park Press, Baltimore.
162. Edwards, G. E., Robinson, S. P., Tyler, N. J. C., and Walker, D. A. (1978). Photosynthesis by isolated protoplasts, protoplast extracts, and chloroplasts of wheat. *Plant Physiol.* **62**, 313–319.
163. Egle, K., and Fock, H. (1967). Light respiration—correlations between CO_2 fixation, O_2 pressure and glycollate concentration. *In* "The Biochemistry of Chloroplasts" (T. W. Goodwin, ed.), Vol. 2, pp. 79–87. Academic Press, New York.
164. Egle, K., and Schenk, W. (1953). Der Einfluss der Temperatur auf die Lage des CO_2-Kompensationspunktes. *Planta* **43**, 83–97.
165. Ehleringer, J., and Björkman, O. (1977). Quantum yields for CO_2 uptake in C_3 and C_4 plants: dependence on temperature, CO_2, and O_2 concentration. *Plant Physiol.* **59**, 86–90.
166. Elkin, L., and Park, R. B. (1975). Chloroplast fluorescence of C_4 plants. III. Fluorescence spectra and relative fluorescence yields of bundle-sheath and mesophyll chloroplasts. *Planta* **127**, 37–47.
167. El-Sharkaway, M. A., Loomis, R. S., and Williams, W. A. (1967). Apparent reassimilation of respiratory carbon dioxide by different plant species. *Physiol. Plant.* **20**, 171–186.
168. Emmerson, R., and Lewis, C. M. (1943). The dependence of the quantum yield of *Chlorella* photosynthesis of the wavelength of light. *Am. J. Bot.* **30**, 165–178.
169. Enns, T. (1967). Facility by carbonic anhydrase of carbon dioxide transport. *Science* **155**, 44–47.
170. Everson, R. G. (1969). Bicarbonate equilibria and the apparent $K_m(HCO_3^-)$ of isolated chloroplasts. *Nature (London)* **222**, 876.
171. Everson, R. G., Cockburn, W., and Gibbs, M. (1967). Sucrose as a product of photosynthesis in isolated spinach chloroplasts. *Plant Physiol.* **42**, 840–844.

172. Everson, R. G., and Slack, C. R. (1968). Distribution of carbonic anhydrase in relation to the C_4 pathway of photosynthesis. *Phytochemistry* **7**, 581–584.
173. Fair, P. (1978). An investigation into the effect of varying nitrogen regimes on the abundance of peroxisomes and activities of nitrate reductase and catalase in barley (*Hordeum vulgare* L.) and maize (*Zea mays* L.). *Ann. Bot. (London)* **42**, 101–107.
174. Fair, P., Tew, J., and Cresswell, C. (1972). The effect of age and leaf position on carbon dioxide compensation point (Γ), and potential photosynthetic capacity, photorespiration and nitrate assimilation in *Hordeum vulgare* L. *J. S. Afr. Bot.* **38**, 81–95.
175. Fair, P., Tew, J., and Cresswell, C. F. (1974). Enzyme activities associated with carbon dioxide exchange in illuminated leaves of *Hordeum vulgare* L. III. Effects of concentration and form of nitrogen supplied on carbon dioxide compensation point. *Ann. Bot. (London)* **38**, 39–43.
176. Farineau, J. (1975). Photoassimilation of CO_2 by isolated bundle-sheath strands of *Zea mays*. I. Stimulation of CO_2 assimilation by adding various intermediates of the photosynthetic cycle; evidence for a deficiency of photosystem II activity. *Physiol. Plant* **33**, 300–309.
177. Feierabend, J. (1976). Temperature sensitivity of chloroplast ribosome formation in higher plants. *In* "Genetics and Biogenesis of Chloroplasts and Mitochondria" (T. Bücher, Neupert, W., Sebald, W., and Werner, S., eds.), pp. 99–102. Elsevier, Amsterdam.
178. Feierabend, J., and Beevers, H. (1972). Developmental studies on microbodies in wheat leaves. II. Ontogeny of particulate enzyme associations. *Plant Physiol.* **49**, 33–39.
179. Filippova, L. A., Mamushina, N. S., and Zubkova, E. K. (1977). On the influence of accumulation of assimilates on photosynthesis in *Chlorella* cells. *Bot. Zh. (Leningrad)* **62**, 179–184.
180. Findenegg, G. R. (1974). Relation between carbonic anhydrase activity and uptake of HCO_3^- and Cl^- in photosynthesis by *Scenedesmus*. *Planta* **11b**, 123–131.
181. Findenegg, G. R. (1976). Correlations between accessibility of carbonic anhydrase for external substrate and regulation of photosynthetic use of CO_2 and HCO_3^- by *Scenedesmus obliquus*. *Z. Pflanzenphysiol.* **79**, 428–437.
182. Fock, H., Bate, G. C., and Egle, K. (1974). Radiogaschromatographische Untersuchungen über die Herkunft der Glykolsäure in photosynthetisch aktiven *Chlorella*-Suspensionen. *Ber. Dtsch. Bot. Ges.* **87**, 239–247.
183. Fock, H., Mahon, J. D., Canvin, D. T., and Grant, B. R. (1974). Estimation of carbon fluxes through photosynthetic and photorespiratory pathways. *In* "Mechanisms of Regulation of Plant Growth" (R. L. Bieleski, A. R. Ferguson, and M. M. Cresswell, eds.), Bull. 12, pp. 235–242. The Royal Society of New Zealand, Wellington.
184. Fock, H., Hilgenberg, W., and Egle, K. (1972). Kohlendioxid- und Sauerstoff-Gaswechsel belichteter Blätter und die CO_2/O_2-quotienten bie normalen und niedrigen O_2-Partialdrucken. *Planta* **106**, 355–361.
185. Fock, H., Klug, K., and Canvin, D. T. (1979). Effect of carbon dioxide and temperature on photosynthetic CO_2 uptake and photorespiratory CO_2 evolution in sunflower leaves. *Planta* **145**, 219–223.
186. Fock, H. Krotkov, G., and Canvin, D. T. (1969). Photorespiration in liverworts and leaves. *In* "Progress in Photosynthesis Research" (H. Metzner, ed.), pp. 482–487. Inst. Chemische Pflanzenphysiologie, Tubingen.
187. Forrester, M. L., Krotkov, G., and Nelson, C. D. (1966). Effect of oxygen on photosynthesis, photorespiration and respiration in detached leaves. I. Soybean, *Plant Physiol.* **41**, 422–427.

188. Forrester, M. L., Krotkov, G., and Nelson, C. D. (1966). Effect of oxygen on photosynthesis, photorespiration, and respiration in detached leaves. II. Corn and other monocotyledons. *Plant Physiol.* **41,** 428–431.
189. Foster, A., and Black, C. C. (1977). *Panicum maximum* photosynthesis. *Plant Cell Physiol.* (special issue), 325–340.
190. Foyer, C. H., and Halliwell, B. (1976). The presence of glutathione and glutathione reductase in chloroplasts: a proposed role of ascorbic acid metabolism. *Planta* **133,** 21–25.
190a. Francis, K. (1979). Photosynthesis by isolated chloroplasts of *Sorghum vulgare. Experientia* **35,** 1324–1326.
191. Fraser, D. E., and Bidwell, R. G. S. (1974). Photosynthesis and photorespiration during the ontogeny of the bean plant. *Can. J. Bot.* **52,** 2561–2570.
192. Frederick, S. E., Gruber, P. J., and Tolbert, N. E. (1973). The occurrence of glycolate dehydrogenase and glycolate oxidase in green plants. An evolutionary survey. *Plant Physiol.* **52,** 318–323.
193. Frederick, S. E., and Newcomb, E. H. (1969). Microbody-like organelles in leaf cells. *Science* **163,** 1353–1355.
194. Frederick, S. E., and Newcomb, E. H. (1971). Ultrastructure and distribution of microbodies in leaves of grasses with and without CO_2-photorespiration. *Planta* **96,** 152–174.
195. Fry, D. J., and Phillips, I. D. J. (1976). Photosynthesis of conifers in relation to annual growth cycles and dry matter production. I. Some C_4 characteristics in photosynthesis of Japanese larch (*Larix leptolepis*). *Physiol. Plant.* **37,** 185–190.
196. Fry, S. C., and Bidwell, R. G. S. (1977). An investigation of photosynthetic sucrose production in bean leaves. *Can. J. Bot.* **55,** 1457–1464.
197. Geiger, D. R. (1976). Effects of translocation and assimilate demand on photosynthesis. *Can. J. Bot.* **54,** 2337–2345.
198. Giaquinta, R. (1978). Source and sink leaf metabolism in relation to phloem translocation. Carbon partitioning and enzymology. *Plant Physiol.* **61,** 380–385.
199. Gibbs, M., Latzko, E., Everson, R. G., and Cockburn, W. (1967). Carbon metabolism: nature and formation of end products. *In* "Harvesting the Sun" (A. San Pietro, F. A. Greer, and T. J. Army, eds.), pp. 111–130. Academic Press, New York.
200. Glidewell, S. M., and Raven, J. A. (1975). Measurement of simultaneous oxygen evolution and uptake in *Hydrodictyon africanum. J. Exp. Bot.* **26,** 479–488.
201. Glover, J. (1973). The dark respiration of sugar-cane and the loss of photosynthate during the growth of a crop. *Ann. Bot. (London)* **37,** 845–852.
202. Gnanam, A., and Kulandaivelu, G. (1969). Photosynthetic studies with leaf cell suspensions from higher plants. *Plant Physiol.* **44,** 1451–1456.
203. Goatly, M. B., Coombs, J., and Smith, H. (1975). Development of C_4 photosynthesis in sugar cane: changes in properties of phosphoenolpyruvate carboxylase during greening. *Planta* **125,** 15–24.
204. Goldsmith, M. H. M., Cataldo, D. A., Karn, J., Brennemann, T., and Trip, P. (1974). The rapid non-polar transport of auxin in the phloem of intact *Coleus* plants. *Planta* **116,** 301–317.
205. Goldstein, L. D., Ray, T. B., Kestler, D. P., Mayne, B. C., and Brown, R. H. (1976). Biochemical characterization of *Panicum* species which are intermediate between C_3 and C_4 photosynthesis plants. *Plant Sci. Lett.* **6,** 85–90.
206. Goldsworthy, A. (1966). Experiments on the origin of CO_2 released by tobacco leaf segments in the light. *Phytochemistry* **5,** 1013–1019.

207. Goto, K. (1979). Modes of control by the circadian oscillator and the hourglass mechanism of the activities of cytoplasmic and chloroplast glyceraldehyde-3-phosphate dehydrogenases in Lemna gibba G3. *Plant Cell Physiol.* **20**, 513–521.
208. Graham, D., and Reed, M. L. (1971). Carbonic anhydrase and the regulation of photosynthesis. *Nature (London) New Biol.* **231**, 81–82.
209. Grodzinski, B. (1978). Glyoxylate decarboxylation during photorespiration. *Planta* **144**, 31–37.
210. Grodzinski, B., and Colman, B. (1976). Intracellular localization of glycolate dehydrogenase in a blue-green alga. *Plant Physiol.* **58**, 199–202.
211. Grossman, D., and Cresswell, C. F. (1973). Influence of nitrogen supply in nutrient media on carbon dioxide compensation point (Γ) of C-4 photosynthetic plants. *S. Afr. J. of Sci.* **69**, 244–246.
212. Gutierrez, M., Gracen, V. E., and Edwards, G. E. (1974). Biochemical and cytological relationships in C_4 plants. *Planta* **119**, 279–300.
213. Halliwell, B. (1974). Oxidation of formate by peroxisomes and mitochondria from spinach leaves. *Biochem. J.* **138**, 77–85.
214. Hanscom, Z., and Ting, I. P. (1977). Physiological responses to irrigation in *Opuntia basilaris* Engelm. & Bigel. *Bot. Gaz. (Chicago)* **138**, 159–167.
215. Hanscom, Z., and Ting, I. P. (1978). Responses of succulents to plant water stress. *Plant Physiol.* **61**, 327–330.
216. Hansen, G. K. (1978). Utilization of photosynthate for growth, respiration, and storage in tops and roots of *Lolium multiflorum*. *Physiol. Plant.* **42**, 5–13.
217. Hansen, G. K., and Jensen, C. R. (1977). Growth and maintenance respiration in whole plants, tops, and roots of *Lolium multiflorum*. *Physiol. Plant.* **39**, 155–164.
218. Harris, G. C., and Stern, A. I. (1977). Isolation and some properties of ribulose-1,5-bisphosphate carboxylase–oxygenase from red kidney bean primary leaves. *Plant Physiol.* **60**, 697–702.
219. Hatch, M. D. (1976). The C_4 pathway of photosynthesis: mechanism and function. *In* "CO_2 Metabolism and Plant Productivity" (R. H. Burris and C. C. Black, eds.), pp. 59–81. Univ. Park Press, Baltimore.
220. Hatch, M. D. (1978). Regulation of enzymes in C_4 photosynthesis. *Curr. Top. Cell. Regul.* **14**, 1–27.
221. Hatch, M. D. (1979). Mechanism of C_4 photosynthesis in *Chloris gayana*: pool sizes and kinetics of $^{14}CO_2$ incorporation into 4-carbon and 3-carbon intermediates. *Arch. Biochem. Biophys.* **194**, 117–127.
222. Hatch, M. D., and Kagawa, T. (1973). Enzymes and functional capacities of mesophyll chloroplasts from plants with C_4-pathway photosynthesis. *Arch. Biochem. Biophys.* **159**, 842–853.
223. Hatch, M. D., Kagawa, T., and Craig, S. (1975). Subdivision of C_4-pathway species based on differing C_4 acid decarboxylating systems and ultrastructural features. *Aust. J. Plant Physiol.* **2**, 111–128.
224. Hatch, M. D., and Mau, S.-L. (1977). Association of NADP- and NAD-linked malic enzyme activities in *Zea mays*: relation to C_4 pathway photosynthesis. *Arch. Biochem. Biophys.* **179**, 361–369.
225. Hatch, M. D., and Oliver, I. R. (1978). Activation and inactivation of phosphoenolpyruvate carboxylase in leaf extracts from C_4 species. *Aust. J. Plant Physiol.* **5**, 571–580.
226. Hatch, M. D., and Osmond, C. B. (1976). Compartmentation and transport in C_4 photosynthesis. *In* "Transport in Plants III" (C. R. Stocking and U. Heber, eds.), pp. 143–184. Springer–Verlag, Berlin.

227. Hatch, M. D., and Slack, C. R. (1966). Photosynthesis by sugar cane leaves. *Biochem. J.* **101**, 103–111.

228. Hatch, M. D., and Slack, C. R. (1968). A new enzyme for the interconversion of pyruvate and phosphopyruvate and its role in the C_4 dicarboxylic acid pathway of photosynthesis. *Biochem. J.* **106**, 141–146.

229. Hatch. M. D., and Slack, C. R. (1970). Photosynthetic CO_2-fixation pathways. *Annu. Rev. Plant Physiol.* **21**, 141–162.

230. Hatch, M. D., and Slack, C. R. (1970). The C_4-carboxylic acid pathway of photosynthesis. *In* "Progress in Phytochemistry" (L. Reinhold and Y. Liwschitz, eds.), pp. 35–106, Wiley–Interscience, London.

231. Hattersley, P. W., Watson, L., and Osmond, C. B. (1977). *In situ* immunofluorescent labelling of ribulose-1,5-bisphosphate carboxylase in leaves of C_3 and C_4 plants. *Aust. J. Plant Physiol.* **4**, 523–539.

232. Hawker, J. S. (1967). Inhibition of sucrose phosphatase by sucrose. *Biochem. J.* **102**, 401–406.

233. Hawker, J. S. (1967). The activity of uridine diphosphate glucose-D-fructose 6-phosphate 2-glucosyltransferase in leaves. *Biochem. J.* **105**, 943–946.

234. Hawker, J. S., Ozbun, J. L., Greenberg, A., and Preiss, J. (1974). Interaction of spinach leaf adenosine diphosphate glucose α-1,4-glucan α-4-glucosyl transferase and α-1,4-glucan α-1,4-glucan-6-glycosyl transferase in synthesis of branched α-glucan. *Arch. Biochem. Biophys.* **160**, 530–551.

235. Heber, U., Egneus, H., Hanck, U., Jensen, M., and Köster, S. (1978). Regulation of photosynthetic electron transport and photophosphorylation in intact chloroplasts and leaves of *Spinacia oleracea* L. *Planta* **143**, 41–49.

236. Heber, U., Kirk, M. R., Gimmler, H., and Schäfer, G. (1974). Uptake and reduction of glycerate by isolated chloroplasts. *Planta* **120**, 31–46.

237. Heber, U. W., and Santarius, K. A. (1970). Direct and indirect transfer of ATP and ADP across the chloroplast envelope. *Z. Naturforsch.* **B25**, 718–728.

238. Heldt, H. W. (1976). Metabolite transport in intact spinach chloroplasts. *In* "The Intact Chloroplast" (J. Barber, ed.), pp. 215–234. Elsevier, Amsterdam.

239. Heldt, H. W., Chon, C. J., Maronde, D., Herold, A., Stankovic, Z. S., Walker, D. A., Kraminer, A., Kirk, M. R., and Heber, U. (1977). Role of orthophosphate and other factors in the regulation of starch formation in leaves and isolated chloroplasts. *Plant Physiol.* **59**, 1146–1155.

240. Heldt, H. W., and Rapley, R. (1970). Specific transport of inorganic phosphate, 3-phosphoglycerate and dihydroxyacetone phosphate, and of dicarboxylates across the inner membrane of spinach chloroplasts. *FEBS Lett.* **10**, 143–148.

241. Hew, C.-S., Krotkov, G., and Canvin, D. T. (1969). Determination of the rate of CO_2 evolution by green leaves in light. *Plant Physiol.* **44**, 662–670.

242. Hew, C.-S., and Wong, Y. S. (1974). Photosynthesis and respiration of ferns in relation to their habitat. *Am. Fern J.* **64**, 40–48.

243. Hill, B. C., and Bown, A. W. (1978). Phosphoenolpyruvate carboxylase activity from *Avena* coleoptile tissue. Regulation by H^+ and malate. *Can. J. Bot.* **56**, 404–407.

244. Ho, L. C. (1976). The relationship between the rates of carbon transport and of photosynthesis in tomato leaves. *J. Exp. Bot.* **96**, 87–97.

245. Hoffmann, P. (1975). Regulative Wechselbeziehungen zwischen Photosynthese und Atmung bie CCC-behandelten Primärblättern von *Triticum aestivum*. *Biochem. Physiol. Pflanz.* **168**, 553–560.

246. Hogetsu, D., and Miyachi, S. (1970). Effect of oxygen on the light-enhanced dark carbon dioxide fixation in *Chlorella* cells. *Plant Physiol.* **45**, 178–182.

246a. Holaday, A. S., and Bowes, G. (1980). C_4 acid metabolism and dark CO_2 fixation in a submersed aquatic macrophyte (*Hydrilla verticillata*). *Plant Physiol.* **65**, 331–335.

247. Holden, M. (1973). Chloroplast pigments in plants with the C_4-dicarboxylic acid pathway of photosynthesis. *Photosynthetica* **7**, 41–49.

248. Holmgren, A., Buchanan, B. B., and Wolosiuk, R. A. (1977). Photosynthetic regulatory protein from rabbit liver is identical with thioredoxin. *FEBS Letters* **82**, 351–354.

249. Holmgren, P., and Jarvis, P. G. (1967). Carbon dioxide efflux from leaves in light and darkness. *Physiol. Plant* **20**, 1045–1051.

250. Huang, A. H. C., and Beevers, H. (1972). Microbody enzymes and carboxylases in sequential extracts from C_3 and C_4 leaves. *Plant Physiol.* **50**, 242–248.

251. Huber, S. C. (1978). Regulation of chloroplast photosynthetic activity by exogenous magnesium. *Plant Physiol.* **62**, 321–325.

252. Huber, S., and Edwards, G. (1975). The effect of oxygen on CO_2 fixation by mesophyll protoplasts of C_3 and C_4 plants. *Biochem. Biophys. Res. Commun.* **67**, 28–34.

253. Huber, S. C., and Edwards, G. E. (1975). An evaluation of some parameters required for the enzymic isolation of cells and protoplasts with CO_2 fixation capacity from C_3 and C_4 grasses. *Plant Physiol.* **55**, 203–209.

254. Huber, S. C., and Edwards, G. E. (1975). C_4 photosynthesis: light-dependent CO_2 fixation by mesophyll cells, protoplasts, and protoplast extracts of *Digitaria sanguinalis*. *Plant Physiol.* **55**, 835–844.

255. Huber, S. C., and Edwards, G. E. (1975). Inhibition of phosphoenolpyruvate carboxylase from C_4 plants by malate and aspartate. *Can. J. Bot.* **53**, 1925–1933.

256. Huber, S. C., and Edwards, G. E. (1976). A high-activity ATP translocator in mesophyll chloroplasts of *Digitaria sanguinalis*, a plant having the C-4 dicarboxylic acid pathway of photosynthesis. *Biochim. Biophys. Acta* **440**, 675–687.

257. Huber, S. C., and Edwards, G. E. (1977). Transport in C_4 mesophyll chloroplasts: characterization of the pyruvate carrier. *Biochim. Biophys. Acta* **462**, 583–602.

258. Huber, S. C., and Edwards, G. E. (1977). Transport in C_4 mesophyll chloroplasts; evidence for an exchange of inorganic phosphate and phosphoenolpyruvate. *Biochim. Biophys. Acta* **462**, 603–612.

259. Huber, S. C., Hall, T. C., and Edwards, G. E. (1977). Light dependent incorporation of $^{14}CO_2$ into protein by mesophyll protoplasts and chloroplasts isolated from *Pisum sativum*. *Z. Pflanzenphysiol.* **85**, 153–163.

260. Humphries, E. C., and Thorne, G. N. (1964). The effect of root formation on photosynthesis of detached leaves. *Ann. Bot. (London)* **28**, 391–400.

261. Husemann, W., and Barz, W. (1977). Photoautotrophic growth and photosynthesis in cell suspension cultures of *Chenopodium rubrum*. *Physiol. Plant* **40**, 77–81.

262. Imafuku, H., and Katoh, T. (1976). Intracellular ATP level and light-induced inhibition of respiration in a blue-green alga, *Anabaena variabilis*. *Plant Cell Physiol.* **17**, 515–524.

263. Ingle, R. K., and Colman, B. (1976). The relationship between carbonic anhydrase activity and glycolate excretion in the blue-green alga *Coccochloris peniocystis*. *Planta* **128**, 217–223.

264. Ishii, R., Takehara, T., Murata, Y., and Miyachi, S. (1977). Effects of light intensity on the rates of photosynthesis and photorespiration in C_3 and C_4 plants. *In* "Biological Solar Energy Conversion" (Mitsui, A., Miyachi, S., san Pietro, A., and Tamura, S., eds.), pp. 265–271. Academic Press, New York.

265. Iwanij, W., Chua, N., and Siekevitz, P. (1975). Synthesis and turnover of ribulose bisphosphate carboxylase and its subunits during the cell cycle of *Chlamydomonas reinhardtii*. *J. Cell Biol.* **64**, 572–585.

438   R. G. S. Bidwell

266. Jackson, W. A., and Volk, R. J. (1970). Photorespiration. *Annu. Rev. Plant Physiol.* **21**, 385–432.
267. Jacobson, B. S., Fong, F., and Heath, R. L. (1975). Carbonic anhydrase of spinach. Studies on its location, inhibition, and physiological function. *Plant Physiol.* **55**, 468–474.
268. Jagendorf, A. T., and Uribe, E. (1966). Photophosphorylation and the chemiosmotic hypothesis. *In* "Energy Conversion of the Photosynthetic Apparatus" (J. M. Olson, G. Hind, H. Lyman and H. W. Siegelman, eds.), pp. 215–245. Brookhaven Nat. Lab., New York.
269. Jensen, R. G., and Bahr, J. T. (1977). Ribulose 1,5-bisphosphate carboxylase-oxygenase. *Annu. Rev. Plant Physiol.* **28**, 379–400.
270. Jensen, R. G., and Bassham, J. A. (1966). Photosynthesis by isolated chloroplasts. *Proc. Natl. Acad. Sci. U.S.* **56**, 1095–1101.
271. Jensen, R. G., and Bassham, J. A. (1968). Photosynthesis by isolated chloroplasts. III. Light activation of the carboxylation reaction. *Biochim. Biophys. Acta* **153**, 227–234.
272. Jensen, R. G., Francki, R. I. B., and Zaitlin, M. (1971). Metabolism of separated leaf cells. I. Preparation of photosynthetically active cells from tobacco. *Plant Physiol.* **48**, 9–13.
273. Jimenez, E., Baldwin, R. L., Tolbert, N. E., and Wood, W. A. (1962). Distribution of C^{14} in sucrose from glycolate-C^{14} and serine-3-C^{14} metabolism. *Arch. Biochem. Biophys.* **98**, 172–175.
274. Joliffe, P. A., and Tregunna, E. B. (1968). Effect of temperature, CO_2 concentration, and light intensity on oxygen inhibition of photosynthesis in wheat leaves. *Plant Physiol.* **43**, 902–906.
275. Joliffe, P. A., and Tregunna, E. B. (1969). Estimation of the CO_2 free exchange pool size in wheat and corn leaves. *Can. J. Bot.* **47**, 1506–1508.
276. Kagawa, T., and Hatch, M. D. (1974). Light-dependent metabolism of carbon compounds by mesophyll chloroplasts from plants with the C_4 pathway of photosynthesis. *Aust. J. Plant Physiol.* **1**, 51–64.
277. Kagawa, T., and Hatch, M. D. (1974). C_4-acids as the source of carbon dioxide for Calvin cycle photosynthesis by bundle sheath cells of the C_4-pathway species *Atriplex spongiosa*. *Biochem. Biophys. Res. Commun.* **59**, 1326–1332.
278. Kagawa, T., and Hatch, M. D. (1975). Mitochondria as a site of C_4 acid decarboxylation in C_4-pathway photosynthesis. *Arch. Biochem. Biophys.* **167**, 687–696.
279. Kanai, R., and Edwards, G. E. (1973). Separation of mesophyll protoplasts and bundle sheath cells from maize leaves for photosynthetic studies. *Plant Physiol.* **51**, 1133–1137.
280. Kapil, R. N., Pugh, T. D., and Newcomb, E. H. (1975). Microbodies and an anomalous "microcylinder" in the ultrastructure of plants with Crassulacean acid metabolism. *Planta* **124**, 231–244.
280a. Kaplan, A., and Björkman, O. (1980). Ratio of CO_2 uptake to O_2 evolution during photosynthesis in higher plants. *Z. Pflanzenphysiol.* **96**, 185–188.
281. Kelly, G. J., and Turner, J. F. (1970). The regulation of pea-seed phosphofructokinase by 6-phosphogluconate, 3-phosphoglycerate, 2-phosphoglycerate and phosphoenol pyruvate. *Biochim. Biophys. Acta* **208**, 360–367.
282. Kelly, G. J., and Gibbs, M. (1973). A mechanism for the indirect transfer of photosynthetically reduced nicotinamide adenine dinucleotide phosphate from chloroplasts to the cytoplasm. *Plant Physiol.* **52**, 674–676.

283. Kelly, G. J., and Latzko, E. (1977). Chloroplast phosphofructokinase. I. Proof of phosphofructokinase activity in chloroplasts. *Plant Physiol.* **60**, 290–294.

284. Kelly, G. J., Latzko, E., and Gibbs, M. (1976). Regulatory aspects of photosynthetic carbon metabolism. *Annu. Rev. Plant Physiol.* **27**, 181–205.

285. Kelly, G. J., Zimmerman, G., and Latzko, E. (1976). Light induced activation of fructose-1,6-bisphosphatase in isolated intact chloroplasts. *Biochim. Biophys. Res. Commun.* **70**, 193–199.

286. Kennedy, R. A. (1976). Relationship between leaf development, carboxylase enzyme activities and photorespiration in the C_4-plant *Portulaca oleracea* L. *Planta* **128**, 149–154.

287. Kennedy, R. A. (1976). Photorespiration in C_3 and C_4 plant tissue cultures: significance of Kranz anatomy to low photorespiration in C_4 plants. *Plant Physiol.* **58**, 573–575.

288. Kennedy, R. A., Barnes, J. E., and Laetsch, W. M. (1977). Photosynthesis in C_4 plant tissue cultures: significance of Kranz anatomy to C_4 acid metabolism in C_4 plants. *Plant Physiol.* **59**, 600–603.

289. Kennedy, R. A., and Laetsch, W. M. (1973). Relationship between leaf development and primary photosynthetic products in the C_4 plant *Portulaca oleracea* L. *Planta* **115**, 113–124.

290. Kennedy, R. A., and Laetsch, W. M. (1974). Plant species intermediate for C_3, C_4 photosynthesis. *Science* **184**, 1087–1089.

291. Kent, S. S. (1977). On the metabolic relationships between the Calvin cycle and the tricarboxylic acid cycle. IV. A plant survey for anomalous acetyl coenzyme A. *Plant Physiol.* **60**, 274–276.

292. Kent, S. S. (1979). Photosynthesis in the higher plant, *Vicia faba*. V. Role of malate as a precursor of the tricarboxylic acid cycle. *Plant Physiol.* **64**, 159–161.

293. Kent, S. S., Pinkerton, F. D., and Strobel, G. A. (1974). Photosynthesis in the higher plant, *Vicia faba*. III. Serine, a precursor of the tricarboxylic cycle. *Plant Physiol.* **53**, 491–495.

294. Keys, A. J., Bird, I. F., Cornelius, M. J., Lea, P. J., Wallsgrove, R. M., and Miflin, B. J. (1978). Photorespiratory nitrogen cycle. *Nature (London)* **275**, 741–743.

295. Kluge, M. (1968). Untersuchungen über den Gaswechsel von *Bryophyllum* während der Lichtperiode. II. Beziehungen zwischen dem Malatgehalt des Gewebes und der CO_2-Aufnahme. *Planta* **80**, 359–377.

296. Kluge, M. (1969). Veränliche Markierungsmuster bei $^{14}CO_2$-Fütterung von *Bryophyllum tubiflorum* zu verschiedenen Zeitpunkten der Hell-Dunkel-Periode. I. Die Fütterung unter Belichtung. *Planta* **88**, 113–129.

296a. Kluge, M., Böcher, M., and Jungnickel, G. (1980). Metabolic control of Crassulacean acid metabolism: evidence for diurnally changing sensitivity against inhibition by malate of PEP carboxylase in *Kalanchoe tubiflora* Hamet. *Z. Pflanzenphysiol.* **97**, 197–204.

297. Kluge, M., and Fischer, K. (1967). Über Zusammenhänge zwischen dem CO_2-Austausch und der Abgabe von Wasserdampf durch *Bryophyllum daigremontianum* Berg. *Planta* **77**, 212–223.

298. Kluge, M., and Heininger, B. (1973). Untersuchungen über den Efflux von Malat aus den Vacuolen der assimilierenden Zellen von *Bryophyllum daigremontianum* und mögliche Einflüsse dieses Vorganges auf den CAM. *Planta* **113**, 333–343.

299. Kluge, M., Lange, O. L., von Eichmann, M., and Schmidt, R. (1973). Diurnaler Säurerhythmus bei *Tillandsia usneoides*. *Planta* **112**, 357–372.

300. Kluge, M., and Osmond, C. B. (1972). Studies on phosphoenolpyruvate carboxylase and other enzymes of Crassulacean acid metabolism of *Bryophyllum tubiflorum* and *Sedum praealtum*. *Z. Pflanzenphysiol.* **66,** 97–105.

301. Kluge, M., and Ting, I. P. (1978). "Crassulacean Acid Metabolism." Springer–Verlag, Berlin.

302. Kirk, M. R., and Heber, U. (1976). Rates of synthesis and source of glycolate in intact chloroplasts. *Planta* **132,** 131–141.

303. Kisaki, T., Hirabayashi, S., and Yano, N. (1973). Effect of the age of tobacco leaves on photosynthesis and photorespiration. *Plant Cell Physiol.* **14,** 505–514.

304. Kisaki, T., Imai, I., and Tolbert, N. E. (1971). Intracellular localization of enzymes related to photorespiration in green leaves. *Plant Cell Physiol.* **12,** 267–273.

305. Kisaki, T., and Tolbert, N. E. (1969). Glycolate and glyoxylate metabolism by isolated peroxisomes or chloroplasts. *Plant Physiol.* **44,** 242–250.

306. Kisaki, T., and Tolbert, N. E. (1970). Glycine as a substrate for photorespiration. *Plant Cell Physiol.* **11,** 247–258.

306a. Koch, K., and Kennedy, R. A. (1980). Characteristics of Crassulacean acid metabolism in the succulent C_4 dicot, *Portulaca oleracea* L. *Plant Physiol.* **65,** 193–197.

307. Kortschak, H. P., Hart, C. E., and Burr, G. O. (1965). Carbon dioxide fixation in sugar cane leaves. *Plant Physiol.* **40,** 209–213.

308. Kortschak, H., and Nickell, L. G. (1970). Calvin-type carbon dioxide fixation in sugarcane stalk parenchyma tissue. *Plant Physiol.* **45,** 515–516.

309. Kow, Y. W., Robinson, J. M., and Gibbs, M. (1977). Influence of pH upon the Warburg effect in isolated intact spinach chloroplasts. II. Interdependency of glycolate synthesis upon pH and Calvin cycle intermediate concentration in the absence of carbon dioxide photoassimilation. *Plant Physiol.* **60,** 492–495.

310. Krause, G. H., Kirk, M., Heber, U., and Osmond, C. B. (1978). O_2-dependent inhibition of photosynthetic capacity in intact isolated chloroplasts and isolated cells from spinach leaves illuminated in the absence of CO_2. *Planta* **142,** 229–233.

311. Krause, G. H., Thorne, S. W., and Lorimer, G. H. (1977). Glycolate synthesis by intact chloroplasts. Studies with inhibitors of photophosphorylation. *Arch. Biochem. Biophys.* **183,** 471–479.

312. Kremer, B. P. (1976). ^{14}C-assimilation patterns and kinetics of photosynthetic $^{14}CO_2$-assimilation of the marine red algae *Bostrichia scorpioides. Planta* **129,** 63–67.

313. Kremer, B. P. (1978). Patterns of photoassimilatory products in Pacific Rhodophyceae. *Can. J. Bot.* **56,** 1655–1659.

314. Kremer, B. P., and Küppers, U. (1977). Carboxylating enzymes and pathway of photosynthetic carbon assimilation in different marine algae—evidence for the C_4-pathway? *Planta* **133,** 191–196.

315. Krenzer, E. G., Moss, D. N., and Crookston, R. K. (1975). Carbon dioxide compensation points of flowering plants. *Plant Physiol.* **56,** 194–206.

316. Kriedemann, P. E. (1971). Photosynthesis and transpiration as a function of gaseous diffusive resistances in orange leaves. *Physiol. Plant.* **24,** 218–225.

317. Kriedemann, P. E., and Lenz, F. (1972). The response of vine leaf photosynthesis to shoot tip excision and stem cincturing. *Vitis* **11,** 193–197.

318. Kriedemann, P. E., and Loveys, B. R. (1975). Hormonal influences on stomatal physiology and photosynthesis. *In* "Environmental and Biological Control of Photosynthesis" (R. Marcelle, ed.), pp. 227–236. Junk, The Hague.

319. Krotkov, G. (1963). Effect of light on respiration. *In* "Photosynthetic Mechanisms in Green Plants" Pub. No. 1145, pp. 452–454. National Academy of Sciences–National Research Council, Washington, D.C.

320. Ku, S.-B., and Edwards, G. E. (1977). Oxygen inhibition of photosynthesis. II. Kinetic characteristics as affected by temperature. *Plant Physiol.* **59**, 991–999.

321. Ku, S.-B., and Edwards, G. E. (1978). Oxygen inhibition of photosynthesis. III. Temperature dependence on quantum yield and its relation to O_2/CO_2 solubility ratio. *Planta* **140**, 1–6.

321a. Ku, M. S. B., Spalding, M. H., and Edwards, G. E. (1980). Intracellular localization of phosphoenolpyruvate carboxykinase in leaves of C_4 and CAM plants. *Plant Sci. Lett.* **19**, 1–8.

322. Kulandaivelu, G., and Gnanam, A. (1974). Physiological studies with isolated leaf cells. *Plant Physiol.* **54**, 569–574.

323. Kumarasinghe, K. S., Keys, A. J., and Whittinham, C. P. (1977). The flux of carbon through the glycolate pathway during photosynthesis by wheat leaves. *J. Exp. Bot.* **28**, 1247–1257.

324. Laber, L. J., Latzko, E., and Gibbs, M. (1974). Photosynthetic path of carbon dioxide in spinach and corn leaves. *J. Biol. Chem.* **249**, 3436–3441.

325. Laetsch, W. M. (1968). Chloroplast specialization in dicotyledons possessing the C_4-dicarboxylic acid pathway of photosynthetic CO_2 fixation. *Am. J. Bot.* **55**, 875–883.

326. Laing, W. A., Ogren, W. L., and Hageman, R. H. (1974). Regulation of soybean net photosynthetic CO_2 fixation by interaction of CO_2, O_2, and ribulose 1,5-diphosphate carboxylase. *Plant Physiol.* **54**, 678–685.

327. Laing, W. A., Ogren, W. L., and R. H. Hageman. (1975). Bicarbonate stabilization of ribulose 1,5-diphosphate carboxylase. *Biochemistry* **14**, 2269–2275.

328. Lambers, H., Noord, R., and Posthumus F. (1979). Respiration of *Senecio* shoots: inhibition during photosynthesis, resistance to cyanide and relation to growth and maintenance. *Physiol. Plant.* **45**, 351–356.

329. Lambers, H., and Steingröver, E. (1978). Growth respiration of a flood-tolerant and a flood-intolerant *Senecio* species: correlation between calculated and experimental values. *Physiol. Plant.* **43**, 219–224.

330. Lambers, H., and Van de Dijk, S. (1979). Cyanide-resistant root respiration and tap root formation in two subspecies of *Hypochaeris radicata*. *Physiol. Plant.* **45**, 235–239.

331. Larsson, C., and Albertson, P.-A. (1974). Photosynthetic $^{14}CO_2$ fixation by chloroplast populations isolated by a polymer two-phase technique. *Biochim. Biophys. Acta* **357**, 412–419.

332. Lawler, D. W., and Fock, H. (1975). Photosynthesis and photorespiratory CO_2 evolution of water-stressed sunflower leaves. *Planta* **126**, 247–258.

333. Lawler, D. W., and Fock, H. (1977). Photosynthetic assimilation of $^{14}CO_2$ by water-stressed sunflower leaves at two O_2 concentrations and the specific activity of the product. *J. Exp. Bot.* **28**, 320–328.

334. LaVergne, D., Bismuth, E., and Champigny, M.-L. (1974). Further studies on phosphoglycerate kinase and ribulose-5-phosphate kinase of the photosynthetic carbon reduction cycle: regulation of the enzymes by the adenine nucleotides. *Plant Sci. Lett.* **3**, 391–397.

335. Lawyer, A. L., and Zelitch, I. (1978). Inhibition of glutamate : glyoxylate aminotransferase activity in tobacco leaves and callus by glycidate, an inhibitor of photorespiration. *Plant Physiol.* **61**, 242–247.

336. Leek, A. E., Halliwell, B., and Butt, V. S. (1972). Oxidation of formate and oxalate in peroxisomal preparations from leaves of spinach beet (*Beta vulgaris* L.). *Biochim. Biophys. Acta* **268**, 299–311.

337. Lehner, K., and Heldt, H. W. (1978). Dicarboxylate transport across the inner membrane of the chloroplast envelope. *Biochim. Biophys. Acta* **501**, 531–544.

338. Lenz, F., Kriedemann, P. E., and Daunicht, H. J. (1972). Effects of roots and emerging axillary shoots on photosynthetic behaviour of *Citrus* cuttings. *Angew. Bot.* **46,** 227–231.

339. Leonard, D. L. (1969). "Photosynthesis and Post-Illumination Fixation of CO_2 by Corn and Barley Leaves." Ph.D. Dissertation, Case Western Reserve Univ., Cleveland, Ohio, 1969.

340. Lester, D. C., Carter, O. G., Kelleher, F. M., and Lasing, D. R. (1972). The effect of gibberellic acid on apparent photosynthesis and dark respiration of simulated swards of *Pennisetum clandestinum* Hochst. *Aust. J. Agric. Res.* **23,** 205–213.

341. Levi, C., Perchorowicz, J. T., and Gibbs, M. (1978). Malate synthesis by dark carbon dioxide fixation in leaves. *Plant Physiol.* **61,** 477–480.

342. Lewenstein, A., and Bachofen, R. (1972). Transient induced oscillations in the level of ATP in *Chlorella fusca. Biochim. Biophys. Acta* **267,** 80–85.

343. Lilley, R. M., Schwenn, J. D., and Walker, D. A. (1973). Inorganic pyrophosphatase and photosynthesis by isolated chloroplasts. II. The controlling influence of orthophosphate. *Biochim. Biophys. Acta* **325,** 596–604.

344. Lilley, R. M., and Walker, D. A. (1975). Carbon dioxide assimilation by leaves, isolated chloropalsts and ribulose bisphosphate carboxylase from spinach. *Plant Physiol.* **55,** 1087–1092.

345. Lilley, R. M., Walker, D. A., and Holborow, K. (1974). The reduction of 3-phosphoglycerate by reconstituted chloroplasts and by chloroplast extracts. *Biochim. Biophys. Acta* **368,** 269–278.

346. Lin, D. C., and Nobel, P. S. (1971). Control of photosynthesis by Mg^{2+}. *Arch. Biochem. Biophys.* **145,** 622–632.

347. Lloyd, N. D. H., and Canvin, D. T. (1977). Photosynthesis and photorespiration in sunflower selections. *Can. J. Bot.* **55,** 3006–3012.

348. Lloyd, N. D. H., Canvin, D. T., and Culver, D. A. (1977). Photosynthesis and photorespiration in algae. *Plant Physiol.* **59,** 936–940.

349. Long, S. P., and Woolhouse, H. W. (1978). The response of net photosynthesis to light and temperature in *Spartina townsendii* (*sensu lato*), a C_4 species from a cool temperate climate. *J. Exp. Bot.* **29,** 803–814.

350. Lord, J. M., Armitage, T. L., and Merret, M. J. (1975). Ribulose 1,5-diphosphate synthesis in *Euglena*. II. Effect of inhibitors on enzyme synthesis during regreening and subsequent transfer to darkness. *Plant Physiol.* **56,** 600–604.

351. Lord, J. M., and Merrett, M. J. (1970). The pathway of glycolate utilization in *Chlorella pyrenoidosa. Biochem. J.* **117,** 929–937.

352. Lor, K.-L., and Cossins, E. A. (1978). Relationships between glycollate and folate metabolism in *Euglena gracilis. Phytochemistry* **17,** 659–665.

353. Lorimer, G. H. (1979). Evidence for the existence of discrete activator and substrate sites for CO_2 on ribulose-1,5-bisphosphate carboxylase. *J. Biol. Chem.* **254,** 5599–5601.

353a. Lorimer, G. H. (1981). The carboxylation and oxygenation of ribulose 1,5-*bis* phosphate carboxylase: The primary events in photosynthesis and photorespiration. *Annu. Rev. Plant Physiol.* **32,** 349–383.

354. Lorimer, G. H., and Andrews, T. J. (1973). Plant photorespiration—an inevitable consequence of the existence of atmospheric oxygen. *Nature (London)* **243,** 359–360.

355. Lorimer, G. H., Andrews, T. J., and Tolbert, N. E. (1973). Ribulose diphosphate oxygenase. II. Further proof of the reaction products and mechanism of action. *Biochemistry* **12,** 18–23.

356. Lorimer, G. H., Badger, M. R., and Andrews, T. J. (1977). D-Ribulose-1,5-bisphosphate carboxylase–oxygenase. Improved methods for the activation and assay of catalytic activities. *Anal. Biochem.* **78,** 66–75.

357. Lorimer, G. H., Badger, M. R., and Andrews, T. J. (1976). The activation of ribulose-1,5-bisphosphate carboxylase by carbon dioxide and magnesium ions. Equilibria, kinetics, a suggested mechanism, and physiological implications. *Biochemistry* **15,** 529–536.

358. Lorimer, G. H., Krause, G. H., and Berry, J. A. (1977). The incorporation of ^{18}O oxygen into glycolate by intact isolated chloroplasts. *FEBS Lett.* **78,** 199–202.

359. Lorimer, G. H., Woo, K. C., Berry. J. A., and Osmond, C. B. (1978). The C_2 photorespiratory carbon oxidation cycle in leaves of higher plants: pathway and consequences. *Proc. Int. Congr. Photosynth. 4th.* 311–322.

360. Loveys, B. R., and Kriedemann, P. E. (1974). Internal control of stomatal physiology and photosynthesis. I. Stomatal regulation and associated changes in endogenous levels of abscisic acid and phaseic acids. *Australian J. Plant Physiol.* **1,** 407–415.

361. Lucas, W. J., Spanswick, R. M., and Dainty, J. (1978). HCO_3^- influx across the plasmalemma of *Chara corallina.* Physiological and biophysical influence of 10 mM k⁺. *Plant Physiol.* **61,** 487–493.

362. Ludwig, L. J., and Canvin, D. T. (1971). The rate of photorespiration during photosynthesis and the relationship of the substrate of light respiration to the products of photosynthesis in sunflower leaves. *Plant Physiol.* **48,** 712–719.

363. Ludwig, L. J., Krotkov, G., and Canvin, D. T. (1969). The relationship of the products of photosynthesis to the substrates for CO_2-evolution in light and darkness. *In* "Progress in Photosynthesis Research" (H. Metzner, ed.), pp. 494–502. Inst. Chem. Pflanzenphysiol., Tubingen.

364. Lüttge, U., and Ball, E. (1977). Water relation parameters of the CAM plant *Kalanchoë daigremontiana* in relation to diurnal malate oscillation. *Oecologia* **31,** 85–94.

365. Lüttge, U., and Ball, E. (1978). Free running oscillations of transpiration and CO_2 exchange in CAM plants without a concomitant rhythm of malate levels. *Z. Pflanzenphysiol.* **90,** 69–77.

366. Lüttge, U., Kluge, M., and Ball, E. (1975). Effects of osmotic gradients on vacuolar malic acid storage. A basic principle in oscillatory behaviour of Crassulacean acid metabolism. *Plant Physiol.* **56,** 613–616.

367. Mangat, B. S., Levin, W. B., and Bidwell, R. G. S. (1974). The extent of dark respiration in illuminated leaves and its control by ATP levels. *Can. J. Bot.* **52,** 673–681.

368. Matsumoto, K., Nishimura, M., and Akazawa, T. (1977). Ribulose-1,5-bisphosphate carboxylase in the bundle sheath cells of maize leaf. *Plant Cell Physiol.* **18,** 1281–1290.

369. McCashin, B., and Canvin, D. T. (1979). Photosynthetic and photorespiratory characteristics of mutants of *Hordeum vulgare* L. *Plant Physiol.* **64,** 354–360.

370. McCree, K. J. (1970). An equation for the rate of respiration of white clover plants grown under controlled conditions. *In* "Prediction and Measurement of Photo-Synthetic Productivity" (I. Setlik, ed.), pp. 221–229. Centre Agr. Publishing & Documentation, Wageningen, The Netherlands.

371. McCree, K. J. (1976). The role of dark respiration in the carbon economy of a plant. *In* "CO_2 Metabolism and Plant Productivity" (R. H. Burris and C. C. Black, eds.), pp. 177–184. Univ. Park Press, Baltimore.

372. McCurry, S. D., and Tolbert, N. E. (1977). Inhibition of ribulose-1,5-bisphosphate carboxylase/oxygenase by xylulose-1,5-bisphosphate. *J. Biol. Chem.* **252,** 8344–8346.

373. Medina, E., Delgado, M., Troughton, J. H., and Medina, J. D. (1977). Physiological ecology of CO_2 fixation in Bromeliaceae. *Flora* **166,** 137–152.
374. Meeuse, B. J. D. (1975). Thermogenic respiration in aroids. *Annu. Rev. Plant Physiol.* **26,** 117–126.
375. Meidner, H. (1962). The minimum intercellular space CO_2 concentration (Γ) of maize leaves and its influence on stomatal movements. *J. Exp. Bot.* **13,** 284–293.
376. Miginiac-Maslow, M., and Champigny, M.-L. (1974). Relationship between the level of adenine nucleotides and the carboxylation activity of illuminated isolated spinach chloroplasts. *Plant Physiol.* **53,** 856–862.
377. Miles, C. D. (1974). Control of the light saturation point for photosynthesis in tomato. *Physiol. Plant.* **31,** 153–158.
378. Milford, G. F. J., and Pearman, I. (1975). The relationship between photosynthesis and the concentration of carbohydrates in the leaves of sugarbeet. *Photosynthetica* **9,** 78–83.
379. Miyachi, S., and Hogetsu, D. (1970). Light-enhanced carbon dioxide fixation in isolated chloroplasts. *Plant Cell Physiol.* **11,** 927–936.
380. Miyachi, S., Kamiya, A., and Miyachi, S. (1977). Wavelength-effects of incident light on carbon metabolism in *Chlorella* cells. *In* "Biological Solar Energy Conversion" (Mitsui, A., Miyachi, S., san Pietro, A., and Tamura, S., eds.), pp. 167–181. Academic Press, New York.
381. Miyachi, S., and Okabe, K. (1976). Oxygen enhancement of photosynthesis in *Anacystis nidulans* cells. *Plant Cell Physiol.* **17,** 973–986.
382. Miziorko, H. M. (1979). Rubulose-1,5-bisphosphate carboxylase. Evidence in support of the existence of distinct CO_2 activator and CO_2 substrate sites. *J. Biol. Chem.* **254,** 270–272.
383. Mollenhauer, H. H., Morre, D. J., and Kelley, A. G. (1966). The widespread occurrence of plant cytosomes resembling animal microbodies. *Protoplasma* **62,** 44–52.
384. Mondal, M. H., Brun, W. A., and Brenner, M. L. (1978). Effects of sink removal on photosynthesis and senescence in leaves of soybean (*Glycine max* L.) plants. *Plant Physiol.* **61,** 394–397.
385. Montes, G., and Bradbeer, J. W. (1976). An association of chloroplasts and mitochondria in *Zea mays* and *Hyptis suaveolens*. *Plant Sci. Lett.* **6,** 35–41.
386. Moore, K. G., Illsley, A., and Lovell, P. H. (1974). Effects of sucrose on petiolar carbohydrate accumulation and photosynthesis in excised *Sinapis* cotyledons. *J. Exp. Bot.* **25,** 887–898.
387. Moore, P. D. (1974). Misunderstandings over C_4 carbon fixation. *Nature (London)* **252,** 438–439.
388. Moore, P. D. (1978). When C_4 plants do best. *Nature (London)* **272,** 400–401.
389. Moss, D. N. (1962). The limiting carbon dioxide concentration for photosynthesis. *Nature (London)* **193,** 587.
390. Moss, D. N., Willmer, C. M., and Crookston, R. K. (1971). CO_2 compensation concentration in maize (*Zea mays* L.) genotypes. *Plant Physiol.* **47,** 847–848.
391. Mousseau, M. (1977). Night respiration in relation to growth, photosynthesis and development of *Chenopodium polyspermum* in long and short days. *Plant Sci. Lett.* **9,** 339–346.
392. Mukerji, S. K. (1977). Corn leaf phosphoenolpyruvate carboxylases: purifications and properties of two isoenzymes. *Arch. Biochem. Biophys.* **182,** 343–351.
393. Mukerji, S. K. (1977). Corn leaf phosphoenolpyruvate carboxylases: the effect of divalent cations on activity. *Arch. Biochem. Biophys.* **182,** 352–359.
394. Mukerji, S. K. (1977). Corn leaf phosphoenolpyruvate carboxylases: % inhibition of

$^{14}CO_2$ fixation by SO_3^{2-} and activation by glucose-6-phosphate. *Arch. Biochem. Biophys.* **182**, 360–365.

395. Muller, W., and Wegmann, K. (1978). Sucrose biosynthesis in *Dunaliella*. III. Regulation by a membrane change. *Planta* **141**, 165–167.

396. Mullins, M. G. (1970). Hormone-directed transport of assimilate in decapitated internodes of *Phaseolus vulgaris*. L. *Ann. Bot. (London)* **34**, 897–909.

397. Nagarajah, S. (1975). Effect of debudding on photosynthesis in leaves of cotton. *Physiol. Plant.* **33**, 28–31.

398. Nagashima, H., Nakamura, S., Nisizawa, K., and Hori, T. (1971). Enzymic synthesis of floridean starch in a red alga *Serraticardia maxima*. *Plant Cell Physiol.* **12**, 243–253.

399. Nagashima, H., Ozaki, H., Nakamura, S., and Nisizawa, K. (1969). Physiological studies on floridean starch, floridoside and trehalose in a red alga, *Serraticardia maxima*. *Shokubntsngakn Zasshi* **82**, 462–473.

400. Nalborczyk, E., LaCroix, L. J., and Hill, R. D. (1975). Environmental influences on light and dark CO_2 fixation by *Kalanchoë daigremontiana*. *Can. J. Bot.* **53**, 1132–1138.

401. Nambudiri, E. M. V., Tidwell, W. D., Smith, B. N., and Hebbert, N. P. (1978). A C_4 plant from the Pliocene. *Nature (London)* **276**, 816–817.

402. Nátr, L., Watson, B. T., and Weatherley, P. E. (1974). Glucose absorption, carbohydrate accumulation, presence of starch, and rate of photosynthesis in barley leaf segments. *Ann. Bot. (London)* **38**, 589–593.

403. Neales, T. F. (1975). The gas exchange patterns of CAM plants. *In* "Environmental and Biological Control of Photosynthesis" (R. Marcelle, ed.), pp. 299–310. Junk, The Hague.

404. Neales, T. F., and Incoll, L. D. (1968). The control of leaf photosynthesis rate by the level of assimilate concentration in the leaf: a review of the hypothesis. *Bot. Rev.* **34**, 107–125.

405. Nelson, E. B., and Tolbert, N. E. (1970). Glycolate dehydrogenase in algae. *Arch. Biochem. Biophys.* **141**, 102–110.

406. Nelson, P. E., and Surzycki, S. J. (1976). Characterization of the oxygenase activity in a mutant of *Chlamydomonas reinhardi* exhibiting altered ribulose-bisphosphate carboxylase. *Eur. J. Biochem.* **61**, 475–480.

407. Nesius, K. K., and Fletcher, J. S. (1975). Contribution of nonautotrophic carbon dioxide fixation to protein synthesis in suspension cultures of Paul's Scarlet rose. *Plant Physiol.* **55**, 643–645.

408. Neyra, C. A., and Hageman, R. H. (1976). Relationships between carbon dioxide, malate, and nitrate accumulation and reduction in corn (*Zea mays* L.) seedlings. *Plant Physiol.* **58**, 726–730.

409. Nishikido, T., and Takanashi, H. (1973). Glycine activation of PEP carboxylase from monocotyledonous C_4 plants. *Biochem. Biophys. Res. Commun.* **53**, 126–133.

410. Nishimura, M., and Akazawa, T. (1978). Biosynthesis of ribulose-1,5-bisphosphate carboxylase in spinach leaf protoplasts. *Plant Physiol.* **62**, 97–100.

411. Nobel, P. S. (1976). Water relations and photosynthesis of a desert CAM plant *Agave deserti*. *Plant Physiol.* **58**, 576–582.

412. Nordhorn, G., Weidner, M., and Willenbrink, J. (1976). Isolation and photosynthetic activities of chloroplasts of the brown alga, *Fucus serratus* L. *Z. Pflanzenphysiol.* **80**, 153–165.

413. Nutbeam, A. R., and Duffus, C. M. (1976). Evidence for C_4 photosynthesis in barley pericarp tissue. *Biochem. Biophys. Res. Commun.* **70**, 1198–1203.

414. Ogren, W. L. (1976). Search for higher plants with modifications of the reductive

446   R. G. S. BIDWELL

pentose phosphate pathway of CO_2 assimilation. *In* "CO_2 Metabolism and Plant Productivity" (C. C. Black and R. H. Burris, eds.), pp. 19–29. Univ. Park Press, Baltimore.

415. Ogren, W. L. (1976). Improving the photosynthetic efficiency of soybean. *World Soybean Res. Proc. World Soybean Res. Conf. 1975,* 255–261.
416. Ogren, W. L., and Bowes, G. (1971). Ribulose diphosphate carboxylase regulates soybean photorespiration. *Nature (London) New Biol.* **230,** 159–160.
417. Okabe, K.-I., Codd, G. A., and Stewart, W. D. P. (1979). Hydroxylamine stimulates carboxylase activity and inhibits oxygenase activity of cyanobacterial RuBP carboxylase/oxygenase. *Nature (London)* **279,** 525–527.
418. Oliver, D. J. (1978). Effect of glyoxylate on the sensitivity of net photosynthesis to oxygen (the Warburg effect) in tobacco. *Plant Physiol.* **62,** 938–940.
419. Oliver, D. J., and Zelitch, I. (1977). Metabolic regulation of glycolate synthesis, photorespiration, and net photosynthesis in tobacco by L-glutamate. *Plant Physiol.* **59,** 688–694.
420. Oliver, D. J., and Zelitch, I. (1977). Increasing photosynthesis by inhibiting photorespiration with glyoxylate. *Science* **196,** 1450–1451.
421. Ong, C. K., and Marshall, C. (1975). Assimilate distribution in *Poa annua* L. *Ann. Bot. (London)* **39,** 413–421.
422. Osmond, C. B. (1971). Metabolite transport in C_4 photosynthesis. *Aust. J. Biol. Sci.* **24,** 159–163.
423. Osmond, C. B. (1976). CO_2 assimilation and dissimilation in the light and dark in CAM plants. *In* "CO_2 Metabolism and Plant Productivity" (R. H. Burris and C. C. Black, eds.), pp. 217–233. Univ. Park Press, Baltimore.
424. Osmond, C. B. (1978). Crassulacean acid metabolism: a curiosity in context. *Annu. Rev. Plant Physiol.* **29,** 379–414.
425. Osmond, C. B., and Allaway, W. G. (1974). Pathway of CO_2 fixation in the CAM plant *Kalanchoë daigremontiana.* I. Patterns of CO_2 fixation in the light. *Aust. J. Plant Physiol.* **1,** 503–511.
426. Osmond, C. B., Allaway, W. G., Sutton, B. G., Troughton, J. H., Queiroz, O., Lüttge, U., and Winter, K. (1973). Carbon isotope descrimination of CAM plants. *Nature (London)* **246,** 41–42.
427. Osmond, C. B., and Avadhani, P. N. (1970). Inhibition of the β-carboxylation pathway of CO_2 fixation by bisulfite compounds. *Plant Physiol.* **45,** 228–230.
428. Osmond, C. B., Bender, M. M., and Burris, R. H. (1973). Pathways of CO_2 fixation in the CAM plant *Kalanchoë daigremontiana.* III. Correlation with $\delta^{13}C$ value during growth and water stress. *Aust. J. Plant Physiol.* **3,** 787–799.
429. Osmond, C. B., and Björkman, O. (1972). Simultaneous measurement of oxygen effects on net photosynthesis and glycolate metabolism in C_3 and C_4 species of *Atriplex. Carnegie Inst. Washington Year Book* **71,** 141–148.
430. Osmond, C. B., and Björkman, O. (1975). Pathways of CO_2 fixation in the CAM plant *Kalanchoë daigremontiana.* II. Effects of O_2 and CO_2 concentration on light and dark CO_2 fixation. *Aust. J. Plant Physiol.* **2,** 155–162.
431. Osmond, C. B., and Harris, B. (1971). Photorespiration during C_4 photosynthesis. *Biochim. Biophys. Acta* **234,** 270–282.
432. Outlaw, W. H., Schmuk, C. L., and Tolbert, N. E. (1976). Photosynthetic carbon metabolism in the palisade parenchyma and spongy parenchyma of *Vicia faba* L. *Plant Physiol.* **58,** 186–189.
433. Ozbun, J. L., Volk, R. J., and Jackson, W. A. (1964). Effects of light and darkness on gaseous exchange of bean leaves. *Plant Physiol.* **39,** 523–527.

434. Pacold, I., and Anderson, L. E. (1973). Energy-charge control of the Calvin cycle enzyme 3-phosphoglyceric acid kinase. *Biochem. Biophys. Res. Commun.* **51**, 139–143.
435. Paech, C., McCurry, S. D., Pierce, J., and Tolbert, N. E. (1978). Active site of ribulose 1,5-bisphosphate carboxylase/oxygenase. In "Photosynthetic Carbon Assimilation" (H. W. Siegelman and G. Hind, eds.), pp. 227–243. Plenum, New York.
436. Parnik, T., Keerberg, O., and Viil, J. (1972). $^{14}CO_2$ evolution from bean leaves at the expense of fast labelled intermediates of photosynthesis. *Photosynthetica* **6**, 66–74.
437. Patrick, J. W., and Wareing, P. F. (1978). Auxin-promoted transport of metabolites in stems of *Phaseolus vulgaris* L. *J. Exp. Bot.* **29**, 359–366.
438. Paul, J. S., and Bassham, J. A. (1977). Maintenance of high photosynthetic rates in mesophyll cells isolated from *Papaver somniferum. Plant Physiol.* **60**, 775–778.
439. Paul, J. S., Krohne, S. D., and Bassham, J. A. (1979). Stimulation of CO_2 incorporation and glutamine synthesis by 2,4-D in photosynthesizing leaf-free mesophyll cells. *Plant Sci. Lett.* **15**, 17–24.
440. Paul, J. S., Sullivan, C. W., and Volcani, B. E. (1975). Photorespiration in diatoms. Mitochondrial glycolate dehydrogenase in *Cylindrotheca fusiformis* and *Nitzschia alba. Arch. Biochem. Biophys.* **169**, 152–159.
441. Paul, J. S., and Volcani, B. E. (1976). Photorespiration in diatoms. IV. Two pathways of glycolate metabolism in synchronized cultures of *Cylindrotheca fusiformis. Arch. Microbiol.* **110**, 247–252.
442. Pearcy, R. W. (1977). Acclimation of photosynthetic and respiratory carbon dioxide exchange to growth temperature in *Atriplex lentiformis* (Torr.) Wats. *Plant Physiol.* **59**, 795–799.
443. Pearson, C. J. (1974). Daily changes in carbon-dioxide exchange and photosynthate translocation in leaves of *Vicia faba. Planta* **119**, 59–70.
444. Peavey, D. G., Steup, M., and Gibbs, M. (1977). Characterization of starch breakdown in the intact spinach chloroplast. *Plant Physiol.* **60**, 305–308.
444a. Peisker, M. (1980). Differential short-time uptake of $^{14}CO_2$ and $^{12}CO_2$ as a method for determining photorespiration. A theoretical approach. *Photosynthetica* **14**, 406–412.
445. Penning de Vries, F. W. T. (1975). The cost of maintenance processes in plant cells. *Ann. Bot. (London)* **39**, 77–92.
446. Penning de Vries, F. W. T. (1975). Use of assimilates in higher plants. In "Photosynthesis and Productivity in Different Environments" (J. P. Cooper, ed.), pp. 459–480. Cambridge Univ. Press, Cambridge.
447. Peterkovsky, A., and Racker, E. (1961). The reductive pentose phosphate cycle. III. Enzyme activities in cell-free extracts of photosynthetic organisms. *Plant Physiol.* **36**, 409–414.
448. Petrov, V. E., Seifullina, N. K., Bairasheva, S. R., and Loseva, N. L. (1974). Energetics of assimilating cells and photosynthesis. I. The relation between energy and gas exchange in photosynthesizing plants. The effect of temperature. *Bot. Zh. (Leningrad)* **59**, 321–331.
449. Pierre, J. N., and Queiroz, O. (1979). Regulation of glycolysis and level of the Crassulacean acid metabolism. *Planta* **144**, 143–151.
450. Poincelot, R. P. (1974). Uptake of bicarbonate ion in darkness by isolated chloroplast envelope membranes and intact chloroplasts of spinach. *Plant Physiol.* **54**, 520–526.
451. Poincelot, R. P. (1975). Transport of metabolites across isolated envelope membranes of spinach chloroplasts. *Plant Physiol.* **55**, 849–852.

452. Poincelot, R. P., and Day, P. R. (1976). Isolation and bicarbonate transport of chloroplast envelope membranes from species of differing net photosynthetic efficiency. *Plant Physiol.* **57**, 334–338.

453. Polevaya, V. S., Smolov, A. R., and Ignatiev, A. R. (1975). Photosynthetic activity of autotrophic tissue cultures of rue. *Dokl. Akad. Nauk SSSR* **225**, 230–231.

454. Pon, N. G., Rabin, B. R., and Calvin, M. (1963). Mechanism of the carboxydismutase reaction. I. The effect of preliminary incubation of substrates, metal ions, and enzyme on activity. *Biochem. Z.* **338**, 7–19.

455. Portis, A. R., Chon, C. J., Mosbach, A., and Heldt, H. W. (1977). Fructose- and sedoheptulosebisphosphatase. The sites of a possible control of CO_2 fixation by light-dependent changes of the stromal Mg^{2+} concentration. *Biochim. Biophys. Acta* **461**, 313–325.

456. Poskuta, J., and Kochańska, K. (1978). The effect of potassium glycidate on the rates of CO_2-exchange and photosynthetic products of bean leaves. *Z. Pflanzenphysiol.* **89**, 393–400.

457. Poskuta, J., Parys, E., Ostrowska, E., and Wolkowa, E. (1975). Photosynthesis, photorespiration, respiration and growth of pea seedlings treated with gibberellic acid (GA_3). *In* "Environmental and Biological Control of Photosynthesis" (R. Marcelle, ed.), pp. 201–209. Junk, The Hague.

458. Poskuta, J., Wróblewska, B., and Mikulska, M. (1976). $^{14}CO_2$ uptake and photosynthetic products of bean leaves after 0 5 or 10 hours pretreatment of plants with 100% O_2 in light. *Z. Pflanzenphysiol.* **78**, 396–402.

458a. Powles, S. B., Chapman, K. S. R., and Osmond, C. B. (1980). Photoinhibition of intact attached leaves of C_4 plants: dependence on CO_2 and O_2 partial pressure. *Aust. J. Plant Physiol.* **7**, 737–747.

459. Powles, S. B., and Osmond, C. B. (1978). Inhibition of the capacity and efficiency of photosynthesis in bean leaflets illuminated in a CO_2-free atmosphere at low oxygen: a possible role for photorespiration. *Aust. J. Plant Physiol.* **5**, 619–629.

460. Preiss, J., Ghosh, H. P., and Wittkop, J. (1967). Regulation of the biosynthesis of starch in spinach leaf chloroplasts. *In* "Biochemistry of Chloroplasts (T. W. Goodwin, ed.), Vol. 2, pp. 131–152. Academic Press, New York.

460a. Prins, H. B. A., Snel, J. F. H., Helder, R. J., and Zanstra, P. E. (1980). Photosynthetic HCO_3^- utilization and excretion in aquatic angiosperms. *Plant Physiol.* **66**, 818–822.

461. Pritchard, G. G., Griffin, W. J., and Whittingham, C. P. (1962). The effect of carbon dioxide concentration, light intensity, and isonicotinyl hydrazide on the photosynthetic production of glycollic acid by *Chlorella*. *J. Exp. Bot.* **13**, 176–184.

462. Pryke, J., and Ap Rees, T. (1977). The pentose phosphate pathway as a source of NADPH for lignin synthesis. *Phytochemistry* **16**, 557–560.

463. Przybylla, K.-R., and Fock, H. (1976). Der Weg des Kohlenstoffs während der CO_2-Freisetzung im Licht und Dunkeln bie Sonnenblumen. *Ber. Dtsch. Bot. Ges.* **89**, 651–661.

464. Pupillo, P., and Giuliani-Piccari, G. (1973). The effect of NADP on the subunit structure and activity of spinach chloroplast glyceraldehyde-3-phosphate dehydrogenase. *Arch. Biochem. Biophys.* **154**, 324–331.

465. Purczeld, P., Chon, C. J., Portis, A. R., Heldt, H. W., and Heber, U. (1978). The mechanism of the control of carbon fixation by the pH in the chloroplast stroma. Studies with nitrite-mediated proton transfer across the envelope. *Biochim. Biophys. Acta* **501**, 488–498.

466. Quebedeaux, B., and Hardy, R. W. F. (1973). Oxygen as a new factor controlling reproductive growth. *Nature (London)* **243**, 477–479.

467. Queiroz, O., and Morel, C. (1974). Photoperiodism and enzyme activity: towards a model for the control of circadian metabolic rhythms in Crassulacean acid metabolism. *Plant Physiol.* **54,** 596–602.
468. Rabinowitch, E. I. (1945). "Photosynthesis and Related Processes," Vol. 1, pp. 567–570. Interscience, New York.
469. Rabson, R., Tolbert, N. E., and Kearney, P. C. (1962). Formation of serine and glyceric acid by the glycolate pathway. *Arch. Biochem. Biophys.* **98,** 154–163.
470. Radin, J. W., Parker, L. L., and Sell, C. R. (1978). Partitioning of sugar between growth and nitrte reduction in cotton roots. *Plant Physiol.* **62,** 550–553.
471. Raghavendra, A. S., and Das, V. S. R. (1977). Aspartate-dependent alanine production by leaf disks of *Amaranthus paniculatus,* an aspartate utilizing NAD-malic enzyme type C$_4$ plant. *New Phytol.* **79,** 481–487.
472. Raghavendra, A. S., and Das, V. S. R. (1978). (Na$^+$–K$^+$-stimulated ATPase in leaves of C$_4$ plants: possible involvement in active transport of C$_4$ acids. *J. Exp. Bot.* **29,** 39–47.
473. Raghavendra, A. S., and Das, V. S. R. (1978). The occurrence of C$_4$ photosynthesis: a supplementary list of C$_4$ plants reported during late 1974–mid 1977. *Photosynthetica* **12,** 200–208.
474. Raghavendra, A. S., Rajendrudu, G., and Das, V. S. R. (1978). Simultaneous occurrence of C$_3$ and C$_4$ photosynthesis in relation to leaf position in *Mollugo nudicaulis. Nature (London)* **273,** 143–144.
475. Randall, D. D., Tolbert, N. E., and Gremel, D. (1971). 3-Phosphoglycerate phosphatase in plants. II. Distribution, physiological considerations, and comparison with P-glycolate phosphatase. *Plant Physiol.* **48,** 480–487.
476. Rathnam, C. K. M. (1978). Heat inactivation of leaf phosphoenolpyruvate carboxylase: protection by aspartate and malate in C$_4$ plants. *Planta* **141,** 289–295.
477. Rathnam, C. K. M. (1979). Metabolic regulation of carbon flux during C$_4$ photosynthesis. II. *In situ* evidence for refixation of photorespiratory CO$_2$ by C$_4$ phosphoenolpyruvate carboxylase. *Planta* **145,** 13–23.
478. Rathnam, C. K. M., and Chollet, R. (1978). CO$_2$ donation by malate and aspartate reduces photorespiration in *Panicum milioides,* a C$_3$–C$_4$ intermediate species. *Biochem. Biophys. Res. Commun.* **85,** 801–808.
479. Rathnam, C. K. M., and Chollet, R. (1979). Phosphoenolpyruvate carboxylase reduces photorespiration in *Panicum milioides,* a C$_3$–C$_4$ intermediate species. *Arch. Biochem. Biophys.* **193,** 346–354.
480. Rathanam, C. K. M., and Das, V. S. R. (1975). Aspartate-type C-4 photosynthetic carbon metabolism in leaves of *Eleusine coracana* Gaertn. *Z. Pflanzenphysiol.* **74,** 377–393.
481. Rathnam, C. K. M., and Edwards, G. E. (1975). Intracellular localization of certain photosynthetic enzymes in bundle sheath cells of plants possessing the C$_4$ pathway of photosynthesis. *Arch. Biochem. Biophys.* **171,** 214–225.
482. Rathnam, C. K. M., and Edwards, G. E. (1976). Distribution of nitrate assimilating enzymes between mesophyll protoplasts and bundle sheath cells in leaves of three groups of C$_4$ plants. *Plant Physiol.* **57,** 881–885.
483. Rathnam, C. K. M., and Edwards, G. E. (1977). C$_4$ acid decarboxylation and CO$_2$ donation to photosynthesis in bundle sheath strands and chloroplasts from species representing three groups of C$_4$ plants. *Arch. Biochem. Biophys.* **182,** 1–13.
484. Rathnam, C. K. M., and Edwards, G. E. (1977). C$_4$-dicarboxylic acid metabolism in bundle-sheath chloroplasts, mitochondria and strands of *Eriochloa borumensis* Hack., a phosphoenolpyruvate-carboxykinase type C$_4$ species. *Planta* **133,** 135–144.

485. Rathnam, C. K. M., and Edwards, G. E. (1977). Use of inhibitors to distinguish between C_4 acid decarboxylation mechanisms in bundle sheath cells of C_4 plants. *Plant Cell Physiol.* **18**, 963–968.

486. Raschke, K. (1977). (^{14}C)Carbon-dioxide fixation by isolated leaf epidermis with stomata closed or open. *Planta* **134**, 69–75.

487. Raven, J. A. (1972). Endogenous inorganic carbon sources in plant photosynthesis. I. Occurrence of the dark respiratory pathways in illuminated green cells. *New Phytol.* **71**, 227–247.

488. Raven, J. A. (1972). Endogenous inorganic carbon sources in plant photosynthesis. II. Comparison of total CO_2 production in the light with measured CO_2 evolution in the light. *New Phytol.* **71**, 995–1014.

489. Raven, J. A. (1976). The rate of cyclic and non-cyclic photophosphorylation and oxidative phosphorylation, and regulation of the rate of ATP consumption in *Hydrodictyon africanum*. *New Phytol.* **76**, 205–212.

490. Raven, J. A. (1976). The quantitative role of "dark" respiratory processes in heterotrophic and photolithotrophic plant growth. *Ann. Bot. (London)* **40**, 587–602.

491. Rawson, H. M., Gifford, R. M., and Bremner, P. M. (1976). Carbon dioxide exchange in relation to sink demand in wheat. *Planta* **132**, 19–23.

492. Ray, T. B., and Black, C. C. (1976). Characterization of phosphoenolpyruvate carboxykinase from *Panicum maximum*. *Plant Physiol.* **58**, 603–607.

493. Ray, T. B., and Black, C. C. (1977). Oxaloacetate as the source of carbon dioxide for photosynthesis in bundle sheath cells of the C_4 species *Panicum maximum*. *Plant Physiol.* **60**, 193–196.

494. Reger, B. J., and Yates, I. E. (1979). Distribution of photosynthetic enzymes between mesophyll, specialized parenchyma and bundle sheath cells of *Arundinella hirta*. *Plant Physiol.* **63**, 209–212.

495. Rehfeld, D. W., Randall, D. D., and Tolbert, N. E. (1970). Enzymes of the glycolate pathway in plants without CO_2-photorespiration. *Can. J. Bot.* **48**, 1219–1226.

496. Reibach, P. H., and Benedict, C. R. (1977). Fractionation of stable carbon isotopes by phosphoenolpyruvate carboxylase from C_4 plants. *Plant Physiol.* **59**, 564–568.

497. Ries, E., and Gauss, V. (1977). D-Glucose as an exogenous substrate of the blue-light enhanced respiration in *Chlorella*. *Z. Pflanzenphysiol.* **82**, 261–273.

498. Rhoades, M. M., and Carvalho, A. (1944). The function and structure of the parenchyma sheath plastids of the maize leaf. *Bull. Torrey Bot. Club* **71**, 335–346.

499. Robinson, J. M., and Gibbs, M. (1974). Photosynthetic intermediates, the Warburg effect, and glycolate synthesis in isolated chloroplasts. *Plant Physiol.* **53**, 790–797.

500. Robinson, J. M., Gibbs, M., and Cotler, D. N. (1977). Influence of pH upon the Warburg effect in isolated intact spinach chloroplasts. I. Carbon dioxide photoassimilation and glycolate synthesis. *Plant Physiol.* **59**, 530–534.

501. Robinson, S. P., McNeil, P. H., and Walker, D. A. (1979). Ribulose bisphosphate carboxylase—lack of dark inactivation of the enzyme in experiments with protoplasts. *FEBS Lett.* **97**, 296–300.

502. Robinson, S. P., Wiskich, J. T., and Paleg, L. G. (1978). Effects of indoleacetic acid on CO_2 fixation, electron transport and phosphorylation in isolated chloroplasts. *Aust. J. Plant Physiol.* **5**, 425–431.

503. Ryan, F. J., Barker, R., and Tolbert, N. E. (1975). Inhibition of ribulose diphosphate carboxylase oxygenase by xylitol-1,5-diphosphate. *Biochem. Biophys. Res. Commun.* **65**, 39–46.

504. Ryan, F. J., Jolly, S. O., and Tolbert, N. E. (1974). Ribulose diphosphate oxygenase. V. Presence in ribulose diphosphate carboxylase from *Rhodospirillum rubrum*. *Biochem. Biophys. Res. Commun.* **59**, 1233–1241.

505. Ryle, G. J. A., and Powell, C. E. (1974). The utilization of recently assimilated carbon in graminaceous plants. *Ann. Appl. Bot.* **77,** 145–158.

506. Sachs, T. (1975). The induction of transport channels by auxin. *Planta* **127,** 201–206.

507. Salerno, G. L., and Pontis, H. G. (1978). Studies on sucrose phosphate synthetase. The inhibitory action of sucrose. *FEBS Lett.* **86,** 263–268.

508. Salin, M. L., and Homann, P. H. (1971). Changes of photorespiratory activity with leaf age. *Plant Physiol.* **48,** 193–196.

509. Samejima, M., and Miyachi, S. (1978). Photosynthetic and light-enhanced dark fixation of $^{14}CO_2$ from the ambient atmosphere and ^{14}C-bicarbonate infiltrated through vascular bundles in maize leaves. *Plant Cell Physiol.* **19,** 907–916.

510. Samish, Y. B., Pallas, J. E., Dornhoff, G. M., and Shibles, R. M. (1972). A reevaluation of soybean leaf photorespiration. *Plant Physiol.* **50,** 28–30.

511. Sankhla, N., and Huber, W. (1975). Regulation of balance between C_3 and C_4 pathway: role of abscisic acid. *Z. Pflanzenphysiol.* **74,** 267–271.

512. Santakumari, M., Reddy, C. S., Reddy, A. R. C., and Das, V. S. R. (1979). CAM behavior in a grain legume, chick pea. *Naturwissenschaftan* **66,** 54–56.

513. Sargent, D. F., and Taylor, C. P. S. (1972). Light induced inhibition of respiration in DCMU-poisoned *Chlorella* caused by photosystem I activity. *Can. J. Bot.* **50,** 13–21.

514. Sargent, D. F., and Taylor, C. P. S. (1975). On the respiratory enhancement in *Chlorella pyrenoidosa* by blue-light. *Planta* **127,** 171–175.

515. Sawhney, S. K., Naik, M. S., and Nicholas, D. J. D. (1978). Regulation of nitrate reduction by light, ATP and mitochondrial respiration in wheat leaves. *Nature (London)* **272,** 647–648.

516. Sayre, R. T., and Kennedy, R. A. (1977). Ecotypic differences in the C_3 and C_4 photosynthetic activity in *Mollugo verticillata*, a C_3–C_4 intermediate. *Planta* **134,** 257–262.

517. Schiebe, R., and Beck, E. (1975). Formation of C-4 dicarboxylic acids by intact spinach chloroplasts. *Planta* **125,** 63–67.

517a. Schnarrenberger, C., Gross, D., Burkhard, Ch., and Herbert, M. (1980). Cell organelles from Crassulacean acid metabolism (CAM) plants. II. Compartmentation of enzymes of the Crassulacean acid metabolism. *Planta* **147,** 477–484.

518. Schötz, F., and Diers, L. (1975). Vergrösserung der Kontaktfläche zwischen Chloroplasten und ihrer cytoplasmichen Umgebung durch tubaläre Ausstülpungen der Plastidenhülle. *Planta* **124,** 277–285.

519. Schulze, E.-D., Ziegler, H., and Stichler, W. (1976). Environmental control of Crassulacean acid metabolism in *Welwitschia mirabilis* Hook. fil. in its natural range of distribution in the Namib Desert. *Oecologia* **24,** 323–334.

520. Semenenko, V. E. (1964). Characteristics of carbon dioxide gas exchange in the transition stages of photosynthesis upon changing from light to darkness, light induced evolution of carbon dioxide. *Sov. Plant Physiol. (Engl. Transl.)* **11,** 375–384.

521. Servaites, J. C., and Ogren, W. L. (1977). Chemical inhibition of the glycolate pathway in soybean leaf cells. *Plant Physiol.* **60,** 461–466.

522. Servaites, J. C., and Ogren, W. L. (1978). Oxygen inhibition of photosynthesis and stimulation of photorespiration in soybean leaf cells. *Plant Physiol.* **61,** 62–67.

523. Setlik, I. (ed.) (1970). "Prediction and Measurement of Photosynthetic Productivity." Centre Agr. Publishing & Documentation, Wageningen, The Netherlands.

524. Shephard, D. C., and Bidwell, R. G. S. (1973). Photosynthesis and carbon metabolism in a chloroplast preparation from *Acetabularia*. *Protoplasma* **76,** 289–307.

525. Shephard, D. C., Levin, W. B., and Bidwell, R. G. S. (1968). Normal photosynthesis in isolated chloroplasts. *Biochem. Biophys. Res. Commun.* **32,** 413–420.

526. Shirahashi, K., Hayakowa, S., and Sugiyama, T. (1978). Cold lability of pyruvate, orthophosphate dikinase in the maize leaf. *Plant Physiol.* **62**, 826–830.

527. Shain, Y., and Gibbs, M. (1971). Formation of glycolate by a reconstituted spinach chloroplast preparation. *Plant Physiol.* **48**, 325–330.

527a. Sicher, R. C., Hatch, A. L., Stumpf, D. K., and Jensen, R. G. (1981). Ribulose-1,5-bisphosphate and activation of the carboxylase in the chloroplast. *Plant Physiol.* **68**, 252–255.

528. Siegel, M., and Lane, D. M. (1972). Interaction of ribulose diphosphate carboxylase with 2-carboxyribitol diphosphate, an analogue of the proposed carboxylated intermediate in the CO_2 fixation reaction. *Biochem. Biophys. Res. Commun.* **48**, 508–516.

529. Singh, R., and Juliano, B. O. (1977). Free sugars in relation to starch accumulation in developing rice grain. *Plant Physiol.* **59**, 417–421.

530. Slabas, A. R., and Walker, D. A. (1976). Enzymic reconstitution of photosynthetic carbon assimilation. *Arch. Biochem. Biophys.* **175**, 590–597.

531. Slack, C. R. (1969). Localization of photosynthetic enzymes in mesophyll and parenchyma sheath chloroplasts of maize and *Amaranthus palmeri. Phytochemistry* **8**, 1387–1391.

532. Slack, C. R., Hatch, M. D., and Goodchild, D. J. (1969). Distribution of enzymes in mesophyll and parenchyma-sheath chloroplasts of maize leaves in relation to the C₄-dicarboxylic acid pathway of photosynthesis. *Biochem. J.* **114**, 489–498.

533. Smillie, R. M., Andersen, K. S., and Bishop, D. G. (1971). Plastocyanin-dependent photoreduction of NADP by agranal chloroplasts from maize. *FEBS Lett.* **13**, 318–320.

534. Smith, B. N., and Brown, W. V. (1973). The Kranz syndrome in the Gramineae as indicated by carbon isotopic ratios. *Am. J. Botany* **60**, 505–513.

535. Smith, B. N., Oliver, J., and McMillan, C. (1976). Influence of carbon source, oxygen concentration, light intensity, and temperature on $^{13}C/^{12}C$ ratios in plant tissues. *Bot. Gaz.* (Chicago) **137**, 99–104.

536. Smith, E. W., Tolbert, N. E., and Ku, H. S. (1976). Variables affecting the CO_2 concentration point. *Plant Physiol.* **58**, 143–146.

537. Shomer-Ilan, A., Beer, S., and Waisel, Y. (1975). *Suaeda monoica,* a C₄ plant without typical bundle sheaths. *Plant Physiol.* **56**, 676–679.

538. Somerville, C. R., and Ogren, W. L. (1981). Photorespiration-deficient mutants of *Arabidopsis thaliana* lacking mitochondrial serine transhydroxymethylase activity. *Plant Physiol.* **67**, 666–671.

539. Spalding, M. H., Schmitt, M. R., Ku, S. B., and Edwards, G. E. (1979). Intracellular localization of some key enzymes of Crassulacean acid metabolism in *Sedum praealtum. Plant Physiol.* **63**, 738–743.

540. Splittstoesser, W. E. (1966). Dark CO_2 fixation and its role in the growth of plant tissue. *Plant Physiol.* **41**, 755–759.

541. Steer, B. T. (1971). The dynamics of leaf growth and photosynthetic capacity in *Capsicum frutescens* L. *Ann. Bot.* (*London*) **35**, 1003–1015.

542. Steer, B. T. (1973). Control of ribulose-1,5-diphosphate carboxylase activity during expansion of leaves of *Capsicum frutescens* L. *Ann. Bot.* (*London*) **37**, 823–829.

543. Stemler, A., and Radmer, R. (1975). Source of photosynthetic oxygen in bicarbonate-stimulated Hill reaction. *Science* **190**, 457–458.

544. Steward, F. C., Craven, G. H., Weerasinghe, S. P. R., and Bidwell, R. G. S. (1971). Effects of prior environmental conditions on the subsequent uptake and release of carbon dioxide in light. *Can. J. Bot.* **49**, 1997–2007.

545. Stocking, C. R., and Larson, S. (1969). A chloroplast cytoplasmic shuttle and the reduction of extraplastid NAD. *Biochem. Biophys. Res. Commun.* **37**, 278–282.

546. Stolwijk, J. A., and Thimann, K. V. (1957). The uptake of carbon dioxide and bicarbonate by roots, and its influence on growth. *Plant Physiol.* **32**, 513–520.

547. Sugiyama, T., Schmitt, M. R., Ku, S. B., and Edwards, G. E. (1979). Differences in cold lability of pyruvate, Pi dikinase among C_4 species. *Plant Cell Physiol.* **20**, 965–971.

548. Sutton, B. G., and Osmond, C. B. (1972). Dark fixation of CO_2 by Crassulacean plants: evidence for a single carboxylation step. *Plant Physiol.* **50**, 360–365.

548a. Szarek, S. R. (1979). The occurrence of Crassulacean acid metabolism: a supplementary list during 1976 to 1979. *Photosynthesis* **13**, 467–473.

549. Szarek, S. R., and Ting, I. P. (1974). Seasonal patterns of acid metabolism and gas exchange in *Opuntia basilaris*. *Plant Physiol.* **54**, 76–81.

550. Szarek, S. R., and Ting, I. P. (1977). The occurrence of Crassulacean acid metabolism among plants. *Photosynthetica* **11**, 330–342.

551. Takabe, T., and Akazawa, T. (1975). Molecular evolution of ribulose-1,5-bisphosphate carboxylase. *Plant Cell Physiol.* **16**, 1049–1060.

552. Tamas, I. A., Atkins, B. D., Ware, S. M., and Bidwell, R. G. S. (1972). Indoleacetic acid stimulation of phosphorylation and bicarbonate fixation by chloroplast preparations in light. *Can. J. Bot.* **50**, 1523–1527.

553. Tamas, I. A., and Bidwell, R. G. S. (1971). Metabolism of glycolic acid-1-^{14}C in barley leaves with or without added CO_2. *Can. J. Bot.* **49**, 299–302.

554. Tamas, I. A., Schwartz, J. W., Hagin, J. M., and Simmonds, R. (1974). Hormonal control of photosynthesis in isolated chloroplasts. *In* "Mechanisms of Regulation of Plant Growth" (R. L. Bieleski, A. R. Ferguson, and M. M. Cresswell, eds.), pp. 261–268. R. Soc. New Zealand, Wellington.

555. Tamas, I. A., Yemm, E. W., and Bidwell, R. G. S. (1970). The development of photosynthesis in dark-grown barley leaves upon illumination. *Can. J. Bot.* **48**, 2313–2317.

556. Theologis, A., and Laties, G. G. (1978). Cyanide-resistant in fresh and aged sweet potato slices. *Plant Physiol.* **62**, 243–248.

557. Theologis, A., and Laties, G. G. (1978). Respiratory contribution of the alternate path during various stages of ripening in avocado and banana fruits. *Plant Physiol.* **62**, 249–255.

558. Thomas, E. A., and Tregunna, E. B. (1968). Bicarbonate ion assimilation in photosynthesis by *Sargassum muticum*. *Can. J. Bot.* **46**, 411–415.

559. Thomas, S. M., Hall, N. P., and Merrett, M. J. (1978). Ribulose 1,5-bisphosphate carboxylase/oxygenase activity and photorespiration during the ageing of flag leaves of wheat. *J. Exp. Bot.* **29**, 1161–1168.

560. Thornley, J. H. M. (1977). Growth, maintenance and respiration: a re-interpretation. *Ann. Bot. (London)* **41**, 1191–1203.

561. Ting, I. P., and Osmond, C. B. (1973). Photosynthetic phosphoenolpyruvate carboxylases: characteristics of alloenzymes from leaves of C_3 and C_4 plants. *Plant Physiol.* **51**, 439–447.

562. Togosaki, R. K., and Gibbs, M. (1967). Enhanced dark CO_2 fixation by preilluminated *Chlorella pyrenoidosa* and *Anacystis nidulans*. *Plant Physiol.* **42**, 991–996.

563. Tolbert, N. E. (1958). Secretion of glycollic acid by chloroplasts. *Brookhaven Symp. Biol.* **11**, 271–290.

564. Tolbert, N. E. (1973). Glycolate biosynthesis. *In* "Current Topics in Cellular Regulation" (B. L. Horecker and E. K. Stadtmann, eds.), pp. 21–49. Academic Press, New York.

565. Tolbert, N. E., Oeser, A., Kisaki, T., Hageman, R. H., and Yamazaki, R. K. (1968). Peroxisomes from spinach leaves containing enzymes related to glycolate metabolism. *J. Biol. Chem.* **243,** 5179–5184.

566. Tolbert, N. E., Oeser, A., Yamazaki, R. K., Hageman, R. H., and Kisaki, T. (1969). A survey of plants for leaf peroxisomes. *Plant Physiol.* **44,** 135–147.

567. Tolbert, N. E., and Osmond, C. B. (eds.) (1976). "Photorespiration in Marine Plants." Univ. Park Press, Baltimore.

568. Tolbert, N. E., and Ryan, F. J. (1976). Glycolate biosynthesis and metabolism during photorespiration. *In* "CO_2 Metabolism and Plant Productivity" (R. H. Burris and C. C. Black, eds.), pp. 141–159. Univ. Park Press, Baltimore.

569. Tolbert, N. E., and Zill, L. P. (1956). Excretion of glycolic acid by algae during photosynthesis. *J. Biol. Chem.* **222,** 895–906.

570. Tomova, N., Setchenska, M., Krusteva, N., Christova, Y., and Detchev, G. (1972). Amino acid activation of glyceraldehyde-3-phosphate dehydrogenase from Chlorella. *Z. Pflanzenphysiol.* **67,** 113–116.

571. Tregunna, E. B., Krotkov, G., and Nelson, C. D. (1961). Evolution of carbon dioxide by tobacco leaves during the dark period following illumination with light of different intensities. *Can. J. Bot.* **39,** 1045–1059.

572. Tregunna, E. B., Krotkov, G., and Nelson, C. D. (1964). Further evidence on the effects of light on respiration during photosynthesis. *Can. J. Bot.* **42,** 989–997.

573. Tregunna, E. B., Krotkov, G., and Nelson, C. D. (1966). Effect of oxygen on the rate of photorespiration in detached tobacco leaves. *Physiol. Plant.* **19,** 723–733.

574. Treharne, K. J., and Cooper, J. P. (1969). The effect of temperature on the activity of carboxylases in tropical and temperate Gramineae. *J. Exp. Bot.* **20,** 170–175.

575. Troughton, J. H. (1971). Aspects of the evolution of the photosynthetic carboxylation in plants. *In* "Photosynthesis and Photorespiration" (M. D. Hatch, C. B. Osmond, and R. O. Slatyer, eds.), pp. 124–129. Wiley–Interscience, New York.

576. Troughton, J. H., and Slatyer, R. O. (1969). Plant water status, leaf temperature, and the calculated mesophyll resistance to carbon dioxide of cotton leaves. *Aust. J. Biol. Sci.* **22,** 815–827.

577. Turner, W. B., and Bidwell, R. G. S. (1965). Rates of photosynthesis in attached and detached bean leaves, and the effect of spraying with indoleacetic acid. *Plant Physiol.* **40,** 446–451.

578. Usuda, H., Kanai, R., and Takeuchi, M. (1971). Comparison of carbon dioxide fixation and the fine structure in various assimilatory tissues of *Amaranthus retroflexus* L. *Plant Cell Physiol.* **12,** 917–930.

579. Usuda, H., and Miyachi, S. (1977). Coupling of malate decarboxylation to CO_2 fixation and the reduction of 3-phosphoglyceric acid in corn bundle sheath cells. *Plant Cell Physiol.* **18,** 1109–1120.

580. Van Bel, A. J. E., and Reinhold, L. (1975). Is the stimulation of sugar transfer by exogenous ATP a pH effect? *Z. Pflanzenphysiol.* **76,** 224–228.

581. Van Norman, R. W., and Brown, A. H. (1952). The relative rates of photosynthetic assimilation of isotopic forms of carbon dioxide. *Plant Physiol.* **27,** 691–709.

582. Vater, J., and Salnikow, J. (1979). Identification of two binding sites of the ribulose-1,5-bisphosphate carboxylase/oxygenase from spinach for ribulose-1,5-bisphosphate and effectors of the carboxylation reaction. *Arch. Biochem. Biophys.* **194,** 190–197.

583. Viil, J., Laisk, A., Oja, V., and Pärnik, T. (1977). Enhancement of photosynthesis caused by oxygen under saturating irradiance and high CO_2 concentrations. *Photosynthetica* **11,** 251–259.

584. Volk, R. J., and Jackson, W. A. (1972). Photorespiratory phenomena in maize: oxygen uptake, isotope discrimination and carbon dioxide efflux. *Plant Physiol.* **49**, 218–223.

585. von Willert, D. J., Kirst, G. O., Treichel, S., and von Willert, K. (1976). The effect of leaf age and salt stress on malate accumulation and phosphoenolpyruvate carboxylase activity in *Mesembryanthemum crystallinum*. *Plant Sci. Lett.* **7**, 341–346.

585a. von Willert, D. J., and von Willert, K. (1979). Light modulation of the activity of the PEP-carboxylase in CAM plants in the Mesembryanthemaceae. *Z. Pflanzenphysiol.* **95**, 43–49.

586. Voskresenskaya, N. P. (1972). Blue light and carbon metabolism. *Annu. Rev. Plant Physiol.* **23**, 219–234.

587. Walker, D. A. (1957). Physiological studies on acid metabolism in green plants. *Biochem. J.* **67**, 73–79.

588. Walker, D. A. (1965). Correlation between photosynthetic activity and membrane integrity in isolated pea chloroplasts. *Plant Physiol.* **40**, 1157–1161.

589. Walker, D. A. (1973). Photosynthetic induction phenomena and the light activation of ribulose diphosphate carboxylase. *New Phytol.* **72**, 209–235.

590. Walker, D. A., Koskiukiewicz, K., and Case, C. (1973). Photosynthesis by isolated chloroplasts: some factors affecting induction in CO_2-dependent oxygen evolution. *New Phytol.* **72**, 237–247.

591. Walker, D. A., McCormick, A. V., and Stokes, D. M. (1971). CO_2-dependent oxygen evolution by envelope-free chloroplasts. *Nature (London)* **233**, 346–347.

592. Walker, D. A., and Lilley, R. M.(1974). Autocatalysis in a reconstituted chloroplast system. *Plant Physiol.* **54**, 950–952.

593. Walker, D. A., and Slabas, A. R. (1976). Stepwise generation of the natural oxidant in a reconstituted chloroplast system. *Plant Physiol.* **57**, 203–208.

594. Wang, D., and Waygood, E. R. (1962). Carbon metabolism of ^{14}C-labelled amino acids in wheat leaves. I. A pathway of glyoxylate–serine metabolism. *Plant Physiol.* **37**, 826–832.

595. Weigl, J., and Calvin, M. (1948). An isotope effect in photosynthesis. *J. Chem. Phys.* **17**, 210 (1948).

596. Werdan, K., and Heldt, H. W. (1972). Accumulation of bicarbonate in intact chloroplasts following a pH gradient. *Biochim. Biophys. Acta* **283**, 430–441.

597. Werdan, K., Heldt, H. W., and Milovancev, M. (1975). The role of pH in the regulation of carbon fixation in the chloroplast stroma. Studies on CO_2 fixation in the light and dark. *Biochim. Biophys. Acta* **396**, 276–292.

598. Widholm, J. M., and Ogren, W. L. (1969). Photorespiratory-induced senescence of plants under conditions of low carbon dioxide. *Proc. Natl. Acad. Sci. U.S.* **63**, 668–675.

599. Wildman, S. G. (1967). The organization of grana-containing chloroplasts in relation to location of some enzymatic systems concerned with photosynthesis, protein synthesis, and ribonucleic acid synthesis. *In* "Biochemistry of Chloroplasts" (T. W. Goodwin, ed.), Vol. 2, pp. 295–319. Academic Press, New York.

600. Wildman, S. G., Jope, C., and Atchison, B. A. (1974). Role of mitochondria in the origin of chloroplast starch grains. Description of the phenomenon. *Plant Physiol.* **54**, 231–237.

601. Wildner, G. F., and Criddle, R. S. (1969). Ribulose diphosphate carboxylase. I. A factor involved in light activation of the enzyme. *Biochem. Biophys. Res. Commun.* **37**, 952–960.

602. Wildner, G. F., and Henkel, J. (1976). Specific inhibition of the oxygenase activity of ribulose-1,5-bisphosphate carboxylase. *Biochem. Biophys. Res. Commun.* **69**, 208–275.

603. Willenbrink, J., and Kremer, B. P. (1973). Lokalisation der Mannitbiosynthese in der marinen Braunalge *Fucus serratus*. *Planta* **113**, 173–178.

604. Williams, L. E., and Kennedy, R. A. (1978). Photosynthetic carbon metabolism during leaf ontogeny in *Zea mays* L.: enzymic studies. *Planta* **142**, 269–274.

605. Willmer, C. M., and Dittrich, P. (1974). Carbon dioxide fixation by epidermal and mesophyll tissue of *Tulipa* and *Commelina*. *Planta* **117**, 123–132.

606. Wilson, D. (1972). Variation in photorespiration in *Lolium*. *J. Exp. Bot.* **23**, 517–524.

607. Winkenbach, F., Parthasarathy, M. V., and Bidwell, R. G. S. (1972). Sites of photosynthetic metabolism in cells and chloroplast preparations of *Acetabularia mediterranea*. *Can. J. Bot.* **50**, 1367–1375.

607a. Winter, K. (1979). CAM in chickpea? *Aust. Natl. Univ. Ann. Rep.* pp. 18–19; and personal communication.

608. Winter, K., Kramer, D., Troughton, J. H., Card, K. A., and Fischer, K. (1977). C_4 pathway of photosynthesis in a member of the Polygonaceae: *Calligonum persicum* (Boiss & Buhse) Boiss. *Z. Pflanzenphysiol.* **81**, 341–346.

609. Winter, K., and Troughton, J. H. (1978). carbon assimilation pathways in *Mesembryanthemum nodiflorum* L. under natural conditions. *Z. Pflanzenphysiol.* **88**, 153–162.

610. Wolosiuk, R. A., and Buchanan, B. B. (1977). Thioredoxin and glutathione regulate photosynthesis in chloroplasts. *Nature (London)* **266**, 565–567.

611. Wolosiuk, R. A., and Buchanan, B. B. (1978). Regulation of chloroplast phosphoribulokinase by the ferredoxin/thioredoxin system. *Arch. Biochem. Biophys.* **189**, 97–101.

612. Wolosiuk, R. A., and Buchanan, B. B. (1978). Activation of chloroplast NADP-linked glyceraldehyde-3-phosphate dehydrogenase by the ferredoxin/thioredoxin system. *Plant Physiol.* **61**, 669–671.

613. Wolpert, J. S., and Ernst-Fonberg, M. L. (1975). A multienzyme complex for CO_2 fixation. *Biochemistry* **14**, 1095–1102.

614. Wolpert, J. S., and Ernst-Fonberg, M. L. (1975). Dissociation and characterization of enzymes from a multienzyme complex involved in CO_2 fixation. *Biochemistry* **14**, 1103–1107.

615. Wong, S. C., and Hew, C. S. (1976). Diffusive resistance, titratable acidity, and CO_2 fixation in two tropical epiphytic ferns. *Am. Fern J.* **66**, 121–124.

616. Wong, W. W., Benedict, C. R., and Kohel, R. J. (1979). Enzymic fractionation of the stable carbon isotopes of carbon dioxide by ribulose-1,5-bisphosphate carboxylase. *Plant Physiol.* **63**, 852–856.

617. Woo, K. C. (1980). Properties and intramitochondrial localization of serine hydroxymethyltransferase in leaves of higher plants. *Plant Physiol.* **63**, 783–787.

618. Woo, K. C., Anderson, J. M., Boardman, N. K., Downton, W. J. S., Osmond, C. B., and Thorne, S. W. (1970). Deficient photosystem II in agranal bundle sheath chloroplasts of C_4 plants. *Proc. Natl. Acad. Sci. U.S.* **67**, 18–25.

619. Woo, K. C., Berry, J. A., and Turner, G. (1978). Release and refixation of ammonia during photorespiration. *Carnegie Inst. Washington Year Book* **78**, 240–245.

620. Woo, K. C., Pyliotis, N. A., and Downton, W. J. S. (1971). Thylakoid aggregation and chlorophyll a/chlorophyll b ratio in C_4 plants. *Z. Pflanzenphysiol.* **64**, 400–413.

621. Wort, J. D. (1976). Mechanism of plant growth stimulation by naphthenic acid. *Plant Physiol.* **58**, 82–86.

622. Yamagouchi, T., Kawa, R. I., and Nisizawa, K. (1966). The incorporation of radio-

active carbon from $HC^{14}O_3^-$ into sugar constituents by a brown alga, *Eisenia bicyclis*, during photosynthesis and its fate in the dark. *Plant Cell Physiol.* **7**, 217–229.

623. Yamamoto, E., Sugiyama, T., and Miyachi, S. (1974). Action spectrum for light activation of pyruvate, phosphate dikinase in maize leaves. *Plant Cell Physiol.* **15**, 987–992.

624. Yemm, E. W., and Bidwell, R. G. S. (1969). Carbon dioxide exchanges in leaves. I. Discrimination between $^{12}CO_2$ and $^{14}CO_2$ in photosynthesis. *Plant Physiol.* **44**, 1328–1334.

624a. Yeoh, H.-H., Badger, M. R., and Watson, L. (1981). Variations in kinetic properties of ribulose-1,5-bisphosphate among plants. *Plant Physiol.* **67**, 1151–1155.

625. Zelitch, I. (1965). The relation of glycolic acid synthesis to the primary photosynthetic carboxylation reaction in leaves. *J. Biol. Chem.* **240**, 1869–1876.

626. Zelitch, I. (1966). Increased rate of net photosynthetic carbon dioxide uptake caused by the inhibition of glycolate oxidase. *Plant Physiol.* **41**, 1623–1631.

627. Zelitch, I. (1968). Investigations on photorespiration with a sensitive ^{14}C-assay. *Plant Physiol.* **43**, 1829–1837.

628. Zelitch, I. (1971). "Photosynthesis, Photorespiration, and Plant Productivity." Academic Press, New York.

629. Zelitch, I. (1972). The photooxidation of glyoxylate by envelope-free spinach chloroplasts and its relation to photorespiration. *Arch. Biochem. Biophys.* **150**, 698–707.

630. Zelitch, I. (1973). Alternate pathways of glycolate synthesis in tobacco and maize leaves in relation to rates of photorespiration. *Plant Physiol.* **51**, 299–305.

631. Zelitch, I. (1974). The effect of glycidate, an inhibitor of glycolate synthesis, on photorespiration and net photosynthesis. *Arch. Biochem. Biophys.* **163**, 367–377.

632. Zelitch, I. (1975). Improving the efficiency of photosynthesis. *Science* **188**, 626–633.

633. Zelitch, I. (1978). Effect of glycidate, an inhibitor of glycolate synthesis in leaves, on the activity of some enzymes of the glycolate pathway. *Plant Physiol.* **61**, 236–241.

634. Zelitch, I., and Day, P. R. (1973). The effect on net photosynthesis of pedigree selection for low and high rates of photorespiration in tobacco. *Plant Physiol.* **52**, 33–37.

635. Ziegler, H., and Ziegler, I. (1965). The influence of light on the $NADP^+$-dependent glycerinaldehyde-3-phosphate dehydrogenase. *Planta* **65**, 369–380.

CHAPTER FOUR

Respiration: A Holistic Approach to Metabolism[1]

DOROTHY F. FORWARD

I. Introduction

Respiration makes available the solar energy fixed by photosynthesis in carbohydrate. Higher plants are obligate aerobes, and transfer of energy is mediated by ATP produced in linkage with the transfer of protons and electrons to O_2 (referred to as oxidative respiration).[2] Respiration is regarded in the following discussion, not merely as this ultimate step, but as the whole process of catabolism of sugar or other substrate leading to the consumption of O_2 and production of CO_2.

Standard pathways, glycolysis, the tricarboxylic acid cycle and the cy-

[1] Abbreviations used in this chapter: ABA, abscisic acid; ADP, adenosine diphosphate; ATP, adenosine triphosphate; CCCP, carbonylcyanide m-chlorophenylhydrazone; CLAM, m-chlorobenzhydroxamic acid; DNP, 2,4-dinitrophenol; F6P, fructose 6-phosphate; FBP, fructose 1,6-bisphosphate; GA, gibberellic acid; IAA, indoleacetic acid; NAD, nicotinamide adenine dinucleotide; NADP, nicotinamide adenine dinucleotide phosphate; P, phosphate; PFK, phosphofructokinase; PK, pyruvate kinase; RNA, ribonucleic acid; RQ, respiratory quotient; SHAM, salicylhydroxamic acid.

[2] This aspect of the process is covered in detail in Chapter 2. (Ed.)

Plant Physiology
A Treatise
Vol. VII: Energy and Carbon Metabolism

tochrome chain of electron transport (with alternatives, the pentose phosphate cycle, cyanide-resistant electron transport and others) have been established, and many control points in the metabolic sequences are known (see Chapter 2). The points at which carbon is diverted from the catabolic stream to provide carbon skeletons for cell constituents are also known. *In vivo* controls of the metabolic processes in the whole organism are harder to discern and are relatively little known.

Basic pathways of respiratory metabolism persist throughout the life of the plant, but variations in intensity and participation of alternate pathways occur spontaneously. Good opportunities to investigate controls come with conspicuous changes in the ontogenetic program, such as seed germination or the onset of senescence. Gaseous exchange provides a means of observing changes in metabolic intensity without destroying the organization of tissues, and changes in respiration rate have been the means of focusing attention on attendant metabolic changes.

Studies of respiration using homogenates or mitochondria deal mainly with biochemical processes and controls. Studies with cells, tissues, organs, and whole organisms introduce a progressive series of organizational controls. Studies of carbon and energy balance in plants or communities of plants are directed toward determining the cost of growth and maintenance in terms of respiration. It is the intention in this chapter to consider what progress has been made by these various approaches toward understanding the regulation of respiration and its role in the metabolism of the plant. No attempt will be made to review all contributions exhaustively, but some examples are considered that have advanced our knowledge.

II. Respiration and Ontogenetic Events

The life cycles of plants, whatever their variations, involve a program of initiation, growth, maturity, senescence, and death. This program is repeated in individual organs, and it may proceed at different rates in different organs of the same plant. Although each organ has its own characteristics, metabolic patterns during ontogenetic development have some features in common in all organs. A knowledge of these and of the intrinsic mechanisms of regulation will contribute to the ultimate interpretation of the role of respiration in the life of the plant. Ontogenetic events that have received a good deal of attention are seed germination, the development and senescence of leaves and flowers, and the ripening of fruit.

A. SEED GERMINATION

The control of seed germination was reviewed several times during the 1970s (58, 59, 111, 123), and in all the reviews the role of respiration was discussed at least incidentally. It is the intent in this section to emphasize the recent work on the earliest stages of germination that provides some insight into the development of respiratory metabolism.

Dry seeds have a very low rate of respiration, but, when water is added, the rate of O_2 uptake begins to increase within a few minutes. It soon reaches a plateau, which lasts through the period of physical water absorption. When the embryo begins to grow and splits the seed coat, O_2 uptake increases rapidly to much higher rates. If the seed is dormant the last phase does not occur. The timing varies widely for seeds of different species, but the pattern is general. For a long time it was assumed that a low rate of respiration before the seed coat split resulted from the low permeability of the latter to O_2. This has now been challenged. Porter and Wareing (86) made direct measurement of the permeability of the seed coats of *Xanthium strumarium* (*X. pensylvanicum*). In this species each burr contains two seeds, the upper one of which remains dormant at a time when the lower seed germinates (after imbibition for 20–24 hr). Not only is seed-coat permeability the same for upper and lower seeds and constant throughout the pregermination period, but it is also high enough to permit a supply of O_2 that is about three times more than that actually consumed by the seeds, both upper and lower. The upper seeds did not germinate when moistened in air, even with the testa removed. The conclusion drawn was that it is not a restricted O_2 supply for respiration that causes dormancy. Porter and Wareing did not find that upper seeds germinated in an atmosphere of high O_2 concentration. They attributed this to destruction of inhibitors, brought about in some unknown way by high O_2 concentration. Edwards (29) reached the same conclusion for charlock [*Brassica kaber* (*Sinapis arvensis*)] seeds.

Others have also found evidence that a restricted O_2 supply is not responsible for dormancy. To cite one more example, Chen (23) found that the seeds of *Phacelia tanacetifolia* respired at the same rate in light and dark, 12 hr after wetting, at which time they germinated in dark, but not in light. The addition of GA induced germination in the light, but increased O_2 tension did not.

In spite of much study the controls of dormancy are but dimly understood. Apparently, in some seeds at least, it is not restricted O_2 supply for respiration that prevents the breaking of dormancy. What, then, controls the rate of respiration in the ungerminated seed? It is known that in mature dry seeds a good deal of degradation of internal structure occurs. The endoplasmic reticulum is disintegrated, Golgi bodies and proplastids disappear. Mitochondrial structure is disorganized, and ribosomes appear largely as monosomes (111). Upon adding water, reorganization begins immediately but may take some time to reach completion. Endoplasmic reticulum reappears, polysomes form and the internal structure of mitochondria is elaborated. Oxygen uptake is one of the first measurable responses to the addition of water. ATP increases early in the imbibition of wheat (*Triticum aestivum*) embryo (77) and radish (*Raphanus sativus*) seeds (66). This is an obvious prerequisite for the synthesis of RNA, polyribosomes, and new proteins.

Most authors deal with one type of reaction in one species and, because the timing of events is very different in different seeds, it is difficult to combine the data to determine a program of metabolic development. Morohashi and Shimokoriyama (68–71) have, however, observed the development of various processes in "mung bean (*Phaseolus mungo*)." Their work indicates that the machinery of glycolysis remains intact in dry seeds and is immediately activated when water is imbibed, whereas mitochondria must be reorganized for efficient electron transport. During the first 3 hr of imbibition at 30°C, CO_2 production is largely by alcoholic fermentation and is strongly inhibited by iodoacetate and fluoride which are inhibitors of glycolysis. However, O_2 uptake is only slightly inhibited until the eighth hour. The activity of certain glycolytic enzymes (aldolase, glyceraldehyde phosphate dehydrogenase, and pyruvate kinase) was high and constant after 20-min imbibition; mitochondrial enzymes (cytochrome oxidase, succinic and malic dehydrogenases) was very low at 20 min and increased substantially during the next hour. Mitochondrial respiration and ADP control also increased during this time.

Solomos *et al.* (108) showed that in pea (*Pisum sativum*) seeds ADP control and phosphorylation develop during 3–12 hr of imbibition at 27°C, while the internal structure of the mitochondria is elaborated and enzymes become active. Nawa and Ahasi (72, 73) found activation of mitochondrial enzymes during the first 3 hr of imbibition of pea cotyledons, whereas protein synthesis began after 3 hr, in parallel with respiration and ADP control. Eldan and Mayer (32) found that in lettuce (*Lactuca sativa*) seeds cytochrome oxidase became active more slowly than NADH-cytochrome c reductase and suggested that its activity may depend on elaboration of cristae. The increase in mitochondrial respiration appears to depend more on the development of functional structure than on the synthesis of new enzymatic proteins.

Respiration in early stages is cyanide sensitive and apparently mediated by cytochrome oxidase in some seeds, but not in all. Wilson and Bonner (125) found no cytochrome c in peanut (*Arachis hypogea*) seeds for 16 hr after wetting. Peanut seeds contain cytochromes a and b, which are reduced when O_2 is deficient, using an aberrant cytochrome chain. After 16 hr, new mitochondria appear and normal respiration occurs. In some seeds an alternate terminal oxidase comes into play that is cyanide resistant but sensitive to hydroxamic acids. Yentur and Leopold (127) reported cyanide-insensitive respiration in several species within the first few hours of imbibition. This was gradually replaced by cyanide-sensitive respiration. Others have made similar observations. The cyanide-insensitive pathway has been called the alternate pathway.[3]

Siedow and Girvin (100) observed that after a 5-hr imbibition the respiration of intact soybean (*Glycine max*) and mung-bean seeds was cyanide sensitive, but the respiration of particles of finely ground seeds was not. Under the experimental conditions, the O_2 uptake of particles was much higher than that of whole seeds, which suggests that substrate reached the mitochondrial enzymes more readily.

Evidence is presented in later sections suggesting that potential for the alternate pathway may exist in various plant organs, coming into play when the electron-transport system is for any reason inadequate to handle the products of glycolysis. In seeds this may be the result of the relatively slow elaboration of mitochondrial function, whereas the glycolytic machinery is intact in the dry seed and is activated immediately when water is added.

Roberts (91) proposed that such a cyanide-resistant chain oxidizes NADPH that is produced in the pentose phosphate cycle and that the operation of this cycle is a necessary step in preparation for germination. Citing ratios of C_6 to C_1 in support of his theory, Roberts claimed that the difference between dormant and nondormant, two-rowed barley (*Hordeum distichon*) seeds lies in greater participation of the pentose phosphate cycle in nondormant seeds before germination. If these ratios give a true indication of the participation of the cycle, participation is highly active in both dormant and nondormant seeds, for the ratios were mostly below 0.3 in both. There is, however, other relevant evidence. For example, Simmonds and Simpson (103) reported that the ratio of C_6 to C_1 in nondormant seeds of wild oats (*Avena fatua*) is much lower (0.5–0.6) than in dormant seeds (0.9) in the first 10 hr of imbibition. In these seeds, GA, which stimulates germination, lowered the ratio to 0.65,

---

[3] For a discussion of the biochemistry of this pathway, see Chapter 2, Section IV,B,3. (Ed.)

whereas it did not affect O_2 uptake. Kovacs and Simpson (48) found that amounts of glucose 6-P dehydrogenase and 6-P-gluconate dehydrogenase, the initial enzymes of the pentose phosphate cycle, increase for 2 days during imbibition of nondormant wild-oat seeds, but decrease in dormant seeds. GA increases the activity of both enzymes in dormant seeds. These authors also concluded that operation of the pentose phosphate cycle is a prerequisite for germination, suggesting that it produces NADPH and the specific metabolites required for essential syntheses. However, the pentose phosphate cycle is not universally a prerequisite for germination, as Morohashi and Shimokoriyama (71) found no evidence for significant participation of the pentose phosphate cycle in imbibed "mung-bean (*Phaseolus mungo*)" seeds.

The programming of ontogenetic events is generally considered to be under hormonal control. It is well known that hormones, including ethylene, may inhibit or promote seed germination, but there is little information about the involvement of respiration in these responses. Variability in responses of different species suggests different immediate mechanisms. Ultimately, these must all be related to integration of the metabolic systems required for germination. Respiration linked with ATP production is a key factor in such integration. Evidence is accumulating that hormones affect the properties of membranes, and this must affect internal communications in the cell and, therefore, metabolic interactions. Knowledge of these controls is scant, but Taylorson and Hendricks (111) rightly emphasize that it is in this realm that knowledge of the regulation of germination should be sought.

Once the embryo splits the seed coat, respiration rates rise rapidly to reach the highest levels attained during the life of the plant. The general characteristics of seedling respiration were discussed in Volume IVA of this treatise by Yemm (126) and need not be further elaborated. The ingredients for respiratory metabolism in the seedling are similar to those in the seed. Malhotra and Spencer (52) showed that mitochondrial structure continues to elaborate for 4 days from wetting in the cotyledons of pea, and James and Spencer (44) reported that cyanide-resistant respiration continues to increase in isolated mitochondria for 6–7 days and finally contributes 15–20% of the succinate oxidation. Ashihara and Komamine (10) found the pentose phosphate cycle to be operative in hypocotyls of "mung bean (*Phaseolus mungo*)" for 4 days after germination, and, in the younger parts, [6-^{14}C]gluconate contributed ^{14}C to building blocks for RNA, lignin, and aromatic amino acids.

As the seedling grows it also differentiates, and attention has been largely directed to the respiration of the individual differentiated organs.

B. LEAF DEVELOPMENT AND SENESCENCE

Each organ of a plant goes through the same cycle as the whole plant—growth, maturity, and senescence. Patterns of respiration have certain features in common, including intense respiration during rapid growth which declines during maturity and is followed by a steadily declining rate. In senescence an increase in respiration, corresponding to the climacteric rise in fruit, may or may not occur. Many aspects of the respiration of leaves have been discussed by Yemm (126). Attention is directed in the following discussion to the importance of position on the plant, to the senescent phase, and to a comparison of attached and detached leaves.

1. Development of Attached Leaves

For study of the ontogenetic drift in attached leaves, two methods have been employed to obtain an age series. Leaves at corresponding nodes on plants of increasing age may be compared to follow the aging of a leaf at a specific node. This may be called a *time* series. An alternative *profile* method has occasionally been used. In this technique the leaves on a plant at any one time are used as an age series, from the youngest at the top to the oldest at the base. Walther (122) combined the two methods in an extensive study of respiration of attached leaves of sunflower (*Helianthus annuus*), in which the initial rate of respiration was measured immediately after excising discs. Using the variety 'Russian Giant' in a controlled environment, Walther found that the *time* drift differs for leaves inserted at different nodes; the *profile* drift differs from the *time* drift, and it varies with the age of the plant. Other varieties and other culture regimes resulted in somewhat modified but essentially similar patterns to those outlined here. These relations are shown in Figs. 1 and 2. Fig. 1 shows the respiration of leaves at the first- and second-formed nodes of plants 10–70 days old. Fig. 2A shows drifts of leaves at individual nodes from 30–70-day-old plants. At 30 days the lower leaves were mature or beginning to senesce, while the upper leaves were young. In the upper leaves the maximum respiration rate was delayed after emergence, and decline from the maximum was steep throughout the mature stage. Leaves at nodes 4–6 were yellowing on the 70-day-old plants, but it is not clear whether any real increase in respiration occurred. Certainly it was not of the magnitude of the increase at node 2.

The age profiles (determined at one time) present a quite different picture (Fig. 2B). In the younger plants there was a steep, linear decline in respiration rate from the youngest to the oldest leaves. In older plants

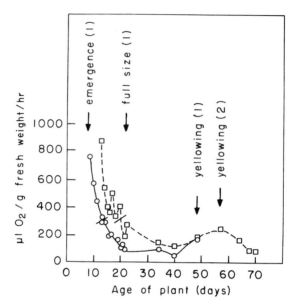

FIG. 1. Drift of respiration rate in primary and secondary leaves of sunflower (*Helianthus annuus*) plants of increasing age, grown at 20 ± 1°C under a 13-hr light regime. Primary leaves, O, secondary leaves, □. Crossbars divide initial values for whole leaves of young plants from initial values for discs from older plants. Emergence, full size, and yellowing of primary leaves (1). Yellowing of secondary leaves (2). From Walther (122).

this changed to a concave form as the respiration rate in the latest-formed leaves decreased, and the rate in the oldest leaves remained constant or increased. The decline in newly formed leaves is doubtlessly the result of aging of the meristem (126) and the development of a terminal inflorescence.

The rate of respiration per unit of nitrogen (R/N) or protein was relatively constant or declined slightly as green leaves at the upper or middle nodes aged, but R/N rose with time in yellowing basal leaves. It was not, however, the same for leaves of different insertion. Indeed, the profiles of R/N were very similar to those based on dry weight, which appear in Fig. 2B. The amount of respiring protein would account at least in part for the drift in time of the respiration rate in green leaves at each node, but it does not account for differences related to position on the plant or for an increased rate in senescing leaves.

A profile of leaf respiration does not represent a simple age series, as Yemm (126) recognized, although it has been used as such. It does

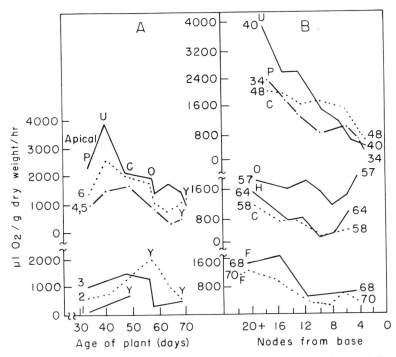

FIG. 2. (A) Drift of respiration rate in leaves at specific nodes on sunflower (*Helianthus annuus*) plants of increasing age, measured as initial rate in discs. Culture conditions as in Fig. 1. Node 1 (basal), ——; node 2, · · · ·; node 3, ——; nodes 4, 5, – · –; node 6, · · · ·; apical nodes below bud, ——. Extensive yellowing, Y. Floral stage: P, inflorescence primordium; U, inflorescence bud visible; C, bud conspicuous; O, bud opening; H, open head; F, fruit setting. (B) Respiration profiles of leaves on the same plants as in part A at different ages. Letter symbols as in part A. Numbers indicate age of the plant. From Walther (122).

provide an account of the distribution of respiratory activity and of changing patterns as the plant ages. The effective physiological age of successive leaves is of complex origin. It is affected by the age of the meristem at the time a leaf is initiated, by irregular intervals of time between the initiation of successive nodes, by different time spans in the life cycle of different leaves, and by coordinating operations within the plant. These last must include, at the least, import and export of nutrients and hormones, which are influenced by the position of the leaf. In short, the respiration of a leaf is dependent on the niche it occupies, both structurally and functionally, in the life of the whole organism.

2. Senescence

a. Attached Leaves. Considerable attention has been focused on the senescent stage. Leaf senescence has been reviewed by Beevers (17) and Thomas and Stoddard (116a), among others. Many genetic and biochemical aspects have been discussed and are not considered in detail here. Leaf senescence has much in common with the senescence of fruits and other plant organs, but the impression is left that the rate of respiration remains constant during yellowing, finally falling in the last stages. However, a rise in respiration that resembles the climacteric rise in fruit is common in detached leaves and may be present, though small, in attached leaves. Observations imply that this respiratory increase, referred to as a climacteric rise, is controlled by ethylene, as in fruit.

Aharoni *et al.* (8) measured respiration rate and ethylene production in primary leaves of pinto bean (*Phaseolus vulgaris*) from plants of increasing age. A shoulder in a declining drift of respiration occurred simultaneously with a peak in declining ethylene production during yellowing. With tobacco (*Nicotiana tabacum*) they used a profile age series. The profile showed a steady decline in leaf respiration rate down the plant, but a peak in ethylene production in yellow leaves near the base. This would appear to imply that in tobacco increased ethylene does not induce a respiratory rise in senescent leaves. But there is no way of knowing whether the yellow leaves had not, in fact, undergone a small climacteric rise that was obscured by the difference inherent in position on the plant, as in sunflower. Macnicol *et al.* (51), using a different

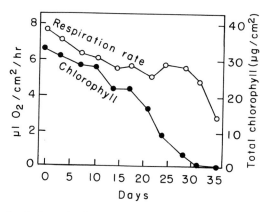

Fig. 3. Drift in the dark-respiration rate and chlorophyll content of an attached tobacco (*Nicotiana tabacum*) leaf subsequent to full expansion, determined as initial rate in discs. From Macnicol *et al.* (51).

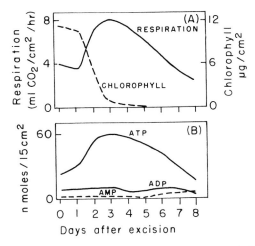

FIG. 4. (A) Drift of respiration rate (whole-leaf basis) and chlorophyll content of excised tobacco (*Nicotiana tabacum*) leaf at 25°C in dark. (B) Adenine nucleotide levels in the same leaf. From Macnicol (50).

variety of tobacco, measured respiration rates of leaves that were inserted at corresponding nodes on plants of increasing age, and they observed a small rise associated with rapid yellowing (Fig. 3).

b. Detached Leaves. When a mature leaf is detached and held in the dark it undergoes a series of changes that resemble senescence, and detached leaves have been used for much of the investigation of leaf senescence. However, the respiration pattern differs quantitatively from that of an attached leaf. The main differences are, usually, a steep decline sooner or later after excision, followed by a much more conspicuous rise and fall than occurs in attached leaves (compare Figs. 3 and 4). The respiratory rise in tobacco leaves was accompanied by an increase in ATP content, and Micnicol (50) observed changes in glycolytic intermediates that indicate an increase in the rate of glycolysis. In excised leaves or discs of sugar beet (*Beta vulgaris*) and tobacco, Aharoni *et al.* (8) found that a rise and fall in respiration is accompanied by a rise and fall in ethylene production (Fig. 5). Exogenous ethylene enhanced respiration before the normal rise, but not significantly at the peak. Both an inhibitor of ethylene synthesis and silver ion, which interferes with the action of ethylene, inhibited the respiratory rise (7). The synthesis of ethylene appears to be under the control of hormones, particularly IAA (7a, 8), but their interaction in effects on senescence is more complex.

Tetley and Thimann (112) found that a dramatic rise and fall of

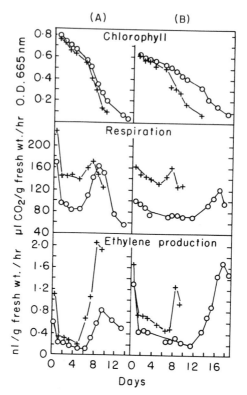

Fig. 5. Changes in ethylene production, respiration rate, and chlorophyll content of (A) sugar beet (*Beta vulgaris*) and (B) tobacco (*Nicotiana tabacum*) leaf discs during senescence in darkness at 20 (○) or 28°C (+). Leaf discs were cut from fully expanded mature leaves. Chlorophyll content was determined in leaf discs that had been senescing in Petri dishes. From Aharoni *et al.* (8).

respiration rate in segments of primary leaves of oat (*Avena sativa*) that occurs 1 day after excision is completely eliminated by adding kinetin to the medium on which the segments are floated. Because DNP had little effect on respiration at the peak, Tetley and Thimann assumed that the enhanced respiration is uncoupled from phosphorylation of ADP and that kinetin acts by maintaining the coupling. However, that a climacteric rise in respiration of tobacco leaves (cf. Fig. 4) and other organs is accompanied by ATP formation makes this explanation of the action of kinetin unlikely. The respiratory rise was accompanied by rapid chlorophyll breakdown, but this also occurred in etiolated leaves, so there is no causal relation.

These observations indicate that the respiration of senescent leaves is similar in kind to that of ripening fruits. A climacteric rise is apparently under the control of ethylene, which interacts with other hormones. The nature of respiratory metabolism and its controls during such a rise have been more extensively investigated with fruit and are more fully discussed in the next section.

There is no reason to believe that respiratory metabolism and its controls are qualitatively different in attached and detached leaves, though the resultant gas exchange is quantitatively different. The nutritional status differs in attached and detached leaves. While net protein loss occurs in both kinds of leaves during senescence, amino acids accumulate in the detached leaf but are massively exported from a senescing, attached leaf to other parts of the plant. There is evidence that protein-degradation products are a major source of respiratory substrate during the climacteric in detached leaves, and their removal must be a factor (though not the only one) in reducing or eliminating an increase in respiration of attached leaves. Furthermore, the rate of senescence is more uniform in detached leaves. It is commonly observed that yellowing does not occur uniformly in attached leaves, but usually begins at the tip and margins and progresses basally. Any accompanying rise in respiration rate would occur at different times in different groups of cells, and the overall rate of gas exchange would be leveled out. The more uniform progress of senescence in a detached leaf suggests that barriers to communication between cells are lowered when a leaf is excised.

An important result of detaching a leaf is that hormone import is cut off, and the hormone balance changes. Thomas and Stoddart (116a) do not consider change in growth regulators to be the primary event in the initiation of leaf senescence but suggest that it could be part of a complex induction system. Although the mechanisms involved in hormonal regulation of senescence remain largely unknown, they may involve structural as well as metabolic changes. There is evidence that intracellular structure undergoes drastic changes in senescing leaves (17, 21, 26, 87, 98, 116a). These follow a general program (17, 21), in which certain organelles and the tonoplast degenerate during yellowing while the mitochondria and nucleus are still intact and apparently functional. The plasma membrane remains visually intact until the terminal stage, when death ensues. Butler and Simon (21) reported that tonoplast breakdown is delayed in detached cucumber (*Cucumis sativus*) cotyledons as compared with attached cotyledons. Ragetli and Weintraub (87), on the other hand, reported earlier breakdown of the tonoplast in detached tobacco leaves than in attached leaves in daylight. Kinetin delays the degenerative structural changes, as may IAA and GA, whereas abscisic acid accelerates these processes (87). The program of degeneration is similar in other organs (17, 21) and could be related to common features of respiratory metabolism in senescence wherever it occurs.

C. FRUIT RIPENING

The most intensively studied aspect of senescence is the ripening of fruit. The phenomenon that led the way into investigation of that process was the pronounced increase in respiration rate, termed the *climacteric*, which is characteristic of many fleshy fruits. The occurrence of a climacteric rise in respiration, its dependence on ethylene, and some aspects of associated changes in fruit were described in Volume IVA of this treatise (33). Since then much study has been devoted to the ripening processes, but basic questions have not yet been fully answered. What initiates the senescent phase of fruit development? What induces an increase in ethylene production? How does ethylene influence the rise in respiration rate? What is the role of respiration in ripening, and how is its rate controlled?

A comprehensive report on the biochemistry of fruits, edited by Hulme (39, 40), appeared in 1970 and 1971. Since then a review of fruit development by Coombe (24) has dealt only incidentally with the role of respiration.

Rhodes (90) proposed a redefinition of the term climacteric to conform to modern usage. He defined it as "a period in the ontogeny of certain fruits, during which a series of biochemical changes is initiated by the autocatalytic production of ethylene, marking the change from growth to senescence and involving an increase in respiration and leading to ripening." Respiration rate is an easily measurable index of the progress of events in this period of ontogeny. Some other events are also measurable in quantitative terms, sometimes involving destruction of the fruit. However, ripening, involving many processes, is not measurable. It is a matter of subjective judgment when a fruit is "ripe," and this may represent different stages of senescence in different fruits.

Not all fruits display a sudden climacteric rise in respiration rate during ripening. Some fruits that were originally classified as nonclimacteric have been shown to undergo a climacteric rise at some stage, under appropriate conditions. There are, however, a few apparent exceptions, notably citrus fruits. There is no clear evidence that the metabolic reactions of nonclimacteric fruit differ from those of climacteric fruit. It has been suggested (90) that the time scale is more extended and, therefore, that the changes are less dramatic. It is also possible that different tissues within the fruit do not respond simultaneously, and so no sharp overall response is observed at one particular time. Evidence of heterogeneity comes from Passam and Bird (79), who found that the inner flesh of honeydew melon (*Cucumis melo*) ripened faster than the outer and that it contributed largely to the climacteric respiration of the whole fruit. Adato and Gazit (6) observed two separate peaks of ethylene production and respiration in developing avocado (*Persea americana*) fruits. The first, a relatively small preclimacteric peak, was produced almost entirely by the seed coat; the second, the climacteric peak, was associated with ripening of the fruit. A different distribution within the fruit could lead to leveling of the overall rate of respiration.

For climacteric fruits it is widely accepted that an increase in respiration rate is initiated by a spontaneous increase in ethylene production, although this has been challenged for some fruits (24, 49). What causes the increase in ethylene or how its consequences are effected are still matters for investigation. The experimenter is faced with a dilemma in this regard. Techniques for analyzing the behavior of whole organs or whole organisms are distinctly limited. Much useful information can be acquired by using detached fruits, tissue slices, mitochondria, or isolated enzyme systems. The question is how far this is directly applicable to the whole organ or organism.

Studies in which some comparison is made are instructive, but rare. One example is the work of Marei and Crane (53) on the fig (*Ficus carica*), which, incidentally, was originally classified as a nonclimacteric fruit. Marei and Crane measured respiration of fruits on the tree as well as after picking and treated the fruits with ethylene by enclosing parts of branches in polyethylene bags. The normal drift of the respiration rate of attached fruits declined during the early phase of rapid growth (I), slowed steadily during a phase of slow growth (II), and rose rapidly to a climacteric peak when the growth rate again increased and ripening began (III). These changes are shown by a solid line in Fig. 6. When fruits were detached at various stages of development, respiration (dotted lines) declined in phases I and II. During the climacteric rise in phase III, isolation made no difference to the rate of respiration. Only at phase III was the fruit independent of the tree with respect to substances and physiological conditions concerned in the regulation of respiration. Ethylene added in the second half of phase II, but not before, brought on an immediate climacteric peak. The endogenous ethylene content was high in phase I, low throughout phase II, and rose rapidly in phase III as the climacteric peak developed. This study provides a clear example of ethylene induction of the climacteric rise in respiration *when the fruit is ready*. Many biochemical changes occur during phase II,

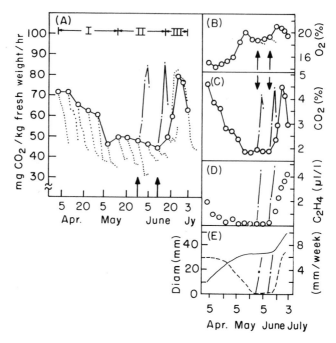

FIG. 6. (A) Respiration rates of Mission fig (*Ficus carica*) fruits. Respiration rate of control (——), ethylene-treated fruits on the tree (–·–), and of control fruits after harvest (· · · ·). Arrows indicate dates of ethylene treatment of attached fruits. (B)–(E) Concentration of O_2, CO_2, and ethylene in the internal atmosphere and diameter of fruits in relation to their growth and development and to ethylene treatment. Arrows indicate the time of ethylene application. From Marei and Crane (53).

but which ones are concerned in the preparation of the fruit for the production of ethylene and the response of respiration rate remain to be determined.

One line of inquiry is into the nature and regulation of respiratory metabolism during the climacteric rise. Laties and collaborators have contributed to this area of investigation. Solomos and Laties (105), finding that cyanide stimulated the respiration of slices of preclimacteric avocado, proposed that the extra respiration in the climacteric proceeds via the alternate cyanide-resistant pathway (see discussion in Chapter 2, Section IV,B,3.). Later (107), they compared the effects of ethylene and cyanide on respiration in green peas, pea seedlings, and cherimoya (*Annona cherimola*) fruits. In seeds and seedlings cyanide inhibited respiration, increased glycolysis and fermentation, and caused a decrease in ATP, whereas ethylene caused none of these changes. In the fruit both ethylene and cyanide increased respiration, glycolysis, and ATP content

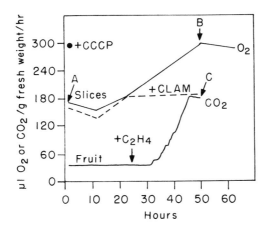

FIG. 7. Rate of respiration in ripening slices (O_2 uptake) and intact fruit (CO_2 output) of avocado (*Persea americana*). A, preclimacteric avocado slices; B, ripened slices at climacteric peak; C, slices prepared from ripened fruit at climacteric peak; +CCCP, uncoupled slices. Fresh slices treated with 3 mM CLAM where indicated; intact fruit treated with ethylene where indicated. From Theologis and Laties (116).

and failed to evoke fermentation. Solomos and Laties concluded that for ethylene to support increased respiration, cyanide-resistant respiration must be present.

However, a more complete investigation of the respiration of both intact fruits and slices of avocado by Theologis and Laties (113) did not support this hypothesis. Figure 7 shows the rates of respiration of intact fruits (with a climacteric rise induced by ethylene), fruit slices held in moist air to undergo ripening, and fruit slices in the presence of a phosphorylation uncoupler, CCCP. Uncoupling increases the respiration rate of intact fruit and fresh slices to that of the climacteric peak value. Uncoupling at the peak did not cause further increase. By a titration technique (11), using cyanide (KCN) and CLAM as inhibitors, Theologis and Laties determined the contribution of electron-transport pathways, as shown in Table I. In the absence of cyanide, the respiration of preclimacteric and climacteric fruits and that of freshly cut slices proceeds entirely by the cytochrome pathway. Higher respiration rates, in uncoupled or aging slices, have an alternate cyanide-resistant-pathway component, and at the maximum rate the alternate pathway is fully engaged. The maximum rate shown in Fig. 7 is equivalent to the total respiratory potential of the tissue, as calculated on the basis of carbon utilization. Respiration rate in the intact, preclimacteric fruit is 12% that of the potential. At the climacteric peak or in fresh slices, whether cut from preclimacteric or peak-climacteric fruit, the respira-

TABLE I

CONTRIBUTION OF CYTOCHROME AND ALTERNATE PATHWAYS TO RESPIRATION OF
AVOCADO (*Persea americana*) FRUIT SLICES[a]

	Control				+KCN
	Total	Cytochrome	Alternate	Residual[b]	Total
Fresh slices from preclimacteric fruits					
Coupled	184	144	0	40	132
Uncoupled	295	156	102	35	120
Preclimacteric slices aged to ripeness					
Coupled	327	169	125	35	125
Uncoupled	320	—	—	—	—
Fresh slices from peak climacteric fruits					
Coupled	165	80	0	77	102
Uncoupled	192	102	0	85	57

[a] Data from Table VI in Theologis and Laties (116); rates of respiration in μl O_2/g fresh weight/hr.
[b] Residual respiration not inhibited by KCN + CLAM.

tory rate is 60%. The normal cytochrome system is capable of this enhanced rate, but further increases require engagement of the alternate pathway. The alternate system is present in the intact fruit, because the total respiration is stimulated by cyanide, but in the absence of cyanide it is not engaged even at the climacteric peak.

Not all fruits are the same in this, although the dwarf banana [*Musa acuminata* (*M. cavendishii*)] is similar (116). Passam and Bird (79) reported that both electron-transport pathways contribute to the climacteric rise in honeydew melon. The capacity of the cytochrome system probably differs in different fruits, but it is reasonable to suppose that it competes favorably for substrate within its capacity (cf. 12) and that the alternate pathway is engaged only when substrate is supplied at a rate that exceeds that capacity. This corresponds with the conclusions drawn by Theologis and Laties from their observations.

Uncoupling increases the respiration of avocado fruit or slices in the preclimacteric stage, but not at the climacteric peak. It is now known that the phosphorylation of ADP continues through the climacteric rise (22, 79, 107), so there must be an adequate supply of ADP for peak respiration, probably through increased turnover of ATP (79). These relations suggest that coupled respiration in the preclimacteric stage is limited by

ADP, but at the peak it is limited by substrate supply. Such a suggestion is supported by Tingwa and Young (118) for avocado slices, by Hulme and Rhodes (41) for apple [*Malus sylvestris* (*Pyrus malus*)], and by Passam and Bird (79) for honeydew melon.

The availability of substrate for the electron-transport system may depend on the rate of catabolism of reserves or on the access of its products to the mitochondria. There is some evidence that the rate of glycolysis increases during the climacteric. It is now fairly clear that glycolysis in plant tissues is under the control of two key enzymes, PFK and PK. Barker and Solomos (14) suggested earlier that the climacteric rise in banana (*Musa acuminata* cv. 'Lacatan') depends on the activation of PFK. Salminen and Young (99) confirmed that FBP increases about 20-fold in banana (*Musa acuminata* cv. 'Valery') during the climacteric rise in respiration that follows application of ethylene and that the enzyme PFK increases in activity by 200%. A crossover between F6P and FBP occurs between the preclimacteric stage and midclimacteric rise, suggesting a change in the properties of PFK. Kinetic and electrophoretic studies indicate activation rather than synthesis of new enzyme, but the mechanism is unknown.

Tomato (*Lycopersicon esculentum*) fruits also accumulate FBP as the climacteric rise in respiration begins (22), and a crossover between F6P and FBP occurs, indicating control by PFK at this time. Similarly, a crossover between phosphoenolpyruvate and pyruvate later in the phase of rising respiration implicates PK control at this stage. The authors attributed increased glycolysis to freer access of vacuolar orthophosphate to PFK in the cytoplasm, where it acts as a positive effector. Combined with the evidence that mitochondrial oxidation is limited by substrate supply during the climacteric, an increase in the rate of glycolysis assumes importance whatever the mechanism involved.

Although present evidence indicates that during the climacteric glycolytic enzymes are activated rather than synthesized *de novo*, protein synthesis is a feature of the climacteric, and some enzymes are synthesized. It appears that these enzymes are not directly in control of respiration rate, because it is possible to inhibit protein synthesis without inhibiting the respiratory rise (35, 62). Ripening, however, requires protein synthesis and is prevented by cycloheximide. Thus it is possible to have a climacteric rise in respiration without ripening, but in climacteric fruit ripening does not occur without the respiratory rise (62, 90). It appears that increased respiration is not, as has been suggested, a response to demand for energy for protein synthesis, but that the latter probably uses the extra energy provided by increased respiration that is coupled with ATP production.

In many fruits the main substrate for respiratory metabolism is sugar, and the respiratory quotient (R.Q.) is close to unity throughout the climacteric rise. Certain pome fruits provide an exception, especially the apple, in which malate is a major reserve. Here the R.Q. may rise from 1 to 1.33 during the climacteric, suggesting the oxidation of malic acid. With added malate the R.Q. may rise still higher, and this is associated with increased activity of NADP-dependent malic enzyme and pyruvate decarboxylase, leading to excess production of CO_2. With this exception the climacteric rise in respiration appears to be an oxidative process.

Oxygen is required for the development of the climacteric and for ethylene synthesis. Wade (121) found that the rate of development of the climacteric rise in respiration of banana (*Musa acuminata* cv. 'William's Hybrid') fruits treated with 2-chloroethylphosponic acid (ethephon) increases with O_2 concentration from 3% upward. Wade carried out a kinetic analysis that indicated that O_2 is an activator of respiration as well as a substrate. Wade's analysis depended on the assumption that O_2 and ethylene bind to a single enzyme, which is unlikely. However, if respiration rate at the climacteric peak is limited by substrate, as the evidence presented suggests, then increasing O_2 concentration could be responsible for increasing supply of substrate at the oxidative sites in the mitochondria.

The role of ethylene in the climacteric is a central problem. Ethylene production increases autonomously in green fruit, and it is fairly commonly believed that a threshold concentration is required to trigger a rise in respiration, which is accompanied by further synthesis of ethylene. A good deal of work has been done in an effort to elucidate the biogenesis of ethylene, its attachment site, and its mode of action at the molecular level. Reviews of the subject include those by McGlasson (61), Abeles (1, 2), and Lieberman (49). The latest of these (49) shows that none of these goals has yet been reached. Lieberman suggested that ethylene is synthesized by a complex enzyme system in or on the plasma membrane, which may be under regulation by auxin. There is no clear evidence that ethylene directly modulates any specific enzyme or that it disrupts membranes in ripening fruit. There is, however, evidence that it may influence transport across membranes (1).

Until the primary action of ethylene is discovered, the question of how it might initiate the climacteric cannot be answered. Two proposals have been made, (*a*) that ethylene changes the properties of membranes so that enzymes become more accessible to substrates and effectors, and (*b*) that ethylene induces synthesis of a new set of enzymes that program the whole ripening process, of which enhanced respiration is one component. There is little definitive evidence to support either theory or, indeed, to show that they are mutually exclusive.

Cells do become leaky as ripening progresses, but probably not before the climacteric is established, and there is no certainty that ethylene is directly responsible (1). However, in senescing cells degeneration of the tonoplast and certain organelles precedes degeneration of the plasma membrane (17, 21), and it remains possible that some internal membrane is changed in such a way as to alter communications and modify the balance of metabolic processes within the cell. Evidence that this occurs in senescing morning glory (*Ipomoea tricolor*) petals is discussed in the next section.

On the other hand, new proteins are synthesized, and these are different in kind in early and late stages of the climacteric (35). RNA synthesis also increases, and it is reasonable to suppose that the climacteric is initiated by a process that controls the specific types of RNA synthesized. Thereafter, various aspects of the climacteric may be under separate controls. Such a scheme was proposed by Frenkel *et al.* (35) and is supported by Rhodes (90) and Lieberman (49), among others. It does not preclude either organizational or biochemical controls for separate processes, and it recognizes that the life cycle of the organism is genetically programmed. It does not answer the question of whether initiation of a new program is implemented by a purely metabolic mechanism or depends on an organizational change that permits the synthesis of new enzymes.

The ontogenetic program is now generally conceived to be mediated by plant hormones, but how they may act remains unknown. Evidence is accumulating that they affect membrane properties as well as nucleic acid synthesis. Evidence of interaction between ethylene and other hormones has been reviewed by Lieberman (49). Information is as yet fragmentary and somewhat chaotic. Abscisic acid may promote ethylene production and ripening. Auxins and cytokinins in some instances antagonize these. But responses differ, not only among species, but in different tissues within the same fruit. It is clear that interactions occur, but until more is known about the mechanisms of hormone action it would be unprofitable to predict how they operate in the control of fruit ripening or, more specifically, in the control of the climacteric rise in respiration.

D. INFLORESCENCE DEVELOPMENT AND SENESCENCE

A respiratory rise has been observed during the development of the inflorescence of some species. A special case is the spadix of Araceae, in which a dramatic rise in respiration rate is accompanied by an equally dramatic rise in tissue temperature. A review of such thermogenic respiration by Meeuse (63) sets out many of its known characteristics. Temperature changes are closely correlated with changes in respiration rate, starch is rapidly depleted, metabolism is generally accelerated, and permeability changes occur. Thermogenic respiration exploits the cyanide-resistant, alternate pathway of electron transport. This may be because, as in slices of some fruits or tubers (113, 116), substrate arrives at the oxidative sites in the mitochondria too rapidly to be dealt with by the cytochrome system.

Ap Rees (9) provided evidence that carbohydrate breakdown during thermogenic respiration in the spadix of *Arum maculatum* is entirely via the Embden–Meyerhof pathway. He reported a large increase (up to 100-fold) in some glycolytic enzymes, including PFK, before the respiratory rise. This was judged to be an essential preliminary, because activities of the appropriate enzymes in early stages of development of the spadix are quite inadequate to sustain the rate of glycolysis prevailing during thermogenesis, as measured by CO_2 production and by starch breakdown. There was, however, no further increase in enzyme activity while the respiration rate was actually rising, so some other control must be operative. At any time the activity of PFK exceeded that required for the prevailing rate of glycolysis. The data do not rule out regulation by PFK and PK but suggest, as an alternative, control by the supply of substrate provided by starch breakdown.

Less is known about the initiation of thermogenic respiration. It depends on the light regime under which the plant is growing, a dark period acting as trigger. During a lag period of about 40 hr, according to Buggeln and Meeuse (20), a hormone, which they called calorigen, is generated and initiates the rise in respiration rate. Ethylene may also be involved (63).

The reason for intense heat production may be related to the fact that respiration is largely supported by the alternate electron transport system, and rapid fuel consumption conserves less energy as ATP and produces more heat. Elevated tissue temperature is of some benefit to the plant. It occurs during pollen production and is accompanied by the dissemination of odors that attract pollinators. *Symplocarpus* produces flowers in early spring, and elevated temperature in the tissue permits development even when the ambient temperature falls below freezing.

Although thermogenic respiration has its unique features, it also has features in common with climacteric respiration in fruits. Other flower parts also exhibit a respiratory rise. Siegelman *et al.* (101) recorded a marked rise in the respiration of hybrid rose (*Rosa*) petals during cell expansion. Uncoupling increased respiration before the rise, but not at the peak. Mayak *et al.* (57) demonstrated a sharp rise in ethylene production, an increase in ABA (see also 56), and a decrease in cytokinin related to the aging of petals of other varieties of roses, but it is not possible to coordinate their timing with the respiratory rise observed by Siegelman *et al.* Nichols (74) found that when carnation (*Dianthus caryophyllus*) was pollinated, ethylene production rose sharply and triggered petal wilting. Intense ethylene production was localized in the style.

Hanson and Kende (36) observed that addition of ethylene to pieces of morning glory petals containing midribs causes rolling up (which occurs naturally in senescing flowers) within 2–3 hr. It simultaneously increases efflux of ions, sucrose, and organic acids to the medium. Kende and Hanson (47) attributed rolling to differential turgor that arises from the inability of adaxial cells of the petal to retain solutes. Membrane phospholipids normally decrease the night before flower opening and decrease rapidly during rolling. The conclusion drawn by Hanson and Kende was that some metabolic process causes a change in membrane composition and permeability which leads to loss of turgor and a surge of ethylene production. These processes accelerate autonomously and hasten senescence.

Matile and Winkenbach (55) showed electron-microscopic evidence that the tonoplast is the first membrane to be modified and to disintegrate in cells of senescing morning glory (*Ipomoea purpurea*) flowers. Suttle and Kende (110) found that leakage of anthocyanin pigment from mature and senescing petals of a *Tradescantia* hybrid was accelerated by ethylene. This implies modification of the tonoplast, because the pigment is confined to the vacuole. A slower efflux of pigment occurred in the presence of an inhibitor of ethylene synthesis, so, apparently, other factors are also concerned in the change in permeability of the tonoplast. Kende and associates have suggested that ethylene does not initiate, but controls the pace of such a change, and of ensuing senescent changes as well. They recognized, however, that all tissues in a flower do not age synchronously and that ethylene production in some cells could trigger senescence in other cells. Suttle and Kende found that in *Tradescantia* flowers more than 70% of the ethylene is produced by the gynoecium and androecium, and endogenous production rises rapidly in the mature flowers before it begins to increase in petals.

Although respiration was not measured in conjunction with other aspects of senescence, it is clear that the senescence of flowers bears a close resemblance to that of fruits, and it is probable that they are under similar controls.

III. Storage Organs and Aging of Slices

In the earlier discussion of respiration of storage organs in Volume IVA of this treatise (33), consideration was given mainly to intact tubers and roots and the characteristics imposed by bulkiness. Since publication of that volume, work has been largely with slices. When storage organs are cut into thin slices, respiration occurs at a much higher rate than in the intact organ. If slices are held in moist air or in an aerated medium, the rate of respiration gradually rises still higher, and metabolic changes occur.[4] Perhaps because of certain similarities to senescence this process is often called aging, although it more nearly resembles a release from dormancy, for it may even involve cell division and differentiation.

Much work has been devoted to attempts to explain the excess respiration induced by "aging." This is discussed in reviews of related topics by Kahl (46), Van Steveninck (120), and Solomos (104) and more recently in papers by Theologis and Laties (113–115), Janes et al. (45), and Rychter et al. (93, 94).

An immediate effect of slicing bulky organs may be reduction and disorganization of endoplasmic reticulum, followed by reorganization during aging (120). Slicing also initiates synthesis of RNA and protein and acceleration of metabolic processes, including glycolysis, the pentose phosphate cycle, the tricarboxylic acid cycle, and lipid metabolism (especially phospholipid synthesis). Membrane permeability and active ion transport increase during aging, as does activity of certain membrane-bound enzymes (120). Rising respiration rates occur within an intricate pattern of structural and functional change in the tissue.

According to Adams and Rowan (3–5) the so-called aging of carrot (*Daucus carota*) slices proceeds in two stages. In the first, synthetic processes are active and the level of endogenous ATP rises. Respiration is limited by the rate of glycolysis, and glycolysis is regulated by the activity of PK, which depends on the availability of ADP. The pentose phosphate cycle operates in this stage, but it does not persist beyond it. In the second stage, syntheses are abated and glycolysis is controlled by PFK, which is activated by ATP. (The concentration of endogenous ATP in these tissues was about 1/50th of the concentration at which it has been shown to act as a negative effector of PFK in some systems.) Changes in the level of intermediates indicated that glycolytic flux increases with aging. Similar results were also found by Theologis and Laties (113) using slices of potato (*Solanum tuberosum*) tuber. The oxidation of fatty

---

[4] For an earlier discussion of these phenomena see Volume II, Chapter 4, pp. 335–360. (Ed.)

TABLE II

CONTRIBUTION OF CYTOCHROME, ALTERNATE, AND RESIDUAL RESPIRATION TO THE TOTAL RESPIRATION OF SLICES OF POTATO (*Solanum tuberosum*)[a]

Slices	Control				$10^{-4}\,M$ KCN added			
	Total	Cyto-chrome	Alter-nate	Residual	Total	Cyto-chrome	Alter-nate	Residual
resh								
Coupled	35	24	0	11	11	0	0	11
Uncoupled[b]	85	75	0	10	11	0	0	11
Aged (24 hr)								
Coupled	112	80	0	32	116	48	36	32
	144[c]	105	0	39	—	—	—	—
Uncoupled[b]	190	116	46	28	154	80	46	26
	206[c]	106	56	44	—	—	—	—
Aged (48 hr)								
Coupled	167	87	50	30	—	—	—	—

[a] From Russet Burbank tubers removed from storage at 7°C. Adapted from Tables I and II and Fig. 2 in Theologis and Laties (113).
[b] The phosphorylation uncoupler CCCP was added.
[c] Data for a second batch of slices.

acids may also increase, providing extra substrate for respiration (104). It is agreed that the induced respiration is essentially mitochondrial (94, 104, 113, 115). The number of mitochondria may increase during aging, but Theologis and Laties (113) offer evidence that this does not explain the increase in respiration rate. In their experiments, fresh potato slices had nearly the same potential capacity for respiration as aged slices.

There has been speculation that induced respiration implies the activation of the cyanide-resistant alternate pathway of electron transport (104, 106). This was based on the observation that slices of potato tuber are sensitive to cyanide inhibition when freshly cut but become increasingly resistant as they age. However, the participation of the alternate pathway proves not to be a requirement. Theologis and Laties (113), using titration with KCN and CLAM, estimated the contribution of the cytochrome and alternate pathways in fresh and aged potato slices. Some of their data are shown in Table II. In the absence of KCN, the respiration rate increased as much as threefold during aging without engaging the alternate pathway. Although the potential for the alternate pathway existed in aged slices, it was not called upon until the normal cytochrome system realized its full potential capacity or was inhibited. In the absence of cyanide this occurred in coupled slices only after 48 hr of aging, or in uncoupled slices after 24 hr.

Cyanide sensitivity in fresh slices is not universal (104, 115). Slices of some other storage organs, like potato, are sensitive; others are resistant to cyanide. Sweet potato (*Ipomoea batatas*) is an example of the latter group. Theologis and Laties (115), using the same technique as with potato, showed that in fresh slices the alternate pathway is partially engaged in the presence of cyanide. In its absence the alternate pathway does not contribute to the respiration of coupled slices, fresh or aged for 24 hr, although both have the capacity for alternate-pathway respiration. The extra respiration elicited by an uncoupler was on some occasions mediated through the alternate pathway; on others it was not. As in potato, the alternate pathway was engaged only when the normal cytochrome pathway was fully saturated, and it was not a necessary ingredient of induced respiration.

The sensitivity of fresh potato slices to cyanide [i.e., the absence of a functional alternate pathway (Table II)] is a matter of interest because Solomos and Laties (106) found that respiration of comparable intact tubers is stimulated by HCN. This suggests that the capacity for alternate-pathway respiration is abolished by slicing, and Solomos (104) and Waring and Laties (124) associated this with breakdown of membrane phospholipids, which gradually reappear during aging. Breakdown does not occur in sweet potato slices, in which the alternate pathway is functional.

If tubers are pretreated with ethylene under appropriate conditions, however, they yield slices that are resistant to cyanide inhibition (27, 45, 93). On the grounds that it required at least 12 hr of pretreatment with ethylene, Rychter *et al.* (93) concluded that the alternate pathway is absent in the tuber until it is induced by ethylene. Once induced it survives slicing.

The situation is not simple. Response to ethylene depends on conditions in the tuber. It differs with storage time and temperature (27). The respiration of tubers immediately after removal from storage at 7°C is stimulated by cyanide (106). If such tubers are treated with ethylene in an atmosphere of 100% O_2 for 24 hr, they yield slices that are inhibited by cyanide. If the tubers are held at room temperature for at least 10 days before treatment, they yield cyanide-resistant slices. After long storage, tubers are less responsive, but if treated with ethylene, 100% O_2, and 10% CO_2, they yield slices resistant to cyanide. The production of cyanide-insensitive slices is not specific to ethylene. It is also brought about by cyanide, acetaldehyde, or ethanol (45). It cannot, therefore, be a specific induction of an enzyme by ethylene. Solomos and Laties (106) were the first to propose that ethylene may cause generation of the alternate pathway, but, later, Solomos (104) cited evidence that the effect is not directly on mitochondrial transport systems, but on cytosomal systems.

The tubers themselves respond to ethylene and the other gases mentioned with a climacteric-like rise in respiration, and high O_2 concentration enhances the response (45, 89, 92, 106). The rate of glycolysis increases (92, 106), as does flux through the tricarboxylic acid cycle (92). The similarity to mature climacteric fruit is apparent, and ethylene action may be similar.

It is not the purpose to discuss here the regulation of apportionment of substrate to the normal cytochrome and alternate pathways in mitochondria. This has been discussed by Bahr and Bonner (12), Ikuma (43), Solomos (104), and others. Suffice it to say that the evidence indicates that in the plant tissues investigated the alternate pathway is operative only when the capacity of the cytochrome pathway is fully engaged. It therefore appears that, as proposed by Theologis and Laties (113, 115), the engagement of the alternate pathway *in vivo* results from access to more substrate than the cytochrome chain can deal with, and the extent to which it operates depends on the excess of substrate, up to the maximum capacity of the alternate pathway. If this is greater than the capacity of the cytochrome pathway, then the total respiration will be stimulated by cyanide.[5]

It is probable that even when the alternate pathway is available it does not contribute to the normal respiration of storage organs or tissue slices. We have seen that, in those tissues that have been investigated, respiration rates are well within the maximum capacity of the cytochrome pathway. We must look to control of the flux of oxidizable substrate to the mitochondrial sites of oxidation for an explanation of induced respiration.

Glycolytic and tricarboxylic acid cycle flux increases during aging of slices, but this is probably not in itself sufficient to account for the increase in respiration. Many other metabolic processes increase in activity, and membrane properties change. Substrate for the electron-transport system could be limited not only by the rate of carbohydrate breakdown, but also by the rate at which substrate can penetrate to the mitochondrial oxidation sites.

Respiration rates in intact storage organs are usually much lower than

---

[5] See discussion in Chapter 2, Section IV,B,3. (Ed.)

in slices and represent only a small fraction of the potential capacity of the normal cytochrome system. It is known that the limitation is not the supply of O_2 as a reactant (33). It is probable that the metabolic controls are of the same kind in the intact organ as in slices. The same cells are involved, and cellular mechanisms must be potentially available in common. But the restraint is more severe in the whole organ, and the question is what overriding control is imposed on metabolic activity. It is a difficult question in view of the multiple effects of slicing and aging.

These effects may be attributed to derepression and genetic direction, or to structural changes, or both. Kahl (46) discussed the problem in terms of partial derepression of the genes that initiate new RNA and protein synthesis, leading to unblocking of metabolic activities. Van Steveninck (120) considered the means of implementation. In view of the changes in membrane structure and properties and in transport systems associated with slicing and aging, Van Steveninck proposed that the endomembrane system is basically responsible for control of cell functions. In the intact organ the symplasm, extending from cell to cell through the plasmadesmata, would be the functional unit, integrating the activities of individual cells. These undergo change when the organ is sliced. Whether the primary effect is genetic or organizational remains to be determined. Certainly, on slicing, a whole new course of metabolism is set in motion by the abrupt change in the environment and the physical condition of the cells.

IV. The Role of Oxygen

Oxygen, as the ultimate acceptor of electrons and protons, is a reactant in respiration, but it is well known that the rate of respiration of plant tissues and organs is not generally limited by its concentration as a reactant. The rate is frequently controlled by the rate of linked phosphorylation of ADP, and situations where respiration rate depends on the provision of products of carbohydrate breakdown to the mitochondria have been discussed. Because most plants and their parts show increasing respiration rates with increasing ambient O_2 concentration, at least up to the concentration found in air, it is clear that O_2 has a secondary role in the regulation of respiration. Some aspects of this were discussed in Volume IVA of this treatise (33, 126), but little progress has been made toward a full explanation of how O_2 influences the respiration of tissues and organs.

A good deal is now known about the metabolic control mechanisms that may affect respiration, including the controls of glycolysis (15, 117, 119). ATP and ADP are required for the key enzymatic reactions in glycolysis, those catalyzed by PFK and PK. ATP may also act as a negative effector of PFK in some systems, but in plant tissues the level of endogenous ATP may be much too low for it to act in this capacity (5). In germinating seeds, detached leaves, climacteric fruits, and aging tissue slices, a rapid rise in respiration is accompanied by increasing levels of ATP. The supply of ADP, however, may be crucial in the control of glycolysis.

Oxygen, by maintaining electron transport, supports the phosphorylation of ADP to ATP and may reduce the ADP available for glycolysis. In some plant tissues, as well as in animal tissues, O_2 may decrease the consumption of sugar in glycolysis, an effect that has come to be known as the Pasteur effect. Alternatively, the Pasteur effect has been defined by plant physiologists as the conservation of carbon in the presence of O_2 (117). This is not necessarily the same thing, although it is sometimes assumed to be (e.g., 106).[6]

For plant tissues the significant comparison is between anoxia and the extinction point, a low external concentration of O_2 at which metabolism is fully aerobic. Below the extinction point carbon loss may increase, with fermentation, toward zero O_2 concentration; above it carbon loss also increases with respiration.

Few attempts have been made at complete analysis of the effects of O_2 since 1965. Effer and Ranson (30) showed by direct measurement that in buckwheat (*Fagopyrum esculentum*) seedlings the consumption of carbohydrates in anoxia was about half that in air. In nitrogen, 60% of the carbohydrate lost appeared in recognized products of fermentation; in air 29% appeared in the respired CO_2. Carbon loss is therefore far from representing the effect of O_2 on glycolysis. Because the conventional carbon loss in nitrogen is about one-third of that lost in air, buckwheat seedlings are said to show no Pasteur effect. However, anoxia produced the same changes in glycolytic intermediates, on transfer from air (31), as in tissues that do show a Pasteur effect (13). These are usually interpreted as indicating an increase in glycolytic flux in the absence of O_2. The changes were essentially the same on transfer to nitrogen from 1.5% O_2, in which the metabolism is wholly aerobic.

Starch breakdown, carbohydrate loss, respiration, and growth were all considerably less in 1.5% O_2 than in air, which indicated a diminution of overall catabolism of sugar. The authors offer two alternative explana-

---

[6] For a definition and discussion of the Pasteur effect, see the author's chapter in Volume IVA of this treatise. (Ed.)

tions of these data. Either glycolysis was decreased with restriction of O_2, or an increase in glycolysis was masked by decreased operation of the pentose phosphate cycle, which must then account for all increase in carbohydrate breakdown as O_2 concentration increased above 1.5%.

A third explanation is possible. Although the key glycolytic enzymes may be activated when O_2 is excluded, the total flux of glycolysis may not be limited by enzymatic activity but by the availability of some substrate or cofactor in the organized seedling. There is no reason to suppose that seedlings do not possess the metabolic systems common to most tissues, but the balance of processes depends on the physiological status. Respiration and growth are active in air. ATP could be rapidly used in the syntheses required for growth, and a reduction of ADP supply, or an excess of ATP, the conditions conventionally regarded as responsible for the Pasteur effect, probably would not occur. This could not in itself account for an actual increase in carbohydrate loss, respiration, and growth in air over that which occurs in nitrogen or 1.5% O_2. The possibility remains that the whole sequence is limited by access of some reactant to the appropriate enzyme, and that this depends on protoplasmic structure and transport systems.

In nongrowing tissues there is less demand for ATP, and, although respiration and phosphorylation are slower, the ADP level may be low enough to limit glycolysis in air. Such tissues may be expected to exhibit a Pasteur effect.

Forward and Cheung (34) investigated the effects of O_2 and a phosphorylation uncoupler on corn (*Zea mays*) coleoptiles that were placed under conditions in which ADP apparently limits the rate of respiration. They provided coleoptile sections with trace amounts of [^{14}C]glucose in atmospheres of air, 5% O_2 (close to the extinction point), and nitrogen in the presence or absence of DNP. A balance' sheet of ^{14}C in various metabolic fractions (Table III) shows that restriction of O_2 and addition of DNP have similar effects, causing increased utilization of the exogenous labeled glucose, increased production of labeled CO_2 and ethanol, and decreased incorporation of ^{14}C in ethanol-insoluble substances, which in this tissue were in all probability derived from products of sugar breakdown.

The greatest differences in distribution of ^{14}C occurred between 5% O_2 and nitrogen (in the absence of DNP) and between air and 5% O_2 (in its presence), that is, between tissues that were virtually totally aerobic and tissues in which fermentation occurred. Diversion of ^{14}C from insoluble compounds to ethanol and CO_2 accounted for about half the increase in fermentation products in nitrogen over those in air. The

TABLE III

DISTRIBUTION OF ^{14}C AS A PERCENTAGE OF ^{14}C ABSORBED AND AS A PERCENTAGE OF ^{14}C GLYCOLYZED, WHEN [U-^{14}C]GLUCOSE WAS SUPPLIED TO SEGMENTS OF CORN (*Zea mays*) COLEOPTILE[a]

Fraction	−DNP			+DNP		
	Air	5% O_2	N_2	Air	5% O_2	N_2
Percentage of total ^{14}C absorbed						
Total sugars	43	36	24	34	22	18
Total glycolyzed	57	64	76	66	78	82
Ethanol soluble[b]	20	19	28	21	20	21
Ethanol insoluble	18	12	3	13	6	2
CO_2 + ethanol	19	33	48	31	52	59
Percentage of total ^{14}C glycolyzed						
Ethanol soluble[b]	35	30	33	32	26	26
Ethanol insoluble	32	19	4	21	7	2
CO_2 + ethanol	33	51	63	47	67	72

[a] From Forward and Cheung (34).
[b] Other than sugars.

other half was accounted for by increased catabolism of sugar. Fermentation products accounted for 63% of the sugar loss in nitrogen. Carbon dioxide accounted for 33% of the sugar loss in air, much as in the buckwheat seedlings studied by Effer and Ranson (30). In corn coleoptiles the Pasteur effect was operating in the metabolism of the exogenous labeled sugar, but its magnitude could not be measured by determining CO_2 and ethanol only.

However, the distribution of ^{14}C did not represent the utilization of endogenous substrate. Although restriction of O_2 to 5% increased the production of respiratory $^{14}CO_2$ from exogenous [^{14}C]glucose by 50%, it reduced the total respiratory CO_2 (mainly from endogenous substrate) by 50%. Lower specific activity of the CO_2 evolved in air (compared to that in 5% O_2) indicated a larger proportion of unlabeled endogenous substrate for respiration in air. The implication is that O_2 is involved in some control over the use of endogenous substrate, overriding controls within the glycolytic sequence. The nature of such a control is unknown, but it is most likely to influence access of native substrate to glycolytic enzymes. There is evidence that sugars (76) and phosphate esters (18) are compartmented in plant cells and that externally applied sugars may enter a metabolically active compartment that is segregated from pools of endogenous sugars (85).

The glycolytic enzymes are no longer believed to be entirely free in a soluble phase of the cytoplasm. Some degree of sequestration has been suggested, earlier by Barker *et al.* (13) and more recently by Dennis and Green (28) and Simcox *et al.* (102), who found enzymes of glycolysis and the pentose phosphate cycle in proplastids of endosperm. Ottaway and Mowbray (78) have assembled a mass of evidence from a variety of animal cells that "the enzymes and intermediates of glycolysis are morphologically or functionally compartmented."

It is becoming increasingly evident that organization of tissues exercises control over metabolic processes, largely by regulating internal communications, and it would be surprising if O_2 were not concerned in maintaining that organization. This would reconcile evidence that the Pasteur effect may operate but that it conserves carbon only by retaining it as carbohydrate with evidence that, as O_2 concentration increases, aerobic plants respire at a higher rate and incorporate more carbon in their cells. In effect, it would mean that the organism releases substrate for catabolism under conditions in which it can use the products and the energy generated for its own growth and maintenance.

Smith and ap Rees (103a) compared the metabolism of [U-^{14}C]sucrose by pea root tips under aerobic and anaerobic conditions. The effect of anoxia on accumulation of various metabolites and on elimination of incorporation of ^{14}C in insoluble compounds corresponded to that in corn coleoptiles (34). The amount of ^{14}C metabolized was reported to be much less in anoxia than in air, but it is not clear how much labeled sucrose was absorbed. It is possible that in root tips O_2 is required for transport of the exogenous sucrose (or hexose derivative) to the glycolytic sites. On the other hand, the glycolytic machinery might have been impaired because the cells deteriorated rapidly in anoxia.

Oxygen has indirect effects on respiration, notably as it interacts with ethylene. Banana fruits respond to ethylene by increased respiration at a minimum O_2 concentration of about 3%, at which metabolism is wholly aerobic, and the rate of development of climacteric respiration increases with O_2 concentration (121). As mentioned earlier, the response of potato tubers to ethylene is greatly enhanced in an atmosphere of 100% O_2, and this involves an enhanced rate of glycolysis. No satisfactory explanation of the interaction of O_2 and ethylene in stimulating respiration has been provided. It must wait elucidation of the action of ethylene. These examples provide additional evidence that, in intact organs, O_2 has a role other than the Pasteur effect in the control of respiratory metabolism.

V. Role of Compartmentation

A prominent feature of the present approach to physiological function is the advancing realization of the importance of cellular organization. In addition to conspicuous organelles, the nucleus, chloroplasts, and mitochondria, other organelles have been characterized, and a complex endomembrane system in the cytoplasm has been recognized. It is obvious that metabolic reactions that occur in such a highly structured medium must be regulated not only by biochemical mechanisms *per se*, but also by the accessibility of the components to one another. This applies in all parts and at all stages in the life of a plant and can never rightly be excluded from consideration of the regulation of metabolic events.

In the foregoing analysis of respiration patterns in organs, tissues, and cells, at various stages of development and under different imposed conditions, the importance of compartmentation and communication within cells has been repeatedly encountered. At the risk of some repetition, further comments on some of the implications of compartmentation in the control of respiratory metabolism are included here.

Some specific instances of the importance of compartmentation have been referred to in reviews dealing with various themes. Oaks and Bidwell (76) produced a valuable review of the compartmentation of intermediary metabolites. Among other things, they reported the existence of separate sugar compartments, and Bieleski (18) reviewed evidence for separate pools of phosphate compounds, including esters. Beevers (15) called particular attention to the role of organelles in controlling the interactions of substrate and enzyme. Van Steveninck (120) suggested that the nucleus and endoplasmic reticulum continuum may be the effective center that controls organelle function through the endomembrane system. Matile (54) reviewed evidence that the vacuole acts as both a storage and lytic compartment for some metabolites, that the tonoplast exercises selective control of the passage of substrates into and out of the vacuole, and that the tonoplast is also an integral part of the internal membrane system, including the endoplasmic reticulum.

Developmental change is often accompanied by a change of pace in respiration. There is continuing discussion of the question, What is the fundamental reason for advance to a new phase? The life cycle is presumably genetically imposed. Thomas and Stoddart (116a) discussed evidence that the primary event initiating senescence in leaves is post-transcriptional and occurs outside the nucleus or chloroplast. They regarded it as the crossing of a metabolic threshold. Entry into a new developmental phase usually involves synthesis of new sorts of RNA and protein. It also usually involves changes in the properties of membranes. It has not been established how these are related. Are new enzymes primarily responsible for all the ensuing changes in metabolism, or does some change in internal communication enable the synthesis or activation of new enzymes? If a key enzyme is induced is it directly involved in a catabolic sequence or in the control of membrane properties and transport? We remain ignorant not only of the initiating event but of the sequence of ensuing events.

Blackman (19), in 1928, introduced the idea that a lowering of *organization resistance* is responsible for the climacteric rise in respiration of ripening fruit. Blackman's concept was that organization of the protoplasm, by regulating the appropriate juxtaposition of reactants, exercises a fundamental control over metabolic reactions. Some part of this control lies in segregation of reactants in what are now called compartments. Oaks and Bidwell (76) proposed that specific changes in compartmentation could permit induction of new enzymes that direct the progress of development. Sacher (98) discussed permeability changes in relation to the climacteric in fruits. There is no clear evidence that gross changes occur before the climacteric, but this does not preclude subtler changes. There is ample evidence that internal membranes are altered during senescence of leaves before the plasmalemma is affected (17, 21). In senescing flowers there is some evidence of change in the permeability of internal membranes before the climacteric rise in respiration (36, 47, 55, 110), and Kende and associates have concluded that a change in membrane properties is the effective cause of a surge in ethylene production (47). Although more evidence is required, changes in cellular organization cannot be dismissed as a possible initiator of the climacteric rise in respiration of senescent organs.

Developmental progress is under genetic control, although it is frequently, if not always, expedited environmentally or hormonally. Much remains to be learned about the action of hormones, but there is increasing evidence that one of their functions is to affect the properties of membranes, thereby controlling communications within cells. The same is true of phytochrome. Taylorson and Hendricks (111) regard such action as important in triggering metabolic activity that leads to seed germination. The possibility that it is important in other developmental events should inspire further research. Beevers (16) has reviewed evidence for the generation of microbodies in seeds and leaves from the endoplasmic reticulum and for hormonal control of the enzymes enclosed in these microbodies.

Apart from normal developmental patterns of metabolism, new patterns can be imposed. Slicing of storage organs causes an immediate increase in respiration rate and leads to further gradual rise in rates of respiration and glycolysis and to other metabolic changes resembling those of senescence, including new protein synthesis. Slicing also causes immediate structural deterioration in the endomembrane system, with gradual reorganization (120), and affects membrane lipids (104). The new course of metabolism might be attributed to derepression of genes or to structural change, or to both. It is certainly set in motion by an abrupt change in physical conditions.

Modern technology has revealed a great deal about the ultrastructure of cells and their organelles, localization of enzymes, membrane structure and properties, and transport across membranes. It seems obvious that within such a system lies the power to control the pace and balance of metabolic reactions occurring within that system. The difficult task that remains is to discover specific mechanisms that may apply controls to the respiratory metabolism of cells in various situations.

VI. The Role of Coordination

The cell is in a sense the universal unit of function in living things, and, in spite of diversity, living cells have many basic mechanisms in common. They do not, however, function independently, but as part of an organism, which is the effectual unit of function. The problems of coordination of activity are among the most difficult to solve, and our knowledge of them is scant as yet. Many of them have been discussed by Steward (109) in relation to the growth of plants. In relation to the regulation of respiration in plants, knowledge is in a primitive state.

Slices of storage tissue respire at a higher rate and pursue a course of respiratory metabolism different from that of the organs from which they were taken. Organs isolated from the plant differ in respiratory behavior from attached organs, at least until they have reached a certain stage of senescence. Organs attached to the plant differ according to their point of attachment. The respiration of vegetative organs is affected by the developmental stage of the plant. Reproductive developments affect the respiration of vegetative organs. These phenomena are recognized, but we can offer little in the way of causal explanations.

One obvious factor is the distribution of organic nutrients and of inorganic ions that act as cofactors in metabolic reactions and as modulators of enzymes as well as being nutrients. An obvious example is the supply of sugar, manufactured mainly in the leaves, to organs that perform inadequate or no photosynthesis. For example, Nichols and Ho (75) demonstrated that the respiration of developing rose petals depends on the import of carbon from leaves. Respiration rate was not directly correlated with current input of carbon, which was in excess of that respired, but was affected indirectly, probably through the prior level of soluble carbohydrates in the petals.

Dependence of root respiration on the supply of sugar from the shoot may be more direct. Hatrick and Bowling (37) found that the respiration rate in roots of very young sunflower plants was highly correlated with the rate of translocation of sucrose from the shoot, to the degree that respiration rate of the roots could be used to measure translocation. In these actively growing plants sugar reaching the roots was completely metabolized. Two-thirds was consumed in respiration, one-third in growth, and none remained as reserve. It is a situation in which respiration may be strictly limited by the supply of substrate that is manufactured in another part of the plant.

Crawford and Huxter (25) found that low respiration rates in 1-cm tips of pea and maize (*Zea mays*) roots, at low temperature, were accompanied by low sugar content and a low ratio of glucose to sucrose. Respiration was increased by added glucose, in preference to sucrose. Their data indicate that root respiration is limited by the supply of sucrose and by the inversion of sucrose to glucose (i.e., by translocation of sucrose and by the activity of invertase at low temperature).

Minchin *et al.* (64) followed the respiration of roots of intact cowpea (*Vigna unguiculata*) plants grown in a special root chamber. The rate of respiration was affected by the vegetative and reproductive status of the plant. It increased during vegetative growth and showed a small peak during senescence and the fall of the first leaf. Respiration rate built up to a maximum at the time when pods were filling at the first reproductive node, and then declined, to rise again during pod ripening. Minchin and co-workers concluded that this course could reflect photosynthetic rates and supply of carbohydrate to the root in the course of plant development. Other effects of the shoot on the root are not, however, excluded. Their proposal finds support in the observation by Hansen and Jensen (35a) that root respiration of *Lolium multiflorum* responded to changes in photosynthesis induced by alternating light intensity over the plants, the response lagging behind changes in photosynthesis.

Apart from movements of nutrients, coordinating influences are very generally attributed to hormones, but there is little information about specific effects of hormones on respiration. The climacteric rise in detached oat leaves can be eliminated by kinetin (112), which retards the whole process of senescence. Cytokinins are synthesized in roots, and their supply to attached leaves may well affect respiration as well as other metabolic systems, but there is at present no way of knowing how. One possible instance of a hormonal effect on respiration appears in observations by Morohashi (67), who examined the influence of the embryonic axis on the development of respiration and respiratory control in cotyledons of pea seeds in the early stages of imbibition. Respiration develops earlier when the cotyledons are attached to the embryonic axis, and its senescent decline is earlier. The evidence suggests that the effect is on structural development in the mitochondria, rather than on the synthesis or activation of enzymes. The effect was completely inhibited by ABA but was not reversed by kinetin or GA.

Interaction of hormones in the initiation and development of a climacteric peak in respiration during fruit ripening has been mentioned previously. Apart from the fact that production of ethylene may be involved, there is no tangible clue to a mechanism for controlling climacteric respiration by hormone action. Because detachment of the fruit, or other organ, generally changes the course of respiration, there must be an exchange of messages within the plant. The messages provide a higher order of control over the activities of cells. Beyond the fact that they are probably carried by hormones, we have as yet little knowledge of what these messages are or how they are implemented.

VII. Respiration and the Carbon Economy of the Plant

In the foregoing sections, aspects of the nature, rate, and control of respiration in cells, tissues, organs, and in the whole plant have been considered. The purpose of this chapter would not be fulfilled without some reference to the relation of respiration to the metabolic and physiological economy of the plant although this subject has been discussed in Chapter 3, Section VI,B.

A different approach may serve to complement the discussion presented there. The attention of experimenters has been largely focused on the yield of crops and attempts to improve production. These matters have been presented and analyzed. Here we may reconsider some of the available information, in terms of the integration of various activities in different parts of the plant to determine the carbon economy of the plant as a whole.

A. RESPIRATORY EFFICIENCY

Photosynthesis provides the carbon that is incorporated in a plant and the potential energy required for the formation and maintenance of its organs and tissues. Respiration makes the energy available. The efficiency of the process may be expressed in terms of the ratio of carbon lost in respiration to that incorporated in tissue (CO_2/C). Estimates of this ratio made by Raven (88) from a variety of sources have been presented in Chapter 3, Table X, together with evidence from Raven's own and other data that in phototrophic growth some respiration may be spared by direct use of ATP and reductant that is generated by photosynthesis. Presumably, this happens within the photosynthetic cells, because the ratio CO_2/C is lower in green organs than in nongreen organs. For higher plants, it would be difficult to estimate the extent of such sparing in quantitative terms, partly because not all cells in "green" tissue contain chlorophyll, also because of exchanges of substances, not only between cells but between different regions of the plant. It is a major feature of the overall economy of the plant that carbon fixed in green cells provides for all the nonphotosynthesizing regions. Carbon-containing substances are transported throughout the plant, especially to the growing regions, which may be remote from the source of photosynthate.

Apart from estimates of respiratory efficiency in isolated instances, some data are available about interrelations within the organism and the effect of the physiological states of the plant. Probably the most complete information about the loss of CO_2 in relation to other fates of assimilated carbon in intact higher plants has been provided by Pate and associates (38, 65, 80). They produced balance sheets of carbon and nitrogen for nodulated legumes, based on measurements of fresh and dry weight, total carbon and nitrogen, respiration in separate organs, and on the composition of xylem sap. Figure 8 reproduces diagrams from Herridge and Pate (38) of three stages of development of cowpea, showing the distribution of carbon as a percentage of net photosynthetic gain of carbon. The interdependence of different regions of the plant is evident, as is the change in carbon economy with advancing development. The carbon lost as CO_2 compared to that incorporated in dry matter (CO_2/C) for the whole plant increased from 0.43 in the first phase of development to 1.33 in the third. But the ratio varied greatly in different parts of the plant, as shown in Table IV. It was much higher in roots and nodules than in leaves. This could be partly because root metabolism is entirely heterotrophic, and also because roots export amides and amino acids to provide partially processed carbon to the shoot, especially to the

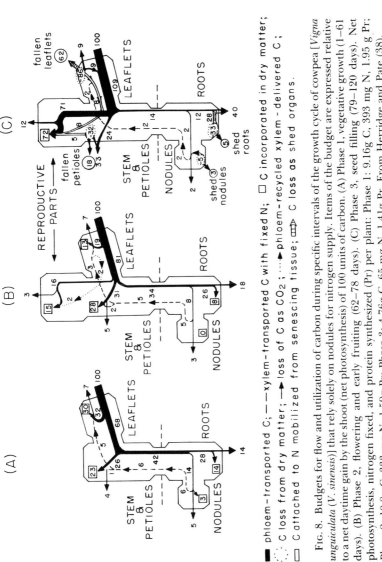

FIG. 8. Budgets for flow and utilization of carbon during specific intervals of the growth cycle of cowpea [*Vigna unguiculata* (*V. sinensis*)] that rely solely on nodules for nitrogen supply. Items of the budget are expressed relative to a net daytime gain by the shoot (net photosynthesis) of 100 units of carbon. (A) Phase 1, vegetative growth (1–61 days). (B) Phase 2, flowering and early fruiting (62–78 days). (C) Phase 3, seed filling (79–120 days). Net photosynthesis, nitrogen fixed, and protein synthesized (Pr) per plant: Phase 1: 9.16g C, 393 mg N, 1.95 g Pr; Phase 2: 10.8g C, 333 mg N, 1.59g Pr; Phase 3: 4.76g C, 65 mg N, 1.41g Pr. From Herridge and Pate (38).

■ phloem-transported C; ── xylem-transported C with fixed N; □ C incorporated in dry matter; ⋮ C loss from dry matter; → loss of C as CO₂; ·····► phloem-recycled xylem-delivered C; ▭▶ C loss as shed tissue; ⬚ C attached to N mobilized from senescing tissue; ⬚ C loss as shed organs.

498   DOROTHY F. FORWARD

TABLE IV

RATIO OF CARBON LOST AS CO_2 TO CARBON INCORPORATED INTO
DRY MATTER IN COWPEA (*Vigna unguiculata*) AT THREE STAGES
OF DEVELOPMENT[a]

Plant Part	CO_2/C		
	Stage I	Stage II	Stage III
Leaflets	0.24	0.54	—
Stem and petioles	0.17	0.18	—
Roots	1.0	2.3	—
Nodules	1.7	—	—
Reproductive parts	—	0.20	0.17
Whole plants	0.43	0.56	1.3

[a] Data derived from Fig. 8 in Herridge and Pate (38).

reproductive parts. Consequently, those parts showed a lower ratio (i.e., higher apparent efficiency). As the plant aged, efficiency declined in the vegetative parts but remained high in reproductive parts. In the last phase, when pods were filling rapidly, vegetative structures were breaking down and contributing organic carbon to growth in reproductive parts as well as squandering it by shedding cells and organs.

It is clear that the ratio CO_2/C does not necessarily represent the respiratory efficiency of cells within a specific region if they are in communication with the rest of the plant. Like cowpea, attached roots of young rapidly growing sunflower plants show a high ratio, CO_2/C = 2, whereas cultured roots of corn show CO_2/C = 0.71 (88). At least part of the difference may represent the export of carbon to the shoot. Roots also absorb salts that serve the metabolism of the whole plant. The uptake of ions is associated with enhanced respiration, and energy generated in "salt respiration" is applied to ion uptake. Veen (120a), as discussed later, assigned as much as half the energy of total respiration of maize roots to ion uptake.

Others attribute inefficiency in some roots to "wasteful respiration" (see Chapter 3, Section VI,B). Lambers and Smakman (48a), for instance, found major participation of the cyanide-resistant pathway in the root respiration of several species of *Senecio*, as indicated by more severe inhibition by SHAM than by cyanide. Ryle *et al.* (95–97) found that in three legumes nodulated roots of plants that were fixing nitrogen respired a higher proportion of the carbon fixed in photosynthesis than did the roots of plants using nitrate, and nodulated plants grew less. This means lower respiratory efficiency, although the energy cost of reducing nitrogen to ammonia is similar to that of reducing nitrate in nongreen tissue or in the dark. It is suggested that some nitrate may be reduced in green tissues, with a contribution of energy generated during photosynthesis, whereas reduction of nitrogen in the nodules must rely entirely on respiration to generate the required energy. Also, the maintenance of the nodule complex may impose an extra respiratory burden. Ryle *et al.* (95) calculated that, in soybean (*Glycine max*), maintenance of nitrate plants left 60% of the daily assimilate available for growth during the vegetative period, declining to 40% in the reproductive period. In nodulated plants relying on nitrogen fixation, 15% less of the assimilate was available in each stage for growth and development.

These studies show that respiratory efficiency is a complex matter. They emphasize the interdependence of the component parts of the organism and show that in an intact plant the relation of respiration to the whole carbon economy in any organ depends very much on what is happening in other parts of the plant as well as on the program of the life cycle. It depends also on the environment. Studies of stress physiology include observations of changes in respiration rate imposed by heat or cold, drought, salt concentration, or waterlogging. Stress effects are multiple, and it is difficult to analyze specific effects on respiration. There is no reason to suppose that intrinsic mechanisms controlling respiration or its role in the economy of the plant under stress are different from those operating under favorable conditions. They are modified by the environment, but not altered in kind.

B. RESPIRATION FOR GROWTH AND MAINTENANCE

Respiration must provide not only carbon skeletons and energy for the synthesis of new plant materials, but also energy for maintenance of the system. Some attempts have been made to resolve total respiration into two components, growth respiration and maintenance respiration. The general approach has been outlined in Chapter 3, Section VI, together with a basic formula, its limitations, and underlying assumptions. The approach of Penning de Vries and associates (81–84) is of interest in the present context because it relates growth efficiency to the metabolism of the plant. It provides a theoretical basis for estimating the respiration associated with the conversion of carbohydrate substrate to structural matter, mainly protein, lipids, cellulose, and lignin. If the composition of the tissue and the chemical nature of the substrate are

500   Dorothy F. Forward

known, the amount of carbon transferred to structural compounds can be calculated together with the energy requirement for the sum of endergonic processes involved. The consumption of O_2 and production of CO_2 in the required energy-releasing processes can then be calculated. Allowance must also be made for the reduction of nitrate and formation of amino acids in the light. According to this analysis, growth respiration is predictable from the increase in dry matter of known composition.

The estimation is necessarily approximate, because some assumptions as to the reactions involved and some simplifications are unavoidable. However, it provides a rational basis for estimates of growth respiration derived from experimental data. A number of these show reasonable conformity to the theoretical values (cf. 35a, 42, 60).

Because the analysis depends on stoichiometrical relations in the conversion reactions, growth respiration should depend on the chemical nature of the substrate and the final products and should not change with rates of the processes under the influence of temperature or other variable conditions.

Determination of the amount and composition of dry matter involves sacrificing plants, and it would be advantageous to use the ratio of growth to respiration in the converse way, to predict yield by measuring respiration. The stumbling block is that the second component of total respiration, maintenance respiration, is difficult to determine with any degree of reliability. Penning de Vries (81, 82) derived theoretical values by analyzing the maintenance processes requiring energy. These include maintenance of cellular structure and of gradients of metabolites and ions, turnover of macromolecules, including enzymes and other proteins, and to a lesser extent other constituents. They also include distribution of materials within the plant and accumulation at specific sites. Too little is known of the dynamics of most of the processes concerned to enable accurate estimation of maintenance respiration from basic data.

Veen (120a) used regression analysis to apportion total respiration of maize roots to growth, ion uptake and maintenance. He assigned 50% of the energy of respiration to ion uptake under normal conditions and 20% of the respiration of the whole plant to uptake and transport of ions. He distinguished respiration used for ion uptake from "maintenance respiration:" During 7 days of culture after root pruning, he apportioned total root respiration to growth (declining from 78 to 25%), ion uptake (increasing from 13 to 60%), and "maintenance respiration" (increasing from 8 to 33%) during the growth period.

Various direct estimates of maintenance respiration have been made using methods mentioned in Chapter 3. Penning de Vries (82) assembled a number of these, and they covered a wide range of values. Evidently, the respiratory cost of maintenance of existing matter is not as predictable as the cost of growth. It differs in different species and changes with the age of the plant and the level of carbohydrates for substrate. It varies with temperature and light intensity and may decrease with prolonged severe water stress or mineral starvation (e.g., 84). As in estimations of total respiratory efficiency, values for attached leaves are invalidated by export of major amounts of photosynthate.

Some values that approach the theoretical prediction have been obtained with whole plants in a steady state over several days. Models based on these have been used to analyze the carbon economy of plants or crops (e.g., 35a, 42, 60). They involve many assumptions and do not necessarily represent actual values, but they provide guidelines for approach to the ultimate goal of understanding the economy of matter and energy in plants. Much knowledge must yet be acquired before the goal is reached.

VIII. Conclusion

In 1965 the author concluded a discussion of respiration with the observation that "as we learn more of the biochemical mechanisms at the disposal of plants, the challenges of the future lie more and more in attempts to find the mechanisms that determine the balance and direction of metabolic processes [33]." Although some progress has been made, the same assertion can be made today. Some important metabolic regulatory mechanisms have been defined, the ultrastructure of cells has been described, and the location of certain reactions within the cell has been determined. Yet we remain ignorant of how the integrated activity of a cell is controlled and how it is conditioned by the organism.

The intensity of respiratory metabolism of cells *in situ* is under normal circumstances much lower than the full capacity of the enzyme machinery. Some of the restriction may be released by detaching an organ, more by slicing it, particularly if the organ or slice is maintained in isolation. Plants have alternate pathways of carbohydrate breakdown and electron transport that can be called into play. The overall rate of respiration cannot, therefore, depend only on the activity of specific enzymes in any one sequence. Respiration rate must depend ultimately on the rate of enzyme reactions, but this is apparently subject to controls of a superior order.

In the studies of respiration that have been considered, it appears that whenever a dramatic increase in rate occurred, whether spontaneously during unfolding of the life cycle or imposed by isolation of organs or tissues, the associated metabolism changed in similar ways. Familiar biochemical control mechanisms are operative but are not in themselves sufficient to explain respiratory behavior. ADP supply is sustained, and ATP level increases but it is not inhibitory. Glycolysis is enhanced, and the supply of substrate to mitochondrial oxidation sites appears to determine maximum rates that are short of the full capacity of the cells. Very generally, there is evidence of change in internal membranes and organelles. Ethylene, interacting with other hormones, is frequently, if not always, involved.

Respiratory metabolism is part of a complex network of reactions, and just as it is necessary to the operation of the whole, it is itself influenced by the processes it supports. The integration of processes is undoubtedly dependent on the organization of cells and organisms. It depends on communication between organelles and compartments within cells, and between cells and organs in the plant.

Hormones are presumed agents of communications, and further knowledge of their action would be most valuable. Apart from ethylene there is at present little evidence of direct linkage of hormones with respiration, and ethylene apparently acts indirectly. Hormones are known to affect membrane properties as well as nucleic acids and, therefore, possibly the implementation of gene action. Both derepression of genes and membrane transport or permeability have been invoked to explain respiratory phenomena such as the climacteric rise, but how either, or more probably both, are effective remains to be discovered.

Not only within the cell but throughout the plant communication is important in regulating the metabolism in any region. Carbohydrates play a prominent part in the nutrition and the structure of a plant. Photosynthesis introduces a special sequence of reactions that are fitted into the metabolic complex. The synthesis of carbohydrates from CO_2 is localized in chloroplastids mainly in the leaves. The products are used for growth and maintenance mainly in nongreen cells, often remote from the leaves. Nitrogen and mineral nutrients are absorbed through the roots. Substrates and cofactors for respiratory and other metabolism must reach the appropriate enzyme sites within the cells in all regions of the plant, and supplies must be apportioned. How all this is regulated still offers the main challenge for the future if we are to understand the respiratory economy of plants.

Hormones are the recognized messengers, and knowledge of their primary actions is of supreme importance. Further knowledge is needed of metabolic controls, also of localization of reactions and compartmentation within the cell, of membrane properties, of transport, as well as of gene derepression and its consequences.

References

1. Abeles, E. B. (1972). Biosynthesis and mechanism of action of ethylene. *Annu. Rev. Plant Physiol.* **23,** 259–292.
2. Abeles, E. B. (1973). "Ethylene in Plant Biology." Academic Press, New York.
3. Adams, P. B. (1970). Effect of adenine nucleotides on the respiration of carrot root slices. *Plant Physiol.* **45,** 495–499.
4. Adams, P. B. (1970). Effect of adenine nucleotides on levels of glycolytic intermediates during the development of induced respiration in carrot root slices. *Plant Physiol.* **45,** 500–503.
5. Adams, P. B., and Rowan, K. S. (1970). Glycolytic control of respiration during aging of carrot root tissue. *Plant Physiol.* **45,** 490–494.
6. Adato, I., and Gazit, S. (1977). Role of ethylene in avocado fruit development and ripening. II. Ethylene production and respiration by harvested fruits. *J. Exp. Bot.* **28,** 644–649.
7. Aharoni, N., and Lieberman, M. (1979). Ethylene as a regulator of senescence in tobacco leaf discs. *Plant Physiol.* **64,** 801–804.
7a. Aharoni, N., Anderson, J. D., and Lieberman, M. (1979). Production and action of ethylene in senescing leaf discs. *Plant Physiol.* **64,** 805–809.
8. Aharoni, N., Lieberman, M., and Sisler, H. D. (1979). Patterns of ethylene production in senescing leaves. *Plant Physiol.* **64,** 796–800.
9. Rees, T. ap. (1977). Conservation of carbohydrate by the non-photosynthetic cells of higher plants. *Symp. Soc. Exp. Biol.* **31,** 7–32.
10. Ashihara, H., and Komamine, A. (1975). The function of the pentose phosphate pathway in *Phaseolus mungo* hypocotyls. *Phytochemistry* **14,** 95–98.
11. Bahr, J. T., and Bonner, W. D. Jr. (1973). Cyanide-insensitive respiration. I. The steady states of skunk cabbage spadix and bean hypocotyl mitochondria. *J. Biol. Chem.* **248,** 3441–3445.
12. Bahr, J. T., and Bonner, W. D. Jr. (1973). Cyanide-insensitive respiration. II. Control of the alternate pathway. *J. Biol. Chem.* **248,** 3446–3450.
13. Barker, J., Kahn, M. A. A., and Solomos, T. (1964). Mechanism of the Pasteur effect. *Nature (London)* **201,** 1126–1127.
14. Barker, J., and Solomos, T. (1962). Mechanism of the "climacteric" rise in respiration in banana fruits. *Nature (London)* **196,** 189.
15. Beevers, H. (1974). Conceptual developments in metabolic control, 1924–1974. *Plant Physiol.* **54,** 437–442.
16. Beevers, H. (1979). Microbodies in higher plants. *Annu. Rev. Plant Physiol.* **30,** 159–193.
17. Beevers, L. (1976). Senescence. *In* "Plant Biochemistry" (J. Bonner and J. E. Varner, eds.), 3rd ed., pp. 771–794. Academic Press, New York.
18. Bieleski, R. L. (1973). Phosphate pools, phosphate transport and phosphate availability. *Annu. Rev. Plant Physiol.* **24,** 225–252.
19. Blackman, F. F. (1954). "Analytic Studies in Plant Respiration." Cambridge Univ. Press, Cambridge.

20. Buggeln, R. G., and Meeuse, B. J. D. (1971). Hormonal control of the respiratory climacteric in *Sauromatum guttatum* (Araceae). *Can. J. Bot.* **49,** 1373–1377.

21. Butler, R. D., and Simon, E. W. (1971). Ultrastructural aspects of senescence in plants. *Adv. Gerontol. Res.* **3,** 73–129.

22. Chalmers, D. J., and Rowan, K. S. (1971). The climacteric in ripening tomato fruit. *Plant Physiol.* **48,** 235–240.

23. Chen, S. S. C. (1970). Influence of factors affecting germination on respiration of *Phacelia tanacetifolia* seeds. *Planta* **95,** 330–335.

24. Coombe, R. G. (1976). The development of fleshy fruits. *Annu. Rev. Plant Physiol.* **27,** 507–528.

25. Crawford, R. M. M., and Huxter, T. J. (1977). Root growth and carbohydrate metabolism at low temperature. *J. Exp. Bot.* **28,** 917–925.

26. Cunninghame, M. E., Bowes, B. G., and Hillman, J. R. (1979). An ultrastructural study of foliar senescence in *Taxus baccata* L. *Ann. Bot. (London)* **43,** 527–528.

27. Day, D. A., Arron, G. P., Christoffersen, R. E., and Laties, G. G. (1978). Effect of ethylene and carbon dioxide on potato metabolism. *Plant Physiol.* **62,** 820–825.

28. Dennis, D. T., and Green, T. R. (1975). Soluble and particulate glycolysis in developing castor bean endosperm. *Biochem. Biophys. Res. Commun.* **64,** 970–974.

29. Edwards, M. M. (1968). Dormancy in seeds of charlock. II. The influence of the seed coat. *J. Exp. Bot.* **19,** 585–600.

30. Effer, W. R., and Ranson, S. L. (1967). Respiratory metabolism in buckwheat seedlings. *Plant Physiol.* **42,** 1042–1052.

31. Effer, W. R., and Ranson, S. L. (1967). Some effects of oxygen concentration on levels of respiratory intermediates in buckwheat seedlings. *Plant Physiol.* **42,** 1053–1058.

32. Eldan, M., and Mayer, A. M. (1972). Evidence for the activation of the NADH-cytochrome C reductase during germination of lettuce. *Physiol. Plant.* **26,** 67–72.

33. Forward, D. F. (1965). The respiration of bulky organs. *In* "Plant Physiology. A Treatise" (F. C. Steward, ed.), Vol. IVA, pp. 311–374. Academic Press, New York.

34. Forward, D. F., and Cheung, K.-W. (1978). The metabolism of glucose in corn coleoptiles in relation to oxygen and 2,4-dinitrophenol. *Can. J. Bot.* **56,** 1444–1452.

35. Frenkel, C., Klein, I., and Dilley, D. R. (1968). Protein synthesis in relation to ripening of pome fruits. *Plant Physiol.* **43,** 1146–1153.

35a. Hansen, G. K., and Jensen, C. R. (1977). Growth and maintenance respiration in whole plants, tops and roots of *Lolium multiflorum*. *Physiol. Plant.* **39,** 155–164.

36. Hanson, A. D., and Kende, H. (1975). Ethylene-enhanced ion and sucrose efflux in morning glory flower tissue. *Plant Physiol.* **55,** 663–669.

37. Hatrick, A. A., and Bowling, D. J. F. (1973). A study of the relationship between root and shoot metabolism. *J. Exp. Bot.* **24,** 607–613.

38. Herridge, D. F., and Pate, J. S. (1977). Utilization of net photosynthate for nitrogen fixation and protein production in an annual legume. *Plant Physiol.* **60,** 759–764.

39. Hulme, A. C. (ed.) (1970). "The Biochemistry of Fruits and their Products," Vol. I. Academic Press, New York/London.

40. Hulme, A. C. (ed.) (1971). "The Biochemistry of Fruits and their Products," Vol. II. Academic Press, New York/London.

41. Hulme, A. C., and Rhodes, M. J. C. (1971). Pome fruits. *In* "The Biochemistry of Fruits and their Products" (A. C. Hulme, ed.), Vol. II, pp. 333–377.

42. Hunt, W. F., and Loomis, R. S. (1979). Respiration modelling and hypothesis testing with a dynamic model of sugar beet growth. *Ann. Bot. (London)* **44,** 5–17.

43. Ikuma, H. (1972). Electron transport in plant respiration. *Annu. Rev. Plant Physiol.* **23,** 419–436.

44. James, T. W., and Spencer, M. S. (1979). Cyanide-insensitive respiration in pea cotyledons. *Plant Physiol.* **64,** 431–434.
45. Janes, H. W., Rychter, A., and Frenkel, C. (1979). Factors influencing the development of cyanide-resistant respiration in potato tissue, *Plant Physiol.* **63,** 837–840.
46. Kahl, G. (1973). Genetic and metabolic regulation in differentiating plant storage tissue cells. *Bot. Rev.* **39,** 274–299.
47. Kende, H., and Hanson, A. D. (1977). On the role of ethylene in aging. *In* "Plant Growth Regulation" (P. E. Pilet, ed.), pp. 172–180. Springer-Verlag, Berlin.
48. Kovacs, M. I., and Simpson, G. M. (1976). Dormancy and enzyme levels in seeds of wild oats. *Phytochemistry* **15,** 455–458.
48a. Lambers, H., and Smakman, G. (1978). Respiration of the roots of flood-tolerant and flood-intolerant *Senecio* species: affinity for oxygen and resistance to cyanide. *Physiol. Plant.* **42,** 163–166.
49. Lieberman, M. (1979). Biosynthesis and action of ethylene. *Annu. Rev. Plant Physiol.* **30,** 533–591.
50. Macnicol, P. K. (1973). Metabolic regulation in the senescing tobacco leaf. II. Changes in glycolytic metabolite levels in the detached leaf. *Plant Physiol.* **51,** 798–801.
51. Macnicol, P. K., Young, R. E., and Biale, J. B. (1973). Metabolic regulation in the senescing tobacco leaf. I. Changes in pattern of ^{32}P incorporation into leaf disc metabolites. *Plant Physiol.* **51,** 793–797.
52. Malhotra, S. S., and Spencer, M. (1973). Structural development during germination of different populations of mitochondria from pea cotyledons. *Plant Physiol.* **52,** 575–579.
53. Marei, N., and Crane, J. C. (1971). Growth and respiratory response of fig (*Ficus carica* L. cv. Mission) fruits to ethylene. *Plant Physiol.* **48,** 249–254.
54. Matile, P. (1978). Biochemistry and function of vacuoles. *Annu. Rev. Plant Physiol.* **29,** 193–213.
55. Matile, P., and Winkenbach, F. (1971). Function of lysosomes and lysosomal enzymes in the senescing corolla of the morning glory (*Ipomoea purpurea*). *J. Exp. Bot.* **22,** 759–771.
56. Mayak, S., and Dilley, D. R. (1976). Regulation of senescence in carnation (*Dianthus caryophyllus*). Effect of abscisic acid and carbon dioxide on ethylene production. *Plant Physiol.* **58,** 663–665.
57. Mayak, S., Halevy, A. H., and Katz, M. (1972). Correlative changes in phytohormones in relation to senescence processes in rose petals. *Physiol. Plant.* **27,** 1–4.
58. Mayer, A. M., and Poljakoff-Mayber, A. (1975). "The Germination of Seeds," 2nd ed. Pergamon, Oxford.
59. Mayer, A. M., and Shain, Y. (1974). Control of seed germination. *Annu. Rev. Plant Physiol.* **25,** 167–193.
60. McCree, K. J. (1976). The role of dark respiration in the carbon economy of a plant. *In* "CO_2 Metabolism and Plant Productivity" (R. H. Burris and C. C. Black, eds.), pp. 177–184. University Park Press, Baltimore, Maryland.
61. McGlasson, W. B. (1970). The ethylene factor. *In* "The Biochemistry of Fruits and their Products" (A. C. Hulme, ed.), Vol. I, pp. 475–519. Academic Press, New York.
62. McGlasson, W. B., Palmer, J. K., Vendrell, M., and Brady, C. J. (1971). Metabolic studies with banana fruit slices. II. Effects of inhibitors on respiration, ethylene production and ripening. *Aust. J. Biol. Sci.* **24,** 1103–1114.
63. Meeuse, B. J. D. (1975). Thermogenic respiration in aroids. *Annu. Rev. Plant Physiol.* **26,** 117–126.
64. Minchin, F. R., Neves, M. C. P., Summerfield, R. J., and Richardson, A. C. (1977). A

chamber designed for continuous, long-term monitoring of legume root respiration. *J. Exp. Bot.* **28,** 507–514.

65. Minchin, F. R., and Pate, J. S. (1973). The carbon balance of a legume and the functional economy of its root nodules. *J. Exp. Bot.* **24,** 259–271.

66. Moreland, D. E., Hussey, G. G., Shriner, C. R., and Farmer, F. S. (1974). Adenosine phosphates in germinating radish (*Raphanus sativus* L.) seeds. *Plant Physiol.* **54,** 560–563.

67. Morohashi, Y. (1980). Development of mitochondrial activity in pea cotyledons following imbibition: influence of the embryonic axis. *J. Exp. Bot.* **31,** 805–812.

68. Morohashi, Y., and Shimokoriyama, M. (1972). Physiological studies on germination of *Phaseolus mungo* seeds. I. Development of respiration and changes in the contents of constituents in the early stages of germination. *J. Exp. Bot.* **23,** 45–53.

69. Morohashi, Y., and Shimokoriyama, M. (1972). Physiological studies on germination of *Phaseolus mungo* seeds. II. Glucose and organic acid metabolism in the early phases of germination. *J. Exp. Bot.* **23,** 54–61.

70. Morohashi, Y., and Shimokoriyama, M. (1975). Further studies on glucose catabolism in the early phases of germination of *Phaseolus mungo* seeds. *J. Exp. Bot.* **26,** 927–931.

71. Morohashi, Y., and Shimokoriyama, M. (1975). Development of glycolytic and mitochondrial activities in the early phase of germination of *Phaseolus mungo* seeds. *J. Exp. Bot.* **26,** 932–938.

72. Nawa, Y., and Asahi, T. (1971). Rapid development of mitochondria in pea cotyledons during the early stage of germination. *Plant Physiol.* **48,** 671–674.

73. Nawa, Y., and Asahi, T. (1973). Biochemical studies on development of mitochondria in pea cotyledons during the early stage of germination. *Plant Physiol.* **51,** 833–838.

74. Nichols, R. (1977). Sites of ethylene production in the pollinated and unpollinated senescing carnation (*Dianthus caryophyllus*) inflorescence. *Planta* **135,** 155–159.

75. Nichols, R., and Ho, L. C. (1979). Respiration, carbon balance, and translocation of dry matter in the corolla of rose flowers. *Ann. Bot.* (*London*) **44,** 19–25.

76. Oaks, A., and Bidwell, R. G. S. (1970). Compartmentation of intermediary metabolites. *Annu. Rev. Plant Physiol.* **21,** 43–66.

77. Obendorf, R. L., and Marcus, A. (1974). Rapid increase in adenosine-5′-triphosphate during early wheat embryo germination. *Plant Physiol.* **53,** 779–781.

78. Ottaway, J. H., and Mowbray, J. (1977). The role of compartmentation in the control of glycolysis. *Current Topics Cellular Reg.* **12,** 107–108.

79. Passam, H. C., and Bird, M. C. (1978). The respiratory activity of honeydew melons during the climacteric. *J. Exp. Bot.* **29,** 325–333.

80. Pate, J. S. (1976). Nutrient mobilization and cycling: Case studies for carbon and nitrogen in organs of a legume. *In* "Transport and Transfer Processes in Plants" (I. F. Wardlaw and G. B. Passioura, eds.), pp. 447–462. Academic Press, New York.

81. Penning de Vries, F. W. T. (1975). Use of assimilates in higher plants. *In* "Photosynthesis and Productivity in Higher Plants" (J. B. Cooper, ed.) pp. 459–480. Cambridge Univ. Press, London.

82. Penning de Vries, F. W. T. (1975). The cost of maintenance processes in plant cells. *Ann. Bot.* (*London*) **39,** 77–92.

83. Penning de Vries, F. W. T., Brunsting, A. H. M., and van Laar, H. H. (1974). Products, requirements and efficiency of biosynthesis: A quantitative approach. *J. Theor. Biol.* **45,** 339–377.

84. Penning de Vries, F. W. T., Witlage, J. M., and Kremer, D. (1979). Rates of respiration and of increase in structural dry matter in young wheat, ryegrass and maize

plants in relation to temperature, to water stress and to their sugar content. *Ann. Bot. (London)* **44**, 595–609.

85. Porter, H. K., and May, L. H. (1955). Metabolism of radioactive sugars in tobacco leaf disks. *J. Exp. Bot.* **6**, 43–63.

86. Porter, N. G., and Wareing, P. F. (1974). The role of the oxygen permeability of the seed coat in the dormancy of seed of *Xanthium pennsylvanicum* Wallr. *J. Exp. Bot.* **25**, 583–594.

87. Ragetli, H. W. J., Weintraub, M., and Lo, E. (1970). Degeneration of leaf cells resulting from starvation after excision. *Can. J. Bot.* **48**, 1913–1922.

88. Raven, J. A. (1976). The quantitative role of "dark" respiratory processes in heterotrophic and photolithotrophic plant growth. *Ann. Bot. (London)* **40**, 587–602.

89. Reid, M. S., and Pratt, H. K. (1972). Effects of ethylene on potato tuber respiration. *Plant Physiol.* **49**, 252–255.

90. Rhodes, M. J. C. (1970). The climacteric and ripening fruit. *In* "The Biochemistry of Fruits and their Products" (A. C. Hulme ed.), Vol. I, pp. 521–533. Academic Press, New York/London.

91. Roberts, E. H. (1969). Seed dormancy and oxidation processes. *Symp. Soc. Exp. Biol.* **23**, 161–192.

92. Rychter, A., Janes, H. W., Chin, C.-K., and Frenkel, C. (1979). Effect of ethanol, acetaldehyde, acetic acid and ethylene on changes in respiration and respiratory metabolites in potato tubers. *Plant Physiol.* **64**, 108–111.

93. Rychter, A., Janes, H. W., and Frenkel, C. (1978). Cyanide-resistant respiration in freshly cut potato slices. *Plant Physiol.* **61**, 667–668.

94. Rychter, A., Janes, H. W., and Frenkel, C. (1979). Effect of ethylene and oxygen on the development of cyanide-resistant respiration in whole potato mitochondria. *Plant Physiol.* **63**, 149–151.

95. Ryle, G. J. A., Powell, C. E., and Gordon, A. L. (1978). Effect of source of nitrogen on the growth of Fiskeby soya bean: the carbon economy of whole plants. *Ann. Bot. (London)* **42**, 637–648.

96. Ryle, G. J. A., Powell, C. E., and Gordon, A. L. (1979a). The respiratory costs of nitrogen fixation in soya bean, cowpea and white clover. I. Nitrogen fixation and the respiration of the nodulated root. *J. Exp. Bot.* **30**, 135–144.

97. Ryle, G. J. A., Powell, C. E., and Gordon, A. L. (1979b). The respiratory costs of nitrogen fixation in soya bean, cowpea and white clover. II. Comparisons of the costs of nitrogen fixation and the utilization of combined nitrogen. *J. Exp. Bot.* **30**, 145–153.

98. Sacher, J. A. S. (1973). Senescence and postharvest physiology. *Annu. Rev. Plant Physiol.* **24**, 197–224.

99. Salminen, S. O., and Young, R. E. (1975). The control of properties of phosphofructokinase in relation to the respiratory climacteric in banana fruit. *Plant Physiol.* **55**, 45–50.

100. Siedow, J. N., and Girvin, M. E. (1980). Alternative respiratory pathway; its role in seed respiration and its inhibition by propylgallate. *Plant Physiol.* **65**, 669–674.

101. Siegelman, H. W., Chow, C. T., and Biale, J. B. (1958). Respiration of developing rose petals. *Plant Physiol.* **33**, 403–407.

102. Simcox, L. P. P., Reed, E. E., Canvin, D. T., and Dennis, D. T. (1977). Enzymes of the glycolytic and pentose phosphate pathways in proplastids from the developing endosperm of *Ricinus communis* L. *Plant Physiol.* **59**, 1128–1132.

103. Simmonds, J. A., and Simpson, G. M. (1971). Increased participation of pentose phosphate pathway in response to after ripening and gibberellic acid treatment in caryopsis of *Avena fatua* L. *Can. J. Bot.* **49**, 1833–1840.

103a. Smith, A. M., and ap Rees, T. (1979). Effects of anaerobiosis on carbohydrate oxidation by roots of *Pisum sativum*. *Phytochemistry* **18**, 1453–1458.

104. Solomos, T. (1977). Cyanide-resistant respiration in higher plants. *Annu. Rev. Plant Physiol.* **28**, 279–297.

105. Solomos, T., and Laties, G. G. (1974). Similarities between the actions of ethylene and cyanide in initiating the climacteric and ripening of avocados. *Plant Physiol.* **54**, 506–511.

106. Solomos, T., and Laties, G. G. (1975). The mechanism of ethylene and cyanide action in triggering the rise of respiration in potato tubers. *Plant Physiol.* **55**, 73–78.

107. Solomos, T., and Laties, G. G. (1976). Effects of cyanide and ethylene on the respiration of cyanide-sensitive and cyanide-resistant plant tissues. *Plant Physiol.* **58**, 47–50.

108. Solomos, T., Malhotra, S. S., Prasad, S., Malhotra, S. K., and Spencer, M. (1972). Biochemical and structural changes in mitochondria and other cellular components of pea cotyledons during germination. *Can. J. Biochem.* **50**, 725–737.

109. Steward, F. C. (1968). "Growth and Organization in Plants." Addison-Wesley Publishing Co. Reading, Massachusetts.

110. Suttle, J. C., and Kende, H. (1978). Ethylene and senescence in petals of *Tradescantia*. *Plant Physiol.* **62**, 267–271.

111. Taylorson, R. B., and Hendricks, S. B. (1977). Dormancy in seeds. *Annu. Rev. Plant Physiol.* **28**, 331–354.

112. Tetley, R. M., and Thimann, K. V. (1974). The metabolism of oat leaves during senescence. I. Respiration, carbohydrate metabolism, and the action of cytokinins. *Plant Physiol.* **54**, 294–303.

113. Theologis, A., and Laties, G. G. (1978). Relative contribution of cytochrome-mediated and cyanide-resistant electron transport in fresh and aged potato slices. *Plant Physiol.* **62**, 232–237.

114. Theologis, A., and Laties, G. G. (1978). Antimycin-insensitive cytochrome-mediated respiration in fresh and aged potato slices. *Plant Physiol.* **62**, 238–242.

115. Theologis, A., and Laties, G. G. (1978). Cyanide-resistant respiration in fresh and aged sweet potato slices. *Plant Physiol.* **62**, 243–248.

116. Theologis, A., and Laties, G. G. (1978). Respiratory contribution of the alternate path during various stages of ripening in avocado and banana fruits. *Plant Physiol.* **62**, 249–255.

116a. Thomas, H., and Stoddart, J. L. (1980). Leaf Senescence. *Annu. Rev. Plant Physiol.* **31**, 83–111.

117. Thomas, M., Ranson, S. L., and Richardson, J. A. (1973). "Plant Physiology," 5th ed. Longman, London.

118. Tingwa, P. O., and Young, R. E. (1974). The effect of tonicity and metabolic inhibitors on respiration and ripening of avocado fruit slices. *Plant Physiol.* **54**, 907–910.

119. Turner, J. F., and Turner, D. H. (1975). The regulation of carbohydrate metabolism. *Annu. Rev. Plant Physiol.* **26**, 159–186.

120. Van Steveninck, R. F. M. (1975). The "washing" phenomenon in plant tissues. *Annu. Rev. Plant Physiol.* **26**, 237–258.

120a. Veen, B. W. (1980). Energy cost of ion transport. *In* "Genetic Engineering of Osmoregulation" (D. W. Rains, R. C. Valentine, and A. Hollaender, eds.), pp. 187–195. Plenum, New York.

121. Wade, N. L. (1974). Effects of oxygen concentration and ethephon upon the respiration and ripening of banana fruits. *J. Exp. Bot.* **25**, 955–964.

122. Walther, A. (1968). Aspects of Aging in Sunflower Leaves. Ph.D. Thesis, Univ. of Toronto, Toronto, Canada.
123. Wareing, P. F., and Saunders, P. F. (1971). Hormones and dormancy. *Annu. Rev. Plant Physiol.* **22,** 261–288.
124. Waring, A. J., and Laties, G. G. (1977). Inhibition of the development of induced respiration and cyanide-insensitive respiration in potato tuber slices by cerulenin and dimethylaminoethanol. *Plant Physiol.* **60,** 11–16.
125. Wilson, S. B., and Bonner, W. D. Jr. (1971). Studies of electron transport in dry and imbibed peanut embryos. *Plant Physiol.* **48,** 340–344.
126. Yemm, E. W. (1965). The respiration of plants and their organs. *In* "Plant Physiology: a Treatise" (F. C. Steward ed.), pp. 231–310. Academic Press, New York.
127. Yentur, S., and Leopold, A. C. (1976). Respiratory transition during seed germination. *Plant Physiol.* **57,** 274–276.

Author Index

A

Abeles, E. B., 62(1), **140,** 478, 479(1), **503**
Ackrell, B. A. C., 188(32, 320), 201(32), **258, 273**
Adams, C. A., 73(2), **141**
Adams, P. B., 482(3–5), 487(5), **503**
Adato, I., 473, **503**
Agrawal, P. K., 315(1), **424**
Aharoni, N., 468, 469(7–8), 470, **503**
Akazawa, T., 298(410, 551), 369(13), 371(368), **424, 443, 445, 453**
Al-Azzawi, M. J., 46(229), 62(229), 83(229), 125(229), 127(229), **152**
Albersheim, P., 33(3), 34(3, 282), 35(12, 140, 282), 37(12, 140, 282), 38, 41(3), **141, 147, 154,** 317(146), **431**
Alberte, R. S., 99(288), **155**
Albertson, P.-A., 317(331), **441**
Albracht, S. P. J., 187(1), **256**
Alexandre, A., 176(285), 187(285), 190, 193(2), 194, 196(3), **256, 271**
Alfonzo, M., 177(395), **277**
Allaway, W. G., 396(425), 399, 402(426), 404, **424, 446**
Allen, J. F., 230(4), 248(4), **256**
Allen, M. B., 246(14, 516), **257, 284**
Allison, A. C., 129(4), **141**
Altamura, N., 176(374), **276**
Altman, A., 313(3), **424**
Amesz, J., 218(507), 228(131), 231(5–7), **256, 263, 283**
Andersen, K. S., 357(8), 375(55), 376(533), **424, 427, 452**
Anderson, J. D., 469(7a), **503**
Anderson, J. M., 236(52), 238(205), 239(8, 52, 205), **257, 259, 267,** 375(618), **456**
Anderson, J. W., 416(7), 423(7), **424**

Anderson, L. E., 305(107), 306(434), 309(434), 315(4), 418(5), **424, 429, 447**
Andreo, C. S., 252(503), **283**
Andrews, T. J., 300(9, 357), 301(357), 303(18, 355, 356), 304(18, 356, 357), 305(357), 341(11, 355), 353(10, 354), 404(19), **424, 425, 442, 443**
Anton-Lamprecht, I., 87(5), **141**
Aparicio, P. J., 230(305), **272**
Apel, P., 380(12), **424**
Armitage, T. L., 299(350), **442**
Armstrong, W. D., 73(297), **155**
Arnon, D. I., 211(10, 13a, 270), 213(9), 215(221), 218(269), 219, 233(10, 11), 236(13), 239(512), 240(269), 246(516), 247(11, 12, 85, 498), 248(11), **257, 260, 268, 270, 271, 282, 283, 284**
Arntzen, C. J., 242(15), **257**
Arosio, P., 235(549), **286**
Arron, G. P., 204(113), 205(113), 206(113), 207(113), 208(113), **262,** 484(27), **504**
Asahi, T., 197(301), 200(301, 317a), **272, 273,** 462, **506**
Asai, J., 195(300), **272**
Asami, S., 369(13), **424**
Ash, J. F., 127(310), **156**
Ashford, A. E., 62(6), **141**
Ashihara, H., 464, **503**
Atchison, B. A., 19(315), 79(315), **156,** 318(600), **455**
Atkins, B. D., 313(552), 314(552), **453**
Atkins, C. A., 297(14), **425**
Atkinson, D. E., 165, **257**
Aung, T., 420(70), **427**
Äuslander, W., 227(18, 248), 245(18), **257, 269**
Austin, B., 404(2), **424**

Index to Plant Names

Numbers in this index designate the pages on which reference is made in the text to the plant in question. No reference is made in the index to plant names included in the titles that appear in the reference lists. In general, when a plant has been referred to in the text sometimes by its common name, sometimes by its scientific name, all such references are listed in the index after the scientific name; cross reference is made, under the common name, to this scientific name. However, in a few instances in which a common name as used cannot be referred with certainty to a particular species, the page numbers follow the common name.

Subject Index

543